NUMERICAL MATHEMATICS AND SCIENTIFIC COMPUTATION

NUMERICAL MATHEMATICS AND SCIENTIFIC COMPUTATION

Books in the series

Monographs marked with an asterix (∗) appeared in the series 'Monographs in Numerical Analysis' which has been folded into, and is continued by, the current series.

*P. Dierckx: *Curve and surface fittings with splines*
*J. H. Wilkinson: *The algebraic eigenvalue problem*
*I. Duff, A. Erisman, and J. Reid: *Direct methods for sparse matrices*
*M. J. Baines: *Moving finite elements*
*J. D. Pryce: *Numerical solution of Sturm–Liouville problems*

K. Burrage: *Parallel and sequential methods for ordinary differential equations*
Y. Censor and S. A. Zenios: *Parallel optimization: theory, algorithms and applications*
M. Ainsworth, J. Levesley, W. Light, and M. Marletta: *Wavelets, multilevel methods and elliptic PDEs*
W. Freeden, T. Gervens, and M. Schreiner: *Constructive approximation on the sphere: theory and applications to geomathematics*
Ch. Schwab: *p- and hp-finite element methods: theory and applications to solid and fluid mechanics*
J. W. Jerome: *Modelling and computation for applications in mathematics, science, and engineering*
Alfio Quarteroni and Alberto Valli: *Domain decomposition methods for partial differential equations*
G. E. Karniadakis and S. J. Sherwin: *Spectral/hp element methods for CFD*
I. Babuška and T. Strouboulis: *The finite element method and its reliability*
B. Mohammadi and O. Pironneau: *Applied shape optimization for fluids*
S. Succi: *The lattice Boltzmann equation for fluid dynamics and beyond*
P. Monk: *Finite element methods for Maxwell's equations*
A. Bellen and M. Zennaro: *Numerical methods for delay differential equations*
J. Modersitzki: *Numerical methods for image registration*
M. Feistauer, J. Felcman, and I. Straškraba: *Mathematical and computational methods for compressible flow*
W. Gautschi: *Orthogonal polynomials: computation and approximation*
M. K. Ng: *Iterative methods for Toeplitz systems*
Michael Metcalf, John Reid, and Malcolm Cohen: *Fortran 95/2003 Explained*
George Em Karniadakis and Spencer Sherwin: *Spectral/hp element methods for CFD, second edition*
Dario A. Bini, Guy Latouche, and Beatrice Meini: *Numerical Methods for Structured Markov Chains*

Numerical Methods for Structured Markov Chains

D. A. Bini

Dipartimento di Matematica, Università di Pisa

G. Latouche

Département d'Informatique, Université Libre de Bruxelles

B. Meini

Dipartimento di Matematica, Università di Pisa

OXFORD
UNIVERSITY PRESS

OXFORD
UNIVERSITY PRESS

Great Clarendon Street, Oxford OX2 6DP

Oxford University Press is a department of the University of Oxford.
It furthers the University's objective of excellence in research, scholarship,
and education by publishing worldwide in

Oxford New York

Auckland Cape Town Dar es Salaam Hong Kong Karachi Kuala Lumpur
Madrid Melbourne Mexico City Nairobi New Delhi Taipei Toronto
Shangai

With offices in

Argentina Austria Brazil Chile Czech Republic France Greece
Guatemala Hungary Italy Japan South Korea Poland Portugal
Singapore Switzerland Thailand Turkey Ukraine Vietnam

Published in the United States
by Oxford University Press Inc., New York

First published 2005

A Catalogue record for this title is availabe from the British Library

Library of Congress Cataloging in Publication Data
(Data available)

ISBN 0 19 852768 3

1 3 5 7 9 10 8 6 4 2

Typeset by the authors using LaTeX
Printed in Great Britain
on acid-free paper by
Biddles Ltd, King's Lynn, Norfolk

PREFACE

The book deals with the numerical solution of structured Markov chains which have a wide applicability in queueing theory and stochastic modeling and include M/G/1 and G/M/1-type Markov chains, quasi-birth–death (QBD) processes, non-skip-free queues and tree-like stochastic processes.

By using a language familiar to both applied probabilists and numerical analysts, the book provides a systematic treatment of the theory and algorithms for important families of structured Markov chains. In fact, the subject is at the intersection of two major research areas: applied probability and numerical analysis which are also of great interest for applications.

The book is addressed to PhD students, researchers in the field of applied probability and numerical analysis but also to engineers and scientists dealing with actual telecommunications and computer systems, and generally with the evaluation of computer system performance. The book is useful both to readers interested in a thorough reference work on algorithms for use in their scientific and applied work, and to those who work in the design and analysis of numerical algorithms and in numerical linear algebra who want to learn more about certain applications and to find more material and motivation for their research.

A comprehensive and self-contained treatment of the Markov chains encountered in the most important queueing problems is performed by means of examples which clarify the applicative importance of the mathematical models. A thorough treatment of the literature on algorithmic procedures for these problems, from the simplest to the most advanced and efficient, is carried out. Functional iterations, doubling methods, logarithmic reduction, cyclic reduction, and subspace iteration are classes of methods which are described and analyzed in detail in the book. Their adaptation to different specific models coming from applications but also interesting in themselves is described and analyzed.

The theory at the basis of the numerical methods developed in the book is treated in a systematic way. Results from the current literature are reported with a unifying language and new results are also provided concerning the convergence and the analysis of the numerical performance of some methods. The book offers also a comprehensive and self-contained treatment of the structured matrix tools which are at the basis of the fastest algorithmic techniques for the solution of structured Markov chains. The main results concerning Toeplitz matrices, displacement operators, and Wiener–Hopf factorizations are reported as far as they are useful for the numerical treatment of Markov chains. Advanced computational techniques like evaluation interpolation at the roots of 1, performed by means of FFT, fast power series arithmetic, Toeplitz matrix computations, and displacement representations, are described in detail. The most up to date algorithms which so far are scattered in diverse papers written with different languages and notation are reported with a unifying language.

Much space is devoted to the theoretical issues concerning Wiener–Hopf factorizations, canonical factorizations and matrix equations trying to relate concepts and tools from the fields of numerical analysis and operator theory with the fields of applied probability. These concepts provide a unifying framework where structured Markov chains can be naturally investigated from a computational point of view. The reader will discover that the Ramaswami formula is nothing else but the canonical factorization of a matrix power series and that FFT-based algorithms allow one to implement this formula in a way that is much faster than the customary implementation. From canonical factorizations applied to more general contexts, the reader will also learn that the Ramaswami formula can be extended to wider classes of Markov chains. Certain transformations, like the reduction of an M/G/1 Markov chain to a QBD process with infinite blocks are viewed from the algebraic point of view in a more general framework.

Each solution method, even the most advanced and apparently complicated one, is reported in detailed algorithmic form so that the reader can translate it into a code in a high-level language with minimum effort.

Besides the major results published in journal papers, the book also contains new achievements concerning the convergence analysis and the computational analysis of certain algorithms. These results become more simple and direct when analyzed in the framework of Wiener algebra and of canonical factorizations.

The book is divided into three parts: Tools, Structured Markov Chains, and Algorithms.

The first part, concerning the tools, is formed of three chapters: Chapter 1 contains a basic description of the fundamental concepts related to Markov chains; Chapter 2 contains a systematic treatment of the structured matrix tools needed throughout the book, including finite Toeplitz matrices, displacement operators, and FFT; the third chapter concerns infinite block Toeplitz matrices, their relationships with matrix power series, and the fundamental problem of solving matrix equations and computing canonical factorizations. Here, the concept of a Wiener algebra is introduced; this provides a natural theoretical framework where the convergence properties of algorithms for solving Markov chains can be easily proved. Also the concept of canonical factorization provides a powerful tool for solving infinite linear systems like the fundamental system $\boldsymbol{\pi}^{\mathrm{T}} P = \boldsymbol{\pi}^{\mathrm{T}}$.

The second part deals with the description and the analysis of structured Markov chains. In Chapter 4, M/G/1-type Markov chains are analyzed; the reduction to solving a nonlinear matrix equation and the role played by canonical factorizations in this regard is pointed out. Phase-type queues are treated in Chapter 5 where G/M/1-type Markov chains, non-skip-free, QBD processes and tree-like processes are described and analyzed.

The third part concerns solution algorithms. In Chapter 6 the major algorithms based on functional iteration techniques for solving nonlinear matrix equations are described, analyzed, and compared. First, the basic concepts on fixed point iterations are recalled, then linearly convergent iterations are treated, and finally Newton's iteration is considered. Chapter 7 concerns quadratically convergent algorithms for solving matrix equations encountered in M/G/1, G/M/1 and QBD processes. The algorithm of logarithmic reduction for solving QBDs is described and analyzed, and an algebraic proof of its convergence is provided. Then the cyclic reduction technique for QBDs is explained and related to logarithmic reduction. Then cyclic reduction is extended to M/G/1 and G/M/1 processes. Here, new convergence and applicability results are proved. Computational issues concerning the implementation of cyclic reduction are treated by relying on the tools introduced in the first part of the book. Chapter 8 deals with some alternative approaches. A general technique for accelerating the convergence of iterative methods is described; its application to accelerating cyclic reduction is analyzed. A functional iteration relying on the combination of cyclic reduction and fixed point iteration is introduced: its convergence is linear but its convergence rate can be arbitrarily large. A doubling method, evaluation interpolation techniques and the invariant subspace method complete the chapter. Finally, in Chapter 9 some specialized structures are investigated and part of the algorithms described in the previous chapters are adapted to the specific cases. Markov chains with limited displacement (non-skip-free processes) are solved by means of functional iterations and cyclic reduction. The special QBD process obtained by reducing an M/G/1-type Markov chain to a QBD with infinite blocks is considered and treated with cyclic reduction. Finally three different algorithms for tree-like stochastic processes, relying on fixed point iterations, Newton's iteration and cyclic reduction, are introduced and analyzed.

Each chapter ends with some bibliographic notes which provide more information about the results treated in the chapter and about further related subjects that the reader might be interested in.

For the sake of completeness, the appendix collects the main general concepts and results used in the book which are part of the numerical background.

A list of the main notation used in the book, the bibliography, the list of algorithms and the index complete the volume.

We wish to thank Albrecht Boettcher, Leiba Rodman, Ilya Spitkovsky and Paolo Tilli for their suggestions and nice discussions which helped us to improve the presentation, and Bruno Iannazzo for carefully reading the manuscript and for pointing out some typos.

This work has been partially carried out under the project "Structured matrix analysis: numerical methods and applications" by the Italian Minister of Research and University, grant number 2002014121, and with the partial support of GNCS, the Italian National Group of Scientific Computing of the INDAM, Istituto Nazionale di Alta Matematica.

CONTENTS

Part I

Tools

1

INTRODUCTION TO MARKOV CHAINS

1.1 Discrete-time Markov chains

Markov chains are used to model systems which evolve in time. They come under various guises but we only consider here *discrete-state* processes, meaning that the total number of states which the process may occupy is either finite or countably infinite. Time may either increase by discrete, constant amounts, as when the modeled system is controlled by a digital clock, or it may increase continuously.

A stochastic process[1] is a family $\{X_t : t \in T\}$ of random variables X_t indexed by some set T and with values in a common set E: $X_t \in E$ for all $t \in T$. Here, E is a denumerable set and is called the *state space*, and T is the *time space*. If T is denumerable, say $T = \mathbb{N}$, the process is said to be *discrete*, otherwise it is *continuous*.

We first define discrete-state, discrete-time Markov chains.

Definition 1.1 (Markov property) *The stochastic process $\{X_n : n \in \mathbb{N}\}$ is a Markov chain if*[2]

$$\mathrm{P}[X_{n+1} = j | X_0, X_1, \ldots, X_{n-1}, X_n] = \mathrm{P}[X_{n+1} = j | X_n], \tag{1.1}$$

for all states $j \in E$, and for all time $n \in \mathbb{N}$.

This means that if one knows the state X_n of the system at time n, then the past history X_0, X_1, ..., X_{n-1} does not help in determining which state might be occupied at time $n+1$. One also usually requires that the laws which govern the evolution of the system be time-invariant; this is formulated as follows.

Definition 1.2 (Homogeneity assumption) *A Markov chain $\{X_n : n \in \mathbb{N}\}$ is homogeneous if*

$$\mathrm{P}[X_{n+1} = j | X_n = i] = \mathrm{P}[X_1 = j | X_0 = i], \tag{1.2}$$

for all states i, $j \in E$ and for all time $n \in \mathbb{N}$.

[1] We give in Section 1.9 a few references for the reader who might wish for a more elaborate presentation than we give here, about stochastic processes in general and Markov chains in particular.

[2] Throughout, we denote by $\mathrm{P}[X = j]$ the probability that the random variable X takes the value j, and by $\mathrm{P}[X = j | Y = i]$ the conditional probability that X takes the value j, given that the random variable Y takes the value i.

A few examples may help the reader get some feeling for the type of processes which we have in mind.

Example 1.3 Queueing models are used to represent systems of resources, traditionally called "servers", which are to be used by several users, called "customers". The terminology stems from applications like bank tellers, hotel reception desks, toll booths on freeways, and so on, where customers actually queue up until they are served by an employee. Queueing theory is used in much more diverse circumstances, however.

Simple queues consist of one server which attends to one customer at a time, in order of their arrivals, with the added assumption that customers are indefinitely patient. We assume that time is discretized into intervals of fixed length, that a random number of customers join the system during each interval, and that the server removes one customer from the queue at the end of each interval, if there is any. Defining α_n as the number of new arrivals during the interval $[n-1, n)$ and X_n as the number of customers in the system at time n, we have

$$X_{n+1} = \begin{cases} X_n + \alpha_{n+1} - 1 & \text{if } X_n + \alpha_{n+1} \geq 1 \\ 0 & \text{if } X_n + \alpha_{n+1} = 0. \end{cases} \tag{1.3}$$

If $\{\alpha_n\}$ is a collection of independent random variables, then X_{n+1} is conditionally independent of $X_0, \ldots, X_{n-1}$ if X_n is known. If, in addition, the α_n's are identically distributed, then $\{X_n\}$ is homogeneous. The state space is $\mathbb{N}$.

Example 1.4 Phase-type random variables are extensively used in Markov modeling. An elementary case is given by the negative binomial distribution: take a Bernoulli sequence of independent trials, each of which results in a success with probability q and failure with probability $1-q$. The trial number V of the kth success has a negative binomial distribution with parameters k and q. We may represent this random variable through a process $\{X_n\}$ on the state space $E = \{0, 1, \ldots, k\}$, where X_n corresponds to the number of successes still to be observed after the nth trial. Starting from $X_0 = k$, we have

$$X_{n+1} = \begin{cases} X_n - \mathcal{I}_{n+1} & \text{if } X_n \geq 1 \\ 0 & \text{if } X_n = 0, \end{cases} \tag{1.4}$$

where $\{\mathcal{I}_n\}$ is a sequence of independent, identically distributed random variables, equal to one with probability q and zero otherwise. The instant when X_n becomes zero determines V.

The Markov and the homogeneity assumptions are extremely powerful, for they automatically induce a number of properties of Markov chains, as we show below, and they allow one to simplify several of the proofs in the remainder of this book. In the sequel, we always assume that Markov chains are *homogeneous.*

Define the matrix $P = (p_{i,j})_{i,j \in E}$ with one row and one column for each state in E and such that

$$p_{i,j} = \mathrm{P}[X_1 = j | X_0 = i], \qquad \text{for all } i,\ j \text{ in } E.$$

This is called the *transition matrix* of the Markov chain; it is a row-stochastic matrix, that is, its elements are nonnegative and its row sums are all equal to one, which we write in short as $P \geq 0$, $P\mathbf{1} = \mathbf{1}$, where $\mathbf{1}$ is the vector with all components equal to one.

Returning to the two examples given earlier, we see that the transition matrix associated with (1.3) is

$$P = \begin{bmatrix} b & q_2 & q_3 & q_4 & \cdots \\ q_0 & q_1 & q_2 & q_3 & \ddots \\ & q_0 & q_1 & q_2 & \ddots \\ 0 & & \ddots & \ddots & \ddots \end{bmatrix}, \tag{1.5}$$

where q_i is the probability $\mathrm{P}[\alpha = i]$ that i new customers join the queue during a unit time interval, α denoting any of the identically distributed random variables α_n, and $b = q_0 + q_1$.

For the Markov chain defined by (1.4), the transition matrix is

$$P = \begin{bmatrix} 1 & & & & 0 \\ q & 1-q & & & \\ & q & 1-q & & \\ & & \ddots & \ddots & \\ 0 & & & q & 1-q \end{bmatrix}. \tag{1.6}$$

For the general phase-type random variables alluded to in Example 1.4, the first row is like in (1.6) but the remainder of the transition matrix may be anything.

As a matter of fact, the matrix (1.5) is a special case of the so-called M/G/1 queue,given by

$$P = \begin{bmatrix} b_0 & b_1 & b_2 & b_3 & \cdots \\ a_{-1} & a_0 & a_1 & a_2 & \ddots \\ & a_{-1} & a_0 & a_1 & \ddots \\ 0 & & \ddots & \ddots & \ddots \end{bmatrix} \tag{1.7}$$

with a_i, b_i nonnegative for all i, $\sum_{i=0}^{+\infty} b_i = \sum_{i=-1}^{+\infty} a_i = 1$. This is one of the structures which we shall extensively investigate. Another one is the G/M/1 queue, the transition matrix of which being

$$P = \begin{bmatrix} b_0 & a_1 & & & 0 \\ b_{-1} & a_0 & a_1 & & \\ b_{-2} & a_{-1} & a_0 & a_1 & \\ b_{-3} & a_{-2} & a_{-1} & a_0 & \ddots \\ \vdots & \vdots & \ddots & \ddots & \ddots \end{bmatrix} \tag{1.8}$$

with a_i, b_i nonnegative for all i, $\sum_{i=-1}^{+\infty} a_{-i} \le 1$, and $b_{-n} = 1 - \sum_{i=-n+1}^{1} a_i$ for $n = 0, 1, 2, \ldots$

It is no surprise that the transition matrix should play an important role since the dynamic behavior of Markov chains is completely characterized by the one-step transition probabilities. The next two propositions are readily proved[3] by induction.

Proposition 1.5 *One has*

$$\mathrm{P}[X_{n+k} = j | X_n = i] = (P^k)_{i,j},$$

for all times $n \ge 0$, all intervals of time $k \ge 0$, and all pairs of states i and j in E.

Let $\boldsymbol{\pi}^{(n)} = (\pi_i^{(n)})_{i \in E}$ be the probability vector of the Markov chain at time n, that is, $\pi_i^{(n)} = \mathrm{P}[X_n = i | X_0]$. Then Proposition 1.5 implies the following vector equation

$$\boldsymbol{\pi}^{(n+1)^{\mathrm{T}}} = \boldsymbol{\pi}^{(n)^{\mathrm{T}}} P, \qquad n \ge 0$$

and also

$$\boldsymbol{\pi}^{(n)^{\mathrm{T}}} = \boldsymbol{\pi}^{(0)^{\mathrm{T}}} P^n, \qquad n \ge 0.$$

This draws attention to the fact that, in addition to the system dynamics, one must choose the starting point X_0. This may be a specific state, or it may be chosen at random, depending on the circumstances. Most of our interest, however, will concentrate on conditional probabilities and we shall seldom explicitly take the initial probability distribution into account.

Proposition 1.6 *One has*

$$\begin{aligned}
&\mathrm{P}[X_{n+1} = j_1, X_{n+2} = j_2, \ldots, X_{n+k} = j_k | X_n = i] \\
&= \mathrm{P}[X_1 = j_1, X_2 = j_2, \ldots, X_k = j_k | X_0 = i] \\
&= p_{i,j_1} p_{j_1,j_2} \cdots p_{j_{k-1},j_k}
\end{aligned}$$

for all times $n \ge 0$, all intervals of time $k \ge 1$, and all sequences of states i, j_1, $\ldots$, j_k in E.

In the definitions and propositions above, the time n is arbitrary but fixed. Proposition 1.6, for instance, states that if we wait n units of time, observe the system and find that it is in state i, then we know that the system will evolve after time n just as if it were a new system which would start in state i with its own clock at zero. Sometimes, similar properties hold if the time chosen for the observation is random, even if it is determined by the system itself. This is true, in particular, for stopping times, as stated by the important strong Markov property, which we introduce below.

[3] The material in this chapter is quite standard and appears widely in the literature on stochastic processes. For that reason, we do not give detailed proofs, except when we feel that the method of proof itself will help in understanding the material in later chapters.

In short, a random variable V is a *stopping time* for the Markov chain $\{X_0, X_1, X_2, \ldots\}$ if it suffices to examine X_0, X_1, ..., X_k, in order to determine whether $V \leq k$ or $V > k$. In other words, we may write that V, if it is finite, is computable, using the Markov chain. To give one example, take a subset A of states and define T as the *first* epoch when the system visits a state in A:

$$T = \min\{n \geq 0 : X_n \in A\},$$

with T being taken to be infinite if $X_n \notin A$ for all n. This is a stopping time. An example of a random variable which is not a stopping time is the *last* visit to A.

Theorem 1.7 (Strong Markov property) *Consider the Markov chain $\{X_n : n \geq 0\}$ with transition matrix P. Let V be a stopping time for the Markov chain. Conditional on $V < \infty$ and $X_V = i$, the process $\{X_{V+n} : n \geq 0\}$ is a Markov chain with transition matrix P and initial state i, and is independent of X_0, X_1, ..., X_V.*

This is a fundamental property, which is used over and over again in the analysis of Markov chains.

1.2 Classification of states

The states of a Markov chain may be transient or positive recurrent or null recurrent, and their characteristics are very different, depending on the category to which they belong. The classification is based on the random variables which count the number of visits to a state and on those which measure intervals of time between successive visits.

Denote by N_j the total number of times the Markov chain visits state j. Formally, we write that

$$N_j = \sum_{n=0}^{+\infty} \mathbb{I}\{X_n = j\}, \tag{1.9}$$

where $\mathbb{I}\{\cdot\}$ is the indicator function: it has the value 1 if the condition inside the braces is true and the value 0 otherwise.

In order to analyze N_j, we need to introduce the *first passage time* θ_j to state j:

$$\theta_j = \min\{n \geq 0 : X_n = j\},$$

taken to be infinite if $X_n \neq j$ for all n. In some cases, we are interested in the first visit to j without taking into account the state at 0. The time T_j until this event is called the *first return time*, defined by

$$T_j = \min\{n \geq 1 : X_n = j\};$$

it differs from θ_j only if $X_0 = j$. We also define

$$f_{i,j} = \mathrm{P}[T_j < \infty | X_0 = i],$$

that is, $f_{i,j}$ is the probability that, starting from i, the Markov chain returns to j in a finite time.

The following property is easily proved. We give the details because they illustrate in a simple manner the use of the strong Markov property.

Theorem 1.8 *The conditional distribution of N_j, given that $X_0 = j$, is*

$$\mathrm{P}[N_j = n | X_0 = j] = f_{j,j}^{n-1}(1 - f_{j,j}), \qquad \text{for } n \geq 1. \tag{1.10}$$

Proof Since $X_0 = j$, the number of visits to j is at least equal to 1. It is exactly 1 if the Markov chain never returns to j after time 0, and thus

$$\mathrm{P}[N_j = 1 | X_0 = j] = \mathrm{P}[T_j = \infty | X_0 = j] = 1 - f_{j,j}.$$

If the Markov chain does return to j, then we may write that $N_j = 1 + \tilde{N}_j$ where $\tilde{N}_j$ is the total number of visits after (and including) the first return, that is, $\tilde{N}_j$ is the total number of visits to j for the process $X_{T_j}, X_{T_j+1}, X_{T_j+2}, \ldots$. By the strong Markov property, this has the same distribution as N_j itself, and we may write that

$$\mathrm{P}[N_j = n | X_0 = j] = f_{j,j}\mathrm{P}[\tilde{N}_j = n - 1 | T_j < \infty] = f_{j,j}\mathrm{P}[N_j = n - 1 | X_0 = j],$$

and (1.10) is easily proved by induction. □

The next property is stated without proof.

Corollary 1.9 *The conditional distribution of N_j, given that $X_0 = i \neq j$, is*

$$\begin{aligned} \mathrm{P}[N_j = 0 | X_0 = i] &= 1 - f_{i,j}, \\ \mathrm{P}[N_j = n | X_0 = i] &= f_{i,j} f_{j,j}^{n-1}(1 - f_{j,j}), \qquad \text{for } n \geq 1. \end{aligned} \tag{1.11}$$

The important observation here is that we identify two categories of states with different characteristics: if $f_{j,j} < 1$, then N_j has a proper distribution and it is finite with probability 1. Such states are called *transient* and may be visited a finite number of times only by the Markov chain. If, on the contrary, $f_{j,j} = 1$, then $\mathrm{P}[N_j = n | X_0 = j] = 0$ for all finite n, which implies that $N_j = \infty$ with probability 1. Such states are called *recurrent* and, once the Markov chain has visited one of these states, it will return to it over and over again.

A further distinction is made between positive recurrent and null recurrent states: a state is *positive recurrent* if the expected[4] return time $\mathrm{E}[T_j | X_0 = j]$ is finite; it is *null recurrent* if the expectation of T_j is infinite, despite the fact that T_j is finite with probability one.

[4]We denote by $\mathrm{E}[X]$ the *expected value* of the random variable X and by $\mathrm{E}[X|A]$ the *conditional expectation* of X, given the event A.

Defining $r_{i,j} = \mathrm{E}[N_j|X_0 = i]$, we find from Theorem 1.8 and Corollary 1.9 that

$$r_{i,j} = \begin{cases} f_{i,j}/(1 - f_{j,j}), & \text{for } i \neq j, \\ 1/(1 - f_{j,j}), & \text{for } i = j, \end{cases} \tag{1.12}$$

where for $i \neq j$ the right-hand side of (1.12) is assumed to be zero if $f_{i,j} = 0$, that is, if it is impossible to reach j from i, and infinity if $f_{i,j} > 0$ and $f_{j,j} = 1$; similarly, for $i = j$ the right-hand side of (1.12) is infinity if $f_{j,j} = 1$.

Thus, another way of making a distinction among the states is by noting that $r_{j,j}$ is finite if and only if state j is transient in which case $r_{i,j}$ is finite for all starting states i; for recurrent states, $r_{i,j}$ is infinite or zero, depending on whether or not it is possible to reach j from i.

As a consequence of this, the matrix P has the following property.

Theorem 1.10 *Consider the power series $S = (s_{i,j})_{i,j \in E} = \sum_{n=0}^{+\infty} P^n$. The element $s_{i,j}$ is the conditional expected number of visits to j, given that the initial state is i.*

If j is a transient state, then $s_{i,j} \leq s_{j,j} < \infty$, for all i. If j is recurrent, then $s_{j,j} = \infty$, and $s_{i,j} = \infty$ or 0, for $i \neq j$.

Proof From (1.9), we have by the monotone convergence theorem[5]

$$\begin{aligned} \mathrm{E}[N_j|X_0 = i] &= \sum_{n=0}^{+\infty} \mathrm{E}[\mathbb{I}\{X_n = j\}|X_0 = i] = \sum_{n=0}^{+\infty} \mathrm{P}[X_n = j|X_0 = i] \\ &= \sum_{n=0}^{+\infty} (P^n)_{i,j} = s_{i,j} \end{aligned}$$

The proof of the theorem is now a consequence of the discussion above. □

1.3 Irreducible classes

One associates in a natural way a *transition graph* with the Markov chain: to each state there corresponds a node of the graph and one defines a directed arc from node i to node j, denoted by (i, j), for each pair of states such that $p_{i,j} > 0$. A finite sequence of directed arcs $(i, i_1), (i_1, i_2), \cdots, (i_k, j)$ is called a *path* from state i to state j. A path from i to i is called a *loop* through i. We also write that state i leads to j if there is a path from i to j in the transition graph, and we write that i and j communicate if i leads to j and j leads to i.

It is useful to adopt the convention that every state communicates with itself, even if there is no loop through it; with this convention, we may regard the relation of communication as an equivalence relation which induces a partition of the state space into equivalence classes, called *communicating classes of states*. In other words, the subset C of E is a communicating class if for every i in C, every j in E which communicates with i belongs to C as well. In the literature,

[5]The monotone convergence theorem is stated as Theorem A.17 in the appendix.

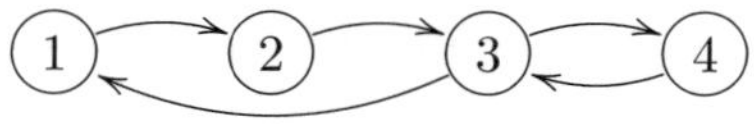

Fig. 1.1 Transition graph of an irreducible Markov chain

communicating classes are also said to be *irreducible.* A Markov chain is called irreducible if all its states communicate, that is, if they form one single communicating class; in this case the transition matrix P is also irreducible.[6] We show in Figure 1.1 an example of a transition graph associated to an irreducible Markov chain.

If a Markov chain with transition matrix P has K communicating classes, denoted by $C_1, C_2, \ldots, C_K$, then the states may be permuted so that the transition matrix $P' = \Pi P \Pi^{\mathrm{T}}$ associated with the permuted states is block triangular:

$$P' = \begin{bmatrix} P_{1,1} & & & 0 \\ P_{2,1} & P_{2,2} & & \\ \vdots & \vdots & \ddots & \\ P_{K,1} & P_{K,2} & \ldots & P_{K,K} \end{bmatrix} \tag{1.13}$$

where $P_{i,j}$ is the submatrix of transition probabilities from the states of C_i to C_j, the diagonal blocks are irreducible square matrices and Π is the permutation matrix associated with the rearrangement.

One also defines *final* and *passage* classes: a communicating class C is a final class if there is no path out of it: for every state i in C, there is no state j out of C such that i leads to j. If, on the contrary, the states in C lead to some state out of C, the class is said to be a passage class. A single state which forms a final class by itself is called *absorbing.* Examples of passage classes and final classes are shown in Figures 1.2 and 1.3. For the Markov chain associated with the transition matrix (1.13), the class C_1 is final and each class C_i, for $2 < i \le K$, is final if and only if $P_{i,j} = 0$ for $j = 1, \ldots, i-1$.

It is clear that, for the Markov chain (1.4) with transition matrix (1.6), the states 1 to k are each in a passage class of their own, if $q > 0$, and that state zero is absorbing. The Markov chain (1.3) with transition matrix (1.7) is irreducible if $a_{-1} > 0$ and $a_{-1} + a_0 < 1$.

In the case where the state space is finite, the directed graph associated with a Markov chain is a useful tool for interpreting the classification of states given in Section 1.2. In fact, it turns out that a state i is transient if it leads to some state j from which there is no path back to i, otherwise it is recurrent. This follows from the two theorems below. Thus, State 3 in both Figures 1.2 and 1.3 is transient.

If the state space is infinite, then the directed graph is not enough to thoroughly identify all transient and recurrent states and we return to this issue in Section 1.5.

[6]See Definition A.8 in the appendix.

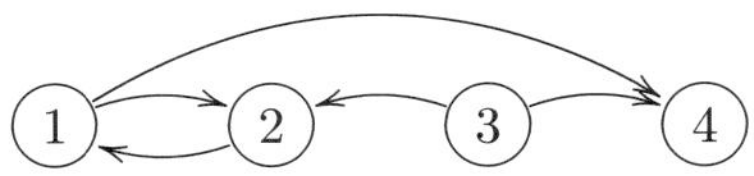

Fig. 1.2 The set $C = \{1, 2\}$ is a passage class, State 3 forms a passage class by itself and State 4 is absorbing.

Fig. 1.3 The set $C = \{1, 2\}$ is a final class, State 3 forms a passage class by itself and State 4 is absorbing.

The theorem below shows that the nature of a state is a class property.

Theorem 1.11 *The states in a passage class are transient. In a final class, the states are either all positive recurrent, or all null recurrent, or all transient.*

The second statement in the theorem only tells us that, if we can somehow determine the nature of one state in a final class, then we can extend our identification to the other states in the same class. In this way, we call a final class positive/null recurrent or transient if all its states are positive/null recurrent or transient, respectively. We can be more specific in the case where a final class only contains finitely many states, as we now state.

Theorem 1.12 *If a final class contains a finite number of states only, then it is positive recurrent.*

A final property worth mentioning is that the interconnection between recurrent states is very strong: once a Markov chain has visited a recurrent state i, it will visit with probability one each state which is on a path from i, and each of these states will be reached in a finite time. That is the meaning of the following theorem.

Theorem 1.13 *If C is a final class of recurrent states, then $f_{i,j} = 1$ for all i and j in C.*

On the basis of these properties, we can make a few general statements about the transition matrix P and the power series S. Assume that there are K communicating classes, denoted by $C_1, C_2, \ldots, C_K$, so that $P' = \Pi P \Pi^{\mathrm{T}}$ has the structure (1.13). The matrix $S' = \sum_{n=0}^{+\infty}(P')^n$ has the same structure as P'. If C_j is a transient class, then $S_{i,j}$ is finite for all i. If C_j is recurrent, then all the elements of $S_{j,j}$ are infinite and, for $i \neq j$, $S_{i,j}$ is identically zero if there is no path from C_i to C_j, or $S_{i,j}$ contains infinite elements only, if there is such a path.

This property holds also in the case where the number K of communicating classes is infinite.

In consequence, once the class structure is elucidated, which is usually very easy, we need only concentrate on the analysis of the final classes. This explains why it is usually assumed that the Markov chain under study is irreducible, or, equivalently, that all states communicate.

1.4 First passages

By Theorem 1.11 we know that, when the Markov chain is irreducible, if one state is transient, then all states are transient and every element in the series $S = \sum_{n=0}^{+\infty} P^n$ is finite. If, on the contrary, one state is recurrent, then every element of S is infinite. Now, let us partition the state space E into two nonempty subsets A and B ($A \cup B = E$, $A \cap B = \emptyset$). Denote by P_A, P_B the submatrices of P made up by the elements of P with indices in A and B, respectively, denote also by $P_{A,B}$ the submatrix of P with row indices in A and column indices in B; similarly for $P_{B,A}$. Arrange the states so that the transition matrix can be partitioned as

$$P = \begin{bmatrix} P_A & P_{A,B} \\ P_{B,A} & P_B \end{bmatrix}. \tag{1.14}$$

Define τ as the first return time to any state out of A:

$$\tau = \min\{n \geq 1 : X_n \notin A\};$$

we have the following property.

Theorem 1.14 *Assume that the Markov chain is irreducible.*

- *For any proper subset A of states, the series $S = \sum_{n=0}^{+\infty} P_A^n$ converges.*
- *For all i and j in A, $S_{i,j}$ is the expected number of visits to state j, starting from i, during the interval of time $[0, \tau]$.*
- *For all i in B and j in A, $(P_{B,A}S)_{i,j}$ is the expected number of visits to state j, starting from i, during the interval of time $[0, \tau]$.*
- *For all i in A, $(S\mathbf{1})_i = \mathrm{E}[\tau | X_0 = i]$, that is, it is the expected first passage time to B, starting from i.*
- *For all i in A and j in B, $(SP_{A,B})_{i,j} = \mathrm{P}[\tau < \infty \text{ and } X_\tau = j | X_0 = i]$, that is, it is the probability that j is the first state visited in B, given that the process starts in i.*

Proof Define the matrix

$$\tilde{P} = \begin{bmatrix} P_A & P_{A,B} \\ 0 & I \end{bmatrix}$$

which corresponds to a new Markov chain where the states in B have been made absorbing: if the Markov chain enters state j in B, then it stays there forever. The important point here is that the first passage time from A to B is identical in the original and in the new processes.

Since P is irreducible, there is for every i in A a path to some j in B, so that the states of A are transient in the new Markov chain, and the expected number of visits to j in A, starting from i in A, is finite. From Theorem 1.10, these expected numbers of visits are given by the upper-left corner of $\tilde{S} = \sum_{n=0}^{+\infty} \tilde{P}^n$, which is clearly given by $\sum_{n=0}^{+\infty} P_A^n$. This proves the first two claims.

The third statement is proved by conditioning on the first transition:

$$\mathrm{E}[N_j | X_0 = i] = \sum_{k \in A} P_{i,k} \mathrm{E}[N_j | X_0 = k] = (P_{B,A}S)_{i,j}.$$

The fourth statement results from the fact that if $X_0 \in A$, then $\tau = \sum_{j \in A} N_j$, where N_j is the total number of visits to A.

Finally, if we decompose the event $[X_\tau = j]$ according to the possible values of τ and the last state visited in A at time $\tau - 1$, we have

$$\begin{aligned}
\mathrm{P}[X_\tau = j | X_0 = i] &= \sum_{n=1}^{+\infty} \sum_{k \in A} \mathrm{P}[X_{n-1} = k, X_n = j, \tau = n | X_0 = i] \\
&= \sum_{n=1}^{+\infty} \sum_{k \in A} (P_A^{n-1})_{i,k} (P_{A,B})_{k,j} = (S P_{A,B})_{i,j},
\end{aligned}$$

which concludes the proof. □

This property has a number of immediate algebraic consequences, which we state as a corollary for future reference.

Corollary 1.15 *The system $(I - P_A)X = I$ always has a solution. The solution is unique and $S = (I - P_A)^{-1}$ if the size of P_A is finite. If the size of P_A is infinite, $S = \sum_{n=0}^{+\infty} P_A^n$ is the minimal nonnegative solution.*

Proof Clearly, $S_1 = I + P_A S$, so that S is a solution of the system $(I - P_A)X = I$. If the size of P_A is finite, that solution is $(I - P_A)^{-1}$. If the size is infinite, there may be multiple solutions. For any nonnegative solution X, we may write

$$\begin{aligned}
X &= I + P_A X \\
&= I + P_A + P_A^2 + \cdots + P_A^n + P_A^{n+1} X \\
&\geq I + P_A + P_A^2 + \cdots + P_A^n
\end{aligned}$$

for all n, so that $X \geq \lim_{n \to +\infty} \sum_{i=0}^{n} P_A^i = S$, which concludes the proof. □

1.5 Stationary distribution

We discuss in this section the asymptotic behavior of the probability distribution of X_n, or, equivalently, the successive powers P^n of the transition matrix, as $n \to +\infty$. Before doing so, we need to introduce the notion of periodicity.

By definition, a state i has *periodicity* $\delta > 1$ if all loops through i in the transition graph have a length which is a multiple of δ. This is equivalent to saying that $\mathrm{P}[X_n = i | X_0 = i] > 0$ only if $n = 0 \mod \delta$. The case of a state i such that $\mathrm{P}[X_n = i | X_0 = i] = 0$ for all n poses technical difficulties only and we do not deal with such cases.

Periodicity is a class property and all states in a communicating class have the same period. Thus, as long as we restrict ourselves to irreducible Markov chains, either all states are non-periodic, or all have the same period δ, which we may call the period of the Markov chain itself.

Figure 1.4 shows an example of a periodic irreducible Markov chain of period 3. The associated transition matrix has the following structure where "$*$" denotes a nonzero element

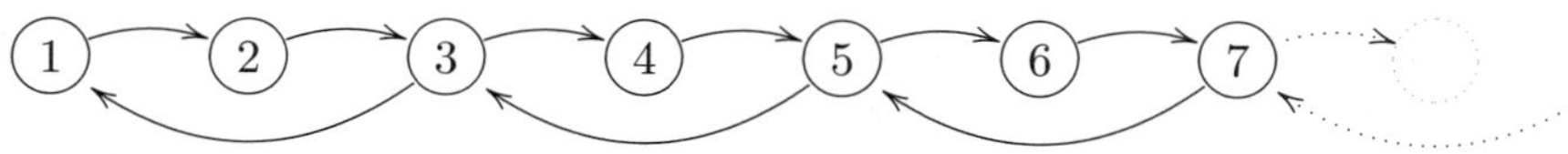

Fig. 1.4 Graph of an infinite irreducible periodic Markov chain of period 3.

$$P = \begin{bmatrix} 0 & * & & & & & & \mathbf{0} \\ 0 & 0 & * & & & & & \\ * & 0 & 0 & * & & & & \\ 0 & 0 & 0 & 0 & * & & & \\ 0 & 0 & * & 0 & 0 & * & & \\ 0 & 0 & 0 & 0 & 0 & 0 & \ddots & \\ 0 & 0 & 0 & 0 & * & 0 & \ddots & \\ \dots & \dots & \dots & \dots & \dots & \dots & \dots & \dots \end{bmatrix}.$$

If P is the transition matrix of an irreducible finite periodic Markov chain of period δ, then there exists a permutation matrix Π such that $P' = \Pi P \Pi^{\mathrm{T}}$ has the form

$$P' = \begin{bmatrix} 0 & 0 & \dots & 0 & P_{1,\delta} \\ P_{2,1} & 0 & \ddots & & 0 \\ & P_{3,2} & \ddots & \ddots & \vdots \\ & & \ddots & 0 & 0 \\ \mathbf{0} & & & P_{\delta,\delta-1} & 0 \end{bmatrix}, \tag{1.15}$$

where the zero diagonal blocks are square. Any irreducible matrix P which can be put in the form (1.15) by means of the transformation $P' = \Pi P \Pi^{\mathrm{T}}$ is called *cyclic* of period (or index of cyclicity) δ.

If the Markov chain is irreducible and nonperiodic, then it is not difficult to verify that for all i and j, there exists n_0 such that $\mathrm{P}[X_n = j | X_0 = i] > 0$ for all $n \geq n_0$. This means that, for every starting state i and every destination state j of interest, there is a strictly positive probability of being in j at any given time, provided that we look far enough in the future. Clearly, this cannot hold if the Markov chain is periodic since successive visits to j are separated by intervals of time which are multiples of the period. Nevertheless, there is a similar property and one shows that for all i and j, there exist r and n_0 such that $\mathrm{P}[X_{n\delta+r} = j | X_0 = i] > 0$ for all $n \geq n_0$, and $\mathrm{P}[X_m = j | X_0 = i] = 0$ for all $m \neq r \mod \delta$.

We are now ready to discuss the limiting distribution, which is easiest to handle in the transient case.

Theorem 1.16 *Assume that the Markov chain is irreducible. If the states are transient, then* $\lim_{n\to+\infty} \mathrm{P}[X_n = j | X_0 = i] = 0$ *for all* i *and* j.

Proof The proof immediately follows from the fact that the series $\sum_{n=0}^{+\infty} (P^n)_{i,j} = \sum_{n=0}^{+\infty} \mathrm{P}[X_n = j | X_0 = i]$ converges for all i and j, by Theorem 1.10. □

For recurrent states, the proofs are more involved and we refer to Norris [93, Sections 1.7 and 1.8]. Let us first discuss the positive recurrent case.

Theorem 1.17 *Assume that the Markov chain is irreducible. The states are positive recurrent if and only if there exists a strictly positive invariant probability vector, that is, a vector* $\boldsymbol{\pi} = (\pi_i)$ *such that* $\pi_i > 0$ *for all* i*, with*

$$\boldsymbol{\pi}^{\mathrm{T}} P = \boldsymbol{\pi}^{\mathrm{T}} \qquad \text{and} \qquad \boldsymbol{\pi}^{\mathrm{T}} \mathbf{1} = 1. \tag{1.16}$$

In that case,

- *if the Markov chain is nonperiodic, then*

$$\lim_{n \to +\infty} \mathrm{P}[X_n = j | X_0 = i] = \pi_j \tag{1.17}$$

for all j*, independently of* i*;*

- *if the Markov chain is periodic, with period* δ*, then*

$$\lim_{n \to +\infty} \mathrm{P}[X_{n\delta} = j | X_0 = j] = \delta \pi_j$$

for all j*.*

- *The invariant vector* $\boldsymbol{\pi}$ *is unique among nonnegative vectors, up to a multiplicative constant.*

One observes that, in the non-periodic case, π_i is the inverse of the expected return time to i, starting from i. Also, we may write (1.17) in the periodic case too.

Null-recurrent states being in some sense at the boundary between transient and positive recurrent states, their properties fall somewhat between Theorems 1.16 and 1.17 as we now state.

Theorem 1.18 *Assume that the Markov chain is irreducible. If the states are null recurrent, then* $\lim_{n \to +\infty} \mathrm{P}[X_n = j | X_0 = i] = 0$ *for all* i *and* j*. In addition, there exists a strictly positive invariant vector, unique up to a multiplicative constant, and such that the sum of its elements is not finite (stated otherwise, this vector is not normalizable).*

Thus, there always exists an invariant vector for the transition matrix of a recurrent Markov chain. Some transient Markov chains also have an invariant vector (with infinite mass, like in the null recurrent case) but some do not.

The following examples illustrate the situations encountered with positive, null recurrent and transient Markov chains.

Example 1.19 For the transition matrix

$$P = \begin{bmatrix} 0 & 1 & & & 0 \\ 1/2 & 0 & 1/2 & & \\ & 1/2 & 0 & 1/2 & \\ 0 & & \ddots & \ddots & \ddots \end{bmatrix}$$

we have $\boldsymbol{\pi}^{\mathrm{T}} P = \boldsymbol{\pi}^{\mathrm{T}}$ with $\boldsymbol{\pi}^{\mathrm{T}} = [1/2, 1, 1, \ldots]$. The vector $\boldsymbol{\pi}$ has "infinite" mass that is, the sum of its components is infinite. Observe that the Markov chain is actually null recurrent but, in order to show this, one needs to verify that the return time to any of its states is finite with probability one.

Example 1.20 For the transition matrix

$$P = \begin{bmatrix} 0 & 1 & & & 0 \\ 1/4 & 0 & 3/4 & & \\ & 1/4 & 0 & 3/4 & \\ & & & & \\ 0 & & \ddots & \ddots & \ddots \end{bmatrix}$$

we have $\boldsymbol{\pi}^{\mathrm{T}} P = \boldsymbol{\pi}^{\mathrm{T}}$ with $\boldsymbol{\pi}^{\mathrm{T}} = [1, 4, 12, 36, 108, \ldots]$. The vector $\boldsymbol{\pi}$ has unbounded elements. In this case the Markov chain is transient.

Example 1.21 For the transition matrix

$$P = \begin{bmatrix} 0 & 1 & & & 0 \\ 3/4 & 0 & 1/4 & & \\ & 3/4 & 0 & 1/4 & \\ & & & & \\ 0 & & \ddots & \ddots & \ddots \end{bmatrix}$$

we have $\boldsymbol{\pi}^{\mathrm{T}} P = \boldsymbol{\pi}^{\mathrm{T}}$ with $\boldsymbol{\pi}^{\mathrm{T}} = (2/3)[1/2, 2/3, 2/9, 2/27, \ldots]$ and $\sum_{i=0}^{+\infty} \pi_i = 1$. In this case the Markov chain is positive recurrent by Theorem 1.17.

1.6 Censoring of a Markov chain

The property in this section is used in later chapters, when we analyze the stationary distribution of structured Markov chains. It is important because it gives a probabilistic meaning to many algorithmic procedures for the stationary distribution of a positive recurrent Markov chain, and, therefore, it illustrates well our approach which relies both on algebraic and on probabilistic arguments.

We partition the state space into two subsets, A and B. We denote by $\{t_0, t_1, t_2, \ldots\}$ the epochs where the Markov chain visits the set A:

$$t_0 = \min\{n \geq 0 : X_n \in A\}, \qquad t_{k+1} = \min\{n \geq t_k + 1 : X_n \in A\},$$

for $k \geq 0$.

Definition 1.22 *The censored process restricted to the subset A is the sequence $\{X_{t_n}\}_{n \geq 0}$ of successive states visited by the Markov chain in A.*

In other words, we make the subset B invisible by censoring out from the original Markov chain all the intervals of time during which it is in B.

Define $Y_n = X_{t_n}$, for $n \geq 0$; by the strong Markov property, the censored process $\{Y_n\}$ is a Markov chain also. If we partition the transition matrix P of $\{X_n\}$ as in (1.14), then the transition matrix of $\{Y_n\}$ is

$$P' = P_A + \sum_{n=0}^{+\infty} (P_{A,B} P_B^n P_{B,A}) \tag{1.18}$$

provided that $\sum_{n=0}^{+\infty}(P_{A,B}P_B^nP_{B,A})$ is convergent. If the series

$$S' = \sum_{n=0}^{+\infty} P_B^n. \tag{1.19}$$

is convergent, then we may rewrite (1.18) as

$$P' = P_A + P_{A,B}S'P_{B,A}. \tag{1.20}$$

The argument goes in three steps.

1. Define $T = S'P_{B,A}$. By Theorem 1.14, $T_{k,j}$ is the probability that, starting from k in B, the Markov chain eventually reaches j in A before any other state in A.
2. Thus, $(P_{A,B}T)_{i,j}$ is the probability that, starting from i in A, the Markov chain moves to B before eventually returning to A, with j being the first state visited in A.
3. Finally, by adding the one-step transition probability matrix P_A, we obtain that $(P')_{i,j}$ is the probability of moving from i to j, either directly or after an interval of unspecified length spent in B.

Observe that there is a strong connection between censoring and Schur complementation:[7] if the state space is finite, then

$$I - P' = I - P_A - P_{A,B}(I - P_B)^{-1}P_{B,A}$$

since $S' = (I - P_B)^{-1}$ by Corollary 1.15, so that $I - P'$ is the Schur complement of $I - P_B$ in the matrix

$$\begin{bmatrix} I - P_A & -P_{A,B} \\ -P_{B,A} & I - P_B \end{bmatrix}.$$

The identification to the Schur complement extends to the case where the set B is infinite since S' is, in that case, the minimal nonnegative inverse of $I - P_B$.

Now, let us partition the stationary vector $\boldsymbol{\pi}$ as $\boldsymbol{\pi}^{\mathrm{T}} = [\boldsymbol{\pi}_A^{\mathrm{T}}, \boldsymbol{\pi}_B^{\mathrm{T}}]$, with $\boldsymbol{\pi}_A = (\pi_i)_{i\in A}$ and $\boldsymbol{\pi}_B = (\pi_i)_{i\in B}$. We have the following property.

Theorem 1.23 *Assume that the Markov chain is irreducible and positive recurrent. Partition the state space E into the subsets A and B. We have*

$$\boldsymbol{\pi}_A^{\mathrm{T}}P' = \boldsymbol{\pi}_A^{\mathrm{T}}, \tag{1.21}$$

$$\boldsymbol{\pi}_B^{\mathrm{T}} = \boldsymbol{\pi}_A^{\mathrm{T}}P_{A,B}S', \tag{1.22}$$

with P' and S' respectively given by (1.20), (1.19).

If the state space is finite, the proof is elementary. If E is infinite, then it is more involved and we refer to Kemeny et al. [72, Proposition 6.4, Lemmas 6.6 and 6.7] and Latouche and Ramaswami [79, Theorems 5.2.1 and 5.3.1] for two different proofs based on probabilistic arguments.

[7] Definition A.6 in the appendix.

In order to make the connection with algorithmic procedures, assume for the time being that $E = \{1, 2, \ldots, M\}$, with M finite, and take $A = \{2, \ldots, M\}$, $B = \{1\}$. It is easy to verify that (1.21, 1.22) are the equations which one obtains after applying the first step of Gaussian elimination to the system $(I-P)^{\mathrm{T}}\boldsymbol{\pi} = 0$. To illustrate the advantages of using both the algebraic and the probabilistic approaches, it suffices to mention that the Grassmann-Taksar-Heyman (GTH) algorithm was based on the observation that Gaussian elimination is identical to censoring [53; 94].

1.7 Continuous-time Markov processes

When it is necessary to model time as a continuous parameter, one uses continuous-time Markov chains, also called *Markov processes.* Their definition below closely mimics Definitions 1.1 and 1.2.

Definition 1.24 *The stochastic process $\{X(t) : t \in \mathbb{R}^+\}$ on the denumerable state space E is a homogeneous Markov process if*

$$\mathrm{P}[X(t+s) = j | X(u) : 0 \le u \le t] = \mathrm{P}[X(t+s) = j | X(t)]$$

and if

$$\mathrm{P}[X(t+s) = j | X(t) = i] = \mathrm{P}[X(s) = j | X(0) = i]$$

for all states i and j in E, for all times $t \ge 0$, and for all intervals $s \ge 0$.

Under suitable regularity assumptions about the stochastic behavior of the Markov process (assumptions which are satisfied in actual applications) one shows that the transition functions $F_{i,j}(t) = \mathrm{P}[X(t) = j | X(0) = i]$ are the minimal nonnegative solutions of the Kolmogorov equations

$$\partial F(t)/\partial t = F(t)Q \tag{1.23}$$

with $F(0) = I$, where the elements of the coefficient matrix Q have the following interpretations.

- For $i \ne j$, $Q_{i,j}$ is the instantaneous transition rate from state i to state j. That is, $Q_{i,j}h$ is the conditional probability that $X(t+h) = j$, given that $X(t) = i$, for an interval of time h small enough. Clearly, $Q_{i,j}$ is nonnegative, and it is strictly positive if it is possible to move from i to j in one jump.
- The diagonal elements are such that $Q_{i,i} = -\sum_{j \in E, j \ne i} Q_{i,j}$. Their interpretation is that the process remains in each state during an exponentially distributed interval of time, with parameter $q_i = -Q_{i,i}$ for state i, before jumping to the next state. If $Q_{i,i} = 0$, then $Q_{i,j} = 0$ for all j, which means that i is an absorbing state: once the process has reached it, it ceases to evolve and remains there forever.

The matrix Q plays exactly the same role for continuous-time Markov processes as the matrix $P - I$ does for discrete-time Markov chains. For instance, the next property is the equivalent of Theorem 1.17.

Theorem 1.25 *Assume that the Markov process is irreducible. It is positive recurrent if and only if there exists a probability vector $\boldsymbol{\pi}$ such that $\pi_i > 0$ for all i, $\boldsymbol{\pi}^{\mathrm{T}}Q = 0$, and $\boldsymbol{\pi}^{\mathrm{T}}\mathbf{1} = 1$. This vector is such that*

$$\lim_{t \to +\infty} \mathrm{P}[X(t) = j | X(0) = i] = \pi_j \tag{1.24}$$

for all j, independently of i, and is unique among nonnegative vectors, up to a multiplicative constant.

Furthermore, if we partition the state space into two proper subsets A and B, and write

$$Q = \begin{bmatrix} Q_A & Q_{A,B} \\ Q_{B,A} & Q_B \end{bmatrix},$$

where Q_A is the submatrix corresponding to the rows and columns in A, and if we define the first passage time out of A as $\tau = \inf\{t \geq 0 : X(t) \notin A\}$ then we have the next property, similar to Theorem 1.14.

Theorem 1.26 *Assume that the Markov process is irreducible.*

- *For any proper subset A of states, the system $-Q_A X = I$ has a minimal nonnegative solution $\tilde{S}$ which is finite.*
- *For all i and j in A, $(\tilde{S})_{i,j}$ is the expected total time spent in state j, starting from i, during the interval of time $[0, \tau]$.*
- *For all i in A, $(\tilde{S}_1\mathbf{1})_i = \mathrm{E}[\tau | X_0 = i]$.*
- *For all i in A and j in B, $(\tilde{S}Q_{A,B})_{i,j} = \mathrm{P}[\tau < \infty$ and $X_\tau = j | X_0 = i]$.*

From an algebraic point of view, these properties reflect the fact that the matrices Q and $P - I$ have exactly the same characteristics. Thus, any procedure which is developed for discrete-time Markov chains is immediately adapted to continuous-time Markov processes. This is the reason why we mostly consider the discrete-time case in the sequel.

1.8 Finite nonnegative matrices

Since the transition matrix of a Markov chain is stochastic, it is in particular nonnegative and, if the state space is finite, one may call upon the Perron–Frobenius theory. This is a consolidated body of many results concerning the *spectral radius* $\rho(A)$ of a nonnegative matrix A and its dominant eigenvectors, that is, the eigenvectors corresponding to the eigenvalues λ such that $|\lambda| = \rho(A)$.

Theorem 1.27. (Perron–Frobenius) *Let $A \geq 0$ be an irreducible matrix of finite size n. Then:*

1. *A has a positive real eigenvalue equal to $\rho(A)$.*
2. *There corresponds to $\rho(A)$ a right eigenvector $\boldsymbol{x} > 0$ and a left eigenvector $\boldsymbol{y}^{\mathrm{T}} > 0$ such that $A\boldsymbol{x} = \rho(A)\boldsymbol{x}$, $\boldsymbol{y}^{\mathrm{T}}A = \rho(A)\boldsymbol{y}^{\mathrm{T}}$.*
3. *$\rho(A)$ is a simple eigenvalue of A, that is, $\rho(A)$ is a simple root of $\det(A - \lambda I)$.*
4. *If A has exactly p eigenvalues of modulus $\rho(A)$ then these eigenvalues are the roots of the equation $\lambda^p - \rho(A)^p = 0$. If $p > 1$ then A is cyclic of index of cyclicity p (see equation (1.15)).*

5. *If B is a nonnegative matrix of size n such that $A - B \geq 0$, $B \neq A$, then $\rho(A) > \rho(B)$.*

Observe that, if the matrix A is not cyclic, then there exists only one eigenvalue of maximum modulus. Matrices which share this property are called *primitive.*

We see that, for finite matrices, the conditions of existence and uniqueness of the invariant probability vector are very simple. Indeed, it immediately follows from Theorem 1.27 that, if P is an irreducible stochastic matrix of finite size, then there exists a unique positive vector such that $\boldsymbol{\pi}^{\mathrm{T}} P = \boldsymbol{\pi}^{\mathrm{T}}$. Therefore, the Markov chain associated with P is positive recurrent and Theorem 1.12 is a direct consequence of the Perron–Frobenius theorem.

If the matrix is nonnegative and not necessarily irreducible then we have the following weaker result:

Theorem 1.28 *Let $A \geq 0$ be a matrix of finite size n. Then:*

1. *A has a nonnegative real eigenvalue equal to $\rho(A)$.*
2. *There correspond to $\rho(A)$ a right eigenvector $\boldsymbol{x} \geq 0$ and a left eigenvector $\boldsymbol{y}^{\mathrm{T}} \geq 0$ such that $A\boldsymbol{x} = \rho(A)\boldsymbol{x}$, $\boldsymbol{y}^{\mathrm{T}} A = \rho(A)\boldsymbol{y}^{\mathrm{T}}$.*
3. *If B is a nonnegative matrix of size n such that $A - B \geq 0$ then $\rho(A) \geq \rho(B)$.*

The spectral radius $\rho(A)$ of a nonnegative matrix A is also called the *Perron–Frobenius eigenvalue* and any corresponding nonnegative right eigenvector $\boldsymbol{x}$ is called *Perron–Frobenius eigenvector.* Similarly any pair $(\rho(A), \boldsymbol{x})$ is called *Perron–Frobenius pair.*

The following result will be used in Chapter 7 in order to prove some convergence properties.

Theorem 1.29 *Let $A \geq 0$ be a matrix of finite size n. If A has only one irreducible final class, then there exists a positive vector $\boldsymbol{x}$ such that $A\boldsymbol{x} = \rho(A)\boldsymbol{x}$.*

Proof For Theorem 1.28 the vector $\boldsymbol{x}$ has nonnegative components. Without loss of generality, in light of Theorem A.9, we may assume that A is in the following form

$$A = \begin{bmatrix} A_{1,1} & & & 0 \\ A_{2,1} & A_{2,2} & & \\ \vdots & \ddots & \ddots & \\ A_{k,1} & \dots & A_{k,k-1} & A_{k,k} \end{bmatrix}$$

where $A_{i,i}$, $i = 1, \dots, k$ are irreducible. Since $A_{1,1}$ corresponds to the final class, then $\rho(A) = \rho(A_{1,1})$. Let us partition $\boldsymbol{v}$ according to the partitioning of A as $\boldsymbol{x} = (\boldsymbol{x}_i)_{i=1,\dots,k}$. We prove the theorem by induction on the number of irreducible classes k. For $k = 1$ the matrix A is irreducible so that for the Perron–Frobenius theorem $\boldsymbol{x} > 0$. Assume that the theorem holds for $k - 1$ irreducible classes and let us prove it for k irreducible classes. Let us denote by A_{k-1} the submatrix of A corresponding to the first $k - 1$ irreducible classes and observe that the vector $\boldsymbol{w}$ with block components $\boldsymbol{x}_i$, $i = 1, \dots, k - 1$ is an eigenvector of A_{k-1}, that is $A_{k-1}\boldsymbol{w} = \rho(A)\boldsymbol{w}$. Therefore for the inductive assumption we have $\boldsymbol{x}_i > 0$ for $i = 1, \ldots, k-1$. We have to prove that $\boldsymbol{x}_k > 0$. Observe that $\boldsymbol{x}_k$ cannot be identically zero, otherwise we have $\sum_{j=1}^{k-1} A_{k,j}\boldsymbol{x}_j = 0$ that is, since $\boldsymbol{x}_j > 0$ for $j = 1, \ldots, k-1$, we would have $A_{k,j} = 0$ for $j = 1, \ldots, k-1$ so that there would be another final class. Assume for a contradiction that $\boldsymbol{x}_k$ has some null componenents, say the first h, and the remaining components are nonzero. In this case, from the inequality $A_{k,k}\boldsymbol{x}_k \le \rho(A)\boldsymbol{x}_k$ we deduce that the elements of $A_{k,k}$ with indices (i,j), where $i \le h$ and $j \ge h+1$, must be zero. This fact implies that $A_{k,k}$ is reducible, which contradicts the assumptions. □

1.9 Bibliographic notes

The interested reader may find in Norris [93] all the proofs which we have not given for properties of Markov chains and Markov processes. The two books by Çinlar [30] and Resnick [102] also offer a very clear introduction to the subject of Markov chains. For a presentation with a clear connection to numerical analysis, one may cite Seneta [105] and Stewart [107]. Two old favorites of ours are Kemeny and Snell [71] and Kemeny *et al.* [72] and one should not forget to mention Chung [33]. With regard to the Perron–Frobenius theory on nonnegative matrices, we direct the reader to the classical books by Berman and Plemmons [9] and Varga [116]. There are plenty of introductory books on probability theory; two recent ones are Grinstead and Snell [54] with an algorithmic slant, and Stirzaker [108].

2

STRUCTURED MATRIX ANALYSIS

2.1 Introduction

We recall and analyse certain structural and computational properties which are the basis of the design and analysis of fast algorithms for the numerical solution of Markov chains. After introducing the definition of discrete Fourier transform and the FFT algorithm for its computation, we describe circulant matrices, z-circulant matrices, and Toeplitz matrices, together with their block analogs, which intervene in the mathematical modeling of structured Markov chains. Then we introduce the concept of displacement operator and of displacement rank which allow one to provide efficient representations and to design efficient algorithms for a wide class of matrices related to Toeplitz matrices.

We first introduce the definition of block matrices, block vectors, Kronecker and Hadamard products of matrices.

Given the positive integers p, q and m_i, n_j, $i = 1, \ldots, p$, $j = 1, \ldots, q$, and given the $m_i \times n_j$ matrices $A_{i,j}$, $i = 1, \ldots, p$, $j = 1, \ldots, q$, the matrix

$$A = \begin{bmatrix} A_{1,1} & A_{1,2} & \ldots & A_{1,q} \\ A_{2,1} & A_{2,2} & \ldots & A_{2,q} \\ \vdots & \vdots & & \vdots \\ A_{p,1} & A_{p,2} & \ldots & A_{p,q} \end{bmatrix}$$

of size $m \times n$, $m = \sum_{i=1}^{p} m_i$, $n = \sum_{j=1}^{q} n_j$, is called a $p \times q$ *block matrix* with *block elements* $A_{i,j}$, $i = 1, \ldots, p$, $j = 1, \ldots, q$. If $q = 1$ the matrix A is called a *block column vector* of block size p, with *block components* $A_{i,1}$, $i = 1, \ldots, p$. Similarly, if $p = 1$ the matrix A is called a *block row vector* of block size q, with *block components* $A_{1,j}$, $j = 1, \ldots, q$. For instance, the matrix

$$A = \left[\begin{array}{ccc|cc} 1 & 1 & 1 & 0 & 0 \\ 1 & 1 & 1 & 0 & 0 \\ \hline 2 & 2 & 2 & 1 & 0 \\ 2 & 2 & 2 & 0 & 1 \end{array}\right]$$

can be rewritten in block form as

$$A = \begin{bmatrix} E & 0_2 \\ 2E & I_2 \end{bmatrix}$$

where E is the 2×3 matrix with all the elements equal to 1, 0_2 is the 2×2 matrix with all the elements equal to 0, and I_2 is the 2×2 identity matrix.

If $A = (a_{i,j})$ and $B = (b_{i,j})$ are matrices of size $m \times n$ and $p \times q$ respectively, we denote by $A \otimes B$ the *Kronecker product* of A and B, that is, the $m \times n$ block matrix with block elements $a_{i,j}B$, $i = 1, \ldots, m$, $j = 1, \ldots, n$, of size $p \times q$. In particular, for an $m \times m$ matrix A we have

$$I_n \otimes A = \begin{bmatrix} A & & 0 \\ & \ddots & \\ 0 & & A \end{bmatrix}, \qquad A \otimes I_n = \begin{bmatrix} a_{1,1}I_n & \ldots & a_{1,m}I_n \\ \vdots & \ddots & \vdots \\ a_{m,1}I_n & \ldots & a_{m,m}I_n \end{bmatrix},$$

where I_n denotes the $n \times n$ identity matrix.

The Kronecker product allows one to write in vector form a product of matrices. For this purpose, introduce the notation

$$\boldsymbol{x} = \mathrm{vec}(X) = [x_{1,1}, \ldots, x_{m,1}, \ldots, \ldots, x_{1,n}, \ldots, x_{m,n}]^{\mathrm{T}} \tag{2.1}$$

which associates with the $m \times n$ matrix X the mn-dimensional vector obtained by concatenating the columns of X. We readily find that

$$\mathrm{vec}(AXB) = (B^{\mathrm{T}} \otimes A)\,\mathrm{vec}(X). \tag{2.2}$$

Similarly, if $\boldsymbol{x}$ is a vector of dimension mn, we define $\mathrm{vec}_m^{-1}(\boldsymbol{x})$ the $m \times n$ matrix such that $\mathrm{vec}(\mathrm{vec}_m^{-1}(\boldsymbol{x})) = \boldsymbol{x}$.

Given the vectors $\boldsymbol{a} = (a_i)_{i=1,n}$, $\boldsymbol{b} = (b_i)_{i=1,n}$ we denote with $\boldsymbol{a} * \boldsymbol{b}$ the *Hadamard, or component-wise* product of the vectors $\boldsymbol{a}$ and $\boldsymbol{b}$, that is,

$$\boldsymbol{a} * \boldsymbol{b} = [a_1 b_1, \ldots, a_n b_n].$$

The Hadamard product is naturally extended to block vectors that is,

$$[A_1, \ldots, A_n] * [B_1, \ldots, B_n] = [A_1 B_1, \ldots, A_n B_n].$$

We denote with $\mathrm{Diag}(A_1, \ldots, A_n)$ the block diagonal matrix having for diagonal blocks the matrices A_i, $i = 1, \ldots, n$. If the blocks are scalars, i.e., $A_i = a_i$, then $\mathrm{Diag}(a_1, \ldots, a_n)$ represents a diagonal matrix.

2.2 Discrete Fourier transform

Let $\mathtt{i}$ be the imaginary unit such that $\mathtt{i}^2 = -1$. For an integer $n \geq 1$, let us define

$$\omega_n = \cos(2\pi/n) + \mathtt{i}\sin(2\pi/n).$$

The complex number ω_n is a *primitive nth root of unity*, i.e., ω_n is such that $\omega_n^n = 1$ and $\omega_n^i \neq \omega_n^j$, $i \neq j$, $i, j = 0, \ldots, n-1$. The entries of the set $\{\omega_n^i, \quad i = 0, \ldots, n-1\}$ are the n solutions of the equation $z^n - 1 = 0$, known also as the *nth roots of unity* or *Fourier points*.

The Vandermonde matrix

$$\Omega_n = (\omega_n^{(i-1)(j-1)})_{i,j=1,n}$$

formed with the Fourier points is called the *Fourier matrix* of order n. It is nonsingular and its inverse is $\Omega_n^{-1} = \frac{1}{n}\overline{\Omega}_n$ where $\overline{\Omega}_n$ is the complex conjugate of the matrix Ω_n. This fact is an immediate consequence of the following property of the nth roots of 1:

$$\sum_{i=0}^{n-1} \omega_n^{ij} = \begin{cases} 0 \text{ if } j \neq 0 \mod n, \\ n \text{ if } j = 0 \mod n. \end{cases}$$

The transformation $\boldsymbol{x} \rightarrow \boldsymbol{y} = \frac{1}{n}\overline{\Omega}_n \boldsymbol{x}$ is called the *discrete Fourier transform* of order n or DFT for short. The inverse transformation $\boldsymbol{y} \rightarrow \boldsymbol{x} = \Omega_n \boldsymbol{y}$ is called the *inverse discrete Fourier transform*, or briefly, IDFT of order n. We also write $\boldsymbol{y} = \mathrm{DFT}_n(\boldsymbol{x})$ and $\boldsymbol{x} = \mathrm{IDFT}_n(\boldsymbol{y})$. In terms of components we have the following relations

$$\begin{aligned} y_i &= \tfrac{1}{n} \textstyle\sum_{j=1}^{n} \overline{\omega}_n^{(i-1)(j-1)} x_j \\ x_i &= \textstyle\sum_{j=1}^{n} \omega_n^{(i-1)(j-1)} y_j \end{aligned} \qquad i = 1, 2, \ldots, n. \tag{2.3}$$

If n is an integer power of 2, the IDFT of a complex vector of length n can be computed by the *fast Fourier transform* (FFT) algorithms with just $5n \log_2 n$ arithmetic operations with real numbers (ops). The same asymptotic number of ops is sufficient for computing the DFT. If $\boldsymbol{x}$ is a real vector, then $\mathrm{IDFT}_n(\boldsymbol{x})$ can be computed in roughly $\frac{5}{2} n \log_2 n$ ops. The same computational cost is sufficient to compute the DFT of a vector $\boldsymbol{y}$ which is the IDFT of a real vector.

Different algorithms and different implementations exist for FFTs. We refer the reader to Section 2.5 for bibliographic notes and references to software.

We may interpret the definition of DFT in terms of polynomials in the following way. Let $y(z) = \sum_{j=0}^{n-1} z^j y_{j+1}$ be the polynomial of degree less than n defined by the vector $\boldsymbol{y}$ and observe from (2.3) that $x_i = y(\omega_n^{i-1})$, $i = 1, \ldots, n$. Therefore we may regard the IDFT of a vector as the values that the polynomial $y(z)$ takes at the nth roots of 1. Similarly, the DFT can be viewed as the coefficients of the polynomial $y(z)$ interpolating the values x_i, $i = 1, \ldots, n$, at the nth roots of 1.

Remark 2.1 From the properties $\Omega_n^{-1} = \frac{1}{n}\overline{\Omega}_n$ and $\Omega_n^T = \Omega_n$, we deduce that the matrix $W_n = \frac{1}{\sqrt{n}}\Omega_n$ is unitary, i.e., $\overline{W}_n^T W_n = I$. This implies that $\|\boldsymbol{x}\|_2 = \|\boldsymbol{y}\|_2$, whenever $\boldsymbol{y} = W_n \boldsymbol{x}$, i.e., $\boldsymbol{y} = \frac{1}{\sqrt{n}} \mathrm{IDFT}_n(\boldsymbol{x})$.

Discrete Fourier transforms can be extended to block vectors. Given the block column vectors $\boldsymbol{x}, \boldsymbol{y}$ of block size n, with $m \times p$ components, the transformation $\boldsymbol{x} \rightarrow \boldsymbol{y} = \frac{1}{n}(\overline{\Omega}_n \otimes I_m)\boldsymbol{x}$ is called *block discrete Fourier transform* of order n. The inverse transformation $\boldsymbol{y} \rightarrow \boldsymbol{x} = (\Omega_n \otimes I_m)\boldsymbol{y}$ is called *block inverse discrete Fourier transform* of order n. As in the scalar case, we write $\boldsymbol{y} = \mathrm{DFT}_n(\boldsymbol{x})$ and $\boldsymbol{x} = \mathrm{IDFT}_n(\boldsymbol{y})$. In terms of block components we have

$$\begin{aligned} Y_i &= \tfrac{1}{n} \textstyle\sum_{j=1}^{n} \overline{\omega}_n^{(i-1)(j-1)} X_j \\ X_i &= \textstyle\sum_{j=1}^{n} \omega_n^{(i-1)(j-1)} Y_j \end{aligned} \qquad i = 1, 2, \ldots, n.$$

Observe that a block DFT (IDFT) consists in computing mp DFTs (IDFTs). More precisely, by denoting $(Y_i)_{h,k}$ the element of Y_i in position (h,k), the vector $\boldsymbol{v}^{(h,k)} = [(Y_1)_{h,k}, (Y_2)_{h,k}, \ldots, (Y_n)_{h,k}]^{\mathrm{T}}$ is such that

$$\boldsymbol{v}^{(h,k)} = \mathrm{DFT}_n([(X_1)_{h,k}, (X_2)_{h,k}, \ldots, (X_n)_{h,k}]^{\mathrm{T}}), \quad h = 1, \ldots, m, \quad k = 1, \ldots, p.$$

A similar relation holds for IDFT_n.

Similarly to the scalar case, we may associate with the block vector $\boldsymbol{y}$ the matrix polynomial $Y(z) = \sum_{i=0}^{n-1} z^i Y_{i+1}$. We call an $m \times p$ *matrix polynomial* any polynomial whose coefficients are $m \times p$ matrices, or equivalently, a matrix whose elements are polynomials. The maximum degree of its elements is called the *degree of the matrix polynomial.* In this way, the IDFT of a block vector can be viewed as the values taken by the matrix polynomial $Y(z)$ at the roots of 1. The DFT of a block vector $\boldsymbol{x}$ can be viewed as the coefficient vector of the matrix polynomial which interpolates X_i, $i = 1, \ldots, n$, at the nth roots of 1.

A nice application of the DFT is Algorithm 2.1 for computing the product $A(z)$ of two $m \times m$ matrix polynomials $B(z)$ and $C(z)$, based on the *evaluation–interpolation* technique. The idea at the basis of this algorithm is to compute the coefficients of $A(z)$ by interpolating to the values of $A(z)$ at the Fourier points by means of a DFT. The latter values are computed by multiplying the values of $B(z)$ and $C(z)$ at the Fourier points which are separately computed by means of two IDFTs. The number N of Fourier points must be greater than the degree of $A(z)$.

The computational effort of this algorithm amounts to $2m^2$ IDFTs of length N, to m^2 DFTs of length N and to N matrix products therefore its cost is $O(m^2 N \log N + m^3 N)$. If the coefficients of $B(z)$ are real, then it follows that U_1 and $U_{N/2+1}$ are real and that $U_i = \overline{U}_{N-i+2}$, $i = 2, \ldots, N/2$. The same property holds for the block components V_i of $\boldsymbol{v}$ if $C(z)$ has real coefficients. Thus, if both $B(z)$ and $C(z)$ have real coefficients, then $\boldsymbol{y}$ also has this property and the computation of $\boldsymbol{w}$, is reduced to computing two products of real matrices and $N/2 - 1$ products of complex matrices. Since a product of complex matrices can be performed with three multiplications and five additions of real matrices by means of (A.3), the overall cost of stage 4 of 2.1 is about $3N/2$ real matrix multiplications, and $5N/2$ real matrix additions. Therefore for a real input the cost of Algorithm 2.1 is

$$N(3m^3 + 2m^2) + \frac{15}{2} m^2 N \log N \tag{2.4}$$

ops up to lower order terms. If m is large enough, the dominant part of the complexity is $3Nm^3$. By using the customary algorithm for polynomial multiplication the cost would be $2m^3(n_1+1)(n_2+1)$ ops. Observe that for moderately large values of n_1 and n_2 one has $N << (n_1+1)(n_2+1)$.

Remark 2.2 From the polynomial interpretation of DFT and IDFT, we deduce a useful property which allows one to save arithmetic operations in the pointwise power series arithmetic which we will describe in Section 3.1.2. In that

---

Algorithm 2.1 Product of matrix polynomials

INPUT: The coefficients of the $m \times m$ matrix polynomials $B(z) = \sum_{i=0}^{n_1} z^i B_i$ and $C(z) = \sum_{i=0}^{n_2} z^i C_i$ of degree n_1 and n_2, respectively.

OUTPUT: The coefficients of the $m \times m$ matrix polynomial $A(z) = B(z)C(z) = \sum_{i=0}^{n} z^i A_i$ of degree at most $n = n_1 + n_2 + 1$.

COMPUTATION:

1. Compute the minimum integer power of two N such that $N > n_1 + n_2$.
2. Let $\boldsymbol{b}$ be the N-dimensional block column vector with elements B_{i-1} for $i = 1, \ldots, n_1 + 1$ and null components elsewhere. Similarly define $\boldsymbol{c}$ as the N-dimensional block column vector with elements C_{i-1} for $i = 1, \ldots, n_2 + 1$ and null components elsewhere.
3. Compute $\boldsymbol{u} = \mathrm{IDFT}_N(\boldsymbol{b})$, $\boldsymbol{v} = \mathrm{IDFT}_N(\boldsymbol{c})$.
4. Compute $\boldsymbol{w} = \boldsymbol{u} * \boldsymbol{v}$.
5. Compute $\boldsymbol{y} = (Y_i)_{i=1,N} = \mathrm{DFT}_N(\boldsymbol{w})$.
6. Output $A_i = Y_{i+1}$, $i = 0, \ldots, n$.

---

context, we have to compute IDFTs of block vectors whose length is doubled at each step by padding them with zeros. More precisely, we need to compute the IDFT_{2n} of the vector $[Y_1, \ldots, Y_n, 0, \ldots, 0]$ of length $2n$ once we have the IDFT_n of $[Y_1, \ldots, Y_n]$ without starting the computation from scratch.

Assume that we are given the block vector $\boldsymbol{y} = (Y_i)_{i=1,n}$ defining the matrix polynomial $Y(z) = \sum_{i=0}^{n-1} z^i Y_{i+1}$ and that we have computed $\mathrm{IDFT}_n(\boldsymbol{y})$, that is, the values $X_j = Y(\omega_n^{j-1})$, $j = 1, \ldots, n$. We wish to compute the values $T_j = Y(\omega_{2n}^{j-1})$ for $j = 1, \ldots, 2n$, that is, $\mathrm{IDFT}_{2n}(\widehat{\boldsymbol{y}})$, $\widehat{\boldsymbol{y}} = (\widehat{Y}_i)_{i=1,2n}$, where $\widehat{Y}_i = Y_i$ for $i = 1, \ldots, n$, and $\widehat{Y}_i = 0$ elsewhere. Since $\omega_{2n}^2 = \omega_n$, we find that $T_{2j-1} = X_j$, $j = 1, \ldots, n$. Therefore, the odd components of $\boldsymbol{t} = (T_j)_{j=1,2n}$ are already available. By direct inspection one proves that the block vector $\boldsymbol{w} = (T_{2j})_{j=1,n}$ of the even components can be computed by means of the expression $\boldsymbol{w} = \mathrm{IDFT}_n(D\boldsymbol{y})$, where $D = \mathrm{Diag}(1, \omega_{2n}, \omega_{2n}^2, \ldots, \omega_{2n}^{n-1}) \otimes I_m$. Therefore, in order to compute T_j, $j = 1, \ldots, 2n$, it is sufficient to compute just one additional IDFT of length n.

The interpolation (a DFT) problem may be stated as follows: given the matrix polynomial $X(z) = \sum_{i=0}^{n-1} z^i X_{i+1}$ of degree less than n, and given $\boldsymbol{y} = (Y_i)_{i=1,n}$, one needs to compute the matrix polynomial $W(z) = \sum_{i=0}^{2n-1} z^i W_{i+1}$ of degree less than $2n$ such that $W(\omega_n^i) = X(\omega_n^i)$, $i = 0, \ldots, n-1$ and $W(\omega_{2n}^{2i-1}) = Y_i$, $i = 1, \ldots, n$. With $w^{(1)} = (W_i)_{i=1,n}$ and $\boldsymbol{w}^{(2)} = (W_i)_{i=n+1,2n}$, one shows that $\boldsymbol{w}^{(1)} = (\boldsymbol{x} + \overline{D} \cdot \mathrm{DFT}_n(\boldsymbol{y}))/2$ and $\boldsymbol{w}^{(2)} = (\boldsymbol{x} - \overline{D} \cdot \mathrm{DFT}_n(\boldsymbol{y}))/2$, so that, here also, it suffices to compute a DFT of length n in order to compute $W_1, \ldots, W_{2n}$.

2.3 Structured matrices

We recall the structural and computational properties of some important classes of matrices among which, circulant, z-circulant and Toeplitz matrices, and their block analogs, that play an important role in the design of algorithms for the solution of structured Markov chains.

2.3.1 Circulant matrices

Circulant matrices are closely related to DFTs.

Definition 2.3 *Given the row vector* $[a_0, a_1, \ldots, a_{n-1}]$, *the* $n \times n$ *matrix*

$$A = (a_{j-i \mod n})_{i,j=1,n} = \begin{bmatrix} a_0 & a_1 & \ldots & a_{n-1} \\ a_{n-1} & a_0 & \ddots & \vdots \\ \vdots & \ddots & \ddots & a_1 \\ a_1 & \ldots & a_{n-1} & a_0 \end{bmatrix}$$

is called the circulant matrix associated with $[a_0, a_1, \ldots, a_{n-1}]$ *and is denoted by* $\mathrm{Circ}(a_0, a_1, \ldots, a_{n-1})$.

A circulant matrix is fully defined by its first row $\boldsymbol{r}^{\mathrm{T}} = [a_0, a_1, \ldots, a_{n-1}]$ or its first column $\boldsymbol{c} = [a_0, a_{n-1}, \ldots, a_1]^{\mathrm{T}}$. Any other row or column is obtained from the preceding one by applying a cyclic permutation to its elements: the last element is moved to the first position and the remaining ones are shifted by one position. With C denoting the circulant matrix associated with $[0, 1, 0, \ldots, 0]$, i.e.,

$$C = \begin{bmatrix} 0 & 1 & 0 & \ldots & 0 \\ 0 & 0 & 1 & \ddots & \vdots \\ \vdots & \ddots & \ddots & \ddots & 0 \\ 0 & \ddots & \ddots & 0 & 1 \\ 1 & 0 & \ldots & 0 & 0 \end{bmatrix}, \tag{2.5}$$

it can easily be verified that

$$A = \sum_{i=0}^{n-1} a_i C^i, \tag{2.6}$$

that is, any circulant matrix can be viewed as a polynomial in C.

By direct inspection we see that

$$C\overline{\Omega}_n = \overline{\Omega}_n \,\mathrm{Diag}(1, \overline{\omega}_n, \overline{\omega}_n^2, \ldots, \overline{\omega}_n^{n-1});$$

multiplying the latter expression on the left by $\frac{1}{n}\Omega_n$ yields

$$\frac{1}{n}\Omega_n C\overline{\Omega}_n = \mathrm{Diag}(1, \overline{\omega}_n, \overline{\omega}_n^2, \ldots, \overline{\omega}_n^{n-1});$$

moreover, taking the conjugate transpose of both sides, we find

$$\frac{1}{n}\Omega_n C^{\mathrm{T}}\overline{\Omega}_n = \mathrm{Diag}(1, \omega_n, \omega_n^2, \ldots, \omega_n^{n-1}),$$

since Ω_n is symmetric. From the above two equations and (2.6) we deduce the following property

Theorem 2.4 *If A is a circulant matrix with first row $\boldsymbol{r}^{\mathrm{T}}$ and first column $\boldsymbol{c}$, then*

$$A = \frac{1}{n}\overline{\Omega}_n \operatorname{Diag}(\boldsymbol{w})\Omega_n,$$

where $\boldsymbol{w} = \Omega_n \boldsymbol{c} = \overline{\Omega}_n \boldsymbol{r}$.

An immediate corollary of the theorem above is that we can compute the product $A\boldsymbol{x}$ of an $n \times n$ circulant matrix A and a vector $\boldsymbol{x}$ by means of two IDFTs of length n and a DFT. In fact, the above result can be rephrased in the form

$$A\boldsymbol{x} = \mathrm{DFT}_n(\mathrm{IDFT}_n(\boldsymbol{c}) * \mathrm{IDFT}_n(\boldsymbol{x})) \tag{2.7}$$

where "$*$" denotes the Hadamard, or component-wise product of vectors.

From Theorem 2.4 we also find that the product of two circulant matrices is still circulant and the inverse of a nonsingular circulant matrix is circulant.

The definition of circulant matrix is naturally extended to block matrices.

Definition 2.5 *Given the block row vector $[A_0, A_1, \ldots, A_{n-1}]$ where A_i, $i = 0, \ldots, n-1$, are $m \times m$ matrices, the $n \times n$ block matrix*

$$A = (A_{j-i \mod n})_{i,j=1,n} = \begin{bmatrix} A_0 & A_1 & \ldots & A_{n-1} \\ A_{n-1} & A_0 & \ddots & \vdots \\ \vdots & \ddots & \ddots & A_1 \\ A_1 & \ldots & A_{n-1} & A_0 \end{bmatrix}$$

is called the block circulant matrix associated with $[A_0, A_1, \ldots, A_{n-1}]$ and is denoted by $\operatorname{Circ}(A_0, A_1, , \ldots, A_{n-1})$.

Similarly to the scalar case we have

$$A = \sum_{i=0}^{n-1} C^i \otimes A_i, \tag{2.8}$$

and Theorem 2.4 is generalized to the following property

Theorem 2.6 *If A is a block circulant matrix with first block row $\boldsymbol{r}^{\mathrm{T}}$ and with first block column $\boldsymbol{c}$ we have*

$$A = \frac{1}{n}(\overline{\Omega}_n \otimes I_m) \operatorname{Diag}(W_1, \ldots, W_n)(\Omega_n \otimes I_m)$$

where

$$[W_1, \ldots, W_n] = \boldsymbol{r}^{\mathrm{T}}(\overline{\Omega}_n \otimes I_m),$$
$$\begin{bmatrix} W_1 \\ \vdots \\ W_n \end{bmatrix} = (\Omega_n \otimes I_m)\boldsymbol{c}.$$

Like circulant matrices, the class of block-circulant matrices is closed under matrix multiplication and inversion.

Equation (2.7) becomes

$$A\boldsymbol{x} = \mathrm{DFT}_n(\mathrm{IDFT}_n(\boldsymbol{c}) * \mathrm{IDFT}_n(\boldsymbol{x})) \tag{2.9}$$

where $\boldsymbol{c}$ is the first block column of A, which shows that one can compute the product $A\boldsymbol{x}$ of an $n \times n$ block circulant matrix A and a block vector $\boldsymbol{x}$ with block components of size $m \times p$ by means of two block IDFTs and one block DFT of length n, and n products of $m \times m$ times $m \times p$ matrices.

We synthesize (2.7) and (2.9) with Algorithm 2.2 for multiplying a block circulant matrix and a block vector. For $m = p = 1$, the algorithm reduces to the scalar case.

Algorithm 2.2 Block circulant matrix-vector product

INPUT: Positive integers m, n, p, where $n = 2^k$, k a positive integer, the n-dimensional block vector $\boldsymbol{c} = (C_i)_{i=1,n}$ with $m \times m$ block components which is the first block column of the block circulant matrix A, and the n-dimensional block vector $\boldsymbol{x} = (X_i)_{i=1,n}$ with $m \times p$ block components.

OUTPUT: The block vector $\boldsymbol{y} = A\boldsymbol{x} = (Y_i)_{i=1,n}$.

COMPUTATION:

1. Compute $\boldsymbol{w} = (W_i)_{i=1,n} = \mathrm{IDFT}_n(\boldsymbol{c})$.
2. Compute $\boldsymbol{v} = (V_i)_{i=1,n} = \mathrm{IDFT}_n(\boldsymbol{x})$.
3. Compute the matrix products $U_i = W_i V_i$, $i = 1, 2, \ldots, n$.
4. Compute $\boldsymbol{y} = \mathrm{DFT}_n(\boldsymbol{u})$ for $\boldsymbol{u} = (U_i)_{i=1,n}$.

The cost of computing $\boldsymbol{y}$, given $\boldsymbol{x}$ and $\boldsymbol{c}$ is clearly $O((m+p)mn \log n + nm^2p)$ ops. If the elements of $\boldsymbol{c}$ are real, then it follows that W_1 and $W_{n/2+1}$ are real and that $W_i = \overline{W}_{n-i+2}$, $i = 2, \ldots, n/2$. The same property holds for the block components V_i of $\boldsymbol{v}$ if $\boldsymbol{x}$ is real. Thus, if both $\boldsymbol{c}$ and $\boldsymbol{x}$ are real, then $\boldsymbol{u}$ also has this property and the computation of U_i, $i = 1, 2, \ldots, n$, is reduced to computing two products of real matrices and $n/2 - 1$ products of complex matrices. Since a product of complex matrices can be performed with three multiplications and five additions of real matrices by means of (A.3), the overall cost of stage 3 is $3(n/2 - 1) + 2$ real matrix multiplications between $m \times m$ and $m \times p$ matrices, $n/2 - 1$ additions of $m \times m$ matrices, and $4(n/2 - 1)$ additions of $m \times p$ matrices. Therefore for a real input the cost of Algorithm 2.2 is

$$n(3m^2p + m(m + 3p)/2)) + \frac{5}{2}(m^2 + 2mp)n \log n \tag{2.10}$$

ops up to lower order terms. If m and p are large enough, the dominant part of the complexity is $3m^2np$. By using the customary algorithm for a matrix-vector product the cost would be $2m^2n^2p$ ops.

2.3.2 z-circulant matrices

A generalization of circulant matrices is provided by the class of z-circulant matrices.

Definition 2.7 *Given a scalar $z \neq 0$ and the row vector $[a_0, a_1, \ldots, a_{n-1}]$, the $n \times n$ matrix*

$$A = \begin{bmatrix} a_0 & a_1 & \ldots & a_{n-1} \\ za_{n-1} & a_0 & \ddots & \vdots \\ \vdots & \ddots & \ddots & a_1 \\ za_1 & \ldots & za_{n-1} & a_0 \end{bmatrix}$$

is called the z-circulant matrix associated with $[a_0, a_1, \ldots, a_{n-1}]$.

Observe that a z-circulant matrix is fully defined by z and by the elements in its first row $\boldsymbol{r}^{\mathrm{T}} = [a_0, a_1, \ldots, a_{n-1}]$ or in its first column $\boldsymbol{c} = [a_0, za_{n-1}, \ldots, za_1]^{\mathrm{T}}$.

We denote by C_z the z-circulant matrix whose first row is $[0, 1, 0, \ldots, 0]$, i.e.,

$$C_z = \begin{bmatrix} 0 & 1 & 0 & \ldots & 0 \\ 0 & 0 & 1 & \ddots & \vdots \\ \vdots & \ddots & \ddots & \ddots & 0 \\ 0 & \ddots & \ddots & 0 & 1 \\ z & 0 & \ldots & 0 & 0 \end{bmatrix},$$

and we easily verify that

$$A = \sum_{i=0}^{n-1} a_i C_z^i. \tag{2.11}$$

That is, any z-circulant matrix can be viewed as a polynomial in C_z.

It is simple to verify that

$$C_{z^n} = zD_z C D_z^{-1}, \qquad D_z = \mathrm{Diag}(1, z, z^2, \ldots, z^{n-1}),$$

where C is the circulant matrix in (2.5). Therefore, if A is z^n-circulant, from (2.11) we deduce that

$$A = D_z \left(\sum_{i=0}^{n-1} a_i z^i C^i \right) D_z^{-1}$$

where $\sum_{i=0}^{n-1} a_i z^i C^i$ is circulant. Whence, from Theorem 2.4 we obtain the following

Theorem 2.8 *If A is the z^n-circulant matrix with first row $\boldsymbol{r}^{\mathrm{T}}$ and first column $\boldsymbol{c}$ then*

$$A = \frac{1}{n} D_z \overline{\Omega}_n \,\mathrm{Diag}(\boldsymbol{w}) \Omega_n D_z^{-1},$$

with $\boldsymbol{w} = \overline{\Omega}_n D_z \boldsymbol{r} = \Omega_n D_z^{-1} \boldsymbol{c}$.

The above theorem states that, like circulant matrices, all the matrices in the z-circulant class can be simultaneously diagonalized by means of a combination of DFT and diagonal scaling with the integer powers of z. Therefore, for any given z, z-circulant matrices are closed under matrix multiplication and inversion.

The extension to block matrices trivially applies to z-circulant matrices.

Definition 2.9 *Given a scalar $z \neq 0$ and the block row vector $[A_0, A_1, \ldots, A_{n-1}]$, the $n \times n$ matrix*

$$A = \begin{bmatrix} A_0 & A_1 & \ldots & A_{n-1} \\ zA_{n-1} & A_0 & \ddots & \vdots \\ \vdots & \ddots & \ddots & A_1 \\ zA_1 & \ldots & zA_{n-1} & A_0 \end{bmatrix}$$

is called the block z-circulant matrix associated with $[A_0, A_1, \ldots, A_{n-1}]$.

The analog of Theorem 2.8 for block z-circulant matrices is stated below.

Theorem 2.10 *If A is the block z^n-circulant matrix with first block column $\mathbf{c}$ and with first block row $[A_0, A_1, \ldots, A_{n-1}]$, then*

$$A = \frac{1}{n}(D_z \otimes I_m)(\overline{\Omega}_n \otimes I_m)\,\mathrm{Diag}(\boldsymbol{w})(\Omega_n \otimes I_m)(D_z^{-1} \otimes I_m),$$

where

$$\boldsymbol{w} = (\overline{\Omega}_n \otimes I_m)(D_z \otimes I_m)\begin{bmatrix} A_0 \\ \vdots \\ A_{n-1} \end{bmatrix} = (\Omega_n \otimes I_m)(D_z^{-1} \otimes I_m)\mathbf{c}.$$

2.3.3 Toeplitz matrices

A *Toeplitz matrix* is a matrix of the form

$$A = \begin{bmatrix} a_0 & a_1 & a_2 & \ldots \\ a_{-1} & a_0 & a_1 & \ddots \\ a_{-2} & a_{-1} & a_0 & \ddots \\ \vdots & \ddots & \ddots & \ddots \end{bmatrix}$$

that is, it has equal elements along each diagonal. The element in position (i, j) of a Toeplitz matrix is a function of the difference $j - i$. For this reason we also write $A = (a_{j-i})_{i,j \in E}$, where the set E of the indices can be $E = \{1, \ldots, n\}$, $E = \mathbb{N}$ or $E = \mathbb{Z}$. In general a matrix is called finite if the set E of indices is $\{1, \ldots, n\}$, is called semi-infinite, or more simply infinite, if $E = \mathbb{N}$ and bi-infinite if $E = \mathbb{Z}$. Similarly, block Toeplitz matrices have the form $A = (A_{j-i})_{i,j \in E}$ where A_k, $k = 0, \pm 1, \pm 2, \ldots$, are $m \times m$ matrices [1].

[1]We should remark that in the classical theory a Toeplitz matrix is defined in the form $A = (a_{i-j})_{i,j \in E}$ so that the elements in the lower triangular part have positive indices and

A simple example of a Toeplitz matrix is obtained by removing the first row of the transition matrix (1.5) of Example 1.3.

An $n \times n$ Toeplitz matrix $A = (a_{j-i})_{i,j=1,n}$, can be embedded into the $2n \times 2n$ circulant matrix B whose first row is $[a_0, a_1, \ldots, a_{n-1}, *, a_{-n+1}, \ldots, a_{-1}]$, where $*$ denotes any number. We observe that the leading $n \times n$ submatrix of B coincides with A. An example with $n = 3$ is shown below

$$B = \left[\begin{array}{ccc|ccc} a_0 & a_1 & a_2 & * & a_{-2} & a_{-1} \\ a_{-1} & a_0 & a_1 & a_2 & * & a_{-2} \\ a_{-2} & a_{-1} & a_0 & a_1 & a_2 & * \\ \hline * & a_{-2} & a_{-1} & a_0 & a_1 & a_2 \\ a_2 & * & a_{-2} & a_{-1} & a_0 & a_1 \\ a_1 & a_2 & * & a_{-2} & a_{-1} & a_0 \end{array}\right].$$

More generally, an $n \times n$ Toeplitz matrix can be embedded into a $q \times q$ circulant matrix for any $q \geq 2n - 1$: it is sufficient to replace $*$ with $q - 2n + 1$ arbitrary elements. If $q = 2n - 1$ there is no extra element. Similarly, an $n \times n$ block Toeplitz matrix A with $m \times m$ blocks can be embedded into a $q \times q$ block circulant matrix B with $m \times m$ blocks for any $q \geq 2n - 1$.

This embedding property allows one to compute the product $\boldsymbol{y} = A\boldsymbol{x}$ of a (block) Toeplitz matrix A and a (block) vector $\boldsymbol{x}$ by means of Algorithm 2.2 in the following manner. First we embed A in a circulant matrix B. Second we define the q-dimensional block vector $\boldsymbol{z} = (Z_i)$ obtained by filling up $\boldsymbol{x}$ with zeros, i.e., $Z_i = X_i$, $i = 1, 2, \ldots, n$, $Z_i = 0$ elsewhere. The first n block components of the block vector $\boldsymbol{w} = B\boldsymbol{z}$ coincide with $\boldsymbol{y}$. If q is chosen as an integer power of 2, then the product $B\boldsymbol{z}$ can be efficiently computed by means of Algorithm 2.2 which is based on the FFT.

We synthesize this computation in Algorithm 2.3 which includes the scalar case when $m = p = 1$.

The complexity analysis of Algorithm 2.3 can be carried out similarly to the case of Algorithm 2.2 and leads to the computational cost of

$$O((m + p)mn \log n + nm^2 p)$$

ops. If $p = m = 1$, that is, A is a Toeplitz matrix and $\boldsymbol{x}$ is a vector, then the asymptotic cost reduces to $O(n \log n)$ ops, versus the $O(n^2)$ cost of the customary algorithm for matrix-vector multiplication. If $m = p$, the asymptotic cost is $O(m^2 n \log n + m^3 n) = O(m^2 n(\log n + m))$; thus, if m is large with respect to $\log n$, the cost of computing FFTs is negligible with respect to the cost of computing the matrix products.

From the complexity bound (2.10) of the product of a circulant matrix and a vector we deduce that for real input the complexity bound of the product of a Toeplitz matrix and a vector is

the elements in the upper triangular part have negative indices. This has specific motivations related to the functional interpretation of Toeplitz matrices. Here we abandon this convention and adopt the definition $A = (a_{j-i})_{i,j \in E}$ which is more suitable for Markov chains applications.

Algorithm 2.3 Block Toeplitz matrix-vector product

INPUT: Positive integers m, n, p, the $m \times m$ matrices A_i, $i = -n+1, \ldots, n-1$, which define the $n \times n$ block Toeplitz matrix $A = (A_{j-i})_{i,j=1,n}$; the n-dimensional block vector $\boldsymbol{x} = (X_i)_{i=1,n}$ with $m \times p$ block components.

OUTPUT: The block vector $\boldsymbol{y} = A\boldsymbol{x} = (Y_i)_{i=1,n}$.

COMPUTATION:

1. Compute the least integer k such that $2^k \geq 2n$; set $q = 2^k$.
2. Define the q-dimensional block column vector $\boldsymbol{v} = (V_i)_{i=1,q}$ such that $V_i = A_{-i+1}$ if $i = 1, \ldots, n$, $V_{q-i+1} = A_i$ if $i = 1, 2, \ldots, n-1$, and $V_i = 0$ elsewhere, and define the $q \times q$ block circulant matrix B having the first block column $\boldsymbol{v}$. The block Toeplitz matrix A is embedded in B.
3. Define the q-dimensional block column vector $\boldsymbol{z} = (Z_i)_{i=1,q}$ such that $Z_i = X_i$ if $i = 1, \ldots, n$, $X_i = 0$ elsewhere.
4. Compute $\boldsymbol{w} = B\boldsymbol{z} = (W_i)_{i=1,q}$ by means of Algorithm 2.2.
5. Set $Y_i = W_i$, $i = 1, \ldots, n$.

$$q(3m^2p + m(m+3p)/2)) + \frac{5}{2}(m^2 + 2mp)q \log q \tag{2.12}$$

up to terms of lower order, where q is the minimum integer power of 2 greater than $2n-1$. In Figure 2.1 we compare the cost (2.12), where $q = q(n)$, and the cost $2n^2m^2p$ of the customary algorithm as a function of n for $m = p = 10$. We see that even for relatively small values of n, Algorithm 2.3 is faster than the customary algorithm.

Algorithm 2.3, formulated for the computation of the product of a block row vector and a block Toeplitz matrix, is reported as Algorithm 2.4.

2.3.4 *Triangular Toeplitz matrices*

Let $Z = (z_{i,j})_{i,j=1,n}$ be the $n \times n$ *lower shift matrix*

$$Z = \begin{bmatrix} 0 & & & 0 \\ 1 & \ddots & & \\ & \ddots & \ddots & \\ 0 & & 1 & 0 \end{bmatrix}, \tag{2.13}$$

with $z_{i+1,1} = 1$ for $i = 1, \ldots, n-1$, $z_{i,j} = 0$ elsewhere.

For a given vector $\boldsymbol{x} = (x_i)_{i=1,n}$, we define the $n \times n$ *lower triangular Toeplitz* matrix

$$L(\boldsymbol{x}) = \sum_{i=0}^{n-1} x_{i+1} Z^i = \begin{bmatrix} x_1 & & & 0 \\ x_2 & x_1 & & \\ \vdots & \ddots & \ddots & \\ x_n & \ldots & x_2 & x_1 \end{bmatrix},$$

having $\boldsymbol{x}$ as first column.

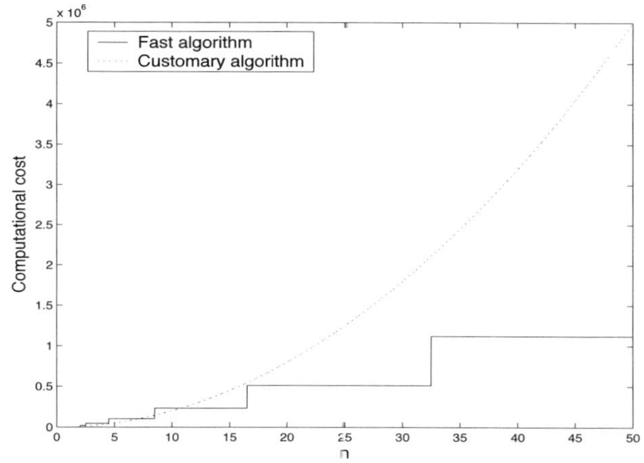

Fig. 2.1 Computational costs of the algorithms for multiplying a block Toeplitz matrix and a block vector for $m = p = 10$ as a function of n.

Similarly, given a row vector $\boldsymbol{y}^{\mathrm{T}} = [y_1, y_2, \ldots, y_n]$, we define the $n \times n$ *upper triangular Toeplitz* matrix

$$U(\boldsymbol{y}^{\mathrm{T}}) = \sum_{i=0}^{n-1} y_{i+1}(Z^{\mathrm{T}})^i = \begin{bmatrix} y_1 & y_2 & \ldots & y_n \\ & y_1 & \ddots & \vdots \\ & & \ddots & y_2 \\ 0 & & & y_1 \end{bmatrix},$$

having $\boldsymbol{y}^{\mathrm{T}}$ as first row.

The above definitions extend to block matrices in the following way: we denote with $L(\boldsymbol{x})$ the *block lower triangular* Toeplitz matrix having first block column $\boldsymbol{x}$, and with $U(\boldsymbol{y}^{\mathrm{T}})$ the *block upper triangular Toeplitz* matrix having first block row $\boldsymbol{y}^{\mathrm{T}}$, that is,

$$L(\boldsymbol{x}) = \begin{bmatrix} X_1 & & & 0 \\ X_2 & X_1 & & \\ \vdots & \ddots & \ddots & \\ X_n & \ldots & X_2 & X_1 \end{bmatrix}, \qquad \boldsymbol{x} = \begin{bmatrix} X_1 \\ X_2 \\ \vdots \\ X_n \end{bmatrix},$$

$$U(\boldsymbol{y}^{\mathrm{T}}) = \begin{bmatrix} Y_1 & Y_2 & \ldots & Y_n \\ & Y_1 & \ddots & \vdots \\ & & \ddots & Y_2 \\ 0 & & & Y_1 \end{bmatrix}, \qquad \boldsymbol{y}^{\mathrm{T}} = [Y_1, Y_2, \ldots, Y_n].$$

Let $T_n = L(\boldsymbol{a})$, $\boldsymbol{a} = (a_{i-1})_{i=1,n}$, be an $n \times n$ lower triangular Toeplitz matrix. Since T_n is a polynomial in Z and $Z^k = 0$ for $k \geq n$, the class of lower

---

Algorithm 2.4 Block vector-Toeplitz matrix product

INPUT: Positive integers m, n, p, the $m \times m$ matrices A_i, $i = -n+1, \ldots, n-1$, which define the $n \times n$ block Toeplitz matrix $A = (A_{j-i})_{i,j=1,n}$; the n-dimensional block row vector $\boldsymbol{x}^{\mathrm{T}} = [X_1, \ldots, X_n]$ with $p \times m$ block components.

OUTPUT: The block row vector $\boldsymbol{y}^{\mathrm{T}} = \boldsymbol{x}^{\mathrm{T}} A = [Y_1, \ldots, Y_n]$

COMPUTATION:

1. Consider the block Toeplitz matrix $H = A^{\mathrm{T}} = (A_{i-j}^{\mathrm{T}})_{i,j=1,n}$ and the block column vector $\boldsymbol{w} = \boldsymbol{x} = (X_i^{\mathrm{T}})_{i=1,n}$.
2. Compute the product $\boldsymbol{v} = H\boldsymbol{w}$ by means of Algorithm 2.3.
3. Output $Y_i = V_i^{\mathrm{T}}$, $i = 1, \ldots, n$.

---

triangular Toeplitz matrices is closed under multiplication. Moreover, by the Cayley–Hamilton theorem (Theorem A.10 in the appendix) the inverse of any nonsingular matrix A can be written as a polynomial in A, therefore T_n^{-1} is still a lower triangular Toeplitz matrix and to compute T_n^{-1} is equivalent to computing the elements in the first column of T_n^{-1}. Similarly, the class of block triangular Toeplitz matrices is closed under matrix product and matrix inversion.

Now, assume $n = 2h$, h a positive integer, and partition T_n into $h \times h$ blocks, writing

$$T_n = \begin{bmatrix} T_h & 0 \\ S_h & T_h \end{bmatrix}, \tag{2.14}$$

where T_h, S_h are $h \times h$ Toeplitz matrices and T_h is lower triangular. If T_n is nonsingular then T_h also is nonsingular and

$$T_n^{-1} = \begin{bmatrix} T_h^{-1} & 0 \\ -T_h^{-1} S_h T_h^{-1} & T_h^{-1} \end{bmatrix}.$$

Thus, the first column $\boldsymbol{v}_n$ of T_n^{-1} is given by

$$\boldsymbol{v}_n = T_n^{-1}\boldsymbol{e}_1 = \begin{bmatrix} \boldsymbol{v}_h \\ -T_h^{-1} S_h \boldsymbol{v}_h \end{bmatrix} = \begin{bmatrix} \boldsymbol{v}_h \\ -L(\boldsymbol{v}_h) S_h \boldsymbol{v}_h \end{bmatrix}, \tag{2.15}$$

where $L(\boldsymbol{v}_h) = T_h^{-1}$ is the lower triangular Toeplitz matrix whose first column is $\boldsymbol{v}_h$.

The same relation holds if T_n is block triangular Toeplitz. In this case the elements $a_0, \ldots, a_{n-1}$ are replaced with the $m \times m$ blocks $A_0, \ldots, A_{n-1}$ and $\boldsymbol{v}_n$ denotes the first block column of T_n^{-1}.

The representation (2.15) of $\boldsymbol{v}_n$ leads to a recursive algorithm for its computation (Algorithm 2.5), which we describe for block triangular Toeplitz matrices of block size $n = 2^k$, for a positive integer k.

The computation of the block vector $\boldsymbol{u}$ at the ith step of Algorithm 2.5 requires the computation of two products of a block Toeplitz matrix and a block vector. Since this can be performed in $O(2^i m^3 + i2^i m^2)$ ops by means of Algorithm 2.3, the overall cost of Algorithm 2.5 is $O(nm^3 + nm^2 \log n)$.

Algorithm 2.5 Block lower triangular Toeplitz matrix inversion

---

INPUT: The positive integer k and the $m \times m$ block elements $A_0, \ldots, A_{n-1}$, $n = 2^k$, of the first block column of the block lower triangular Toeplitz matrix T_n, where $\det A_0 \neq 0$.

OUTPUT: The first block column $\boldsymbol{v}_n$ of T_n^{-1}.

COMPUTATION:

1. Set $\boldsymbol{v}_1 = A_0^{-1}$
2. For $i = 0, \ldots, k-1$, given $\boldsymbol{v}_h$, $h = 2^i$:
   (a) Compute the block Toeplitz matrix-vector products $\boldsymbol{w} = S_h \boldsymbol{v}_h$ and $\boldsymbol{u} = -L(\boldsymbol{v}_h)\boldsymbol{w}$.
   (b) Set
$$\boldsymbol{v}_{2h} = \begin{bmatrix} \boldsymbol{v}_h \\ \boldsymbol{u} \end{bmatrix}.$$

More precisely, if the input is real, from the complexity estimate (2.12) with $p = m$, we deduce that the complexity bounds of the ith stage reduce to $2(3m^3q + \frac{15}{2}m^2q\log q + 2m^2q)$ for $q = 2^{i+1}$. Moreover, in this complexity estimate, we counted twice the cost of the computation of the DFT of the vector $\boldsymbol{v}_h$ (filled with zeros) which appears in both the Toeplitz matrix-vector products $S_h\boldsymbol{v}_h$ and $L(\boldsymbol{v}_h)\boldsymbol{w}$ (see Steps 2 and 3 of Algorithm 2.3). Taking this into consideration, the complexity bound is reduced to $\frac{25}{2}m^2h\log(h) + 6m^3h + 4m^2h$. Therefore, since $\sum_{i=0}^{k-1} 2^{i+1} = 2(2^k - 1)$, the overall cost is less than
$$25m^2n\log(2n) - (12m^3 + 8m^2)n \tag{2.16}$$
where we have bounded $\log 2^i$ with $\log n$.

The block triangular Toeplitz system $T_n\boldsymbol{x} = \boldsymbol{b}$ can be solved in $O(m^2n\log n + m^3n)$ ops, by first computing the inverse matrix T_n^{-1} by means of Algorithm 2.5 and then computing the product $T_n^{-1}\boldsymbol{b}$ by means of Algorithm 2.3.

Alternatively, the computation of the first block column of the inverse of T_n might be performed by using the customary approach, i.e, by inverting the diagonal block A_0 and by applying forward substitution. This amounts to computing $n(n+1)/2$ matrix products and $n(n-1)/2$ matrix sums. The cost in the case of real input is n^2m^3 ops, up to terms of lower order. A plot of the latter cost estimate versus the bound (2.16) as functions of n is reported in Figure 2.2, with $m = 10$.

Algorithm 2.5 can be easily adjusted to invert a block upper triangular Toeplitz matrix at the same cost. This is described as Algorithm 2.6.

If T_n is block upper triangular Toeplitz, then the system $T_n\boldsymbol{x} = \boldsymbol{b}$ can be solved in $O(m^2n\log n + m^3n)$ ops, by first computing the inverse matrix T_n^{-1} by means of Algorithm 2.6 and then computing the product $T_n^{-1}\boldsymbol{x}$ by means of Algorithm 2.3.

If the block Toeplitz matrix T_n is not block triangular, its inverse is not block Toeplitz, as the following simple example shows:

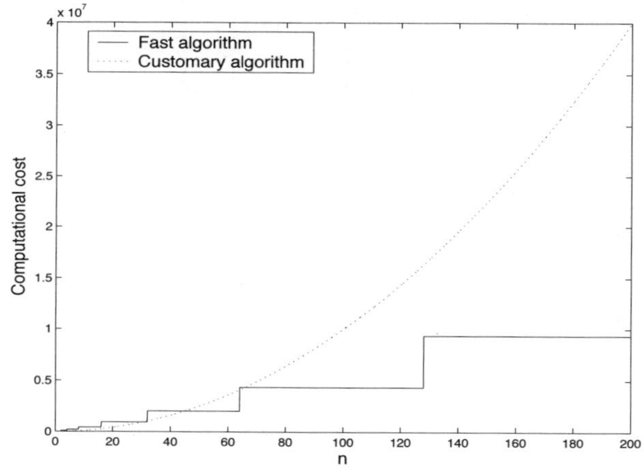

Fig. 2.2 Computational costs of the algorithms for inverting a block triangular Toeplitz matrix for $m = 10$ as a function of n.

Example 2.11 Let $n = 4$ and

$$A = \begin{bmatrix} 4 & 3 & 2 & 1 \\ 0 & 4 & 3 & 2 \\ 1 & 0 & 4 & 3 \\ 0 & 1 & 0 & 4 \end{bmatrix}.$$

We have

$$A^{-1} = \frac{1}{265} \begin{bmatrix} 65 & -50 & 5 & 5 \\ 12 & 56 & -48 & 5 \\ -14 & 23 & 56 & -50 \\ -3 & -14 & 12 & 65 \end{bmatrix}.$$

However, it is possible to introduce a more general structure which is preserved under inversion. This is the subject of the next section.

2.4 Displacement operators

The concept of displacement operator and displacement rank is a powerful tool for dealing with Toeplitz matrices. We refer the reader to Section 2.5 for historical and bibliographic notes. Here, we recall the main results concerning displacement rank, and prepare the tools that we will use later on. Throughout this section we refer to section 2.11 of [12].

Define the *displacement operator*

$$\Delta(A) = AZ - ZA, \tag{2.17}$$

applied to an $n \times n$ matrix A, where Z is the lower shift matrix of (2.13). Multiplying the matrix A on the left by Z shifts down each row of A by one position. Similarly, multiplying the matrix A on the right by Z shifts each column

Algorithm 2.6 Block upper triangular Toeplitz matrix inversion

INPUT: The positive integer k and the $m \times m$ block elements $A_0, \ldots, A_{n-1}$, $n = 2^k$, of the first block row of the upper block triangular Toeplitz matrix T_n, where $\det A_0 \neq 0$.

OUTPUT: The first block row $\boldsymbol{v}_n^{\mathrm{T}}$ of T_n^{-1}.

COMPUTATION:

1. Set $B_i = A_i^{\mathrm{T}}$, $i = 0, \ldots, n-1$, and $\boldsymbol{b} = (B_{i-1})_{i=1,n}$.
2. Apply Algorithm 2.5 to the block lower triangular matrix $L(\boldsymbol{b}) = T_n^{\mathrm{T}}$ and compute the first block column $\boldsymbol{v}_n$ of $(T_n^{\mathrm{T}})^{-1}$.
3. Output the block row vector $\boldsymbol{v}_n^{\mathrm{T}}$.

of A by one position to the left. In particular, if $A = (a_{j-i})_{i,j=1,n}$ is Toeplitz then

$$\Delta(A) = \left[\begin{array}{c|c} a_1\ a_2\ \ldots\ a_{n-1} & 0 \\ \hline & -a_{n-1} \\ 0 & \vdots \\ & -a_2 \\ & -a_1 \end{array}\right] = \boldsymbol{e}_1\boldsymbol{e}_1^{\mathrm{T}}AZ - ZA\boldsymbol{e}_n\boldsymbol{e}_n^{\mathrm{T}}, \tag{2.18}$$

where $\boldsymbol{e}_1$ and $\boldsymbol{e}_n$ denote the first and the last column of the $n \times n$ identity matrix. Therefore, $\Delta(A)$ has at most rank 2. We say that a matrix A has *displacement rank* (at most) k with respect to the operator Δ if $\operatorname{rank}\Delta(A) = k$ ($\operatorname{rank}\Delta(A) \leq k$). As a particular case, Toeplitz matrices have displacement rank at most 2, so that the class of matrices with "low" displacement rank are a generalization of Toeplitz matrices.

It is important to recall that an $n \times n$ matrix X has rank k if and only if there exist $n \times k$ matrices V, W of full rank such that $X = VW^{\mathrm{T}}$. Therefore, if $\Delta(A)$ has rank k then there exist two $n \times k$ matrices V and W of full rank such that $\Delta(A) = VW^{\mathrm{T}}$ and vice versa. We call a *displacement generator* of A with respect to the operator Δ any pair (V, W) of such matrices. For instance, in the case of (2.18) one has $\Delta(A) = VW^{\mathrm{T}}$ where

$$V = \left[\begin{array}{c|c} 1 & 0 \\ 0 & a_{n-1} \\ \vdots & \vdots \\ 0 & a_1 \end{array}\right], \quad W^{\mathrm{T}} = \left[\begin{array}{ccc} a_1 \ \ldots & a_{n-1} & 0 \\ \hline 0 \ \ldots & 0 & -1 \end{array}\right].$$

Observe that we might have a pair of $n \times h$ matrices (V, W) such that $h > k$, $VW^{\mathrm{T}} = \Delta(A)$ and $\operatorname{rank}(V) = \operatorname{rank}(W) = \operatorname{rank}\Delta(A) = k$. We call such a pair a *displacement generator of nonminimal rank* of A. A generator of nonminimal rank stores the information about $\Delta(A)$ in a redundant way. Using numerical linear algebra tools like the singular value decomposition any generator can be reduced to minimal rank.

The displacement generator of a matrix A, together with the first column of A, contains all the information which allows one to represent all the elements of A, as shown in the next theorem.

Theorem 2.12 *Let A be an $n \times n$ matrix having first column $\boldsymbol{a}$ and displacement rank k. If (V, W) is a displacement generator of A, then*

$$A = L(\boldsymbol{a}) + \sum_{i=1}^{k} L(\boldsymbol{v}_i) U(\boldsymbol{w}_i^{\mathrm{T}} Z^{\mathrm{T}})$$

where $\boldsymbol{v}_i$ and $\boldsymbol{w}_i$ are the ith column of V and W, respectively.

Equivalent representations can be given in terms of different operators. Besides the Sylvester type operator Δ we may consider operators of the Stein type like $\nabla(A) = A - ZAZ^{\mathrm{T}}$. More generally, we may consider operators like $AZ_1 - Z_2A$ or $A - Z_1AZ_2^{\mathrm{T}}$ where Z_1 and Z_2 can be different, say z-circulant, or $(-z)$-circulant. We refer the reader to [70] and to [12, Section 2.11] for more details on this regard.

A simple but important consequence of Theorem 2.12 is that any matrix with displacement rank k can be decomposed as the sum of at most $k+1$ matrices, each of them the product of a lower and an upper triangular Toeplitz matrix. Therefore, the product $\boldsymbol{y} = A\boldsymbol{x}$ can be split into at most $2k+1$ products of triangular Toeplitz matrices and vectors. Each one of these products can be efficiently performed by using Algorithm 2.3 in $O(n \log n)$ ops. The overall cost of the algorithm for computing $\boldsymbol{y}$ in this way is $O(kn \log n)$.

Another nice consequence of Theorem 2.12 concerns the inverse matrix of A. Observe that if A is nonsingular then pre- and post-multiplying (2.17) by A^{-1} yields the simple relation

$$\Delta(A^{-1}) = -A^{-1}\Delta(A)A^{-1}, \tag{2.19}$$

from which we conclude that the displacement ranks of A and of A^{-1} coincide. Moreover, given a displacement generator (V, W) of the nonsingular matrix A, the pair $(-A^{-1}V, (A^{-1})^{\mathrm{T}}W)$ is a displacement generator for A^{-1}. This allows one to represent A^{-1} in a compact way by means of Theorem 2.12 as

$$A^{-1} = L(A^{-1}\boldsymbol{e}_1) - \sum_{i=1}^{k} L(A^{-1}\boldsymbol{v}_i) U(\boldsymbol{w}_i^{\mathrm{T}} A^{-1} Z^{\mathrm{T}}). \tag{2.20}$$

Observe that even though the inverse of a Toeplitz matrix A is not generally Toeplitz, its displacement rank is at most 2.

Example 2.13 For the Toeplitz matrix A of Example 2.11 we have $\Delta(A^{-1}) = VW^{\mathrm{T}}$ where

$$V = \frac{1}{265}\begin{bmatrix} -65 & 25 \\ -12 & 25 \\ 14 & 15 \\ 3 & -205 \end{bmatrix}, \quad W^{\mathrm{T}} = \frac{1}{265}\begin{bmatrix} 205 & -15 & -25 & -25 \\ 3 & 14 & -12 & -65 \end{bmatrix}.$$

Computing the displacement representation of A^{-1} is reduced to solving at most $2k+1$ linear systems. This is particularly convenient when k is small with respect to n. Moreover, once A^{-1} is represented by means of its displacement generator, solving any additional system of the kind $A\boldsymbol{x} = \boldsymbol{b}$ is reduced to computing the product $\boldsymbol{x} = A^{-1}\boldsymbol{b}$, with the cost of $O(kn \log n)$ ops.

Displacement representations are also useful for computing products of matrices with low displacement rank. It is a simple matter to prove that

$$\Delta(AB) = A\Delta(B) + \Delta(A)B \tag{2.21}$$

so that a displacement generator (possibly of nonminimal rank) of $C = AB$ is given by (V_C, W_C) where

$$V_C = \left[AV_B \middle| V_A \right], \quad W_C = \left[W_B \middle| B^{\mathrm{T}} W_A \right],$$

and (V_A, W_A) and (V_B, W_B) are displacement generators of A and B, respectively. Therefore, to compute the displacement generator (possibly of nonminimal rank) of the product AB given the displacement generators of A and B, one only needs to compute the products AV_B and $B^{\mathrm{T}}W_A$, at the cost $O(kn \log n)$, where k is the maximum displacement rank of A and B, if we use the displacement representation of A and B and Algorithm 2.3.

The concept of displacement rank is easily extended to block matrices. Let $\mathcal{Z} = Z \otimes I$, where I is the $m \times m$ identity matrix, and consider the operator $A \to A\mathcal{Z} - \mathcal{Z}A$ applied to an $n \times n$ block matrix with $m \times m$ blocks. Observe that this operator coincides with Δ if $m = 1$. Therefore, for notational simplicity, we will denote it with the same symbol and write that $\Delta(A) = A\mathcal{Z} - \mathcal{Z}A$.

It is easy to see that, if $A = (A_{j-i})_{i,j=1,n}$ is block Toeplitz, then

$$\Delta(A) = \left[\begin{array}{c|c} A_1 \; A_2 \; \dots \; A_{n-1} & 0 \\ \hline & -A_{n-1} \\ 0 & \vdots \\ & -A_2 \\ & -A_1 \end{array}\right]$$

$$= (\boldsymbol{e}_1 \otimes I)(\boldsymbol{e}_1 \otimes I)^{\mathrm{T}} A\mathcal{Z} - \mathcal{Z}A(\boldsymbol{e}_n \otimes I)(\boldsymbol{e}_n \otimes I)^{\mathrm{T}}.$$

We say that the block matrix A has *block displacement rank* k if k is the minimum integer such that there exist $n \times k$ block matrices V and W with $m \times m$ blocks satisfying $\Delta(A) = VW^{\mathrm{T}}$. Any such pair (V, W) is called a *block displacement generator* of A. The representation theorem 2.12 still holds as well as equations (2.19), (2.20) and (2.21) suitably adjusted to the block notation. We synthesize these properties in the following

Theorem 2.14 *Let $\Delta(A) = A\mathcal{Z} - \mathcal{Z}A$. If A is nonsingular then*

$$\Delta(A^{-1}) = -A^{-1}\Delta(A)A^{-1}.$$

Given matrices A, B, C such that $A = BC$ then

$$\Delta(A) = B\Delta(C) + \Delta(B)C.$$

Moreover, if $\Delta(A) = VW^{\mathrm{T}}$, *where* V *and* W *are* $n \times k$ *block matrices with* $m \times m$ *blocks, then*

$$A = L(\boldsymbol{a}) + \sum_{i=1}^{k} L(\boldsymbol{v}_i)U(\boldsymbol{w}_i^{\mathrm{T}} \mathcal{Z}^{\mathrm{T}})$$

where $\boldsymbol{a}$ *is the first block column of* A, $\boldsymbol{v}_i$ *and* $\boldsymbol{w}_i$ *denote the* i*th block column of* V *and* W, *respectively, for* $i = 1, \ldots, k$. *In particular, if* A *is nonsingular, then*

$$A^{-1} = L(\boldsymbol{a}') - \sum_{i=1}^{k} L(A^{-1}\boldsymbol{v}_i)U(\boldsymbol{w}_i^{\mathrm{T}} A^{-1} \mathcal{Z}^{\mathrm{T}})$$

where $\boldsymbol{a}'$ *is the first block column of* A^{-1}.

It is interesting to point out that, $\Delta(A) = 0$ if and only if A is block lower triangular Toeplitz. Observe also that the "dual" operator $\Delta_2(A) = A\mathcal{Z}^{\mathrm{T}} - \mathcal{Z}^{\mathrm{T}} A$, which satisfies the same formal properties of Δ, is such that $\Delta_2(A) = 0$ if and only if A is block upper triangular Toeplitz. In Chapter 9 we will use both operators Δ and Δ_2 and we will use the notation Δ_1 instead of Δ. The symbol Δ will be used only to denote a general displacement operator in the expressions which are valid for both Δ_1 and Δ_2.

We summarize the properties of $\Delta_2(A)$ in the following theorem.

Theorem 2.15 *Let* $\Delta_2(A) = A\mathcal{Z}^{\mathrm{T}} - \mathcal{Z}^{\mathrm{T}} A$. *If* A *is nonsingular then*

$$\Delta_2(A^{-1}) = -A^{-1}\Delta_2(A)A^{-1}.$$

Given matrices A, B, C *such that* $A = BC$ *then*

$$\Delta_2(A) = B\Delta_2(C) + \Delta_2(B)C.$$

Moreover, if $\Delta_2(A) = VW^{\mathrm{T}}$, *where* V *and* W *are* $n \times k$ *block matrices with* $m \times m$ *blocks, then*

$$A = U(\boldsymbol{a}^{\mathrm{T}}) - \sum_{i=1}^{k} L(\mathcal{Z}\boldsymbol{v}_i)U(\boldsymbol{w}_i^{\mathrm{T}})$$

where $\boldsymbol{v}_i$ *and* $\boldsymbol{w}_i$ *denote the* i*th block column of* V *and* W, *respectively, for* $i = 1, \ldots, k$, *and* $\boldsymbol{a}^{\mathrm{T}}$ *is the first block row of* A. *In particular, if* A *is nonsingular, then*

$$A^{-1} = U(\boldsymbol{a}''^{\mathrm{T}}) + \sum_{i=1}^{k} L(\mathcal{Z}A^{-1}\boldsymbol{v}_i)U(\boldsymbol{w}_i^{\mathrm{T}} A^{-1}).$$

where $\boldsymbol{a}''^{\mathrm{T}}$ *is the first block row of* A^{-1}.

Another interesting property which relates the operators $\Delta_1 = \Delta$ and Δ_2 is expressed by the following

Theorem 2.16 *If* $\Delta_1(A) = A\mathcal{Z} - \mathcal{Z}A$ *and* $\Delta_2(A) = A\mathcal{Z}^{\mathrm{T}} - \mathcal{Z}^{\mathrm{T}}A$ *then*

$$\Delta_1(A) = -\mathcal{Z}\Delta_2(A)\mathcal{Z} - \mathcal{Z}A(\boldsymbol{e}_m \otimes I)(\boldsymbol{e}_m \otimes I)^{\mathrm{T}} + (\boldsymbol{e}_1 \otimes I)(\boldsymbol{e}_1 \otimes I)^{\mathrm{T}}A\mathcal{Z}.$$

Proof Multiply both sides of $\Delta_2(A) = A\mathcal{Z}^{\mathrm{T}} - \mathcal{Z}^{\mathrm{T}}A$ by $\mathcal{Z}$ on the right and on the left by $\mathcal{Z}$ and obtain

$$\mathcal{Z}\Delta_2(A)\mathcal{Z} = \mathcal{Z}A\mathcal{Z}^{\mathrm{T}}\mathcal{Z} - \mathcal{Z}\mathcal{Z}^{\mathrm{T}}A\mathcal{Z}.$$

Use that fact that $I - \mathcal{Z}\mathcal{Z}^{\mathrm{T}} = (\boldsymbol{e}_1 \otimes I)(\boldsymbol{e}_1 \otimes I)^{\mathrm{T}}$ and $I - \mathcal{Z}^{\mathrm{T}}\mathcal{Z} = (\boldsymbol{e}_m \otimes I)(\boldsymbol{e}_m \otimes I)^{\mathrm{T}}$ and obtain the claim. □

The Toeplitz structure, and more generally, the displacement structure, can be effectively used for computing matrix inverses. For solving general block Toeplitz systems there are algorithms based on Schur complementation and displacement rank. We refer the reader to Section 2.5 for references both on theoretical analysis and on the available software.

An alternative to direct algorithms are iterative algorithms which provide a sequence of successive approximations to the solution of the linear system $A\boldsymbol{x} = \boldsymbol{b}$. For positive definite systems, particular attention is paid to conjugate gradient iteration which provides the exact solution after mn steps, but which may provide reasonable approximations after just a few iterations. Furthermore, convergence can be accelerated by means of preconditioning techniques.

Each step of the conjugate gradient requires the multiplication of a Toeplitz matrix and a vector. This computation is inexpensive if performed by means of Algorithm 2.3. For nonsymmetric matrices, iterative algorithms like GMRES and Arnoldi methods should be used. All these algorithms for Toeplitz inversion can be extended to the case of matrices having a low displacement rank.

References are given in Section 2.5.

2.5 Bibliographic notes

Analysis of algorithms for FFT can be found in Van Loan [115]; discrete fast transforms are analyzed in Elliot and Rao [36]. A treatise on circulant matrices is in Davis [35], α-circulant matrices have been introduced in Cline, Plemmons, and Worm [34], and some properties of circulant and α-circulant matrices can be found in Bini and Pan [12]. A package for computing fast trigonometric transforms has been developed by Swarztrauber in [109]. It may be downloaded from `http://www.netlib.org/fftpack`. A very efficient package in C (the "best FFT in the West": FFTW) , with a hardware-dependent optimization, has been designed at MIT by Matteo Frigo and Steven G. Johnson [39]. It may be downloaded from `http://www.fftw.org`.

Algorithms for Toeplitz and polynomial computations are analyzed in [12]. Other important algorithmic issues concerning Toeplitz and Toeplitz-like matrices are treated in Kailath and Sayed [70] and in Heinig and Rost [62]. Concerning preconditioning techniques and conjugate gradients we refer the reader to the books [66] and [70] and to the survey papers [32] and [31]. Arnoldi-like methods and GMRES can be found in [50] and [103]. Software for Toeplitz and block Toeplitz matrices has been developed in [8] and can be downloaded from `http://www.win.tue.nl/wgs`.

The theory of displacement operators dates back to the papers by Kailath *et al.* [68], [67]. A basic treatment, with many historical notes, is given in Chapter 1 of [70] and in [69].

3

MATRIX EQUATIONS AND CANONICAL FACTORIZATIONS

3.1 Matrix power series and structured matrices

Let $\{A_i\}_{i\in\mathbb{Z}}$ be a sequence of $m \times m$ matrices. We call a *matrix power series* any formal expression of the kind $A(z) = \sum_{i=0}^{+\infty} z^i A_i$, where "formal" means that at the moment we ignore issues like the convergence of the power series for given values of z. If $A_i = 0$ for $i > n$ and $A_n \neq 0$ for a nonnegative integer n, then $A(z)$ is a *matrix polynomial* of degree n.

We call a *matrix Laurent power series* any formal expression of the kind $A(z) = \sum_{i=-\infty}^{+\infty} z^i A_i$. Given $n_1, n_2 \geq 0$, if $A_i = 0$ for $i < -n_1$ and for $i > n_2$, and if $A_{-n_1} \neq 0$, $A_{n_2} \neq 0$, then we call $A(z) = \sum_{i=-n_1}^{n_2} z^i A_i$ a *matrix Laurent polynomial* of degree (n_1, n_2).

Matrix power series and matrix Laurent power series are closely related to block Toeplitz matrices. Here we point out the strict relationship between structured matrices and matrix power series and show the interplay between algorithms for structured matrix computations and algorithms for matrix power series computations.

Given a matrix Laurent power series $A(z) = \sum_{i=-\infty}^{+\infty} z^i A_i$ we denote by

$$
\begin{aligned}
T_n[A(z)] &= (A_{j-i})_{i,j=1,n},\\
T_\infty[A(z)] &= (A_{j-i})_{i,j\in\mathbb{N}},\\
T_{\pm\infty}[A(z)] &= (A_{j-i})_{i,j\in\mathbb{Z}},
\end{aligned}
$$

the $n \times n$, semi-infinite and bi-infinite block Toeplitz matrices, respectively, associated with $A(z)$.

If $A(z)$ is a matrix Laurent polynomial, such that $A_i = 0$ for $|i| > k$, then $T_n[A(z)]$, $T_\infty[A(z)]$, $T_{\pm\infty}[A(z)]$ are block banded block Toeplitz matrices. If $A(z)$ is a matrix power series in z then $T_n[A(z)]$, $T_\infty[A(z)]$ and $T_{\pm\infty}[A(z)]$ are block upper triangular Toeplitz matrices. Similarly, if $A(z)$ is a matrix power series in z^{-1} then the associated block Toeplitz matrices are block lower triangular.

3.1.1 Power series and infinite Toeplitz matrices

Wiener algebra plays an important role in the analysis of matrix Laurent power series. We call a *Wiener algebra* the set $\mathcal{W}$ of complex $m \times m$ matrix valued functions $A(z) = \sum_{i=-\infty}^{+\infty} z^i A_i$ such that $\sum_{i=-\infty}^{+\infty} |A_i|$ is finite. We recall (see, e.g., [27]) that, if $A(z) \in \mathcal{W}$ is nonsingular for $|z| = 1$ then $A(z)^{-1}$ exists for

$|z| = 1$ and $A(z)^{-1} \in \mathcal{W}$; moreover, if $A(z), B(z) \in \mathcal{W}$, then also $A(z) + B(z)$ and $A(z)B(z)$ are in $\mathcal{W}$. Thus $\mathcal{W}$ is closed under addition, multiplication and inversion, so that it has the structure of an algebra.

We denote by $\mathcal{W}_+$ and $\mathcal{W}_-$ the subalgebras of $\mathcal{W}$ made up by power series of the kind $\sum_{i=0}^{+\infty} z^i A_i$ and $\sum_{i=0}^{+\infty} z^{-i} A_i$, respectively; the former are analytic functions in the open unit disk $\mathbb{D} = \{z \in \mathbb{C} : \quad |z| < 1\}$, the latter are analytic outside the closed unit disk. Observe that a function in $\mathcal{W}_+$ may be invertible in $\mathcal{W}$ but not in $\mathcal{W}_+$. A simple example of this fact is the function $zI \in \mathcal{W}_+$; its inverse $z^{-1}I \in \mathcal{W}_-$ does not belong to $\mathcal{W}_+$.

The Wiener condition $\sum_{i=-\infty}^{+\infty} |A_i| < +\infty$ is specific to the matrix functions encountered in Markov chains where, typically, $A_i \geq 0$ and $\sum_{i=-\infty}^{+\infty} A_i$ is stochastic so that $\sum_{i=-\infty}^{+\infty} |A_i| = \sum_{i=-\infty}^{+\infty} A_i$ is finite. A general example of this situation is represented by the M/G/1 Markov chains which will be studied in Chapter 4 where the transition matrix P is the block Hessenberg stochastic matrix of (4.3).

In order to deal with semi-infinite and bi-infinite matrices we have to regard them as linear operators on $\ell^2(\mathbb{N})$ and $\ell^2(\mathbb{Z})$ respectively, that is, on the space of infinite vectors $\boldsymbol{x} = (x_i)$ indexed by $\mathbb{N}$ and $\mathbb{Z}$, respectively, such that $\sum_i |x_i|^2$ is finite. In this context it is useful to use also the notation $\ell^1(\mathbb{N})$ and $\ell^1(\mathbb{Z})$ for the vector spaces $\boldsymbol{x} = (x_i)$ such that $\sum_i |x_i|$ is finite. The properties $\ell^1(\mathbb{N}) \subset \ell^2(\mathbb{N})$ and $\ell^1(\mathbb{Z}) \subset \ell^2(\mathbb{Z})$ are easily verified. For $\boldsymbol{x} = (x_i) \in \ell^2$ we define the norm $\|\boldsymbol{x}\|_2 = (\sum_i |x_i|^2)^{1/2}$, similarly, for $\boldsymbol{x} \in \ell^1$ we define the norm $\|\boldsymbol{x}\|_1 = \sum_i |x_i|$. Other important vector spaces are the spaces $\ell^\infty(\mathbb{N})$ and $\ell^\infty(\mathbb{Z})$ of vectors indexed by $\mathbb{N}$, $\mathbb{Z}$, respectively, such that $\|\boldsymbol{x}\|_\infty = \sup_i |x_i|$ is finite.

A linear operator $A : \ell^2 \to \ell^2$ is *bounded* if and only if

$$\sup_{\boldsymbol{x} \in \ell^2, \boldsymbol{x} \neq 0} \frac{\|A\boldsymbol{x}\|_2}{\|\boldsymbol{x}\|_2} < +\infty,$$

and we have the following classical result

Theorem 3.1 *The bi-infinite block Toeplitz matrix $(A_{j-i})_{i,j \in \mathbb{Z}}$ defines a bounded operator on $\ell^2(\mathbb{Z})$ if and only if there exists a function $A(z)$ defined for $|z| = 1$ such that $\sup_{|z|=1} |A(z)|$ is finite, and such that $\{A_j\}_{j \in \mathbb{Z}}$ is the sequence of the Fourier coefficients of $A(z)$:*

$$A_j = \frac{1}{2\pi} \int_0^{2\pi} A(e^{\mathrm{i}\theta}) e^{-\mathrm{i}j\theta} d\theta, \quad j \in \mathbb{Z}.$$

The same property holds for semi-infinite matrices.

As a consequence of the above theorem we conclude that if $A(z) \in \mathcal{W}$, then the associated semi- or bi-infinite block Toeplitz matrix defines a bounded operator.

Concerning the invertibility of a function in $\mathcal{W}_+$ we have the following result (see, e.g., [27]), a similar property holding for functions in $\mathcal{W}_-$.

Theorem 3.2 *The function $A(z) \in \mathcal{W}_+$ is invertible in $\mathcal{W}_+$ if and only if one of the following equivalent conditions hold:*

1. $\det A(z) \neq 0$ *for* $|z| \leq 1$;
2. $T_\infty[A(z)]$ *is invertible.*

Given matrix Laurent power series $A(z), B(z), C(z) \in \mathcal{W}$ such that $C(z) = A(z) + B(z)$, it is immediate to verify that

$$T_{\pm\infty}[C(z)] = T_{\pm\infty}[A(z)] + T_{\pm\infty}[B(z)], \quad T_\infty[C(z)] = T_\infty[A(z)] + T_\infty[B(z)].$$

Likewise, if $A(z) = \alpha B(z)$ then

$$T_{\pm\infty}[A(z)] = \alpha T_{\pm\infty}[B(z)], \quad T_\infty[A(z)] = \alpha T_\infty[B(z)],$$

where α is any complex number.

Concerning the product of matrix Laurent power series we may prove a similar property. Let us first consider the case of semi-infinite block Toeplitz matrices and of matrix power series. Consider the formal matrix power series $A(z) = \sum_{i=0}^{+\infty} z^i A_i$, $B(z) = \sum_{i=0}^{+\infty} z^i B_i$, $C(z) = \sum_{i=0}^{+\infty} z^i C_i$, such that $A(z) = B(z)C(z)$. By equating the coefficients of z^i in the above expression for $i = 0, 1, \ldots$, we get

$$\begin{aligned} A_0 &= B_0C_0, \\ A_1 &= B_0C_1 + B_1C_0, \\ A_2 &= B_0C_2 + B_1C_1 + B_2C_0, \\ &\vdots \end{aligned}$$

which we may rewrite in matrix form as

$$\begin{bmatrix} A_0 & A_1 & A_2 & \ldots \\ & A_0 & A_1 & \ddots \\ & & A_0 & \ddots \\ 0 & & & \ddots \end{bmatrix} = \begin{bmatrix} B_0 & B_1 & B_2 & \ldots \\ & B_0 & B_1 & \ddots \\ & & B_0 & \ddots \\ 0 & & & \ddots \end{bmatrix} \begin{bmatrix} C_0 & C_1 & C_2 & \ldots \\ & C_0 & C_1 & \ddots \\ & & C_0 & \ddots \\ 0 & & & \ddots \end{bmatrix}, \tag{3.1}$$

where the block triangular Toeplitz matrices in the above expression are semi-infinite, that is, their subscripts range in $\mathbb{N}$. In this way we may regard the product of matrix power series as a product of block upper triangular Toeplitz matrices.

Similarly, we may look for the inverse of $A(z)$, that is, a matrix power series $H(z) = \sum_{i=0}^{+\infty} z^i H_i$ such that $A(z)H(z) = I$. If the matrix function $A(z)$ is invertible for $|z| \leq 1$, then the inverse of $T_\infty[A(z)]$ exists and is block upper triangular in light of Theorem 3.2. Once again, equating the coefficients of z^i on both sides of $A(z)H(z) = I$ we obtain that

$$\begin{bmatrix} A_0 & A_1 & A_2 & \ldots \\ & A_0 & A_1 & \ddots \\ & & A_0 & \ddots \\ 0 & & & \ddots \end{bmatrix}^{-1} = \begin{bmatrix} H_0 & H_1 & H_2 & \ldots \\ & H_0 & H_1 & \ddots \\ & & H_0 & \ddots \\ 0 & & & \ddots \end{bmatrix}. \tag{3.2}$$

Summarizing this property we have

$$\begin{aligned} &T_\infty[A(z)] = T_\infty[B(z)]T_\infty[C(z)], \quad \text{if } A(z) = B(z)C(z), \\ &T_\infty[A(z)]^{-1} = T_\infty[H(z)], \quad \text{if } H(z) = A(z)^{-1}. \end{aligned} \tag{3.3}$$

In general, however, if $A(z)$, $B(z)$ and $C(z)$ are matrix Laurent power series, then $T_\infty[A(z)] \neq T_\infty[B(z)]T_\infty[C(z)]$, and $T_\infty[A(z)]^{-1} \neq T_\infty[H(z)]$ as the following example shows.

Example 3.3 Let $A(z) = z^{-1}I + I + zI$, $B(z) = A(z)$ so that

$$T_\infty[A(z)] = T_\infty[B(z)] = \begin{bmatrix} I & I & & 0 \\ I & I & I & \\ & I & I & \ddots \\ 0 & & \ddots & \ddots \end{bmatrix},$$

and

$$T_\infty[A(z)B(z)] = \begin{bmatrix} 3I & 2I & I & & & 0 \\ 2I & 3I & 2I & I & & \\ I & 2I & 3I & 2I & I & \\ & I & 2I & 3I & 2I & \ddots \\ 0 & & \ddots & \ddots & \ddots & \ddots \end{bmatrix}.$$

We have

$$T_\infty[A(z)]T_\infty[B(z)] = \begin{bmatrix} 2I & 2I & I & & & 0 \\ 2I & 3I & 2I & I & & \\ I & 2I & 3I & 2I & I & \\ & I & 2I & 3I & 2I & \ddots \\ 0 & & \ddots & \ddots & \ddots & \ddots \end{bmatrix},$$

that is, $T_\infty[A(z)]T_\infty[B(z)]$ is not block Toeplitz and differs from the matrix $T_\infty[A(z)B(z)]$ for the block element in position $(1, 1)$.

Nevertheless, formula (3.3) is still valid if we replace semi-infinite matrices with bi-infinite matrices.

To see this, assume that $B(z) = \sum_{i=-\infty}^{+\infty} z^i B_i$ and $C(z) = \sum_{i=-\infty}^{+\infty} z^i C_i$ are two matrix Laurent power series in the Wiener algebra, and define $A(z) = \sum_{i=-\infty}^{+\infty} z^i A_i$, $A_k = \sum_{i=-\infty}^{+\infty} B_i C_{k-i}$, so that $A(z) = B(z)C(z)$, Consider the matrix $S = T_{\pm\infty}[B(z)]T_{\pm\infty}[C(z)]$ and observe that the element $S_{i,j}$ is given by

$$\sum_{p=-\infty}^{+\infty} B_{p-i}C_{j-p} = \sum_{q=-\infty}^{+\infty} B_q C_{j-i-q}$$

so that $S_{i,j} = A_{j-i}$. Then S is the block Toeplitz matrix associated with $A(z)$, and

$$T_{\pm\infty}[A(z)] = T_{\pm\infty}[B(z)]T_{\pm\infty}[C(z)], \quad \text{if } A(z) = B(z)C(z), \tag{3.4}$$

from which we conclude that

$$T_{\pm\infty}[A(z)]^{-1} = T_{\pm\infty}[H(z)], \quad \text{if } H(z) = A(z)^{-1}, \tag{3.5}$$

provided that $A(z) \in \mathcal{W}$ is invertible.

The above properties clearly show the interplay between bi-infinite block Toeplitz matrices and matrix Laurent power series.

In Chapter 7 we will encounter the problem of inverting an infinite block triangular Toeplitz matrix $T_\infty[A(z)]$, where $A(z) = \sum_{i=0}^{k-1} z^i A_i$, $A_i \le 0$ for $i \ge 1$, A_0 is an M-matrix and $\sum_{i=0}^{k-1} A_i$ is a nonsingular M-matrix (for the definition of M-matrix we refer to Definition A.12). Under this assumption $A(z)^{-1}$ is a matrix power series and its coefficients can be easily computed relying on the following

Theorem 3.4 *Let $A(z) = \sum_{i=0}^{k-1} z^i A_i$, be such that $A_i \le 0$ for $i \ge 1$, and $\sum_{i=0}^{k-1} A_i$ is a nonsingular M-matrix. Then $A(z)$ is invertible in $\mathcal{W}_+$ and $V(z) = A(z)^{-1} = \sum_{i=0}^{+\infty} z^i V_i$ has nonnegative block coefficients. Moreover, if $A(1)^{-1} - \sum_{i=0}^{n-1} V_i \le \epsilon E$ then $\sum_{i=n}^{+\infty} V_i \le \epsilon E$, where E is the $m \times m$ matrix with elements equal to 1.*

Proof Since $\sum_{i=0}^{k-1} A_i$ is a nonsingular M-matrix, and since $A_i \le 0$ for $i \ge 1$, then $A(z)$ is in $\mathcal{W}_+$ and it is nonsingular for $|z| \le 1$. Therefore it is invertible in $\mathcal{W}_+$ for Theorem 3.2. Since for any $n \ge 1$ one has $T_n[A(z)]^{-1} = T_n[V(z)]$, and since $T_n[A(z)]$ is a nonsingular M-matrix, then for Theorem A.13 one has $T_n[A(z)]^{-1} \ge 0$ so that $V_i \ge 0$ for $i = 0, \ldots, n-1$, for any n. Since $A(1)^{-1} = V(1) = \sum_{i=0}^{+\infty} V_i$ then $0 \le A(1)^{-1} - \sum_{i=0}^{n-1} V_i = \sum_{i=n}^{+\infty} V_i$. □

According to the above theorem, the matrix $(T_\infty[A(z)])^{-1}$ can be approximated with the infinite banded Toeplitz matrix associated with the matrix polynomial $V_n(z) = \sum_{i=0}^{n-1} z^i V_i$, and the approximation error is at most ϵ in infinity norm if $A(1)^{-1} - \sum_{i=0}^{n-1} V_i \le \epsilon E$.

Relying on the nonnegativity properties of the coefficients of $V(z)$ one may design a procedure for computing a value of n such that $\sum_{i=n}^{+\infty} V_i \le \epsilon E$ and for computing the block coefficients $V_0, V_1, \ldots, V_{n-1}$. Indeed, by using the doubling technique described in Algorithm 2.5 we may compute the sequences of inverses $(T_{2^i}[A(z)])^{-1}$, $i = 0, 1, \ldots$, having first block row with block elements V_j, $j = 0, 1, \ldots, 2^i - 1$, until $A(1)^{-1} - \sum_{j=0}^{2^i-1} V_j \le \epsilon E$. In fact, in light of Theorem 3.4, the latter inequality implies that $\sum_{j=2^i}^{+\infty} V_j \le \epsilon E$.

Algorithm 3.1 synthesizes the above computation, i.e., given the error bound ϵ, it computes the first N block elements of the first block row of $(T_\infty[A(z)])^{-1}$, where N is such that $\sum_{i=N}^{+\infty} V_i \le \epsilon E$. Here we follow the notation of Section 2.3.4, in particular $L(\boldsymbol{v})$ is the block lower block triangular Toeplitz matrix whose first block column is $\boldsymbol{v}$, and the matrix S_N is the Toeplitz matrix defined in formula (2.14).

Algorithm 3.1 is the extension to the infinite case of Algorithm 2.6. Concerning comments on the computational cost, and on the computation of the Toeplitz

---

Algorithm 3.1 Infinite block triangular Toeplitz matrix inversion

INPUT: The positive integer k and the $m \times m$ block elements $A_0, \ldots, A_{k-1}$, of the first block row of the block triangular Toeplitz matrix $T_\infty[A(z)]$, such that $A_i \le 0$ for $i \ge 1$, and $\sum_{i=0}^{k-1} A_i$ is a nonsingular M-matrix; an error bound $\epsilon > 0$.

OUTPUT: An integer N and the first N block elements V_i, $i = 0, \ldots, N-1$, of the first block row of $T_\infty[A(z)]^{-1}$, where N is such that $\sum_{i=N}^{+\infty} V_i \le \epsilon E$.

COMPUTATION:

1. Set $B_i = A_i^{\mathrm{T}}$, $i = 0, \ldots, k-1$.
2. Compute $W = (\sum_{i=0}^{k-1} B_i)^{-1}$.
3. Set $N = 1$, $\boldsymbol{y}_1 = B_0^{-1}$ and $Y_0 = \boldsymbol{y}_1$.
4. Given $\boldsymbol{y}_N = (Y_i)_{i=0,N-1}$
   (a) compute the Toeplitz matrix-vector products $\boldsymbol{w} = S_N \boldsymbol{y}_N$, $\boldsymbol{u} = -L(\boldsymbol{y}_N)\boldsymbol{w}$, where S_N is the Toeplitz matrix defined in formula (2.14);
   (b) set
   $$\boldsymbol{y}_{2N} = \begin{bmatrix} \boldsymbol{y}_N \\ \boldsymbol{u} \end{bmatrix};$$
   (c) set $N = 2N$.
5. If $W - \sum_{i=0}^{N-1} Y_i \le \epsilon E$ then output N and $V_i = Y_i^{\mathrm{T}}$, $i = 0, \ldots, N-1$; else continue from step 4.

---

matrix-vector products $\boldsymbol{w} = S_N \boldsymbol{y}_N$ and $\boldsymbol{u} = -L(\boldsymbol{y}_N)\boldsymbol{w}$ we refer the reader to the comments after Algorithm 2.5. In particular, the overall computation cost is less than $25m^2 N \log(2N) + (12m^3 + 8m^2)N$.

3.1.2 *Power series arithmetic*

In this section we first examine some algorithmic issues related to computing the first n block coefficients of the product of two matrix power series and of the inverse of a given matrix power series; later, we consider the case of matrix Laurent power series.

Let $A(z)$, $B(z)$ and $C(z)$ be matrix power series such that $A(z) = B(z)C(z)$. The first n coefficients of $A(z)$ are the coefficients of the polynomial

$$A(z) \mod z^n = \sum_{i=0}^{n-1} z^i A_i,$$

which is the matrix polynomial obtained by applying element-wise the modular relation to $A(z)$. Here, $p(z) \mod z^n$ is the remainder on division of $p(z)$ by z^n.

Observe that the relation which we obtain by cutting (3.1) to the finite block size n, can be rewritten in polynomial form as[1]

[1] Another interesting remark which relates structured matrices with matrix polynomials is that equating the coefficients of the same degree in both sides of $A(z) = B(z)C(z) \mod z^n - 1$ yields the matrix equation $\mathrm{Circ}(A_0, \ldots, A_{n-1}) = \mathrm{Circ}(B_0, \ldots, B_{n-1})\,\mathrm{Circ}(C_0, \ldots, C_{n-1})$, where $\mathrm{Circ}(\cdot)$ denotes the circulant matrix of Definition 2.5.

$$A(z) = B(z)C(z) \mod z^n.$$

Therefore, we may compute the first n block coefficients of the product $A(z) = B(z)C(z)$ by performing this computation modulo z^n and applying Algorithm 2.4 to the product $[A_0, A_1, \dots, A_{n-1}] = [B_0, B_1, \dots, B_{n-1}]T_n[C(z)]$: since block upper triangular Toeplitz matrices are uniquely defined by their first block row, the first block row of $T_n[A(z)]$ is the product of the first block row of $T_n[B(z)]$ and $T_n[C(z)]$.

Observe that if the matrix power series $B(z)$ and $C(z)$ reduce to matrix polynomials of degree n_1 and n_2, and if $A(z) = B(z)C(z)$ then the relation between the coefficients of $A(z)$, $B(z)$ and $C(z)$ is fully determined by the product of the upper block triangular Toeplitz matrices of size $n = n_1 + n_2 + 1$ obtained by truncating (3.1), i.e., $T_n[A(z)] = T_n[B(z)]T_n[C(z)]$.

The computation of this product of block upper triangular Toeplitz matrices can be done by reducing it to computing the product $A(z)$ of the two matrix polynomial $B(z)$ and $C(z)$ relying on Algorithm 2.1.

Similarly, the computation of the first n block coefficients of $H(z) = A(z)^{-1}$, is obtained by reducing the power series modulo z^n. This corresponds to truncating equation (3.2) to block size n. In this way the computation is reduced to inverting the $n \times n$ block triangular Toeplitz matrix $T_n[A(z)]$. For this computation we may apply Algorithm 2.6.

Observe also that $H(z)$ can be computed by formally applying the Newton iteration to the equation $H(z)^{-1} - A(z) = 0$ where the unknown is $H(z)$. We recall that the Newton method for approximating the zeros of a function $f(x) : \mathbb{R} \to \mathbb{R}$ consists in generating the sequence $x_{k+1} = x_k - f(x_k)/f'(x_k)$, $k = 0, 1, \dots$, starting from a given x_0. For the function $f(x) = x^{-1} - a$ this recurrence turns into $x_{k+1} = 2x_k - x_k^2 a$ so that we formally obtain the functional iteration $\phi^{(k+1)}(z) = 2\phi^{(k)}(z) - \phi^{(k)}(z)^2 A(z)$, $k = 0, 1, \dots$, where $\phi^{(0)}(z) = A_0^{-1}$. It is easy to show that $\phi^{(k)}(z) = H(z) \mod z^{2^k}$. Whence the algorithm obtained by rewriting the above formula as $\phi^{(k+1)}(z) = 2\phi^{(k)}(z) - \phi^{(k)}(z)^2 A(z) \mod z^{2^{k+1}}$ and by implementing the latter equation by means of the evaluation–interpolation at the roots of 1, is equivalent to Algorithm 2.6.

A more general approach which applies to matrix Laurent power series can be followed if the matrix power series are in the Wiener algebra.

The coefficients of matrix Laurent power series which are analytic in the annulus

$$\mathbb{A} = \{z \in \mathbb{C} : \quad r < |z| < R\}, \quad 0 < r < 1 < R, \tag{3.6}$$

have a useful decay property as stated by the following classical result concerning Laurent series which are analytic in an annulus (see Theorem 4.4c of the book by Henrici [63]):

Theorem 3.5 *Let $a(z) = \sum_{i=-\infty}^{+\infty} z^i a_i$ be a Laurent power series analytic in the annulus $\mathbb{A}$ of (3.6). Then for any ρ such that $r < \rho < R$, the coefficients a_n satisfy*

$$|a_n| \le M(\rho)\rho^{-n}, \quad n = 0, \pm 1, \pm 2, \dots,$$

where $M(\rho) = \max_{|z|=\rho} |a(z)|$.

Applying the above theorem to all the elements of the matrix Laurent power series $A(z)$ we deduce the following result.

Theorem 3.6 *Let* $A(z) = \sum_{i=-\infty}^{+\infty} z^i A_i$ *be a matrix Laurent power series analytic in the annulus* $\mathbb{A}$ *of* (3.6). *Then for any* ρ *such that* $r < \rho < R$, *the coefficients* A_n *satisfy*

$$|A_n| \le M(\rho)\rho^{-n}, \quad n = 0, \pm 1, \pm 2, \ldots,$$

where $M(\rho)$ *is the* $m \times m$ *matrix with elements* $\max_{|z|=\rho} |a_{i,j}(z)|$. *In particular, if* $r < 1 < R$ *then the elements of* A_i *converge exponentially to 0 for* $n \to \pm\infty$. *Moreover, if* $A(z)$ *is nonsingular for* $z \in \mathbb{A}$ *then there exists* $H(z) = \sum_{i=-\infty}^{+\infty} z^i H_i$ *analytic in* $\mathbb{A}$ *such that* $A(z)H(z) = I$, *and the coefficients* H_n *satisfy*

$$|H_n| \le M(\rho)\rho^{-n}, \quad n = 0, \pm 1, \pm 2, \ldots,$$

where $M(\rho)$ *is the* $m \times m$ *matrix with elements* $\max_{|z|=\rho} |h_{i,j}(z)|$. *In particular, if* $r < 1 < R$ *then the elements of* H_n *converge exponentially to 0 for* $n \to \pm\infty$.

A particular case of Theorem 3.5 restricted to matrix power series is the following.

Theorem 3.7 *Let* $a(z) = \sum_{i=0}^{+\infty} z^i a_i$ *be analytic for* $|z| < R$. *Then for any* ρ *such that* $0 < \rho < R$ *one has*

$$|a_n| \le M(\rho)\rho^{-n}, \quad n = 0, 1, 2, \ldots,$$

where $M(\rho) = \max_{|z|=\rho} |a(z)|$.

The decay property allows one to look at a matrix Laurent power series as a matrix Laurent polynomial plus a term which has negligible coefficients. More precisely, for a fixed $\epsilon > 0$ let n_1, n_2 be nonnegative integers such that the matrix $\sum_{i<-n_1} |A_i| + \sum_{i>n_2} |A_i|$ has elements at most ϵ. Then we have

$$A(z) = \sum_{i=-n_1}^{n_2} z^i A_i + R_\epsilon(z)$$

where $R_\epsilon(z) = \sum_{i<-n_1} z^i A_i + \sum_{i>n_2} z^i A_i$ has coefficients with small moduli. In this case we say that the matrix Laurent power series has ϵ-*degree* at most (n_1, n_2). If ϵ is the machine precision of floating point arithmetic, we say that the *numerical degree* is at most (n_1, n_2).

Similarly, we define the ϵ-*degree* and *numerical degree* of a matrix power series. More precisely, we say that the matrix power series $A(z) = \sum_{i=0}^{+\infty} z^i A_i$ has ϵ degree at most n if the matrix $\sum_{i>n} |A_i|$ has elements at most ϵ.

The following result can be proved by direct inspection.

Theorem 3.8 *Let* $A(z) = \sum_{i=-\infty}^{+\infty} z^i A_i$ *be convergent for* $|z| = 1$. *Let* n_1 *and* N *be nonnegative integers such that* $N \ge n_1 + 1$. *Let* ω_N^i, $i = 0, \ldots, N-1$ *be the*

Nth roots of 1. Consider the matrix Laurent polynomial $P(z)=\sum_{i=-n_1}^{N-n_1-1} z^i P_i$ *of degree at most* $(n_1, N-n_1-1)$ *such that* $P(\omega_N^i)=A(\omega_N^i)$, $i=0,\ldots,N-1$. *Then*

$$P_i = A_i + \sum_{j=1}^{+\infty}(A_{i+Nj}+A_{i-Nj}), \quad \text{for } i=-n_1,\ldots,N-n_1-1.$$

According to the above theorem, if $A(z)$ has ϵ-degree at most (n_1, n_2) and we choose $N \geq n_1+n_2+1$, then the coefficients of the matrix Laurent polynomial $P(z)$ of degree at most $(n_1, N-n_1-1)$, which interpolates $A(z)$ at the Nth roots of 1, are such that $|P_i - A_i|$ has elements less than ϵ, that is P_i provide approximations to A_i within the absolute error ϵ. Observe in fact that $N-n_1-1 \geq n_2$.

As a first application of the decay property stated in Theorem 3.6 and of Theorem 3.8 we describe an algorithm for approximating the coefficients H_i, $i=-n_1,\ldots,n_2$, of $H(z)=A(z)^{-1}$, where the matrix Laurent power series $A(z)$ is analytic and invertible in the annulus $\mathbb{A}$ and $H(z)$ has ϵ-degree at most (n_1, n_2). The algorithm is based on the evaluation–interpolation technique. Our goal is to compute the coefficients of the matrix Laurent polynomial $\widehat{H}(z) = \sum_{i=-n_1}^{N-n_1-1} z^i \widehat{H}_i$ such that $\widehat{H}(\omega_N^i) = H(\omega_N^i)$ for $i=0,\ldots,N-1$, where $N \geq n_1+n_2+1$. In fact, for Theorem 3.8 these coefficients provide approximations to H_i for $i=-n_1,\ldots,n_2$ within the absolute error ϵ. In the case where the ϵ-degree is not known, it is possible to apply an adaptive strategy which we discuss after the description of Algorithm 3.3.

Let $P(z)=\sum_{i=-n_1}^{n_2} z^i P_i$ be a matrix Laurent polynomial of degree (n_1, n_2) and let $N \geq n_1+n_2+1$ be an integer (say, a power of 2). Consider the vector

$$\boldsymbol{u} = \begin{bmatrix} P_0 \\ \vdots \\ P_{N-n_1-1} \\ P_{-n_1} \\ \vdots \\ P_{-1} \end{bmatrix}. \tag{3.7}$$

where $P_j = 0$ for $j=n_2+1,\ldots,N-n_1-1$ From the definition of IDFT, since $\omega_N^{N-i} = \omega_N^{-i}$ we conclude that $\boldsymbol{v}=(V_i)_{i=1,N} = \mathrm{IDFT}_N(\boldsymbol{u})$ is such that $V_i = P(\omega_N^{i-1})$, $i=1,\ldots,N$. Assuming that $P(z)$ is nonsingular for $|z|=1$, we have that V_i is nonsingular for $i=1,\ldots,N$. Therefore, denoting $\boldsymbol{w}=(V_i^{-1})_{i=1,N}$ and $\boldsymbol{y}=(Y_i)_{i=1,N}=\mathrm{DFT}_N(\boldsymbol{w})$ we find that the matrix Laurent polynomial $Q(z)=\sum_{i=-n_1}^{N-n_1-1} z^i Q_i$, of degree at most $(n_1, N-n_1-1)$, defined by

$$\begin{aligned} Q_i &= Y_{i+1}, && \text{for } i=0,\ldots,N-n_1-1, \\ Q_i &= Y_{N+i+1}, && \text{for } i=-n_1,\ldots,-1, \end{aligned}$$

is such that

$$Q(\omega_N^i) = P(\omega_N^i)^{-1}, \quad i = 0, \dots, N-1.$$

The same property holds if $\boldsymbol{v}$ is the vector with the values that a matrix Laurent power series $A(z)$ takes at the Nth roots of 1, i.e., $V_i = A(\omega_N^{i-1})$, $i = 1, \dots, N$. In this case $A(\omega_N^i) = P(\omega_N^i)^{-1}$, $i = 0, \dots, N-1$.

This is the basis of Algorithm 3.2.

Algorithm 3.2 Matrix Laurent power series inversion

INPUT: The $m \times m$ matrix Laurent power series $A(z)$, analytic and nonsingular for $z \in \mathbb{A}$ of (3.6), a positive number ϵ, integers $n_1, n_2 \geq 0$ such that the ϵ-degree of $H(z) = A(z)^{-1}$ is at most (n_1, n_2).

OUTPUT: Approximations $\widehat{H}_i$, within the absolute error ϵ, to the coefficients H_i, $i = -n_1, \dots, n_2$, of $H(z)$.

COMPUTATION: Let $N = 2^q$ be the minimum integer power of 2 such that $N \geq n_1 + n_2 + 1$.

1. Compute $\boldsymbol{v} = (V_i)_{i=1,N}$ such that $V_i = A(\omega_N^{i-1})$, $i = 1, \dots, N$.
2. Compute $\boldsymbol{w} = (W_i)_{i=1,N}$, where $W_i = V_i^{-1}$, $i = 1, \dots, N$, which provide the values of $H(\omega_N^{i-1})$.
3. Compute $\boldsymbol{y} = (Y_i)_{i=1,N} = \mathrm{DFT}_N(\boldsymbol{w})$.
4. Output the approximations $\widehat{H}_i = Y_{i+1}$, for $i = 0, \dots, n_2$, $\widehat{H}_i = Y_{N+i+1}$, $i = -n_1, \dots, -1$.

Observe that the computation of $V_i = A(\omega_N^{i-1})$, $i = 1, \dots, N$, at stage 1 of Algorithm 3.2 can be carried out within the absolute error ϵ if one knows an upper bound (m_1, m_2) on the ϵ-degree of the input Laurent series $A(z)$. By Theorem 3.8, if M is the least integer power of 2 such that $M \geq \max\{N, m_1 + m_2 + 1\}$, then an IDFT of order M provides the V_i's within the absolute error bound ϵ.

The cost of Algorithm 3.2 is dominated by the inversion of V_i, $i = 1, \dots, N$, that is $O(Nm^3)$ ops, while the costs of stages 1 and 3 are $O(Mm^2 \log M)$ and $O(Nm^2 \log N)$ ops, respectively. Moreover, in view of Theorem 3.6, the farther r and R are from 1, the lower is the number N of interpolation points needed to reach a required accuracy.

It is interesting to point out that the evaluation–interpolation technique at the basis of Algorithm 3.2 can be applied to solve more general problems. Assume that we are given a rational function $F(X_1, \dots, X_h)$ of the matrices $X_1, \dots, X_h$. Given matrix functions $X_i(z) \in \mathcal{W}$, $i = 1, \dots, h$, in the Wiener algebra, assume also that $H(z) = F(X_1(z), \dots, X_h(z))$ is analytic in the annulus $\mathbb{A}$ of (3.6). Our goal is to approximate the coefficients of $H(z)$ given the functions $X_i(z)$, $i = 1, \dots, h$. Relying on the evaluation–interpolation technique, we describe in Algorithm 3.3 a paradigm for performing this computation where we assume we are given the ϵ-degree (n_1, n_2) of $H(z)$.

If the evaluation at stage 1 of Algorithm 3.3 is not affected by errors, then from Theorem 3.8 it follows that $|P_i - H_i| < \epsilon E$, $i = -n_1, \dots, n_2$, where E is the matrix with all the elements equal to 1. Possible errors introduced at

Algorithm 3.3 Point-wise Laurent power series computation—1

INPUT: The matrix functions $X_i(z) \in \mathcal{W}$, $i = 1, \ldots, h$, and a rational matrix function $F(X_1, X_2, \ldots, X_h)$ such that $H(z) = F(X_1(z), \ldots, X_h(z))$ is analytic for $z \in \mathbb{A}$ of (3.6). Integers $n_1, n_2 \geq 0$ and a positive number ϵ such that the ϵ-degree of $H(z) = F(X_1(z), \ldots, X_h(z))$ is at most (n_1, n_2).

OUTPUT: Approximations $\widehat{H}_i$, within the absolute error ϵ, to the coefficients H_i, $i = -n_1, \ldots, n_2$, of $H(z)$.

COMPUTATION: Let $N = 2^q$ be the minimum integer power of 2 such that $N \geq n_1 + n_2 + 1$.

1. Compute $W_{i,j} = X_i(\omega_N^j)$, $j = 0, \ldots, N-1$, $i = 1, \ldots, h$
2. Compute $F_j = F(W_{1,j}, \ldots, W_{h,j})$, $j = 0, \ldots, N-1$.
3. Compute the coefficients P_i of the matrix Laurent polynomial $P(z) = \sum_{i=-n_1}^{N-n_1-1} z^i P_i$ such that $P(\omega_N^j) = F_j$, $j = 0, \ldots, N-1$
4. Output approximations $\widehat{H}_i = P_i$ to H_i, for $i = -n_1, \ldots, n_2$.

stage 1, generated by truncating the power series $X_j(z)$ to their ϵ-degree, are not amplified by the discrete Fourier transform (see Remark 2.1).

The above algorithm reduces the computation of a given function of matrix power series to the evaluation of the same function to a set of N numerical matrix values. In particular, it can be applied for the computation of the product $A(z)B(z)$ of matrix Laurent power series and of the inverse $A(z)^{-1}$, by choosing $F(X_1, X_2) = X_1 X_2$ and $F(X_1) = X_1^{-1}$, respectively.

If the ϵ-degree of $H(z)$ is not known it is possible to apply a doubling strategy provided that we are given a test for detecting if

$$\sum_{i=-N/2}^{N/2-1} |H_i - P_i| + \sum_{i<-N/2} |H_i| + \sum_{i>N/2-1} |H_i| < \epsilon E, \tag{3.8}$$

where $P(z) = \sum_{i=-N/2}^{N/2-1} z^i P_i$ is the matrix Laurent polynomial which interpolates $H(z)$ at the Nth roots of 1. Observe that, since

$$\sum_{i<-N/2} |H_i| + \sum_{i>N/2-1} |H_i| \leq \sum_{i=-N/2}^{N/2-1} |H_i - P_i| + \sum_{i<-N/2} |H_i| + \sum_{i>N/2-1} |H_i|,$$

the condition (3.8) implies that the ϵ-degree (n_1, n_2) of $H(z)$ is such that $N/2 \geq n_1$ and $N/2 - 1 \geq n_2$.

Algorithm 3.4 relies on the evaluation–interpolation strategy at the Nth roots of 1 where N doubles at each step. An efficient way for implementing this doubling technique, which does not require us to compute from scratch each IDFT and DFT in the evaluation stage, has been described in Remark 2.2.

In the particular case where $H(z)$ is a matrix power series, then Algorithms 3.2 and 3.3 are simpler since we can set $n_1 = 0$ so that $H(z)$ is approximated by

Algorithm 3.4 Point-wise Laurent power series computation—2

INPUT: The matrix functions $X_i(z) \in \mathcal{W}$, $i = 1, \dots, h$, and a rational matrix function $F(X_1, X_2, \dots, X_h)$ such that $H(z) = F(X_1(z), \dots, X_h(z))$ is analytic for $z \in \mathbb{A}$ of (3.6). A positive number ϵ and a test TEST$(H(z), P(z))$ which is true if $\sum_{i=-N/2}^{N/2-1} |H_i - P_i| + \sum_{i<-N/2} |H_i| + \sum_{i>N/2-1} |H_i| < \epsilon E$, where E is the matrix with all the elements equal to 1.

OUTPUT: Approximations P_i, within the absolute error ϵ, to the coefficients H_i, $i = -N/2, \dots, N/2-1$, of $H(z) = F(X_1(z), \dots, X_h(z))$ where N is such that the ϵ-degree of $H(z)$ is at most $(N/2, N/2-1)$.

COMPUTATION: Set $N = 2$.

1. Compute $W_{i,j} = X_i(\omega_N^j)$, $j = 0, \dots, N-1$, $i = 1, \dots, h$.
2. Compute $F_j = F(W_{1,j}, \dots, W_{h,j})$, $j = 0, \dots, N-1$.
3. Compute the coefficients $P_{-N/2}, \dots, P_{N/2-1}$ of the matrix Laurent polynomial $P(z)$ such that $P(\omega_N^j) = F_j$, $j = 0, \dots, N-1$.
4. If TEST$(H(z), P(z))$ is false then set $N = 2N$ and continue from step 1.
5. Output the approximations P_j to the coefficients H_j, $j = -N/2, \dots, N/2-1$.

a matrix polynomial of degree at most $N-1$. Concerning Algorithm 3.4, setting $n_1 = 0$ does not reduce to its minimum value the length of the discrete Fourier transforms. Algorithm 3.5 is an optimized version of Algorithm 3.4 in the case where $H(z)$ is a matrix power series when we are given a test for detecting if

$$\sum_{i=0}^{N-1} |H_i - P_i| + \sum_{i=N}^{+\infty} |H_i| < \epsilon E, \tag{3.9}$$

where $P(z) = \sum_{i=0}^{N-1} z^i P_i$ is the polynomial which interpolates $H(z)$ at the Nth roots of 1. Observe that, since

$$\sum_{i=N}^{+\infty} |H_i| \le \sum_{i=0}^{N-1} |H_i - P_i| + \sum_{i=N}^{+\infty} |H_i|,$$

the condition (3.9) implies that the ϵ-degree of $H(z)$ is at most N.

The following example shows the effectiveness of the evaluation–interpolation technique with the automatic detection of the ϵ-degree.

Example 3.9 Consider the function $F(X) = (I - X)^{-1}$ with $X(z) = zB + z^2C$, where $B, C \ge 0$ and $\|B\| + \|C\| < 1$ for some operator norm $\|\cdot\|$. From the expression $H(z) = F(X(z)) = \sum_{i=0}^{+\infty}(zB + z^2C)^i$, we find that $H(z)$ is convergent for $|z| \le 1$ and $H_i \ge 0$. Observe that

$$H'(z) = \left[(I - zB - z^2C)^{-1}\right]' = (I - zB - z^2C)^{-1}(B + 2zC)(I - zB - z^2C)^{-1}$$

so that $H'(1) = (I - B - C)^{-1}(B + 2C)(I - B - C)^{-1}$ is given explicitly and is computable without knowing the coefficients of $H(z)$.

Algorithm 3.5 Point-wise power series computation

INPUT: The matrix Laurent power series $X_i(z)$, $i = 1, \ldots, h$, and a rational matrix function $F(X_1, X_2, \ldots, X_h)$, such that $H(z) = F(X_1(z), \ldots, X_h(z))$ is a matrix power series convergent for $|z| \le 1$. A positive number ϵ and a test $\mathrm{TEST}(H(z), P(z))$ which is true if $\sum_{i=0}^{N-1} |H_i - P_i| + \sum_{i=N}^{+\infty} |H_i| < \epsilon E$.

OUTPUT: Approximations P_i, within the absolute error ϵ, to the coefficients H_i, $i = 0, \ldots, N-1$, of $H(z) = F(X_1(z), \ldots, X_h(z))$ where N is greater than or equal to the ϵ-degree of $H(z)$.

COMPUTATION: Set $N = 2$.

1. Compute $W_{i,j} = X_i(\omega_N^j)$, $j = 0, \ldots, N-1$, $i = 1, \ldots, h$.
2. Compute $F_j = F(W_{1,j}, \ldots, W_{h,j})$, $j = 0, \ldots, N-1$.
3. Compute the coefficients $P_0, \ldots, P_{N-1}$ of the matrix polynomial $P(z)$ of degree at most $N-1$ such that $P(\omega_N^j) = F_j$, $j = 0, \ldots, N-1$.
4. If $\mathrm{TEST}(H(z), P(z))$ is false then set $N = 2N$ and continue from step 1.
5. Output the approximations $P_0, \ldots, P_{N-1}$ to the values $H_0, \ldots, H_{N-1}$.

Let $P(z)$ be the polynomial of degree at most $N-1$ such that $P(\omega_N^j) = F_j$, $j = 0, \ldots, N-1$. Then, in light of Theorem 3.8 applied to $H(z)$ we have

$$P'(1) = \sum_{i=1}^{N-1} iH_i + \sum_{i=1}^{N-1} \sum_{j=1}^{+\infty} iH_{i+Nj}.$$

Therefore,

$$H'(1) - P'(1) = \sum_{i=N}^{+\infty} iH_i - \sum_{i=1}^{N-1} \sum_{j=1}^{+\infty} iH_{i+Nj} \ge \sum_{i=N}^{-\infty} H_i,$$

and so, $H'(1) - P'(1) \le \epsilon E$ implies that $\sum_{i=N}^{+\infty} H_i \le \epsilon E$, i.e., $H(z)$ has ϵ-degree at most N. Since both $H'(1)$ and $P'(1)$ are explicitly computable, the condition $H'(1) - P'(1) \le \epsilon E$ is a guaranteed test for dynamically checking the ϵ-degree of $H(z)$.

3.2 Wiener–Hopf factorization

Let $a(z) = \sum_{i=-\infty}^{+\infty} z^i a_i$ be a complex-valued function in the Wiener algebra $\mathcal{W}$, such that $a(z) \neq 0$ for $|z| = 1$. A *Wiener–Hopf factorization* of $a(z)$ is a decomposition

$$a(z) = u(z) z^{\kappa} l(z), \qquad |z| = 1 \tag{3.10}$$

where κ is an integer, $u(z) = \sum_{i=0}^{+\infty} z^i u_i$ and $l(z) = \sum_{i=0}^{+\infty} z^{-i} l_{-i}$ belong to the Wiener algebra and are different from zero for $|z| \le 1$ and $1 \le |z| \le \infty$, respectively.

It is well known [27] that for functions in the Wiener algebra which are nonsingular for $|z| = 1$ the Wiener–Hopf factorization always exists. In particular, l_0 and u_0 are nonzero.

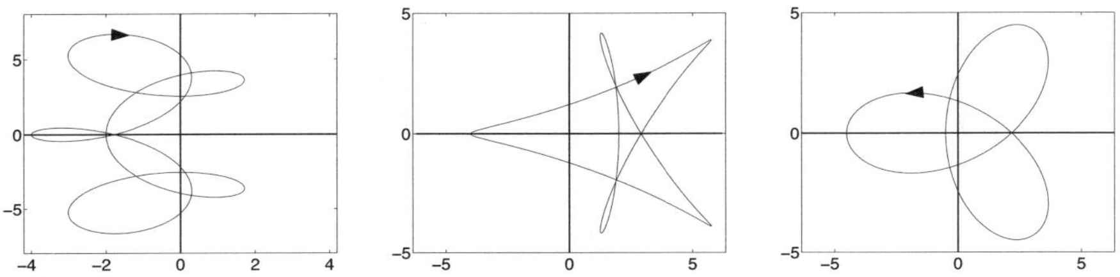

(a) $a_1(z)$, $\kappa = 0$ (b) $a_2(z)$, $\kappa = -1$ (c) $a_3(z)$, $\kappa = 2$

Fig. 3.1 Winding numbers of $a_1(z) = 2z^{-4} - 2z^{-1} + 3z + 1$, $a_2(z) = 3z^{-2} + z^{-1} - 2 + 2z^3$ and $a_3(z) = 2z^{-1} - 1/2 + 3z^2$.

If the function $a(z)$ is analytic in the annulus $\mathbb{A}$ of (3.6), and if $\xi \in \mathbb{A}$ is such that $a(\xi) = 0$, then

$$\begin{cases} u(\xi) = 0 & \text{if } |\xi| > 1 \\ l(\xi) = 0 & \text{if } |\xi| < 1. \end{cases}$$

that is, the zeros of $a(z)$ of modulus greater than 1 coincide with the zeros of $u(z)$, and the zeros of $a(z)$ of modulus less than 1 coincide with the zeros of $l(z)$. In other words, the Wiener–Hopf factorization provides a splitting of the zeros of $a(z)$ with respect to the unit circle.

The number κ coincides with the *winding number* of $a(z)$. Its formal definition, related to the polar representation $a(z) = \rho(z)e^{\mathrm{i}\theta(z)}$, is

$$\kappa = \frac{1}{2\pi}\left(\lim_{t\to 2\pi^-} \theta(\cos t + \mathrm{i}\sin t) - \lim_{t\to 0^+} \theta(\cos t + \mathrm{i}\sin t)\right).$$

In other words the winding number of $a(z)$ is the number of times that the oriented curve $\Gamma = \{a(e^{\mathrm{i}\theta}) : \quad \theta \in [0, 2\pi]\}$ winds round the origin in the complex plane. In Figure 3.1 we show the curve Γ for different functions $a(z)$.

We may describe the Wiener–Hopf factorization in matrix form in terms of bi-infinite Toeplitz matrices. For simplicity, consider the most interesting case where $\kappa = 0$. For the matrices $T_{\pm\infty}[a(z)] = (a_{j-i})_{i,j\in\mathbb{Z}}$, $T_{\pm\infty}[u(z)] = (u_{j-i})_{i,j\in\mathbb{Z}}$, $T_{\pm\infty}[l(z)] = (l_{j-i})_{i,j\in\mathbb{Z}}$, where $u_i = l_{-i} = 0$ if $i < 0$,

$$T_{\pm\infty}[a(z)] = T_{\pm\infty}[u(z)]T_{\pm\infty}[l(z)] = T_{\pm\infty}[l(z)]T_{\pm\infty}[u(z)],$$

where $T_{\pm\infty}[u(z)]$ is upper triangular and $T_{\pm\infty}[l(z)]$ is lower triangular.

A similar factorization holds for semi-infinite matrices: if

$$T_\infty[a(z)] = (a_{j-i})_{i,j\in\mathbb{N}}, \quad T_\infty[u(z)] = (u_{j-i})_{i,j\in\mathbb{N}}, \quad T_\infty[l(z)] = (l_{j-i})_{i,j\in\mathbb{N}},$$

then $T_\infty[a(z)] = T_\infty[u(z)]T_\infty[l(z)]$, that is,

$$\begin{bmatrix} a_0 & a_1 & a_2 & \cdots \\ a_{-1} & a_0 & a_1 & \ddots \\ a_{-2} & a_{-1} & a_0 & \ddots \\ \vdots & \ddots & \ddots & \ddots \end{bmatrix} = \begin{bmatrix} u_0 & u_1 & u_2 & \cdots \\ & u_0 & u_1 & \ddots \\ 0 & & \ddots & \ddots \end{bmatrix} \begin{bmatrix} l_0 & & & 0 \\ l_{-1} & l_0 & & \\ l_{-2} & l_{-1} & l_0 & \\ \vdots & \ddots & \ddots & \ddots \end{bmatrix}.$$

That is, a Wiener–Hopf factorization is equivalent to a UL factorization of a semi-infinite Toeplitz matrix into an upper triangular Toeplitz matrix $U = T_\infty[u(z)]$ and a lower triangular Toeplitz matrix $L = T_\infty[l(z)]$.

Wiener–Hopf factorizations may also be defined for matrix valued functions. Let $A(z) = \sum_{i=-\infty}^{+\infty} z^i A_i$ be an $m \times m$ complex matrix valued function in the Wiener algebra such that $\det A(z) \neq 0$ for $|z| = 1$. A *Wiener–Hopf factorization* of $A(z)$ is a decomposition

$$A(z) = U(z)\,\mathrm{Diag}(z^{\kappa_1}, \dots, z^{\kappa_m})L(z), \qquad |z| = 1, \tag{3.11}$$

where $\kappa_1 \leq \cdots \leq \kappa_m$ are integers called the *partial indices* of $A(z)$, $U(z) = \sum_{i=0}^{+\infty} z^i U_i$ and $L(z) = \sum_{i=0}^{+\infty} z^{-i} L_{-i}$, belong to the Wiener algebra and are invertible for $|z| \leq 1$ and $1 \leq |z| \leq \infty$, respectively. Note that in particular U_0 and L_0 are nonsingular.

If $A(z)$ is analytic in the annulus $\mathbb{A}$ of (3.6), we say that $\lambda \in \mathbb{A}$ is a *root* of $A(z)$ if $\det A(\lambda) = 0$. The roots of $U(z)$ in $\mathbb{A}$ coincide with the roots of $A(z)$ of modulus greater than 1 and the roots of $L(z)$ in $\mathbb{A}$ coincide with the roots of $A(z)$ of modulus less than 1.

Similarly, for matrix valued functions we have the following existence condition:

Theorem 3.10 *Let $A(z) = \sum_{i=-\infty}^{+\infty} z^i A_i$ be in the Wiener algebra such that $\det A(z) \neq 0$ for $|z| = 1$. Then there exists a Wiener–Hopf factorization* (3.11) *of $A(z)$.*

Also in the block case, if $\kappa_1 = \cdots = \kappa_m = 0$ we may provide a matrix interpretation given in terms of block Toeplitz matrices. In that case, $T_{\pm\infty}[A(z)] = T_{\pm\infty}[U(z)]T_{\pm\infty}[L(z)]$ where $T_{\pm\infty}[L(z)]$ is block lower triangular and $T_{\pm\infty}[U(z)]$ is block upper triangular. A similar factorization holds for semi-infinite block matrices, i.e., $T_\infty[A(z)] = T_\infty[U(z)]T_\infty[L(z)]$.

Remark 3.11 Let $a(z) = u(z)l(z)$ be a Wiener–Hopf factorization of $a(z)$ with $\kappa = 0$. If $a_i = 0$ for $i < -p$, where $p \geq 0$, then $l(z^{-1})$ is a polynomial of degree at most p and $a(z)$ has at most p zeros in the open unit disk $\mathbb{D}$. If $a_i = 0$ for $i > q$, where $q \geq 0$, then $u(z)$ is a polynomial of degree at most q and $a(z)$ has at most q zeros outside the closure of $\mathbb{D}$. The same property holds for matrix valued functions where the coefficients are $m \times m$ matrices provided that there exists a Wiener–Hopf factorization $A(z) = U(z)L(z)$ with null partial indices. In this case $\det A(z)$ has at most mp zeros of modulus less than 1 and mq zeros of modulus greater than 1.

A Wiener–Hopf factorization with null partial indices is called *canonical factorization.*

It is interesting to observe that the Wiener–Hopf factorization of a function $a(z)$ or of a matrix function $A(z)$ provides an expression for the inverse of $a(z)$ and $A(z)$, respectively: we obtain from (3.11) that

$$A(z)^{-1} = L(z)^{-1} \operatorname{Diag}(z^{-\kappa_1}, \ldots, z^{-\kappa_m}) U(z)^{-1}, \quad |z| = 1,$$

where $L(z^{-1})^{-1}$ and $U(z)^{-1}$ are matrix power series analytic for $|z| < 1$.

Remark 3.12 Let $A(z) = U(z)L(z)$ be a canonical factorization. Since the functions $A(z)$, $U(z)$ and $L(z)$ belong to the Wiener algebra and are nonsingular for $|z| = 1$, then from Theorem 3.1 the matrices $T_{\pm\infty}[A(z)]$, $T_{\pm\infty}[L(z)]$ and $T_{\pm\infty}[U(z)]$ define bounded operators together with their inverses. The same property holds also for $T_\infty[A(z)]$, $T_\infty[L(z)]$, and $T_\infty[U(z)]$. In particular we have

$$T_{\pm\infty}[A(z)]^{-1} = T_{\pm\infty}[L(z)]^{-1} T_{\pm\infty}[U(z)]^{-1}$$
$$T_\infty[A(z)]^{-1} = T_\infty[L(z)]^{-1} T_\infty[U(z)]^{-1}.$$

Remark 3.13 In the framework of Markov chains we encounter matrix Laurent power series $A(z)$ such that $\det A(z) = 0$ for $|z| = 1$. In this case the assumption of invertibility of $A(z)$ for $|z| = 1$ does not hold, and a Wiener–Hopf factorization does not exist in the form defined in this section. However, we may still consider factorizations of the kind $A(z) = U(z) \operatorname{Diag}(z^{\kappa_1}, \ldots, z^{\kappa_m}) L(z)$ where $U(z)$ and $L(z^{-1})$ are nonsingular for $|z| < 1$ and may be singular for $|z| = 1$. In this case, the nice property of invertibility of $A(z)$, $L(z)$ and $U(z)$ no longer holds for $|z| = 1$.

In the following, we will call *weak Wiener–Hopf factorization* a factorization of the kind (3.11) where $U(z)$ and $L(z)$ are in the Wiener algebra $\mathcal{W}$ and $U(z)$ and $L(z^{-1})$ are invertible for $|z| < 1$. That is, we allow $U(z)$ or $L(z)$, or both, to be singular for some values z of modulus 1. Similarly, we call *weak canonical factorization* a weak Wiener–Hopf factorization with null partial indices.

For instance, the factorization $A(z) = (I - zH)(I - z^{-1}\boldsymbol{u}\boldsymbol{u}^{\mathrm{T}})$, where $\rho(H) \leq 1$ and $\boldsymbol{u}^{\mathrm{T}}\boldsymbol{u} = 1$, is a weak canonical factorization. In fact, $L(z) = I - z^{-1}\boldsymbol{u}\boldsymbol{u}^{\mathrm{T}}$ is singular for $z = 1$ and one shows that $L(z^{-1})^{-1} = I + \boldsymbol{u}\boldsymbol{u}^{\mathrm{T}} \sum_{i=1}^{+\infty} z^i$. Observe that $L(z)^{-1}$ is convergent for $|z| > 1$ but not for $z = 1$. However, its coefficients are bounded.

Conditions for the existence of a (weak) canonical factorization of a matrix Laurent power series of the kind $\sum_{i=-1}^{+\infty} z^i A_i$ can be given in terms of solutions of certain nonlinear matrix equations. This is the subject of the next section.

3.3 Nonlinear matrix equations

Let us consider the $m \times m$ matrix Laurent power series

$$S(z) = \sum_{i=-1}^{+\infty} z^i A_i$$

which we assume to be in the Wiener algebra, and define the matrix power series

$$A(z) = zS(z).$$

Since $S(z) \in \mathcal{W}$, $A(z)$ is analytic in the open unit disk $\mathbb{D}$ and continuous in its closure. We associate with $A(z)$ the matrix equation

$$A_{-1} + A_0 X + A_1 X^2 + \cdots = 0 \tag{3.12}$$

where the unknown X is an $m \times m$ matrix, and we are interested in solutions X of (3.12) which are *power bounded*, i.e., such that $|X^k|$ is bounded from above by a constant for any positive integer k. This boundedness condition is satisfied if $\rho(X) < 1$ since X^k converges to zero for $k \to +\infty$. Conversely, if X is power bounded then $\rho(X) \leq 1$. However, if $\rho(X) = 1$, then X is not necessarily power bounded, a simple example being $X = \left[\begin{smallmatrix} 1 & 1 \\ 0 & 1 \end{smallmatrix}\right]$, for which $X^k = \left[\begin{smallmatrix} 1 & k \\ 0 & 1 \end{smallmatrix}\right]$. In particular, among the power bounded solutions, we are interested in those solutions (if they exist) which have minimal spectral radius.

Definition 3.14 *A solution X of* (3.12) *is called a spectral minimal solution or more simply, a minimal solution, if $\rho(X) \leq \rho(Y)$ for any other solution Y. A solution X is called a minimal nonnegative solution if $0 \leq X \leq Y$ for any other nonnegative solution Y, where the inequalities hold component-wise.*

Under suitable conditions, which are naturally satisfied in the context of Markov chains, the minimal nonnegative solution exists and coincides with the spectral minimal solution.

Let G be a power bounded solution of (3.12), and assume that λ is an eigenvalue of G corresponding to the eigenvector $\boldsymbol{u}$, i.e., such that $G\boldsymbol{u} = \lambda\boldsymbol{u}$, $\boldsymbol{u} \neq 0$. Since $G^i\boldsymbol{u} = \lambda^i\boldsymbol{u}$, multiplying (3.12) on the right by $\boldsymbol{u}$ yields

$$A_{-1}\boldsymbol{u} + \lambda A_0\boldsymbol{u} + \lambda^2 A_1\boldsymbol{u} + \cdots = A(\lambda)\boldsymbol{u} = 0,$$

where the left-hand side is convergent since $|\lambda| \leq \rho(G) \leq 1$ and $S(z) \in \mathcal{W}$. In other words, for any solution G such that $\rho(G) \leq 1$, if $(\lambda, \boldsymbol{u})$ is an eigenpair of G, i.e., $G\boldsymbol{u} = \lambda\boldsymbol{u}$ then $A(\lambda)\boldsymbol{u} = 0$. This property extends to Jordan chains as we state in the next theorem. First, we have to introduce some definitions.

Let $J = U^{-1}GU$ be the Jordan canonical form of G. If λ is an eigenvalue of G belonging to a Jordan block of size k,

$$\begin{bmatrix} \lambda & 1 & 0 & \dots & 0 \\ & \lambda & 1 & \ddots & \vdots \\ & & \ddots & \ddots & 0 \\ & & & \ddots & 1 \\ 0 & & & & \lambda \end{bmatrix}$$

and if $\boldsymbol{u}_1, \dots, \boldsymbol{u}_k$ are the corresponding columns of U such that

$$\begin{aligned} G\boldsymbol{u}_1 &= \lambda\boldsymbol{u}_1, \\ G\boldsymbol{u}_i &= \lambda\boldsymbol{u}_i + \boldsymbol{u}_{i-1}, \quad i = 2, \dots, k, \end{aligned}$$

we say that the vectors $\boldsymbol{u}_1 = \boldsymbol{u}, \boldsymbol{u}_2, \dots, \boldsymbol{u}_k$, form a *cycle of length* k corresponding to the eigenvalue λ of G.

If λ is a root of the matrix power series $A(z)$ we say that the vectors $\boldsymbol{v}_1, \ldots, \boldsymbol{v}_h$ form a *Jordan chain of length* h for $A(z)$ corresponding to λ if

$$\sum_{j=0}^{i} \frac{1}{j!} A^{(j)}(\lambda)\boldsymbol{v}_{i-j+1} = 0, \quad \text{for } i = 0, \ldots, h-1,$$

where $A^{(j)}(z)$ denotes the jth derivative of $A(z)$ with respect to z.

The following result which relates the Jordan chains of $A(z)$ to the Jordan cannical form of G has been proved in [40].

Theorem 3.15 *Let G be a power bounded solution of* (3.12), *and let $\boldsymbol{u}_1, \ldots, \boldsymbol{u}_k$ be a cycle of length k corresponding to the eigenvalue λ of G. Then the vectors $\boldsymbol{u}_1, \ldots, \boldsymbol{u}_k$ form a Jordan chain for the matrix power series $A(z)$ corresponding to λ. Conversely, let λ_j, $|\lambda_j| \le 1$, $j = 1, \ldots, \nu$, be roots of $A(z)$ with Jordan chains $\boldsymbol{u}_i^{(j)}$, $i = 1, \ldots, k_j$ for $j = 1, \ldots, \nu$; define the matrix U having columns $\boldsymbol{u}_i^{(j)}$, $i = 1, \ldots, k_j$, $j = 1, \ldots, \nu$, and the block diagonal matrix J having Jordan blocks J_j of size $k_j \times k_j$ associated with λ_j. Assume that $\sum_{i=1}^{\nu} k_i = m$ and that $\det U \neq 0$. Then the matrix $G = UJU^{-1}$ solves* (3.12) *and* $\rho(G) \le 1$.

An immediate consequence of Theorem 3.15 follows.

Corollary 3.16 *If $A(z)$ has exactly m roots in the closed unit disk, counted with their multiplicities, and if G is a solution of* (3.12) *such that $\rho(G) \le 1$, then G is the unique minimal solution of* (3.12).

Canonical factorization is a useful tool which allows one to characterize the minimal solutions of equation (3.12). In order to show this we need a preliminary result which generalizes to matrix power series Corollary 1 on page 252 of [74], valid for matrix polynomials.

Lemma 3.17 *Let $P(z) = \sum_{i=0}^{+\infty} z^i P_i$ and $Q(z) = \sum_{i=0}^{+\infty} z^i Q_i$ be $m \times m$ matrix power series and assume that there exists a matrix G such that $P(z) = Q(z)(zI - G)$, and $\lim_n Q_n G^{n+1} = 0$. Then $\sum_{i=0}^{+\infty} P_i G^i = 0$. Similarly, if $P(z) = (zI - R)C(z)$ and $\lim_n R^{n+1} C_n = 0$, where $C(z) = \sum_{i=0}^{+\infty} z^i C_i$ is an $m \times m$ power series and R is an $m \times m$ matrix, then $\sum_{i=0}^{+\infty} R^i P_i = 0$.*

Proof From $P(z) = Q(z)(zI - G)$ we find that $P_0 = -Q_0 G$ and that $P_i = Q_{i-1} - Q_i G$, for $i > 0$. Therefore,

$$\sum_{i=0}^{n} P_i G^i = -Q_0 G + (Q_0 - Q_1 G)G + (Q_1 - Q_2 G)G^2 + \cdots$$
$$+ (Q_{n-1} - Q_n G)G^n = -Q_n G^{n+1},$$

and we conclude that $\sum_{i=0}^{+\infty} P_i G^i = \lim_n \sum_{i=0}^{n} P_i G^i = -\lim_n Q_n G^{n+1} = 0$. A similar argument is used for the second claim. □

We are ready now to prove the following theorem.

Theorem 3.18 *Let $S(z) = \sum_{i=-1}^{+\infty} z^i A_i$ be an $m \times m$ matrix valued function in the Wiener algebra $\mathcal{W}$. If there exists a canonical factorization*

$$S(z) = U(z)L(z), \qquad L(z) = L_0 + z^{-1}L_{-1}, \qquad |z| = 1,$$

then $G = -L_0^{-1}L_{-1}$ is the unique solution of (3.12) such that $\rho(G) < 1$. In particular, G is the minimal solution. Conversely, if there exists a solution G of (3.12) such that $\rho(G) < 1$ and if $A(z) = zS(z)$ has exactly m roots in the open unit disk and is nonsingular for $|z| = 1$, then $S(z)$ has a canonical factorization

$$S(z) = U(z)L(z) = (U_0 + zU_1 + \cdots)(I - z^{-1}G), \qquad |z| = 1 \tag{3.13}$$

where $U_i = \sum_{j=i}^{+\infty} A_j G^{j-i}, \quad i \geq 0$.

Proof If there exists a canonical factorization for $S(z)$, since $A_i = 0$ for $i < -1$ then, in light of Remark 3.11, $L(z) = L_0 + z^{-1}L_{-1}$. Moreover, from the properties of canonical factorizations L_0 is nonsingular and we may write $L(z) = L_0(I - z^{-1}G)$ with $G = -L_0^{-1}L_{-1}$ so that

$$A(z) = (U_0 + zU_1 + z^2U_2 + \cdots)L_0(zI - G).$$

Since $L(z)$ is nonsingular for $|z| \geq 1$, $\rho(G) < 1$ and $\det A(z)$ has exactly m zeros in the open unit disk. Since $\rho(G) < 1$ and $\lim_i U_i = 0$, we conclude from Lemma 3.17 that $X = G$ is a solution of (3.12). In particular, G is the unique minimal solution in light of Corollary 3.16. Conversely, assume that G is a solution of (3.12) such that $\rho(G) < 1$ and $A(z)$ is nonsingular for $|z| = 1$. Consider the formal product

$$S(z)\sum_{i=0}^{+\infty} z^{-i}G^i = \sum_{i=-\infty}^{+\infty} z^i U_i,$$

with

$$U_i = \begin{cases} \left(\sum_{j=-1}^{+\infty} A_j G^{j+1}\right) G^{-i-1} & \text{for } i < 0 \\ \sum_{j=i}^{+\infty} A_j G^{j-i} & \text{for } i \geq 0. \end{cases} \tag{3.14}$$

Observe that the matrices U_i are well defined since the series $A(z)$ is convergent for $|z| \leq 1$ and $\rho(G) < 1$. Since G is such that $\sum_{j=-1}^{+\infty} A_j G^{j+1} = 0$, then $U_i = 0$ for $i < 0$.

Clearly, $L(z) = I - z^{-1}G$ is in the Wiener algebra and is nonsingular for $1 \leq |z| \leq \infty$. Concerning $U(z) = \sum_{i=0}^{+\infty} z^i U_i$, observe that for any matrix norm $\|\cdot\|$, the condition $\sum_{i=0}^{+\infty} |U_i| < +\infty$ (that is, $U(z) \in \mathcal{W}$) is equivalent to $\sum_{i=0}^{+\infty} \|U_i\| < +\infty$. Moreover, since $\rho(G) < 1$, there exists a matrix norm $\|\cdot\|$ such that $\sigma = \|G\| < 1$ (see Theorem A.2). In this way, from (3.14) we have

$$\sum_{i=0}^{+\infty} \|U_i\| \leq \sum_{i=0}^{+\infty}\sum_{j=i}^{+\infty} \sigma^{j-i}\|A_j\| = \sum_{i=0}^{+\infty} \sigma^i \sum_{j=i}^{+\infty} \|A_j\| \leq \frac{1}{1-\sigma}\sum_{i=0}^{+\infty} \|A_i\| < +\infty,$$

which shows that $U(z)$ is in the Wiener algebra, so that we may write

$$S(z) = \left(\sum_{i=0}^{+\infty} U_i z^i \right) (I - z^{-1}G), \quad |z| = 1.$$

Since $A(z)$ is nonsingular for $|z| = 1$ and since the roots of $A(z)$ in the open unit disk are the eigenvalues of G, from the relation $\det A(z) = \det U(z) \det(zI - G)$ one concludes that $\det U(z)$ cannot have zeros of modulus less than or equal to 1. Therefore (3.13) is a canonical factorization. □

Similarly, the following result is valid for weak canonical factorizations.

Theorem 3.19 *Let $S(z) = \sum_{i=-1}^{+\infty} z^i A_i$ be an $m \times m$ matrix valued function in the Wiener algebra $\mathcal{W}$. If there exists a weak canonical factorization*

$$S(z) = U(z)L(z), \quad L(z) = L_0 + z^{-1}L_{-1}, \quad |z| = 1,$$

such that $G = -L_0^{-1}L_{-1}$ is power bounded, then G is a minimal solution of (3.12) such that $\rho(G) \le 1$. Conversely, if $\sum_{i=0}^{+\infty}(i+1)|A_i| < +\infty$, there exists a power bounded solution G of (3.12) such that $\rho(G) = 1$, and if all the zeros of $\det A(z)$ in the open unit disk are eigenvalues of G then there exists a weak canonical factorization of the form (3.13).

Proof We do not give the detailed proof of the first part, since it follows the same lines of the proof of Theorem 3.18. For the second part, by following the same argument as in the proof of Theorem 3.18, we obtain (3.14). Now we prove that $U(z) \in \mathcal{W}$. Since G is power bounded there exists a constant $\sigma > 0$ such that $|G^j| \le \sigma E$ for any $j \ge 0$ where E is the matrix with all the elements equal to 1. Therefore we have

$$\sum_{i=0}^{+\infty} |U_i| \le \sigma \sum_{i=0}^{+\infty} \sum_{j=i}^{+\infty} |A_j| E = \sigma \sum_{i=0}^{+\infty} (i+1)|A_i| E$$

which is bounded from above, therefore $U(z) \in \mathcal{W}$. Moreover $\det U(z)$ cannot have zeros in the open unit disk since $\det(zS(z)) = \det U(z) \det(zI - G)$ and the zeros of $\det(zS(z))$ in $\mathbb{D}$ are eigenvalues of G. □

3.3.1 Linearization

Solving the matrix equation (3.12) is equivalent to solving a semi-infinite linear system. Assume that there exists a canonical factorization of $S(z)$, denote by G the minimal solution of (3.12) and recall that $\rho(G) < 1$. From the property $\sum_{i=-1}^{+\infty} A_i G^{i+1} = 0$ we deduce that

$$\begin{bmatrix} A_0 & A_1 & A_2 & \dots \\ A_{-1} & A_0 & A_1 & \dots \\ & A_{-1} & A_0 & \dots \\ 0 & & \ddots & \ddots \end{bmatrix} \begin{bmatrix} G \\ G^2 \\ G^3 \\ \vdots \end{bmatrix} = \begin{bmatrix} -A_{-1} \\ 0 \\ 0 \\ \vdots \end{bmatrix} \tag{3.15}$$

where the above system is semi-infinite. In other words, the infinite block vector with components $G, G^2, \dots$, is a solution of the above semi-infinite system, and its columns are in $\ell^2(\mathbb{N})$. Moreover, from the existence of the canonical factorization, the semi-infinite matrix $(A_{j-i})_{i,j\in\mathbb{N}}$ is invertible (as we mentioned in Remark 3.12), and this implies the uniqueness in $\ell^2(\mathbb{N})$ of the solution of the above system.

In the context of structured Markov chains, the function $S(z) = z^{-1}A(z)$ has a weak canonical factorization $S(z) = U(z)L(z)$ where $L(z) = I - z^{-1}G$, G is power bounded and $U(z)$ is nonsingular for $|z| \le 1$. Then G is a solution of the matrix equation (3.12) and the system (3.15) still holds, but the matrix does not represents an operator with bounded inverse in $\ell^2(\mathbb{N})$. Still, $T_\infty[U(z)]$ is invertible and $T_\infty[L(z)]$ has a block lower triangular inverse with bounded elements G^{i-j}, $i \ge j$.

3.3.2 Quadratic matrix equations

The special case

$$A_{-1} + A_0X + A_1X^2 = 0 \tag{3.16}$$

where (3.12) is quadratic deserves special attention since the particular features of this problem allow one to give stronger properties of the solutions. Furthermore, as we will show in Section 3.4, more general matrix equations can be reduced to the quadratic case.

Let us consider the easier case where the function $S(z) = z^{-1}A_{-1}+A_0+zA_1$ is nonsingular for $|z| = 1$ and denote $A(z) = zS(z)$. Since $\det A(z)$ is a polynomial, then a necessary condition for the existence of a canonical factorization of $S(z)$ is that $\det A(z)$ has exactly m zeros in the open unit disk (see Remark 3.11).

We have the following:

Theorem 3.20 *Define $S(z) = z^{-1}A_{-1} + A_0 + zA_1$ and $A(z) = zS(z)$, where A_i, $i = -1, 0, 1$, are $m \times m$ matrices. Assume that the polynomial $\det A(z)$ has zeros ξ_i, $i = 1, \ldots, 2m$, such that*

$$|\xi_1| \le \cdots \le |\xi_m| < 1 < |\xi_{m+1}| \le \cdots \le |\xi_{2m}|$$

where we assume zeros at ∞ if the degree of $\det A(z)$ is less than $2m$. If there exists a matrix G which solves (3.16) such that $\rho(G) < 1$ then the following properties hold:

1. *$S(z)$ has a canonical factorization*

$$S(z) = (U_0 + zU_1)(I - z^{-1}G), \quad |z| = 1, \tag{3.17}$$

with

$$U_0 = A_0 + A_1G, \quad U_1 = A_1;$$

moreover, $\det U_0 \ne 0$, G is the minimal solution of (3.16) and $\rho(G) = |\xi_m|$.

2. *The matrix $R = -U_1U_0^{-1} = -A_1(A_1G + A_0)^{-1}$ is the minimal solution of the matrix equation*

$$A_1 + XA_0 + X^2A_{-1} = 0; \tag{3.18}$$

moreover, $\rho(R) = 1/|\xi_{m+1}|$ and $A_1G = RA_{-1}$.

3. *The matrix* U_0 *solves the equation*

$$X = A_0 - A_1 X^{-1} A_{-1};$$

moreover $U_0 = A_0 + RA_{-1}$.

4. *The matrix Laurent power series* $S(z)$ *is invertible in the annulus* $\mathbb{A} = \{z \in \mathbb{C} : \quad |\xi_m| < |z| < 1/|\xi_{m+1}|\}$ *and* $H(z) = S(z)^{-1} = \sum_{i=-\infty}^{+\infty} z^i H_i$ *is such that*

$$H_i = \begin{cases} G^{-i} H_0, & i < 0, \\ \sum_{j=0}^{+\infty} G^j U_0^{-1} R^j, & i = 0, \\ H_0 R^i, & i > 0. \end{cases}$$

5. *If the matrix equation*

$$A_{-1} X^2 + A_0 X + A_1 = 0 \tag{3.19}$$

has a solution $\widehat{G}$ *such that* $\rho(\widehat{G}) < 1$, *then* $\widehat{G}$ *is the minimal solution,* $\det H_0 \neq 0$, $\det(A_{-1}\widehat{G} + A_0) \neq 0$ *and* $\widehat{R} = -A_{-1}(A_{-1}\widehat{G} + A_0)^{-1}$ *is the minimal solution of the equation*

$$X^2 A_1 + X A_0 + A_{-1} = 0.$$

Moreover, we have the following representation of H_i*:*

$$H_i = \begin{cases} G^{-i} H_0 = H_0 \widehat{R}^{-i}, & i \le 0, \\ \widehat{G}^i H_0 \;\; = H_0 R^i, & i \ge 0. \end{cases}$$

Therefore, $G = H_{-1} H_0^{-1}$, $\widehat{G} = H_1 H_0^{-1}$, $R = H_0^{-1} H_1$, $\widehat{R} = H_0^{-1} H_{-1}$.

Proof Part 1 follows from Theorem 3.18. To prove part 2, we rewrite (3.17) as $S(z) = (I - zR)(U_0 + z^{-1} A_{-1})$, and we find from Lemma 3.17 that R solves (3.18). By using the same argument as in the proof of Theorem 3.18 we prove that R is the minimal solution and that $\rho(R) = 1/|\xi_{m+1}|$. Finally, from the definition of R, we have $A_1 G = -R(A_1 G + A_0)G = RA_{-1}$. Part 3 can be proved by direct inspection. Concerning part 4, since the matrix functions $S(z)$, $U(z)$ and $L(z) = I - z^{-1}G$ are analytic and invertible in the annulus $\mathbb{A}$, so are their inverses and $H(z) = S(z)^{-1} = (I - z^{-1}G)^{-1} U_0^{-1} (I - zR)^{-1}$, from which we conclude that

$$H_i = \begin{cases} (\sum_{j=0}^{+\infty} G^j U_0^{-1} R^j) R^i = H_0 R^i & \text{for } i > 0, \\ \sum_{j=0}^{+\infty} G^j U_0^{-1} R^j & \text{for } i = 0, \\ G^{-i} (\sum_{j=0}^{+\infty} G^j U_0^{-1} R^j) = G^{-i} H_0 & \text{for } i < 0. \end{cases} \tag{3.20}$$

Concerning part 5, if there exists a solution $\widehat{G}$ of $A_{-1} X^2 + A_0 X + A_1 = 0$ such that $\rho(\widehat{G}) < 1$ then, by following the same arguments as before, one shows that

$A_{-1}\widehat{G}+A_0$ is nonsingular and that the matrix $\widehat{R} = -A_{-1}(A_{-1}\widehat{G}+A_0)^{-1}$ solves the equation $X^2A_1 + XA_0 + A_{-1} = 0$. In particular, we obtain that $H_i = \widehat{G}^i H_0$ for $i \geq 0$ and $H_i = H_0\widehat{R}^{-i}$ for $i < 0$, so that

$$H(z) = \left(I + \sum_{i=1}^{+\infty}(z^{-i}G^i + z^i\widehat{G}^i) \right) H_0 = H_0 \left(I + \sum_{i=1}^{+\infty}(z^{-i}\widehat{R}^i + z^iR^i) \right).$$

From the latter equation, the singularity of H_0 would imply that $\det H(z) = 0$ for any $z \in \mathbb{A}$ which is absurd since $\det H(z) = 1/\det S(z)$ is nonzero for any $z \in \mathbb{A}$. □

Observe that the existence of a solution G of (3.16) with $\rho(G) < 1$ is not sufficient to guarantee the existence of the solution $\widehat{G}$ with $\rho(\widehat{G}) < 1$ of the reversed matrix equation (3.19). In fact, the existence of a canonical factorization of $S(z)$ does not imply the existence of a canonical factorization of the reversed matrix Laurent polynomial $zS(z^{-1})$ as shown by the following example.

Example 3.21 Consider the polynomial $S(z) = z^{-1}A_{-1} + A_0 + zA_1$ with

$$A_{-1} = \begin{bmatrix} \frac{1}{3} & 0 \\ 1 & 2 \end{bmatrix}, \quad A_0 = I, \quad A_1 = \begin{bmatrix} 2 & 0 \\ 0 & \frac{1}{20} \end{bmatrix}.$$

One has $a(z) = \det zS(z) = (2z^2 + z - \frac{1}{3})(\frac{1}{20}z^2 + z + 2)$ so that $a(z)$ has two zeros of modulus less than 1 and two zeros of modulus greater than 1. Moreover, there exists a canonical factorization $S(z) = (I - zR)U_0(z^{-1}G - I)$ with

$$G = \begin{bmatrix} \frac{361}{234} & \frac{45421}{14040} \\ -\frac{40}{39} & -\frac{239}{117} \end{bmatrix}, \quad R = \begin{bmatrix} -\frac{2101}{4680} & \frac{45421}{14040} \\ -\frac{1}{1560} & -\frac{239}{4680} \end{bmatrix}, \quad U_0 = \begin{bmatrix} \frac{478}{117} & \frac{45421}{7020} \\ -\frac{2}{39} & \frac{2101}{2340} \end{bmatrix},$$

in fact G and R are the minimal solutions of (3.16) and (3.18), respectively. However, the equation (3.19) does not have a minimal solution so that there is no canonical factorization of the matrix Laurent polynomial $z^{-1}A_1 + A_0 + zA_{-1}$. In fact, the block coefficient H_0 of $H(z)$ is singular.

An immediate consequence of the above results is the following relation concerning the linearization of a quadratic matrix equation, given in terms of bi-infinite matrices, which holds if there exist G and $\widehat{G}$ with $\rho(G) < 1$ and $\rho(\widehat{G}) < 1$ which solve (3.16) and (3.19), respectively:

$$\begin{bmatrix} \ddots & \ddots & \ddots & & & 0 \\ & A_{-1} & A_0 & A_1 & & \\ & & A_{-1} & A_0 & A_1 & \\ & & & A_{-1} & A_0 & A_1 \\ 0 & & & & \ddots & \ddots & \ddots \end{bmatrix} \begin{bmatrix} \vdots \\ \widehat{G}^2 \\ \widehat{G} \\ I \\ G \\ G^2 \\ \vdots \end{bmatrix} H_0 = \begin{bmatrix} \vdots \\ 0 \\ 0 \\ I \\ 0 \\ 0 \\ \vdots \end{bmatrix}.$$

A sufficient condition for the invertibility of H_0 for general matrix Laurent power series of the kind $F(z) = \sum_{i=-1}^{+\infty} z^i F_i$ is expressed by the following theorem, shown in [26], which extends part of the results of Theorem 3.20.

Theorem 3.22 *Let $F(z) = \sum_{i=-1}^{+\infty} z^i F_i \in \mathcal{W}$ be invertible for $|z| = 1$ and denote $H(z) = F(z)^{-1} \in \mathcal{W}$, $H(z) = \sum_{i=-\infty}^{+\infty} z^i H_i$. If there exist canonical factorizations of $F(z)$ and of $F(z^{-1})$, then H_0 is invertible.*

Proof Canonical factorization of $F(z)$, if exists, has the form

$$F(z) = \left(\sum_{j=0}^{+\infty} z^j U_j \right) (I - z^{-1} G),$$

with the matrix G having spectral radius strictly less than 1. Taking the inverses in both sides yields

$$H(z) = \left(\sum_{k=0}^{+\infty} z^{-k} G^k \right) \left(\sum_{j=0}^{+\infty} z^j V_j \right).$$

From here it follows, in particular, that

$$H_0 = \sum_{k=0}^{+\infty} G^k V_k$$

and

$$H_{-i} = \sum_{k=0}^{+\infty} G^{k+i} V_k = G^i H_0, \quad i = 1, 2, \ldots \tag{3.21}$$

Suppose now that the coefficient H_0 is singular. Then there exists a non-zero vector $\boldsymbol{x}$ for which $H_0 \boldsymbol{x} = 0$. From (3.21) it follows that also $H_{-i}\boldsymbol{x} = 0$ for all $i = 1, 2, \ldots$, so that the vector function

$$\phi^+(z) := z^{-1} H(z) \boldsymbol{x} = \sum_{j=1}^{+\infty} z^{j-1} (H_j \boldsymbol{x})$$

is analytic in the unit disk. On the other hand, the vector function

$$\phi^-(z) = -z^{-1} \boldsymbol{x}$$

is analytic outside the unit disk, vanishes at infinity, and together with $\phi^+(z)$ satisfies the boundary condition

$$\phi^+(z) + H(z)\phi^-(z) = 0. \tag{3.22}$$

Thus, the homogeneous Riemann-Hilbert problem (3.22) with the matrix coefficient $H(z)$ has non-trivial solutions. This means (see, e.g., [46]) that $H(z)$ does not admit a canonical factorization. Equivalently, its inverse $F(z)$ is such that $F(z^{-1})$ does not admit a canonical factorization. The contradiction obtained shows that H_0 is invertible. □

Observe for completeness that condition on the Laurent series decomposition of $F(z)$ in Theorem 3.22 is essential. Namely, there even exist scalar functions $F(z)$ in $\mathcal{W}$ admitting a canonical factorization for which $F(z)^{-1}$ has a vanishing constant term. As a concrete example, let $H(z) = -2z^{-4} + 2z^{-1} - 3z$. Being a trigonometric polynomial, this function belongs to $\mathcal{W}$. It is also non-vanishing on the unit circle, with the winding number equal to zero. Thus, this function (and therefore its inverse $F(z)$) admits a canonical factorization. Nevertheless, the constant term of $H(z) = F(z)^{-1}$ is indeed zero.

3.3.3 Power series matrix equations

Let us assume that the $m \times m$ matrix Laurent power series $S(z) = \sum_{i=-N}^{+\infty} z^i A_i$ belongs to the Wiener algebra, where $N \geq 1$, and define the matrix power series $A(z) = z^N S(z)$. Observe that the semi-infinite block Toeplitz matrix associated with $S(z)$ is in upper generalized block Hessenberg form, that is,

$$T_\infty[S(z)] = \begin{bmatrix} A_0 & A_1 & A_2 & A_3 & A_4 \dots \\ A_{-1} & A_0 & A_1 & A_2 & A_3 & \ddots \\ \vdots & \ddots & \ddots & \ddots & \ddots & \ddots \\ A_{-N} & \dots & A_{-1} & A_0 & A_1 & \ddots \\ & A_{-N} & \dots & A_{-1} & A_0 & \ddots \\ 0 & & \ddots & \ddots & \ddots & \ddots \end{bmatrix}. \tag{3.23}$$

If we partition the above matrix into $mN \times mN$ blocks $\mathcal{A}_i$, $i = -1, 0, 2, \dots$, we may look at (3.23) as a block matrix in block upper Hessenberg form, i.e.,

$$\begin{bmatrix} \mathcal{A}_0 & \mathcal{A}_1 & \mathcal{A}_2 & \mathcal{A}_3 \dots \\ \mathcal{A}_{-1} & \mathcal{A}_0 & \mathcal{A}_1 & \mathcal{A}_2 \dots \\ & \mathcal{A}_{-1} & \mathcal{A}_0 & \mathcal{A}_1 & \ddots \\ & & \mathcal{A}_{-1} & \mathcal{A}_0 & \ddots \\ 0 & & & \ddots & \ddots \end{bmatrix}, \tag{3.24}$$

where

$$\mathcal{A}_i = \begin{bmatrix} A_{iN} & A_{iN+1} & \dots & A_{iN+N-1} \\ A_{iN-1} & A_{iN} & \ddots & \vdots \\ \vdots & \ddots & \ddots & A_{iN+1} \\ A_{iN-N+1} & \dots & A_{iN-1} & A_{iN} \end{bmatrix}, \quad i = -1, 0, 1, \dots . \tag{3.25}$$

Defining $\mathcal{S}(z) = \sum_{i=-1}^{+\infty} z^i \mathcal{A}_i$ we may denote by $T_\infty[\mathcal{S}(z)]$ the matrix in (3.24) so that $T_\infty[\mathcal{S}(z)] = T_\infty[S(z)]$.

We now discuss the roots of $\alpha(z) = \det \mathcal{A}(z)$ where $\mathcal{A}(z) = z\mathcal{S}(z)$. Observe that

$$\mathcal{A}(z) = \begin{bmatrix} \phi_0(z) & \phi_1(z) & \dots & \phi_{N-1}(z) \\ z\phi_{N-1}(z) & \phi_0(z) & \ddots & \vdots \\ \vdots & \ddots & \ddots & \phi_1(z) \\ z\phi_1(z) & \dots & z\phi_{N-1}(z) & \phi_0(z) \end{bmatrix} \tag{3.26}$$

where $\phi_j(z) = \sum_{i=-1}^{+\infty} z^{i+1} A_{iN+j}$, $j = 0, \dots, N-1$. Therefore $\mathcal{A}(z)$ is a block z-circulant matrix (see Section 2.3.2). The following result, which has been proved in Gail, Hantler and Taylor [42] in the context of Markov chains, is a direct consequence of Theorem 2.10:

Theorem 3.23 *The matrix power series $\mathcal{A}(z^N)$ can be block diagonalized as follows:*

$$\begin{aligned} &\frac{1}{N}(\Omega_N \otimes I_m)\,(D(z) \otimes I_m)^{-1}\, \mathcal{A}(z^N)\,(D(z) \otimes I_m)\,(\overline{\Omega}_N \otimes I_m) \\ &= \begin{bmatrix} A(z) & & & 0 \\ & A(z\overline{\omega}_N) & & \\ & & \ddots & \\ 0 & & & A(z\overline{\omega}_N^{N-1}) \end{bmatrix}, \end{aligned}$$

for $z \neq 0$, $|z| \leq 1$, where $D(z) = \mathrm{Diag}(1, z, \dots, z^{N-1})$, Ω_N is the $N \times N$ Fourier matrix and I_m is the $m \times m$ identity matrix.

From the above theorem it follows that

$$\det \mathcal{A}(z^N) = \prod_{i=0}^{N-1} \det A(z\omega_N^i). \tag{3.27}$$

Therefore, if w is a zero of $\det A(z)$ then w^N is a zero of $\det \mathcal{A}(z)$.

Now we show that if there exists a weak canonical factorization

$$S(z) = U(z)L(z), \quad |z| = 1, \tag{3.28}$$

where

$$U(z) = \sum_{i=0}^{+\infty} z^i U_i, \quad L(z) = I_m + \sum_{i=1}^{N} z^{-i} L_{-i}, \tag{3.29}$$

then there exists a closely related weak canonical factorization of $\mathcal{S}(z)$.

Theorem 3.24 *Let $S(z) = \sum_{i=-N}^{+\infty} z^i A_i$ be an $m \times m$ matrix valued function in the Wiener algebra which has a weak canonical factorization* (3.28) *where $U(z)$*

and $L(z)$ *are given in* (3.29)*. The matrix Laurent power series* $\mathcal{S}(z)$ *has the weak canonical factorization*

$$\mathcal{S}(z) = \mathcal{U}(z)\mathcal{L}(z), \quad |z| = 1, \tag{3.30}$$

where

$$\mathcal{U}(z) = \sum_{i=0}^{+\infty} z^i \mathcal{U}_i, \quad \mathcal{L}(z) = I_{mN} - z^{-1}\mathcal{G},$$

and

$$\mathcal{G} = \begin{bmatrix} 0 & I_m & 0 & \dots & 0 \\ 0 & 0 & I_m & \ddots & \vdots \\ \vdots & \ddots & \ddots & \ddots & 0 \\ 0 & \dots & 0 & 0 & I_m \\ -L_{-N} & -L_{-N+1} & \dots & \dots & -L_{-1} \end{bmatrix}^N . \tag{3.31}$$

Moreover, $\mathcal{G}$ *is a minimal solution of the matrix equation*

$$\mathcal{A}_{-1} + \mathcal{A}_0\mathcal{X} + \mathcal{A}_1\mathcal{X}^2 + \cdots = 0. \tag{3.32}$$

Given the block row vector $\boldsymbol{r}^{\mathrm{T}} = [R_1, R_2, \dots, R_N]$ we define

$$C(\boldsymbol{r}^{\mathrm{T}}) = \begin{bmatrix} 0 & I & 0 & \dots & 0 \\ 0 & 0 & I & \ddots & \vdots \\ \vdots & \ddots & \ddots & \ddots & 0 \\ 0 & \dots & 0 & 0 & I \\ R_1 & R_2 & \dots & \dots & R_N \end{bmatrix}$$

the *block companion matrix* associated with $\boldsymbol{r}^{\mathrm{T}}$. In this way, the result expressed in the above theorem can be rewritten as $\mathcal{G} = C(\boldsymbol{g}^{\mathrm{T}})^N$ where $\boldsymbol{g}^{\mathrm{T}} = [-L_{-N}, -L_{-N+1}, \dots, -L_{-1}]$.

The N-th power of an $N \times N$ block companion matrix can be factorized as products of two block triangular Toeplitz matrices as stated by the following classical result known as Barnett factorization [7].

Theorem 3.25 *The matrix* $\mathcal{G}$ *of* (3.31) *can be factorized as*

$$\mathcal{G} = - \begin{bmatrix} I_m & & & 0 \\ L_{-1} & I_m & & \\ \vdots & \ddots & \ddots & \\ L_{-N+1} & \dots & L_{-1} & I_m \end{bmatrix}^{-1} \begin{bmatrix} L_{-N} & L_{-N+1} & \dots & L_{-1} \\ & \ddots & \ddots & \vdots \\ & & L_{-N} & L_{-N+1} \\ 0 & & & L_{-N} \end{bmatrix} . \tag{3.33}$$

Besides providing a useful representation of the matrix $\mathcal{G}$, the Barnett factorization can be used for proving Theorem 3.24

Proof of Theorem 3.24 By rewriting the factorization $S(z) = U(z)L(z)$ in matrix form we get

$$T_\infty[S(z)] = \begin{bmatrix} U_0 & U_1 & U_2 & \dots \\ & U_0 & U_1 & \ddots \\ & & U_0 & \ddots \\ 0 & & & \ddots \end{bmatrix} \begin{bmatrix} I_m & & & & & 0 \\ L_{-1} & I_m & & & & \\ \vdots & \ddots & \ddots & & & \\ L_{-N} & \dots & L_{-1} & I_m & & \\ & L_{-N} & \dots & L_{-1} & I_m & \\ 0 & & \ddots & \ddots & \ddots & \ddots \end{bmatrix}.$$

By reblocking the above semi-infinite matrices into $mN \times mN$ blocks we obtain

$$T_\infty[S(z)] = T_\infty[\mathcal{S}(z)] = \begin{bmatrix} \widetilde{\mathcal{U}}_0 & \widetilde{\mathcal{U}}_1 & \widetilde{\mathcal{U}}_2 & \dots \\ & \widetilde{\mathcal{U}}_0 & \widetilde{\mathcal{U}}_1 & \ddots \\ & & \widetilde{\mathcal{U}}_0 & \ddots \\ 0 & & & \ddots \end{bmatrix} \begin{bmatrix} \widetilde{\mathcal{L}}_0 & & & 0 \\ \widetilde{\mathcal{L}}_{-1} & \widetilde{\mathcal{L}}_0 & & \\ & \widetilde{\mathcal{L}}_{-1} & \widetilde{\mathcal{L}}_0 & \\ 0 & & \ddots & \ddots \end{bmatrix} = \mathcal{U}\mathcal{L} \quad (3.34)$$

where

$$\widetilde{\mathcal{L}}_0 = \begin{bmatrix} I_m & & & 0 \\ L_{-1} & I_m & & \\ \vdots & \ddots & \ddots & \\ L_{-N+1} & \dots & L_{-1} & I_m \end{bmatrix}, \quad \widetilde{\mathcal{L}}_{-1} = \begin{bmatrix} L_{-N} & L_{-N+1} & \dots & L_{-1} \\ & \ddots & \ddots & \vdots \\ & & L_{-N} & L_{-N+1} \\ 0 & & & L_{-N} \end{bmatrix}.$$

Since $\widetilde{\mathcal{L}}_0$ is nonsingular we may scale to the left $\mathcal{L}$ by $\widetilde{\mathcal{L}}_0^{-1}$ and $\mathcal{U}$ to the right by $\widetilde{\mathcal{L}}_0$, so that (3.34) can be rewritten as

$$T_\infty[\mathcal{S}(z)] = \begin{bmatrix} \mathcal{U}_0 & \mathcal{U}_1 & \mathcal{U}_2 & \dots \\ & \mathcal{U}_0 & \mathcal{U}_1 & \ddots \\ & & \mathcal{U}_0 & \ddots \\ 0 & & & \ddots \end{bmatrix} \begin{bmatrix} I_{mN} & & & 0 \\ \mathcal{L}_{-1} & I_{mN} & & \\ & \mathcal{L}_{-1} & I_{mN} & \\ 0 & & \ddots & \ddots \end{bmatrix},$$

where $\mathcal{U}_i = \widetilde{\mathcal{U}}_i\widetilde{\mathcal{L}}_0$, $i = 0, 1, \dots$, $\mathcal{L}_{-1} = -\mathcal{G}$, and $\mathcal{G} = -\widetilde{\mathcal{L}}_0^{-1}\widetilde{\mathcal{L}}_{-1}$. The above matrix factorization provides the functional factorization $\mathcal{S}(z) = \mathcal{U}(z)(I_{mN} + z^{-1}\mathcal{L}_1)$. We now show that this is a weak canonical factorization. Observe that $\mathcal{U}(z)$ and $\mathcal{L}(z)$ belong to the Wiener algebra since $U(z), L(z) \in \mathcal{W}$. By Theorem 3.25, $\mathcal{G} = C(\boldsymbol{g}^{\mathrm{T}})^N$, therefore the eigenvalues of $\mathcal{G}$ are the Nth powers of the zeros of the polynomial $\det(z^N I_m + \sum_{i=1}^N z^{N-i}L_{-i})$, see Theorem A.11 in the appendix, which are the zeros of $\det L(z)$. Thus ξ is a zero of $\det L(z)$ if and only if ξ^N is a zero of $\det \mathcal{L}(z) = \det(I_{mN} - z^{-1}\mathcal{G})$, and $\det \mathcal{L}(z)$ has zeros with modulus less than or equal to 1 only. Since, by (3.27), the roots of $\mathcal{A}(z)$ are the Nth powers of the roots of $A(z)$, we conclude that (3.30) is a weak canonical factorization. From Theorem 3.19 we deduce that $\mathcal{G}$ is a minimal solution of (3.32). $\square$

Remark 3.26 From the structure (3.33) it follows that the first block row of $-\mathcal{G}$, i.e., $[L_{-N}, \ldots, L_{-1}]$ defines all the block coefficients of $L(z)$. Therefore, the knowledge of $\mathcal{G}$ provides the factors $L(z)$ and $\mathcal{L}(z)$ of the canonical factorization of $S(z)$ and $\mathcal{S}(z)$, respectively, at no cost The factors $U(z)$ and $\mathcal{U}(z)$ can be computed from the equations $S(z) = U(z)L(z)$, and $\mathcal{S}(z) = \mathcal{U}(z)\mathcal{L}(z)$, respectively. Conversely, the knowledge of the factor $\mathcal{L}(z)$ provides immediately the solution of the matrix equation (3.32). Moreover, the factor $L(z)$ provides the first block row of $\mathcal{G}$ which uniquely defines $\mathcal{G}$ through (3.33).

Theorem 3.27 *Let $S(z) = \sum_{i=-N}^{+\infty} z^i A_i$ be an $m \times m$ matrix Laurent power series in the Wiener algebra such that $\det S(z) \neq 0$ for $|z| = 1$, and such that there exists a canonical factorization $S(z) = U(z)L(z)$, $|z| = 1$, where $L(z) = I_m + \sum_{i=1}^{N} z^{-i} L_{-i}$, $U(z) = \sum_{i=0}^{+\infty} z^i U_i$. Let $H(z) = \sum_{i=-\infty}^{+\infty} z^i H_i$ be the inverse of $S(z)$. Consider the $q \times q$ block Toeplitz matrix $T_q[H(z)] = (H_{j-i})_{i,j=1,q}$, for $q > N$. The coefficients of $L(z)$ satisfy the following equation*

$$\boldsymbol{e}_q^{\mathrm{T}} \otimes I_m = U_0[L_{-q+1}, L_{-q+2}, \ldots, L_{-1}, I_m] T_q[H(z)]$$

where $L_i = 0$ for $i < -N$ and $\boldsymbol{e}_q$ is the q-th column of the $q \times q$ identity matrix.

Proof We use a similar argument as in Theorem 3.24, adapted to bi-infinite matrices. Let $T_{\pm\infty}[S(z)] = (A_{j-i})_{i,j\in\mathbb{Z}}$, where $A_i = 0$ for $i < -N$. Partition $T_{\pm\infty}[S(z)]$ into $mq \times mq$ blocks and from the canonical factorization of $S(z)$ obtain the UL decomposition similar to (3.34) where the matrices are bi-infinite and the blocks $\widetilde{\mathcal{L}}_i$ and $\widetilde{\mathcal{U}}_i$ have size mq. Scale on the right the factor $\mathcal{L}$ by multiplying it by $\widetilde{\mathcal{L}}_0^{-1}$ and obtain the decomposition

$$T_{\pm\infty}[S(z)] = \begin{bmatrix} \ddots & \ddots & \ddots & \ddots & \ddots \\ & \widetilde{\mathcal{U}}_0 & \widetilde{\mathcal{U}}_1 & \widetilde{\mathcal{U}}_2 & \ldots \\ & & \widetilde{\mathcal{U}}_0 & \widetilde{\mathcal{U}}_1 & \ddots \\ & & & \widetilde{\mathcal{U}}_0 & \ddots \\ 0 & & & & \ddots \end{bmatrix} \begin{bmatrix} \ddots & & & & 0 \\ \ddots & I_{mN} & & & \\ & \mathcal{L}_{-1} & I_{mN} & & \\ & & \mathcal{L}_{-1} & I_{mN} & \\ 0 & & & \ddots & \ddots \end{bmatrix} \begin{bmatrix} \ddots & & & 0 \\ & \widetilde{\mathcal{L}}_0 & & \\ & & \widetilde{\mathcal{L}}_0 & \\ 0 & & & \ddots \end{bmatrix}$$

where $\mathcal{L}_{-1} = \widetilde{\mathcal{L}}_{-1}\widetilde{\mathcal{L}}_0^{-1}$. Compute the inverses on both sides of the above relation, observe that $T_{\pm\infty}[S(z)]^{-1} = T_{\pm\infty}[H(z)] = (H_{j-i})_{i,j\in\mathbb{Z}}$, and compare the diagonal blocks in both sides of the equation obtained in this way. We find that $T_q[H(z)] = \widetilde{\mathcal{L}}_0^{-1} \sum_{i=0}^{+\infty} (-\mathcal{L}_{-1})^i \mathcal{K}_i$, where $\sum_{i=0}^{+\infty} z^i \mathcal{K}_i = (\sum_{i=0}^{+\infty} z^i \widetilde{\mathcal{U}}_i)^{-1}$. Now observe that the last block row of $\mathcal{L}_{-1}$ is zero since $q > N$, and deduce that multiplying to the left $T_q[H(z)]$ by $(\boldsymbol{e}_q^{\mathrm{T}} \otimes I_m)\widetilde{\mathcal{L}}_0$ yields $(\boldsymbol{e}_q^{\mathrm{T}} \otimes I_m)\widetilde{\mathcal{L}}_0 T_q[H(z)] = (\boldsymbol{e}_q^{\mathrm{T}} \otimes m)\mathcal{K}_0 = (\boldsymbol{e}_q^{\mathrm{T}} \otimes I_m)\widetilde{\mathcal{U}}_0^{-1} = U_0^{-1}(\boldsymbol{e}_q^{\mathrm{T}} \otimes I_m)$. □

The above theorem relates the inverse of $S(z)$ with the canonical factorization of $S(z)$. Observe that, in order to compute the canonical factorization of $S(z)$ it is sufficient to compute the $2N + 1$ central coefficients of the inverse $S(z)^{-1}$

and then to solve a block Toeplitz system. Concerning the latter problem we refer the reader to Section 2.4, and to Section 3.1.2 for the former. Conversely, if the canonical factorization of $S(z)$ is known, then the coefficients H_i of $H(z) = S(z)^{-1}$ can be computed by means of the equation

$$H_j = \sum_{i=0}^{+\infty} U_i^{(-1)} L_{-i-j}^{(-1)}, \quad j = 0, \pm 1, \pm 2, \ldots,$$

derived from (3.28) where $\sum_{i=0}^{+\infty} z^i U_i^{(-1)} = U(z)^{-1}$, $\sum_{i=0}^{+\infty} z^{-i} L_{-i}^{(-1)} = L(z)^{-1}$ and $U_{-i}^{(-1)} = L_i^{(-1)} = 0$ for $i > 0$.

We recall that, if $S(z)$ is analytic and invertible in a suitable annulus containing the unit circle, then the coefficients $U_i^{(-1)}$ and $L_{-i}^{(-1)}$ decay exponentially to zero for $i \to \infty$ by Theorem 3.6. In this way the matrices H_j can be approximated by truncating the infinite summations to a finite sum.

3.4 Reduction of matrix equations

The technique of reblocking introduced in the previous section may be used to reduce a polynomial matrix equation of the kind

$$A_{-1} + A_0 X + A_1 X^2 + \cdots + A_{N-1} X^N = 0 \tag{3.35}$$

into a quadratic matrix equation. More precisely, defining the matrices $\mathcal{A}_{-1}, \mathcal{A}_0$, and $\mathcal{A}_1$ as in (3.24), (3.25), where we set $A_i = 0$ if $i < -1$ or $i > N-1$, we find that the matrix

$$\mathcal{X} = \begin{bmatrix} 0 & \ldots & 0 & X \\ 0 & \ldots & 0 & X^2 \\ \vdots & & \vdots & \vdots \\ 0 & \ldots & 0 & X^N \end{bmatrix}$$

solves the quadratic matrix equation

$$\mathcal{A}_{-1} + \mathcal{A}_0 \mathcal{X} + \mathcal{A}_1 \mathcal{X}^2 = 0. \tag{3.36}$$

Observe that the nonnull eigenvalues of $\mathcal{X}$ coincide with the nonnull eigenvalues of X^N, i.e., with the Nth power of the eigenvalues of X. In this way, X is a minimal solution of (3.35) if and only if $\mathcal{X}$ is a minimal solution of (3.36) and $\rho(\mathcal{X}) = \rho(X)^N$.

A different strategy, which allows one to reduce any power series matrix equation into a quadratic matrix equation, has been introduced by Ramaswami in the framework of Markov chains [100]. Here we recall the main result given in terms of algebraic properties.

Let G be a solution of the matrix equation (3.12), and define the semi-infinite matrices

$$\mathcal{A}_{-1} = \begin{bmatrix} A_{-1} & 0 & \dots \\ 0 & 0 & \dots \\ \vdots & \vdots & \end{bmatrix}, \quad \mathcal{A}_0 = \begin{bmatrix} A_0 & A_1 & A_2 & \dots \\ & I_m & 0 & \dots \\ & & I_m & \ddots \\ 0 & & & \ddots \end{bmatrix}, \quad \mathcal{A}_1 = \begin{bmatrix} 0 & & & 0 \\ -I_m & 0 & & \\ & -I_m & \ddots & \\ 0 & & \ddots & \ddots \end{bmatrix}.$$

Then the semi-infinite matrix

$$\mathcal{X} = \begin{bmatrix} X & 0 & \dots \\ X^2 & 0 & \dots \\ \vdots & \vdots & \ddots \end{bmatrix},$$

solves the quadratic matrix equation with semi-infinite blocks

$$\mathcal{A}_{-1} + \mathcal{A}_0\mathcal{X} + \mathcal{A}_1\mathcal{X}^2 = 0.$$

Observe that the semi-infinite matrix $\mathcal{X} - \lambda I$ is not invertible if λ is zero or equal to an eigenvalue of X. Therefore, the eigenvalues of $\mathcal{X}$, defined as the values of λ which make $\mathcal{X} - \lambda I$ noninvertible, coincide with the eigenvalues of X or with zero.

3.5 Infinite systems and canonical factorization

In this section we analyze the problem of computing a vector $\boldsymbol{x} = (x_i)_{i\in\mathbb{N}}$ which solves the semi-infinite system

$$\boldsymbol{x}^{\mathrm{T}} T_\infty[S(z)] = \boldsymbol{b}^{\mathrm{T}} \tag{3.37}$$

given the vector $\boldsymbol{b} = (b_i)_{i\in\mathbb{N}}$ and the matrix $T_\infty[S(z)]$ associated with the $m \times m$ matrix Laurent power series $S(z) = \sum_{i=-\infty}^{+\infty} z^i A_i \in \mathcal{W}$.

This problem is crucial in the context of Markov chains where we have to compute a vector $\boldsymbol{\pi}$ such that $\boldsymbol{\pi}^{\mathrm{T}}(I - P) = 0$ and where $T_\infty[S(z)]$ is a submatrix of the semi-infinite matrix $I - P$.

We partition the vectors $\boldsymbol{x}$ and $\boldsymbol{b}$ into vectors $\boldsymbol{x}_i$, $\boldsymbol{b}_i$, $i = 1, 2, \dots$, of dimension m, and rewrite the above system as

$$[\boldsymbol{x}_1^{\mathrm{T}}, \boldsymbol{x}_2^{\mathrm{T}}, \dots] \begin{bmatrix} A_0 & A_1 & A_2 & A_3 & \dots \\ A_{-1} & A_0 & A_1 & A_2 & \ddots \\ A_{-2} & A_{-1} & A_0 & A_1 & \ddots \\ \vdots & \ddots & \ddots & \ddots & \ddots \end{bmatrix} = [\boldsymbol{b}_1^{\mathrm{T}}, \boldsymbol{b}_2^{\mathrm{T}}, \dots]. \tag{3.38}$$

If $\det S(z) \neq 0$ for $|z| = 1$ and if there exists a canonical factorization $S(z) = U(z)L(z)$ of $S(z)$, then the above system takes the form

$$\boldsymbol{x}^{\mathrm{T}} T_\infty[U(z)] T_\infty[L(z)] = \boldsymbol{b}^{\mathrm{T}} \tag{3.39}$$

where $T_\infty[U(z)]$ is the block upper triangular Toeplitz matrix

$$T_\infty[U(z)] = \begin{bmatrix} U_0 & U_1 & U_2 & \dots \\ & U_0 & U_1 & \ddots \\ 0 & & \ddots & \ddots \end{bmatrix}$$

and $T_\infty[L(z)]$ is the block lower triangular Toeplitz matrix

$$T_\infty[L(z)] = \begin{bmatrix} L_0 & & & 0 \\ L_{-1} & L_0 & & \\ L_{-2} & L_{-1} & L_0 & \\ \vdots & \ddots & \ddots & \ddots \end{bmatrix}.$$

Since $L(z)$ is in $\mathcal{W}_-$ and is nonsingular for $1 \le |z| \le \infty$, $K(z) = L(z)^{-1} = \sum_{i=0}^{+\infty} z^{-i}K_{-i}$ exists and belongs to $\mathcal{W}_-$, that is $\sum_{i=0}^{+\infty} |K_{-i}| < +\infty$ so that $\sup_{|z|=1} |K(z)|$ is a matrix with finite elements. Therefore, by Remark 3.12 $T_\infty[L(z)]^{-1} = T_\infty[K(z)]$ is a bounded operator on $\ell^2(\mathbb{N})$ and

$$\boldsymbol{c}^{\mathrm{T}} = \boldsymbol{b}^{\mathrm{T}} T_\infty[K(z)] \tag{3.40}$$

is in $\ell^2(\mathbb{N})$ if $\boldsymbol{b} \in \ell^2(\mathbb{N})$.

Multiplying on the right both sides of (3.39) by $T_\infty[L(z)]^{-1}$ yields

$$\boldsymbol{x}^{\mathrm{T}} T_\infty[U(z)] = \boldsymbol{c}^{\mathrm{T}}. \tag{3.41}$$

Since $T_\infty[U(z)]$ also is a bounded operator, as well as its inverse, $\boldsymbol{x} \in \ell^2(\mathbb{N})$ if $\boldsymbol{b} \in \ell^2(\mathbb{N})$.

We thus obtain the following property.

Theorem 3.28 *Let $S(z) = \sum_{i=-\infty}^{+\infty} z^i A_i \in \mathcal{W}$ be an $m \times m$ matrix Laurent power series such that $\det S(z) \neq 0$ for $|z| = 1$, and such that there exists a canonical factorization $S(z) = U(z)L(z)$, $U(z) = \sum_{i=0}^{+\infty} z^i U_i$, $L(z) = \sum_{i=0}^{+\infty} z^{-i} L_{-i}$. If $\boldsymbol{b} \in \ell^2(\mathbb{N})$ then the system (3.37) has a solution $\boldsymbol{x} \in \ell^2(\mathbb{N})$. Partition $\boldsymbol{x}$ and $\boldsymbol{b}$ into m-dimensional subvectors $\boldsymbol{x}_i$, $\boldsymbol{b}_i$, $i = 1, 2, \dots$, so that $\boldsymbol{x} = (\boldsymbol{x}_i)_{i=1,2,\dots}$, $\boldsymbol{b} = (\boldsymbol{b}_i)_{i=1,2,\dots}$. Then*

$$\begin{aligned} \boldsymbol{x}_1^{\mathrm{T}} &= \boldsymbol{c}_1^{\mathrm{T}} U_0^{-1}, \\ \boldsymbol{x}_i^{\mathrm{T}} &= \left(\boldsymbol{c}_i^{\mathrm{T}} - \sum_{j=1}^{i-1} \boldsymbol{x}_j^{\mathrm{T}} U_{i-j} \right) U_0^{-1}, \quad i = 2, 3, \dots, \end{aligned} \tag{3.42}$$

where

$$\begin{aligned} \boldsymbol{c}_i^{\mathrm{T}} &= \sum_{j=i}^{+\infty} \boldsymbol{b}_j^{\mathrm{T}} K_{i-j}, \quad i = 1, 2, \dots \\ \sum_{i=0}^{+\infty} & z^i K_{-i} = L(z^{-1})^{-1}. \end{aligned} \tag{3.43}$$

Observe that the expressions for $\boldsymbol{x}_i$, $i = 1, \ldots, n$, in the above equations are obtained by solving the block upper triangular finite system

$$[\boldsymbol{x}_1^{\mathrm{T}}, \ldots, \boldsymbol{x}_n^{\mathrm{T}}]T_n[U(z)] = [\boldsymbol{c}_1^{\mathrm{T}}, \ldots, \boldsymbol{c}_n^{\mathrm{T}}]$$

by forward substitution. The same system can be solved with a lower asymptotic complexity by means of Algorithm 2.6.

The first M block components of $\boldsymbol{c}$ can be approximated in the following way. Using Algorithm 2.5, we compute $T_M[K(z)]$ where M is large enough so that $\sum_{i>M} |\boldsymbol{c}_i|$ is negligible; then we compute $[\boldsymbol{b}_1^{\mathrm{T}}, \ldots, \boldsymbol{b}_M^{\mathrm{T}}]T_M[K(z)]$ with Algorithm 2.4.

A simpler situation, which is often encountered in practice, occurs if $\boldsymbol{b}_i = 0$ for i sufficiently large, say, for $i > M$. In this case $\boldsymbol{c}_i = 0$ for $i > M$ and the computation is reduced to solving the finite system $[\boldsymbol{c}_1^{\mathrm{T}}, \ldots, \boldsymbol{c}_M^{\mathrm{T}}]T_M[L(z)] = [\boldsymbol{b}_1^{\mathrm{T}}, \ldots, \boldsymbol{b}_M^{\mathrm{T}}]$. In practice, this situation holds for $\boldsymbol{b} \in \ell^2(\mathbb{N})$ since $\boldsymbol{b}_i$ is numerically zero for i large enough. This system can be solved by means of Algorithms 2.5 and 2.4.

In the case where $S(z) = \sum_{i=-1}^{+\infty} z^i A_i$, we may assume that $L(z)$ is of the form $L(z) = I - z^{-1}G$, so that $K(z) = \sum_{i=0}^{+\infty} z^{-i}G^i$ and we arrive at the equation

$$\boldsymbol{c}_i^{\mathrm{T}} = \sum_{j=i}^{+\infty} \boldsymbol{b}_j^{\mathrm{T}} G^{j-i}, \quad i = 1, 2, \ldots$$

If $\boldsymbol{b}_i$ is zero (or numerically negligible) for $i > M$, then the vectors $\boldsymbol{c}_i$, $i = 1, 2, \ldots$, can be computed (or approximated) by means of the following Horner-like scheme:

$$\begin{aligned} \boldsymbol{c}_M^{\mathrm{T}} &= \boldsymbol{b}_M^{\mathrm{T}}, \\ \boldsymbol{c}_i^{\mathrm{T}} &= \boldsymbol{c}_{i+1}^{\mathrm{T}} G + \boldsymbol{b}_i^{\mathrm{T}}, \qquad i = M-1, \ldots, 1. \end{aligned}$$

If $S(z) = U(z)L(z)$ is a weak canonical factorization, in principle we cannot apply the above technique since the matrix power series $U(z)^{-1}$ and $L(z)^{-1}$ may have unbounded coefficients. However, under additional assumptions on the vector $\boldsymbol{b}$, we may still apply the above machinery as stated by the following theorem.

Theorem 3.29 *Let $S(z) = \sum_{i=-\infty}^{+\infty} z^i A_i \in \mathcal{W}$ be an $m \times m$ matrix Laurent power series such that there exists a weak canonical factorization $S(z) = U(z)L(z)$, $U(z) = \sum_{i=0}^{+\infty} z^i U_i$, $L(z) = \sum_{i=0}^{+\infty} z^{-i} L_{-i}$. If $U(z)$ is nonsingular for $|z| = 1$, the coefficients of $L(z)^{-1}$ are uniformly bounded in norm by a constant, and $\sum_{i=1}^{+\infty} i\|\boldsymbol{b}_i\|_1$ is finite, then (3.42) and (3.43) provide a solution $\boldsymbol{x} \in \ell^2(\mathbb{N})$ of the system (3.37).*

If $L(z)$ is nonsingular for $|z| = 1$, the coefficients of $U(z)^{-1}$ are uniformly bounded in norm by a constant, and $\sum_{i=1}^{+\infty} \|\boldsymbol{b}_i\|_1$ is finite, then (3.42) and (3.43) provide a solution $\boldsymbol{x} \in \ell^\infty(\mathbb{N})$ of the system (3.37).

Proof Assume that, in the weak canonical factorization of $S(z)$, the matrix $U(z)$ is nonsingular for $|z| = 1$, and that the coefficients of $K(z) = \sum_{i=0}^{+\infty} z^{-i}K_{-i} = L(z)^{-1}$ are uniformly bounded in norm by a constant, more specifically let us assume that $\|K_{-i}^{\mathrm{T}}\|_1 \le \gamma$, $i \ge 0$. If $\sum_{i=1}^{+\infty} i\|\boldsymbol{b}_i\|_1 < +\infty$, then we have from (3.43)

$$\|\boldsymbol{c}_i\|_1 \le \sum_{j=i}^{+\infty} \|\boldsymbol{b}_j\|_1 \cdot \|K_{i-j}^{\mathrm{T}}\|_1 \le \gamma \sum_{j=i}^{+\infty} \|\boldsymbol{b}_j\|_1 < +\infty, \quad i = 1, 2, \ldots,$$

so that $\|\boldsymbol{c}\|_1 = \sum_{i=1}^{+\infty} \|\boldsymbol{c}_i\|_1 \le \gamma \sum_{i=1}^{+\infty} i\|\boldsymbol{b}_i\|_1 < +\infty$, that is, $\boldsymbol{c} \in \ell^1(\mathbb{N})$. This implies that $\boldsymbol{c} \in \ell^2(\mathbb{N})$ and, since $U(z)^{-1}$ is in the Wiener algebra, we find that $\boldsymbol{x}^{\mathrm{T}} = \boldsymbol{c}^{\mathrm{T}} T_\infty[U(z)]^{-1}$ is in $\ell^2(\mathbb{N})$. In this way we may apply (3.42) to compute $\boldsymbol{x}$.

We proceed in a similar manner if $L(z)^{-1}$ is in the Wiener algebra and $U(z)^{-1}$ has bounded coefficients. In this case, assuming that $\sum_{i=1}^{+\infty} \|\boldsymbol{b}_i\|_1 < +\infty$, we deduce that $b(z) = \sum_{i=0}^{+\infty} z^i \boldsymbol{b}_{i+1}$ is in the Wiener algebra, that the function $c(z) = \sum_{i=0}^{+\infty} z^i \boldsymbol{c}_{i+1}$ is such that $c(z) = L(z^{-1})^{-1} b(z)$ and, being a product of functions in $\mathcal{W}$, that $c(z)$ belongs to $\mathcal{W}$. In particular, $\sum_{i=1}^{+\infty} \|\boldsymbol{c}_i\|_1$ is finite, that is $\boldsymbol{c} \in \ell^1(\mathbb{N})$. From the boundedness of the coefficients of $R(z) = U(z)^{-1}$ it follows that $\boldsymbol{x}^{\mathrm{T}} = \boldsymbol{c}^{\mathrm{T}} T_\infty[R(z)]$ has bounded coefficients, i.e., $\boldsymbol{x} \in \ell^\infty(\mathbb{N})$. Once again we may apply equations (3.42) for computing $\boldsymbol{x}$. □

The above technique for solving semi-infinite block Toeplitz systems can be used for solving the more general system

$$\boldsymbol{x}^{\mathrm{T}} \left[\begin{array}{c|ccc} V_0 & V_1 & V_2 & \cdots \\ \hline V_{-1} & & & \\ V_{-2} & & T_\infty[S(z)] & \\ \vdots & & & \end{array}\right] = \boldsymbol{b}^{\mathrm{T}} \tag{3.44}$$

where V_0 is an $mN \times mN$ matrix, V_i, $i \ge 1$, are $mN \times m$ matrices, V_i, $i \le -1$, are $m \times mN$ matrices, for N positive integer, and $T_\infty[S(z)]$ is the matrix in (3.38).

In fact, we have the following

Theorem 3.30 *Let $S(z) = \sum_{i=-\infty}^{+\infty} z^i A_i \in \mathcal{W}$ be an $m \times m$ matrix Laurent power series such that $\det S(z) \ne 0$ for $|z| = 1$ and such that there exists a canonical factorization $S(z) = U(z)L(z)$, $U(z) = \sum_{i=0}^{+\infty} z^i U_i$, $L(z) = \sum_{i=0}^{+\infty} z^{-i} L_{-i}$. Consider the system (3.44) where $\boldsymbol{b} \in \ell^2(\mathbb{N})$ and $\sum_{i=1}^{+\infty} \|V_i\|^2$, $\sum_{i=1}^{+\infty} \|V_{-i}\|^2$, are finite for a given matrix norm $\|\cdot\|$. Partition $\boldsymbol{x}$ and $\boldsymbol{b}$ into subvectors $\boldsymbol{x}_i$, $\boldsymbol{b}_i$, $i = 0, 1, 2, \ldots$, respectively, where $\boldsymbol{x}_0$ and $\boldsymbol{b}_0$ have dimension mN, while $\boldsymbol{x}_i$, $\boldsymbol{b}_i$, $i = 1, 2, \ldots$, have dimension m. Then there exists finite the matrices*

$$Y = T_\infty[S(z)]^{-1} \begin{bmatrix} V_{-1} \\ V_{-2} \\ \vdots \end{bmatrix}, \qquad W = V_0 - [V_1, V_2, \ldots]Y,$$

and the vector $\boldsymbol{f}^{\mathrm{T}} = \boldsymbol{b}_0^{\mathrm{T}} - [\boldsymbol{b}_1^{\mathrm{T}}, \boldsymbol{b}_2^{\mathrm{T}}, \ldots]Y$. If W is nonsingular, then there exists a solution $\boldsymbol{x} \in \ell^2(\mathbb{N})$ of (3.44) where $\boldsymbol{x}_0$ solves the $mN \times mN$ linear system $\boldsymbol{x}_0^{\mathrm{T}} W = \boldsymbol{f}^{\mathrm{T}}$, and the vector $\widehat{\boldsymbol{x}} = (\boldsymbol{x}_i)_{i \ge 1}$ solves the system

$$\widehat{\boldsymbol{x}}^{\mathrm{T}} T_{\infty}[S(z)] = \widehat{\boldsymbol{b}}^{\mathrm{T}}$$

where $\widehat{\boldsymbol{b}} = (\widehat{\boldsymbol{b}}_i)_{i\geq 1}$, $\widehat{\boldsymbol{b}}_i^{\mathrm{T}} = \boldsymbol{b}_i^{\mathrm{T}} - \boldsymbol{x}_0^{\mathrm{T}} V_i$, $i = 1, 2, \ldots$.

Proof By following the same argument used to prove Theorem 3.30 we deduce that the columns of Y are in $\ell^2(\mathbb{N})$ so that W exists finite as well as $\boldsymbol{f}^{\mathrm{T}}$. From (3.44) one has that

$$\begin{aligned} &\boldsymbol{x}_0^{\mathrm{T}} V_0 + \widehat{\boldsymbol{x}}^{\mathrm{T}} \begin{bmatrix} V_{-1} \\ V_{-2} \\ \vdots \end{bmatrix} = \boldsymbol{b}_0^{\mathrm{T}}, \\ &\boldsymbol{x}_0^{\mathrm{T}}[V_1, V_2, \ldots] + \widehat{\boldsymbol{x}}^{\mathrm{T}} T_{\infty}[S(z)] = [\boldsymbol{b}_1^{\mathrm{T}}, \boldsymbol{b}_2^{\mathrm{T}}, \ldots]. \end{aligned} \tag{3.45}$$

Therefore, from the second expression we deduce that if $\boldsymbol{x}_0$ is known, then the vector $\widehat{\boldsymbol{x}} = (\boldsymbol{x}_i)_{i\geq 1}$ solves the system

$$\widehat{\boldsymbol{x}}^{\mathrm{T}} T_{\infty}[S(z)] = \widehat{\boldsymbol{b}}^{\mathrm{T}},$$

where $\widehat{\boldsymbol{b}}^{\mathrm{T}} \in \ell^2(\mathbb{N})$, so that by Theorem 3.28 there exists a solution $\widehat{\boldsymbol{x}}^{\mathrm{T}} \in \ell^2(\mathbb{N})$. By recovering $\widehat{\boldsymbol{x}}^{\mathrm{T}}$ from the second equation of (3.45) and by substituting $\widehat{\boldsymbol{x}}^{\mathrm{T}}$ in the first equation of (3.45) we obtain $\boldsymbol{x}_0^{\mathrm{T}} W = \boldsymbol{f}^{\mathrm{T}}$. □

For the above theorem the first component $\boldsymbol{x}_0$ of the solution can be computed by solving a finite linear system, while the successive components can be computed by applying the formulas of Theorem 3.28.

Remark 3.31 The above theorem is still valid under the weaker assumptions of Theorem 3.29 on the matrix function $S(z)$ under the additional condition that $\sum_{i=1}^{+\infty} i\|V_i\|_1 < +\infty$ and $\sum_{i=1}^{+\infty} \|V_{-i}\|_1 < +\infty$ if $L(z)$ is singular for some z of modulus 1, and that $\sum_{i=1}^{+\infty} \|V_i\|_1 < +\infty$, $\sum_{i=1}^{+\infty} i\|V_{-i}\|_1 < +\infty$ if $U(z)$ is singular for some z of modulus 1.

The assumptions of Theorem 3.29 are satisfied in a wide class of problems encountered in Markov chains where $S(z)$ has the form $S(z) = I - \sum_{i=-1}^{+\infty} z^i P_i$, with $P = \sum_{i=-1}^{+\infty} P_i$ being stochastic, and where $L(z) = I - z^{-1}G$, with G stochastic. The computation of the stationary vector $\boldsymbol{\pi}$ by using Theorem 3.29 leads to the well-known Ramaswami formula, which was first obtained by using a probabilistic argument, given in Section 4.2. The linear system for $\boldsymbol{\pi}_0$ in the Ramaswami formula can be derived from Theorem 3.30.

3.6 Shifting technique

In this section we describe a suitable technique for transforming a weak canonical factorization of a function $S(z)$ into a canonical factorization of a slightly modified function $\widetilde{S}(z)$ which is nonsingular on the unit circle. This technique consists in shifting into zero any λ such that $\det S(\lambda) = 0$ and $|\lambda| = 1$. In order to apply this strategy we need to know, besides λ, the corresponding vectors $\boldsymbol{v} \neq 0$ such that $S(\lambda)\boldsymbol{v} = 0$. In the context of Markov chains this information is readily available, since $S(1) = I - P$, where P is a stochastic matrix, and therefore $\lambda = 1$ and $\boldsymbol{v} = \boldsymbol{1}$; moreover, under weak assumptions, $\lambda = 1$ is the only root of modulus 1 of $S(z)$ and is simple.

Let $S(z) = \sum_{i=-N}^{+\infty} z^i A_i$ be an $m \times m$ matrix Laurent power series in the Wiener algebra and let $A(z) = z^N S(z)$. Let us, in addition, assume that the derivative $S'(z)$ of $S(z)$ belongs to the Wiener algebra as well, that is,

$$\sum_{i=-N}^{+\infty} |iA_i| < +\infty. \tag{3.46}$$

Let us, finally, assume that there is only one simple zero λ of $\det S(\lambda)$ which has modulus 1 and that $\boldsymbol{v}$ is a vector such that $S(\lambda)\boldsymbol{v} = 0$, $\boldsymbol{v} \neq 0$. The general case with more zeros on the unit circle will be treated later on.

We introduce the following matrix function

$$\widetilde{S}(z) = S(z)(I - z^{-1}\lambda Q)^{-1}, \quad Q = \boldsymbol{v}\boldsymbol{u}^{\mathrm{T}} \tag{3.47}$$

where $\boldsymbol{u}$ is any fixed vector such that $\boldsymbol{v}^{\mathrm{T}}\boldsymbol{u} = 1$.

Observe that $Q^i = Q$ for any $i > 0$ and, consequently, the inverse of $I - z^{-1}\lambda Q$ is formally

$$(I - z^{-1}\lambda Q)^{-1} = I + Q\sum_{i=1}^{+\infty} z^{-i}\lambda^i.$$

Therefore, from (3.47), we obtain that

$$\widetilde{S}(z) = S(z) + S(z)Q\sum_{i=1}^{+\infty} z^{-i}\lambda^i.$$

By denoting $\widetilde{S}(z) = \sum_{i=-\infty}^{+\infty} z^i \widetilde{A}_i$, from the above relation we find that

$$\widetilde{A}_i = A_i + \left(\sum_{j=i+1}^{+\infty} \lambda^{j-i} A_j\right) Q = A_i + \lambda^{-i}\left(\sum_{j=i+1}^{+\infty} \lambda^j A_j\right) Q, \tag{3.48}$$

for $i = 0, \pm 1, \pm 2, \ldots$. We assumed that $A_i = 0$ if $i < -N$. Since $S(\lambda)\boldsymbol{v} = 0$, therefore $\sum_{j=-N}^{+\infty} \lambda^j A_j Q = 0$, and $\widetilde{A}_i = 0$ for $i < -N$ as well. The coefficients of the matrix Laurent power series $\widetilde{S}(z) = \sum_{i=-N}^{+\infty} z^i \widetilde{A}_i$ are given in (3.48); we may also write

$$\widetilde{A}_i = A_i - \left(\sum_{j=-N}^{i} \lambda^{j-i} A_j\right) Q, \quad i = -N, -N+1, \ldots. \tag{3.49}$$

Both the functions $A(z) = z^N S(z)$ and $\widetilde{A}(z) = z^N \widetilde{S}(z)$ are matrix power series, and

$$\widetilde{A}(z) = A(z)(I - z^{-1}\lambda Q)^{-1}. \tag{3.50}$$

Observe that $\widetilde{S}(z)$ is defined for $z = \lambda$, where it takes the value

$$\widetilde{S}(\lambda) = \sum_{i=-N}^{+\infty} \lambda^i A_i + \left(\sum_{i=-N+1}^{+\infty} (i+N)\lambda^i A_i \right) Q,$$

which is finite by (3.46).

The transformation (3.47) has the effect of moving the root λ to zero leaving unchanged the remaining roots as stated by the following theorem.

Theorem 3.32 *Assume that the $m \times m$ matrix Laurent power series $S(z) = \sum_{i=-N}^{+\infty} z^i A_i$ is in the Wiener algebra together with $S'(z)$, and define $A(z) = z^N S(z)$. Let λ be the only zero of $\det A(z)$ with modulus 1, assumed to be simple, and let $\boldsymbol{v} \neq 0$ be such that $S(\lambda)\boldsymbol{v} = 0$. Then the matrix function $\widetilde{S}(z)$ defined in (3.47) is in the Wiener algebra and, with $\widetilde{A}(z) = z^N \widetilde{S}(z)$, one has:*

1. *if $z \notin \{0, \lambda\}$, then $\det \widetilde{A}(z) = 0$ if and only if $\det A(z) = 0$;*
2. *$\det \widetilde{A}(0) = 0$ and $\widetilde{A}(0)\boldsymbol{v} = 0$;*
3. *$\det \widetilde{A}(\lambda) \neq 0$ and $\widetilde{A}(z)$ is nonsingular for $|z| = 1$;*
4. *if $\mu \notin \{0, \lambda\}$ is a root of $A(z)$, and if $\boldsymbol{w} \neq 0$ is such that $A(\mu)\boldsymbol{w} = 0$, then μ is a root of $\widetilde{A}(z)$ and $\widetilde{A}(\mu)\boldsymbol{r} = 0$, for $\boldsymbol{r} = (I - \frac{\lambda}{\mu}Q)\boldsymbol{w} = \boldsymbol{w} - \frac{\lambda}{\mu}(\boldsymbol{u}^{\mathrm{T}}\boldsymbol{w})\boldsymbol{v}$;*
5. *if 0 is a root of $A(z)$, $\boldsymbol{w} \neq 0$ is such that $A(0)\boldsymbol{w} = 0$, and if $\boldsymbol{v}$ and $\boldsymbol{w}$ are linearly dependent, then the vectors $\boldsymbol{v}$, $\boldsymbol{w}$ form a Jordan chain of length 2 corresponding to 0; otherwise, if $\boldsymbol{v}$ and $\boldsymbol{w}$ are linearly independent, and if $\boldsymbol{u}$ is such that $\boldsymbol{u}^{\mathrm{T}}\boldsymbol{w} = 0$, then $\widetilde{A}(0)\boldsymbol{w} = 0$.*

Proof From (3.49), we have

$$\sum_{i=-N}^{+\infty} |\widetilde{A}_i| \leq \sum_{i=-N}^{+\infty} |A_i| + \sum_{i=-N}^{+\infty} \left(\sum_{j=i+1}^{+\infty} |A_j| \right) Q$$

which is bounded in view of (3.46). Concerning the zeros of $\det \widetilde{A}(z) = 0$, observe that $\det(I - z^{-1}\lambda Q)^{-1} = z/(z - \lambda)$ so that

$$\det \widetilde{A}(z) = \frac{z}{z - \lambda} \det A(z).$$

Therefore, if $z \notin \{0, \lambda\}$, then $\det \widetilde{A}(z) = 0$ if and only if $\det A(z) = 0$. In particular $\det \widetilde{A}(z) \neq 0$ if $|z| = 1$ and $z \neq \lambda$. Since λ is a simple zero of $\det A(z)$, then also $\det \widetilde{A}(\lambda) \neq 0$. Moreover, $\det \widetilde{A}(0) = 0$ and, from (3.49) we have $\widetilde{A}(0)\boldsymbol{v} = \widetilde{A}_{-N}\boldsymbol{v} = A_{-N}(I - Q)\boldsymbol{v} = 0$.

If $\mu \notin \{0, \lambda\}$ is a root of $A(z)$, and if $\boldsymbol{w} \neq 0$ is such that $A(\mu)\boldsymbol{w} = 0$, then from (3.50) we have

$$\widetilde{A}(\mu)\left(I - \frac{\lambda}{\mu}Q\right)\boldsymbol{w} = A(\mu)\boldsymbol{w} = 0,$$

which shows that $\widetilde{A}(\mu)\boldsymbol{r} = 0$.

Suppose that $A(0)\boldsymbol{w} = 0$, $\boldsymbol{w} \neq 0$; if $\boldsymbol{w} = \alpha\boldsymbol{v}$ for some scalar α, then the vectors $\boldsymbol{v}$ are $\boldsymbol{w}$ form a Jordan chain of length 2 corresponding to 0 since $\widetilde{A}(0)\boldsymbol{v} = 0$ and

$$\widetilde{A}'(0)\boldsymbol{v} + \widetilde{A}(0)\boldsymbol{w} = (A_{-N+1}(I - Q) - \lambda^{-1}A_{-N}Q)\boldsymbol{v} + A_{-N}(I - Q)\boldsymbol{w} = 0;$$

if $\boldsymbol{w} \neq \alpha\boldsymbol{v}$, then we may choose $\boldsymbol{u}$ such that $\boldsymbol{u}^{\mathrm{T}}\boldsymbol{w} = 0$, and then $\widetilde{A}(0)\boldsymbol{w} = A_{-N}(I - Q)\boldsymbol{w} = A_{-N}\boldsymbol{w} = 0$. □

Observe that if $A(z)$ has k nonzero roots $\mu_1 = \lambda, \mu_2, \ldots, \mu_k$ and linearly independent vectors $\boldsymbol{w}_1 = \boldsymbol{v}, \boldsymbol{w}_2, \ldots, \boldsymbol{w}_k$ such that $A(\mu_i)\boldsymbol{w}_i = 0$, for $i = 1, \ldots, k$, then $\widetilde{A}(z)$ has the roots $0, \mu_2, \ldots, \mu_k$ such that $\widetilde{A}(0)\boldsymbol{v} = 0$, $\widetilde{A}(\mu_i)\boldsymbol{r}_i = 0$, for $i = 2, \ldots, k$, where $\boldsymbol{r}_i = (I - \frac{\lambda}{\mu_i}Q)\boldsymbol{w}_i = \boldsymbol{w}_i - \frac{\lambda}{\mu_i}(\boldsymbol{u}^{\mathrm{T}}\boldsymbol{w}_i)\boldsymbol{v}$, $i = 2, \ldots, k$; moreover, the vectors $\boldsymbol{v}, \boldsymbol{r}_2, \ldots, \boldsymbol{r}_k$, are linearly independent.

The theorem above allows one to pass from a weak canonical factorization of $S(z)$ to a canonical factorization of $\widetilde{S}(z)$, and vice-versa, so that we may apply algorithms for the canonical factorization in order to compute a weak canonical factorization. Let us assume that there exists a weak canonical factorization

$$S(z) = U(z)L(z)$$

where $U(z)$ is nonsingular for $|z| = 1$ and λ is the only zero of $\det L(z)$ of modulus 1. Let $\widetilde{S}(z)$ be the "shifted" matrix Laurent power series of (3.47). One easily shows that

$$\widetilde{S}(z) = \widetilde{U}(z)\widetilde{L}(z),$$

where

$$\begin{aligned} \widetilde{U}(z) &= U(z), \\ \widetilde{L}(z) &= L(z)(I - z^{-1}\lambda Q)^{-1} \end{aligned} \tag{3.51}$$

is a canonical factorization of $\widetilde{S}(z)$; $\widetilde{L}(z^{-1})$ is a matrix polynomial of degree at most N, and the coefficients of $\widetilde{L}(z) = \sum_{i=0}^{N} z^{-i}\widetilde{L}_{-i}$ and of $L(z) = \sum_{i=0}^{N} z^{-i}L_{-i}$ are related by

$$\begin{aligned} L_{-i} &= \widetilde{L}_{-i} - \lambda\widetilde{L}_{-i+1}Q, \quad i = 1, 2, \ldots, N, \\ L_0 &= \widetilde{L}_0, \\ \widetilde{L}_{-i} &= L_{-i} + \left(\sum_{j=0}^{i-1} \lambda^{i-j}L_{-j}\right)Q, \quad i = 1, 2, \ldots, N. \end{aligned} \tag{3.52}$$

In fact, the first and the second equations above are obtained by comparing the coefficients in the expression $\widetilde{L}(z)(I - z^{-1}\lambda Q) = L(z)$; the third equation can be inductly proved by using the first one.

The above relations between canonical factorizations lead to the following.

Theorem 3.33 *Let $S(z) = \sum_{i=-1}^{+\infty} z^i A_i$ be an $m \times m$ matrix valued function in the Wiener algebra $\mathcal{W}$, together with $S'(z)$. Let λ be the only zero of $\det S(z)$ of modulus 1 and let $\boldsymbol{v} \neq 0$ be such that $S(\lambda)\boldsymbol{v} = 0$. If there exists a weak canonical factorization*

$$S(z) = U(z)L(z), \qquad L(z) = I - z^{-1}G, \quad |z| = 1,$$

such that G is power bounded and $\rho(G) = \lambda$, then G is the minimal solution of the matrix equation $\sum_{i=-1}^{+\infty} A_i X^{i+1} = 0$. Moreover, the matrix $\widetilde{G} = G - \lambda Q$ is the minimal solution of the matrix equation $\sum_{i=-1}^{+\infty} \widetilde{A}_i X^{i+1} = 0$ and $\rho(\widetilde{G}) < 1$.

Proof The first part of the theorem follows from Theorem 3.19. Concerning the second part, observe that $\widetilde{S}(z)$ has a canonical factorization $\widetilde{S}(z) = \widetilde{U}(z)\widetilde{L}(z)$, $\widetilde{U}(z) = U(z)$, and $\widetilde{L}(z) = (I - z^{-1}G)(I - z^{-1}\lambda Q)^{-1} = I - z^{-1}\widetilde{G}$, since $G\boldsymbol{v} = \lambda \boldsymbol{v}$. Therefore, in light of Theorem 3.18, $\widetilde{G}$ is the minimal solution of the matrix equation $\sum_{i=-1}^{+\infty} \widetilde{A}_i X^{i+1} = 0$, and is the unique solution with spectral radius less than one. □

We can also provide an explicit relation between the formal inverse $H(z) = S(z)^{-1} = \sum_{i=-\infty}^{+\infty} z^i H_i$ and the inverse $\widetilde{H}(z) = \widetilde{S}(z)^{-1} = \sum_{i=-\infty}^{+\infty} z^i \widetilde{H}_i$. From (3.47), we obtain

$$H(z) = (I - z^{-1}\lambda Q)^{-1}\widetilde{H}(z)$$

which, expressed in terms of the coefficients, turns into

$$H_i = \widetilde{H}_i + \lambda^{-i} Q \sum_{j=i+1}^{+\infty} \lambda^j \widetilde{H}_j.$$

Furthermore, if we define

$$H^* = \sum_{i=-\infty}^{+\infty} \lambda^i \widetilde{H}_i = \widetilde{H}(\lambda) = \widetilde{S}(\lambda)^{-1}$$

then we find that

$$H_i = \widetilde{H}_i - \lambda^{-i} Q\Big(\sum_{j=-\infty}^{i} \lambda^j \widetilde{H}_j - H^*\Big). \tag{3.53}$$

The shifting technique also applies in the case where we are given a left eigenvector $\boldsymbol{w}$ such that $\boldsymbol{w}^{\mathrm{T}} S(\lambda) = 0$, and λ is the only zero of $\det S(z)$ of modulus 1: in that case, we define

$$\widetilde{S}(z) = (I - z^{-1}\lambda Q)^{-1} S(z), \quad Q = \boldsymbol{u}\boldsymbol{w}^{\mathrm{T}},$$

where $\boldsymbol{u}$ is any vector such that $\boldsymbol{w}^{\mathrm{T}}\boldsymbol{u} = 1$. For the transformed matrix Laurent power series $\widetilde{S}(z)$ we may write a theorem analogous to Theorem 3.32. In this case also the (weak) canonical factorizations of $S(z)$ and of $\widetilde{S}(z)$ can be related: if $S(z) = U(z)L(z)$ is a weak canonical factorization of $S(z)$, where $\det U(\lambda) = 0$ and $L(z)$ is nonsingular for $|z| = 1$, then $\widetilde{S}(z)$ has a canonical factorization $\widetilde{S}(z) = \widetilde{U}(z)\widetilde{L}(z)$, where

$$\begin{aligned} \widetilde{U}(z) &= (I - z^{-1}\lambda Q)^{-1}U(z), \\ \widetilde{L}(z) &= L(z). \end{aligned} \tag{3.54}$$

Formally, since $(I - z^{-1}\lambda Q)^{-1} = I + Q\sum_{i=1}^{+\infty} z^{-i}\lambda^i$, if we write $\sum_{i=-\infty}^{+\infty} z^i\widetilde{U}_i = (I - z^{-1}\lambda Q)^{-1}U(z)$, then we find that $\widetilde{U}_i = 0$ for $i < 0$ and

$$\widetilde{U}_i = U_i + Q\sum_{j=i+1}^{+\infty} \lambda^{j-i}U_j, \quad i = 0, 1, \ldots.$$

In the case where $S(z) = \sum_{i=-\infty}^{M} z^iA_i$, for a nonnegative integer M, and where λ is the only zero of $\det S(z)$ on the unit circle, we may move the root λ to infinity by applying a similar shifting technique. Let us assume that $\boldsymbol{v}$ is such that $S(\lambda)\boldsymbol{v} = 0$. We observe that $\mu = \lambda^{-1}$ is a root of the "reversed" matrix Laurent power series $T(z) = \sum_{i=-\infty}^{M} z^{-i}A_i = \sum_{i=-M}^{+\infty} z^iA_{-i}$, and it is the only zero of $\det T(z)$ on the unit circle. Define the "shifted" matrix Laurent power series

$$\widetilde{T}(z) = T(z)(I - z^{-1}\lambda^{-1}Q)^{-1}, \quad Q = \boldsymbol{v}\boldsymbol{u}^{\mathrm{T}}$$

where $\boldsymbol{u}^{\mathrm{T}}\boldsymbol{v} = 1$; the function

$$\widetilde{S}(z) = \widetilde{T}(z^{-1}) = S(z)(I - z\lambda^{-1}Q)^{-1}$$

has the same roots of $S(z)$ except for the point $z = \lambda$ which is moved to infinity. In this case also we may relate as follows the (weak) canonical factorizations of $S(z)$ and $\widetilde{S}(z)$. Assuming that $S(z) = U(z)L(z)$ is a weak canonical factorization of $S(z)$, where $\det L(\lambda) = 0$ and $U(z)$ is nonsingular for $|z| = 1$, $\widetilde{S}(z)$ has a canonical factorization $\widetilde{S}(z) = \widetilde{U}(z)\widetilde{L}(z)$ where $\widetilde{U}(z) = U(z)$, $\widetilde{L}(z) = L(z)(I - z\lambda^{-1}Q)^{-1}$. Moreover, $\widetilde{L}(z) = \sum_{i=0}^{+\infty} z^{-i}\widetilde{L}_{-i}$ is such that

$$\widetilde{L}_i = L_i + \sum_{j=-\infty}^{i-1} \lambda^{-i+j}L_jQ, \quad i = 0, -1, -2, \ldots.$$

In the case where $S(z)$ has more than one root of modulus 1, the shift technique can be repeatedly applied in order to shift these roots either to zero or to infinity.

3.7 Bibliographic notes

Relationships between matrix equations, Wiener–Hopf factorizations, infinite Toeplitz matrices and matrix (Laurent) power series are investigated in [14], [15], [24]. The results expressed in Theorems 3.18 and 3.19 are new.

There is a wide literature on the theory of Wiener–Hopf factorizations. The books [27] and [48] contain all the material on Toeplitz matrices which is needed for our purposes. Theorem 3.1 is a classic result by Toeplitz [111]. The notion of the Wiener–Hopf factorization as it is used here was introduced by Kreĭn in the scalar case [73] and by Gohberg and Kreĭn [49] in the matrix case. Theorem 3.10 was proved by Gohberg and Kreĭn [49]. A comprehensive and self-contained survey on factorization of matrix function of the Wiener–Hopf type over L_p from the operator theory point of view, with extensive comments on related results and with a wide bibliography, can be found in [85] and in [46].

Matrix polynomials and related topics are treated in the book [47], where relations between inverses of matrix Laurent polynomials and Wiener–Hopf factorizations are investigated. In particular, concerning monic matrix polynomials, chapter 4 contains necessary and sufficient conditions for the existence of a canonical factorization, together with representation formulas for the spectral minimal solution of matrix equations. Theorems 3.20 and 3.22 have been proved in [14] and [26], respectively.

Linearizations of matrix polynomials are introduced in the book [47]; linearizations of matrix equations are introduced in the papers [100], [25], [11], and [15].

The shift technique has been introduced in the paper [59] and generalized in [15] and in [24]. A result related to this technique concerns the deflation of a root of an analytic matrix function and is described in section 1.2 of [46].

Part II

Structured Markov chains

4

M/G/1-TYPE MARKOV CHAINS

4.1 Introduction

We assume from now on that in our Markov chains the state space is two-dimensional. One illustrative example is given by a buffer in a telecommunications system. The buffer content follows the evolution equation

$$X_{n+1} = \max(X_n + \alpha_n - 1, 0) \tag{4.1}$$

just as in Example 1.3, where X_n is the buffer occupancy at time n and α_n is the number of new packets which arrive during the nth transmission interval. Often, it is not legitimate to assume, like we did earlier, that the α_n's are independent and identically distributed random variables. Instead, they depend on various factors which vary slowly in time, such as the number of active customers or the congestion of the network.

These external factors often are modeled as a Markov chain $\{\varphi_n\}$ on some state space S, and one lets the distribution of α_n depend on φ_n. For historical reasons, S is called the set of *phases*, the process $\{\varphi_n\}$ itself being called the *environmental* Markov chain; the structure of S may be more or less complicated, according to circumstances.

Under this assumption, the two-dimensional process $\{(X_n, \varphi_n)\}$ is a Markov chain on the state space $E = \mathbb{N} \times S$ and the transition matrix has the structure

$$P = \begin{bmatrix} B & A_2 & A_3 & A_4 & \dots \\ A_0 & A_1 & A_2 & A_3 & \ddots \\ & A_0 & A_1 & A_2 & \ddots \\ & & A_0 & A_1 & \ddots \\ 0 & & & \ddots & \ddots \end{bmatrix} \tag{4.2}$$

similar to (1.5); here, however, B and the A_k's are matrices, with

$$(A_k)_{i,j} = \mathrm{P}[\alpha_n = k, \varphi_{n+1} = j | \varphi_n = i] \qquad \text{for } k \geq 0,\ i, j \in S,$$

being the probability that k packets arrive during a packet transmission interval, and that the next environmental phase is j, given that the current one is i.

This is called a Markov chain of M/G/1 type, to emphasize the resemblance between (4.2) and (1.5). The component X_n of the Markov chain is generally called *the level*, and we also say that the structure of (4.2) is *skip-free to lower levels*, in reference to the dynamic evolution of the process: starting in level n at time t, the process may move without hindrance to higher levels at time $t+1$, but it may not move in one step beyond $n-1$ in the direction of lower levels. For similar reasons, Markov chains with the structure (1.8) where X_{t+1} is at most equal to $X_t + 1$ are called *skip-free to higher levels.* Furthermore, since the transition probabilities from $X_t = i$ to $X_{t+1} = j$ depend on $j - i$ only and not specifically on i and j (for i and j at least equal to 1), we also say that the transition matrix is *homogeneous* away from level 0. Observe that the skip-free to lower level property of Markov chains corresponds to an upper block Hessenberg transition matrix. Similarly, the skip-free to higher level property generates lower block Hessenberg transition matrices. Finally a transition matrix which is homogeneous away from level 0 has the block Toeplitz structure except for its first block row.

Applications are numerous and varied and we give in Section 4.8 a few pointers to the literature. Our analysis, however, is based on the high-level structure displayed in (4.2) only, and not on the finer structure of the sub-blocks; we may, therefore, assume without loss of generality that the phase set S is simply $\{1, 2, \ldots, m\}$. Furthermore, we assume that m is finite, since most of the developments in this book require that assumption; nevertheless, we shall indicate it, when a property holds even for $m = \infty$.

More properties of the case where $m = \infty$ can be found in the specific chapters of the book [79].

4.2 Markov chains skip-free to lower levels

Our first family of structured processes have transition matrices given by

$$P = \begin{bmatrix} B_0 & B_1 & B_2 & B_3 & \ldots \\ A_{-1} & A_0 & A_1 & A_2 & \ldots \\ & A_{-1} & A_0 & A_1 & \ddots \\ & & A_{-1} & A_0 & \ddots \\ 0 & & & \ddots & \ddots \end{bmatrix} \tag{4.3}$$

where A_i, for $i \geq -1$, and B_i, for $i \geq 0$, are nonnegative matrices in $\mathbb{R}^{m \times m}$ such that $\sum_{i=-1}^{+\infty} A_i$, $\sum_{i=0}^{+\infty} B_i$, are stochastic. That is P is in upper block Hessenberg form and it is block Toeplitz except for its first block row.

The two-dimensional state space is $E = \mathbb{N} \times \{1, 2, \ldots, m\}$, where m is finite, and we have

$$(A_k)_{j,j'} = \mathrm{P}[X_{n+1} = i + k, \varphi_{n+1} = j' | X_n = i, \varphi_n = j],$$

for $k \geq -1$, $i \geq 1$, and $1 \leq j, j' \leq m$, and

$$(B_k)_{j,j'} = \mathrm{P}[X_{n+1} = k, \varphi_{n+1} = j' | X_n = 0, \varphi_n = j],$$

for $k \geq 0$. In words, if the Markov chain is in the nonboundary level $i \geq 1$, then A_{-1} gives the transition probabilities down by one level and A_k, for $k \geq 0$, gives the transition probabilities up by k levels, while B_k gives the transition probabilities up to level k, starting from the boundary level 0.

We shall see that the key to analyze such Markov chains is the matrix equation

$$X = \sum_{i=-1}^{+\infty} A_i X^{i+1}, \tag{4.4}$$

where the unknown X is an $m \times m$ matrix. We need to make some assumptions, to ensure that the stochastic process is sufficiently well-behaved. The first one is as follows.

Condition 4.1 *The Markov chain with transition matrix* (4.3) *is irreducible and nonperiodic.*

This is a traditional hypothesis. It guarantees, by Theorems 1.16 and 1.17, that the state probabilities at time t have a limit, as $t \to +\infty$, independently of the initial state. If, in addition, the Markov chain is positive recurrent, then the limits are strictly positive. We denote by $\boldsymbol{\pi} \in \mathbb{R}^{\mathbb{N}}$ the stationary probability vector, that is, the unique solution of the system (1.16), and we partition it as $\boldsymbol{\pi} = (\boldsymbol{\pi}_n)_{n=0,1,\ldots}$, with $\boldsymbol{\pi}_n = (\pi_{n,j})_{j=1,\ldots,m}$, and $\pi_{n,j} = \lim_{t\to+\infty} \mathrm{P}[X_t = n, \varphi_t = j | X_0 = n', \varphi_0 = j']$, independently of n' and j'.

We determine $\boldsymbol{\pi}$ in three steps. First, we analyze the probabilities of first passage to lower levels; we define θ as the first return time to the level 0:

$$\theta = \min\{n \geq 0 : X_n = 0\}$$

and we define the matrices $G^{(n)}$, with

$$G^{(n)}_{j,j'} = \mathrm{P}[\theta < \infty, \varphi_\theta = j' | X_0 = n, \varphi_0 = j]$$

being the probability that, starting from the state (n, j) in level n, the process reaches down to the level 0 in a finite time, and that $(0, j')$ is the first state visited in level 0.

Lemma 4.2 *One has*

$$H \begin{bmatrix} G^{(1)} \\ G^{(2)} \\ G^{(3)} \\ \vdots \end{bmatrix} = \begin{bmatrix} A_{-1} \\ 0 \\ 0 \\ \vdots \end{bmatrix}, \tag{4.5}$$

where

$$H = \begin{bmatrix} I - A_0 & -A_1 & -A_2 & -A_3 & \ldots \\ -A_{-1} & I - A_0 & -A_1 & -A_2 & \ddots \\ & -A_{-1} & I - A_0 & -A_1 & \ddots \\ & & -A_{-1} & I - A_0 & \ddots \\ 0 & & & \ddots & \ddots \end{bmatrix}. \tag{4.6}$$

Furthermore, the matrices $G^{(1)}$, $G^{(2)}$, $G^{(3)}$,..., form the minimal nonnegative solution of the system (4.5) *according to the element-wise ordering.*

Proof This is a direct consequence of the last statement of Theorem 1.14 and of Corollary 1.15, where we take the set A to be the level 0, and the set B to be made up of all the other levels. □

Next, we observe that any sequence of displacements which takes the process from level n to level 0 also takes the process from level $n+1$ to level 1, and from level $n+2$ to level 2, etc., so that we actually have

$$G_{j,j'}^{(n)} = \mathrm{P}[\theta(k) < \infty, \varphi_{\theta(k)} = j' | X_0 = n+k, \varphi_0 = j], \qquad \text{for all } k \geq 0, \quad (4.7)$$

where $\theta(k) = \min\{n \geq 0 : X_n = k\}$ is the first return time to level k. This leads to the following result.

Theorem 4.3 *Denote by $G_{\min}$ the matrix of first passage probabilities from level $k+1$ to level k, independently of $k \geq 0$.*

The matrix $G_{\min}$ is the minimal nonnegative solution of (4.4) *in the sense of Definition 3.14, and one has $G^{(n)} = G_{\min}^n$, for all $n \geq 0$. If the Markov chain is recurrent, then $G_{\min}$ is stochastic, otherwise it is substochastic with $G_{\min}\mathbf{1} \leq \mathbf{1}$, $G_{\min}\mathbf{1} \neq \mathbf{1}$.*

Proof We decompose the trajectory from level n to level 0 into a first passage from level n to $n-1$ followed by a first passage from level $n-1$ to 0. In so doing, we obtain that $G^{(n)} = G_{\min}G^{(n-1)}$, from which one proves by induction that $G^{(n)} = G_{\min}^n$.

With this, (4.5) becomes

$$H \begin{bmatrix} G_{\min} \\ G_{\min}^2 \\ G_{\min}^3 \\ \vdots \end{bmatrix} = \begin{bmatrix} A_{-1} \\ 0 \\ 0 \\ \vdots \end{bmatrix},$$

the first row of which is equivalent to $G_{\min} = \sum_{i=-1}^{+\infty} A_i G_{\min}^{i+1}$. This shows that $G_{\min}$ is one solution of (4.4).

Now, take another nonnegative solution Z of (4.4). We have

$$H \begin{bmatrix} Z \\ Z^2 \\ Z^3 \\ \vdots \end{bmatrix} = \begin{bmatrix} A_{-1} \\ 0 \\ 0 \\ \vdots \end{bmatrix},$$

so that, by Lemma 4.2, $Z^n \geq G^{(n)}$, which shows that $Z \geq G_{\min}$ and that $G_{\min}$ is the minimal nonnegative solution.

If the Markov chain is recurrent, then the process reaches in finite time any state in level 0 (with probability one), starting from any state in level 1 and, therefore, the probability $(G_{\min}\mathbf{1})_j$ of reaching in finite time *some* state in level 0, starting from $(1,j)$ is also equal to 1 for all j. Conversely, if the Markov chain is transient, then the probability of returning to the states in level 0 is strictly less than one, which implies that $(G_{\min}\mathbf{1})_j < 1$ for at least one value of j. □

In the next theorem, proved in [98], we apply the censoring principle described in Section 1.6 and obtain a recurrence equation for the subvectors $\boldsymbol{\pi}_n$, for $n \geq 1$. The same result can also be obtained by means of a purely algebraic approach as an immediate consequence of Theorems 3.28, 3.29 and 3.30. In order to give more details of this algebraic approach we need further properties which are explained in Section 4.4, therefore we pospone the algebraic interpretation of this theorem to Section 4.5.

Theorem 4.4 *Assume that the Markov chain of M/G/1 type is positive recurrent, and let*

$$A_n^* = \sum_{i=n}^{+\infty} A_i G_{\min}^{i-n}, \qquad B_n^* = \sum_{i=n}^{+\infty} B_i G_{\min}^{i-n}, \qquad \text{for } n \geq 0. \tag{4.8}$$

Then $I - A_0^*$ is nonsingular,

$$\boldsymbol{\pi}_n^{\mathrm{T}} = \left(\boldsymbol{\pi}_0^{\mathrm{T}} B_n^* + \sum_{i=1}^{n-1} \boldsymbol{\pi}_i^{\mathrm{T}} A_{n-i}^*\right)(I - A_0^*)^{-1}, \qquad \text{for } n \geq 1, \tag{4.9}$$

and $\boldsymbol{\pi}_0$ is such that

$$\boldsymbol{\pi}_0^{\mathrm{T}} B_0^* = \boldsymbol{\pi}_0^{\mathrm{T}}. \tag{4.10}$$

Proof Choose an arbitrary level n and partition E into the subsets $A = \{(i,j) : 0 \leq i \leq n, 1 \leq j \leq m\}$ and $B = \{(i,j) : i > n, 1 \leq j \leq m\}$. We show that, in the notations of Section 1.6,

$$S' P_{B,A} = \begin{bmatrix} 0 & \dots & 0 & G_{\min} \\ 0 & \dots & 0 & G_{\min}^2 \\ 0 & \dots & 0 & G_{\min}^3 \\ \vdots & & \vdots & \vdots \end{bmatrix}$$

by Theorem 1.14. The argument goes as follows: the (i,i')th block of $S'P_{B,A}$ gives the first passage probabilities from level $n+i$ in B to level i' in A, before any other level in A. In view of the skip-free property of the Markov chain, the first level visited in A is of necessity the level n, so that the last block of columns is the only one which may be different from 0. Then we recall from Theorem 4.3 that the first passage probabilities from level $n+i$ to level n are given by $G_{\min}^i$.

Simple algebraic manipulations show that the matrix $P_A + P_{A,B} S' P_{B,A}$, which we denote by P', is given by

$$P' = \begin{bmatrix} B_0 & B_1 & \dots & B_{n-1} & B_n^* \\ A_{-1} & A_0 & \dots & A_{n-2} & A_{n-1}^* \\ & A_{-1} & \ddots & \vdots & \vdots \\ & & \ddots & A_0 & A_1^* \\ 0 & & & A_{-1} & A_0^* \end{bmatrix}.$$

With $\boldsymbol{\pi}_A^{\mathrm{T}} = [\boldsymbol{\pi}_0^{\mathrm{T}}, \boldsymbol{\pi}_1^{\mathrm{T}}, \dots, \boldsymbol{\pi}_n^{\mathrm{T}}]$, (4.9) directly results from (1.21). The fact that $I - A_0^*$ is nonsingular is justified as follows: the matrix P' is irreducible since, otherwise, the states in A would break into separate communicating classes, in violation of Condition 4.1, then we apply Corollary 1.15.

Incidentally, we may write that

$$G_{\min} = (I - A_0^*)^{-1} A_{-1}, \tag{4.11}$$

which readily results from (4.4).

If we take $A = \{(0, j) : 1 \le j \le m\}$, then $P' = B_0^*$, so that the last statement follows, and the proof is complete. □

It is worth observing that Lemma 4.2 and Theorems 4.3 and 4.4 hold even if $m = \infty$, in which case we only need to take care of replacing the matrix $(I - A_0^*)^{-1}$ in (4.9) by the series $\sum_{\nu=0}^{+\infty} (A_0^*)^\nu$. In the sequel, however, the assumption that $m < \infty$ becomes important.

Remark 4.5 There exist M/G/1-type systems with a slightly more general transition structure at level zero, because the behavior at the boundary is markedly different from its behavior in the other levels. In such cases, we have

$$P = \begin{bmatrix} B_0 & B_1 & B_2 & B_3 & \dots \\ C_{-1} & A_0 & A_1 & A_2 & \dots \\ & A_{-1} & A_0 & A_1 & \ddots \\ & & A_{-1} & A_0 & \ddots \\ 0 & & & \ddots & \ddots \end{bmatrix} \tag{4.12}$$

where B_0 is now a matrix in $\mathbb{R}^{m_0 \times m_0}$, with m_0 possibly different from m, C_{-1} is in $\mathbb{R}^{m \times m_0}$ and the blocks B_i, for $i \ge 1$, are in $\mathbb{R}^{m_0 \times m}$.

It is not very difficult to adapt Theorem 4.4 to these circumstances, the only difference being that B_0^* is now given by

$$B_0^* = B_0 + B_1^* (I - A_0^*)^{-1} C_{-1}.$$

4.3 Ergodicity

We determine now the conditions under which Theorem 4.4 applies, that is, the conditions under which the Markov chain is positive recurrent. We will need to impose some constraint on the blocks A_{-1}, A_0, A_1, ... The simplest is the condition given below. It is not very restrictive in practice, but we defer the discussion of this question to Section 4.7.

Condition 4.6 *The matrix*

$$A = \sum_{i=-1}^{+\infty} A_i \tag{4.13}$$

is irreducible, in addition to being stochastic.

Since $m < \infty$, Condition 4.6 implies in light of Theorem 1.27 that A has an invariant probability vector $\boldsymbol{\alpha}$, that is, a vector such that $\boldsymbol{\alpha}^{\mathrm{T}} A = \boldsymbol{\alpha}^{\mathrm{T}}$, $\boldsymbol{\alpha}^{\mathrm{T}}\mathbf{1} = 1$.

Theorem 4.4 is incomplete in that it does not indicate when the Markov chain is positive recurrent, and it lacks a normalizing equation for $\boldsymbol{\pi}_0$, as the solution of (4.10) is given up to a multiplicative constant only.

Before stating the next theorem it is useful to introduce the following matrix functions

$$\begin{aligned} S(z) &= z^{-1}A_{-1} + A_0 + zA_1 + z^2 A_2 + \cdots = \sum_{i=-1}^{+\infty} z^i A_i, \\ A(z) &= zS(z). \end{aligned} \tag{4.14}$$

Under our assumptions, $A(z)$ is a matrix power series whose elements are analytic in the open unit disk $\mathbb{D}$ and convergent in the closed unit disk. That series plays an important role and we call it the *generating function* of the Markov chain.

Theorem 4.7 *Assume that $m < \infty$, that Conditions 4.1 and 4.6 hold and that*

$$\sum_{i=-1}^{+\infty} (i+1)A_i < +\infty, \tag{4.15}$$

so that we may define the vector

$$\boldsymbol{a} = \sum_{i=-1}^{+\infty} iA_i\mathbf{1}; \tag{4.16}$$

moreover, set

$$\mu = \boldsymbol{\alpha}^{\mathrm{T}}\boldsymbol{a}, \tag{4.17}$$

where $\boldsymbol{\alpha}$ is the vector such that $\boldsymbol{\alpha}^{\mathrm{T}} A = \boldsymbol{\alpha}^T$, $\boldsymbol{\alpha}^{\mathrm{T}}\mathbf{1} = 1$, for $A = \sum_{i=-1}^{+\infty} A_i$. Then the Markov chain with transition matrix (4.3) is recurrent if and only if $\mu \le 0$ and transient if and only if $\mu > 0$. It is positive recurrent if and only if $\mu < 0$ and $\boldsymbol{b} = \sum_{i=1}^{+\infty} iB_i\mathbf{1} < \infty$.

We do not give the detailed proof, because it is quite technical (see [4, Proposition 4.6], [92, Theorem 1.3.1] or [41, Corollary 3], for instance). Nevertheless, the property is simple to interpret. Assume that the state is (n, i) at some time t; at time $t+1$ it will be (n', j), with n' and j random. The component a_i of the vector $\boldsymbol{a}$ is the expectation of the jump size $n' - n$, when the phase is i, and μ is obtained by averaging over the phases with the probability distribution $\boldsymbol{\alpha}$. If $\mu < 0$, then the jumps are negative on the average and the process is attracted to the level 0 at the bottom of the state space.

The constant μ in (4.17) is called the *drift* of the Markov chain.

Similarly, the component b_i of the vector $\boldsymbol{b}$ is the expected level reached in one jump from the state $(0, i)$. If $\boldsymbol{b} < \infty$ and if the drift is negative, then the cycles from level 0 back to level 0 take a finite amount of time, on average. Building upon the definitions in Section 1.2, one eventually reaches the conclusion that the states are positive recurrent.

If $\mu = 0$, or if $\mu < 0$ and $b_i = \infty$ for some i, then the cycles have an infinite expectation, so that the process is null recurrent. If $\mu > 0$, then the drift is away from level 0 and there is a strictly positive probability that the process never returns to level 0, which makes the Markov chain transient.

In order to fully characterize the stationary distribution, we need an additional equation. The one given below is adapted from [104].

Theorem 4.8 *Assume that $m < \infty$ and that Conditions 4.1 and 4.6 hold. If the Markov chain with transition matrix (4.3) is positive recurrent, then $\boldsymbol{\pi}_0$ is such that*

$$\boldsymbol{\pi}_0^{\mathrm{T}} \boldsymbol{b} - \mu \boldsymbol{\pi}_0^{\mathrm{T}} \mathbf{1} + \boldsymbol{\pi}_0^{\mathrm{T}} (I - B)(I - A)^{\#} \boldsymbol{a} = -\mu,$$

where $B = \sum_{n=0}^{+\infty} B_n$ and the operator $(\cdot)^{\#}$ denotes the group inverse[1] *of a matrix.*

Proof In addition to (4.14), define

$$B(z) = \sum_{i=0}^{+\infty} z^i B_i \qquad \text{and} \qquad \boldsymbol{\pi}(z) = \sum_{i=0}^{+\infty} z^i \boldsymbol{\pi}_i = \boldsymbol{\pi}_0 + z\boldsymbol{\psi}(z).$$

[1] The group inverse of a matrix M is the unique matrix $M^{\#}$ such that $MM^{\#}M = M$, $M^{\#}MM^{\#} = M^{\#}$ and $M^{\#}M = MM^{\#}$. We refer to [29, Chapters 7 and 8] for details and only recall here that, if P is the transition matrix of an irreducible Markov chain, and if $M = I - P$, then $M^{\#}$ is the unique solution of the linear system

$$MX = I - \mathbf{1}\boldsymbol{\pi}^{\mathrm{T}}, \qquad \boldsymbol{\pi}^{\mathrm{T}} X = \mathbf{0},$$

or equivalently, the unique solution of

$$XM = I - \mathbf{1}\boldsymbol{\pi}^{\mathrm{T}}, \qquad X\mathbf{1} = \mathbf{0},$$

where $\boldsymbol{\pi}$ is the stationary probability vector of P [29, Theorem 8.5.5]. The second system is interesting because it has the same coefficient matrix as the system $\boldsymbol{\pi}^{\mathrm{T}} M = \mathbf{0}$, $\boldsymbol{\pi}^{\mathrm{T}} \mathbf{1} = 1$ for the stationary distribution and, therefore, may be solved by the same procedure.

One may verify by direct substitution in either of the systems above that

$$M^{\#} = (M + \mathbf{1}\boldsymbol{\pi}^{\mathrm{T}})^{-1} - \mathbf{1}\boldsymbol{\pi}^{\mathrm{T}}.$$

We write the system $\boldsymbol{\pi}^{\mathrm{T}} P = \boldsymbol{\pi}^{\mathrm{T}}$ as

$$\boldsymbol{\pi}_n^{\mathrm{T}} = \boldsymbol{\pi}_0^{\mathrm{T}} B_n + \sum_{i=1}^{n+1} \boldsymbol{\pi}_i^{\mathrm{T}} A_{n-i},$$

for $n \geq 0$, multiply the nth equation by z^n, sum over all n, and obtain that

$$\boldsymbol{\psi}(z)^{\mathrm{T}}(zI - A(z)) + \boldsymbol{\pi}_0^{\mathrm{T}}(I - B(z)) = 0. \tag{4.18}$$

The latter equation is a functional interpretation of $\boldsymbol{\pi}^{\mathrm{T}} P = \boldsymbol{\pi}^{\mathrm{T}}$, where P is represented by $\psi(z)$ and P by $A(z)$ and $B(z)$.

Now, for all z in $(0, 1]$, $A(z)$ is finite, nonnegative and irreducible, therefore, we may define its Perron–Frobenius eigenvector/eigenvalue pair $\boldsymbol{u}(z)$, $\chi(z)$, with

$$A(z)\boldsymbol{u}(z) = \chi(z)\boldsymbol{u}(z). \tag{4.19}$$

We post-multiply (4.18) by $\boldsymbol{u}(z)$ and obtain that

$$\boldsymbol{\psi}^{\mathrm{T}}(z)\boldsymbol{u}(z) = -\boldsymbol{\pi}_0^{\mathrm{T}}(I - B(z))\boldsymbol{u}(z)/(z - \chi(z)).$$

At $z = 1$, $\chi(1) = 1$ and $\boldsymbol{u}(1) = \mathbf{1}$ by continuity, the left-hand side is $\boldsymbol{\psi}^{\mathrm{T}}(1)\mathbf{1} = 1 - \boldsymbol{\pi}_0^{\mathrm{T}}\mathbf{1}$ and, therefore, the right-hand side has a limit as z tends to 1. We find by L'Hospital's rule that

$$1 - \boldsymbol{\pi}_0^{\mathrm{T}}\mathbf{1} = [-\boldsymbol{\pi}_0^{\mathrm{T}}(I - B)\boldsymbol{u}'(1) + \boldsymbol{\pi}_0^{\mathrm{T}} B'(1)\mathbf{1}]/(1 - \chi'(1)). \tag{4.20}$$

Taking the derivative of (4.19) with respect to z and evaluating it at $z = 1$, we obtain that

$$(A - I)\boldsymbol{u}'(1) = \chi'(1)\mathbf{1} - A'(1)\mathbf{1} = \chi'(1)\mathbf{1} - (\boldsymbol{a} + \mathbf{1}) \tag{4.21}$$

by (4.16). Premultiplying this equation by $\boldsymbol{\alpha}^{\mathrm{T}}$ gives us $\chi'(1) = \boldsymbol{\alpha}^{\mathrm{T}}\boldsymbol{a} = 1 + \mu$. Furthermore, it results from [86, Lemma 2.3] that the solution of (4.21) is

$$\boldsymbol{u}'(1) = (A - I)^{\#}\boldsymbol{a} + c\mathbf{1}$$

where c is an unknown scalar.

Altogether, (4.20) becomes

$$1 - \boldsymbol{\pi}_0^{\mathrm{T}}\mathbf{1} = [\boldsymbol{\pi}_0^{\mathrm{T}}(I - B)(I - A)^{\#}\boldsymbol{a} + \boldsymbol{\pi}_0^{\mathrm{T}}\boldsymbol{b}]/(-\mu),$$

from which the statement follows. □

4.4 Properties of the generating function

It is clear that the whole stationary distribution of the M/G/1-type Markov chain is easily determined once the minimal nonnegative solution of (4.4) is identified and one is, therefore, interested in knowing as much as possible about the solutions of that equation. We briefly present in this section, and in Section

4.6, properties which have been proved in [40; 41], through a detailed analysis of their spectral characteristics.

Assume that G is any solution of (4.4) with eigenvalue λ and associated eigenvector $\boldsymbol{v}$. We post-multiply both sides of (4.4) by $\boldsymbol{v}$ and find that

$$G\boldsymbol{v} = \lambda\boldsymbol{v} = \sum_{i=-1}^{+\infty} \lambda^{i+1} A_i \boldsymbol{v} = A(\lambda)\boldsymbol{v}, \tag{4.22}$$

where the generating function $A(z)$ is defined in (4.14). Observe that equation (4.22) is a particular case of the more general Theorem 3.15.

In particular, $G_{\min}$ being a substochastic matrix, its eigenvalues have modulus at most one, and we are led to determine the values of z in the closed unit disk for which $zI - A(z)$ is singular; in other words, we are looking for the zeros in the closed unit disk of the scalar function

$$a(z) = \det(zI - A(z)). \tag{4.23}$$

This function inherits the properties of $A(z)$: it is a power series, analytic in the open unit disk $\mathbb{D}$ and convergent for $|z| = 1$.

We define a new Markov chain, on the state space $E' = \mathbb{Z} \times \{1, 2, \ldots, m\}$ with transition matrix

$$P' = T_{\pm\infty}[A(z)] = \begin{bmatrix} \ddots & \ddots & \ddots & \ddots & & \\ \ddots & A_0 & A_1 & A_2 & A_3 & \\ & A_{-1} & A_0 & A_1 & A_2 & \ddots \\ & & A_{-1} & A_0 & A_1 & \ddots \\ & & & A_{-1} & A_0 & \ddots \\ 0 & & & & \ddots & \ddots \end{bmatrix}; \tag{4.24}$$

that is, we remove the boundary at level 0 and we allow the Markov chain to move freely to any negative level, as well as to the positive ones. This we shall call the *doubly infinite* or *bi-infinite* Markov chain.

Even if we assume that Conditions 4.1 and 4.6 are satisfied, it may happen that (4.24) is not irreducible. For instance, take

$$S(z) = \begin{bmatrix} 0 & zr_1 & 0 & z^{-1}p_1 \\ z^{-1}p_2 & 0 & q_2 & zr_2 \\ z^{-1}p_3 & q_3 & 0 & 0 \\ zr_4 & z^{-1}p_4 & 0 & 0 \end{bmatrix}. \tag{4.25}$$

It is irreducible if all the parameters are strictly positive, nevertheless, inspection of the transition graph of P' shows that there are three disjoint final classes: E'_0, E'_1 and E'_2, with

$$E'_k = \{(n, 1), (n+1, 2), (n+1, 3), (n+2, 4) : n = k \mod 3\},$$

for $k = 0, 1$ and 2 (see Figure 4.1).

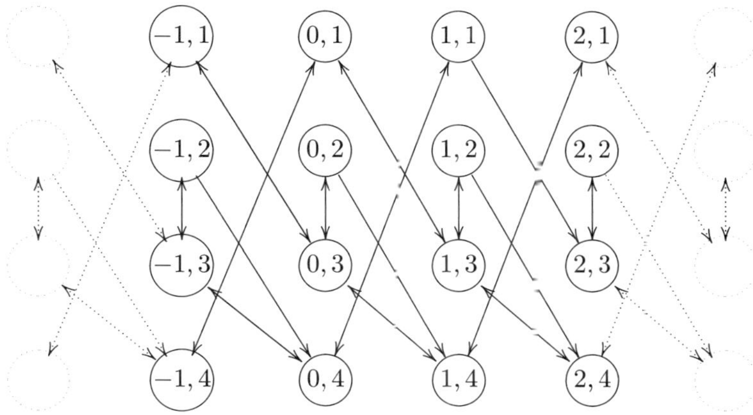

Fig. 4.1 Graph of the bi-infinite Markov chain defined by (4.25).

This is an important observation because the class structure of P' influences the roots of $a(z)$ in the following way. For every phase j, define $\mathcal{D}_j$ as the set of integers d for which there is a path from $(0, j)$ to (d, j). Define $\mathcal{D}$ as the union of the $\mathcal{D}_j$'s, for $j = 1, \ldots, m$, and define

$$\kappa = \gcd \mathcal{D}. \tag{4.26}$$

An equivalent definition of κ is

$$\kappa = \max\{k : \quad z^{-m/k} a(z^{1/k}) \text{ is a (single valued) function in } |z| \le 1\}. \tag{4.27}$$

The following property is proved in [41].

Theorem 4.9 *Assume that m is finite, that Condition 4.6 holds, and that $\boldsymbol{a}$ is finite. Define the drift μ as in Theorem 4.7.*

- *If $\mu < 0$, then $a(z)$ has $m - \kappa$ zeros in the open unit disk, and κ simple zeros on the unit circle at the κth roots of 1;*
- *if $\mu = 0$ and $\sum_{i=-1}^{+\infty}(i+1)^2 A_i$ is finite, then $a(z)$ has $m - \kappa$ zeros in the open unit disk, and κ zeros of multiplicity 2 on the unit circle at the κth roots of 1;*
- *if $\mu > 0$, then $a(z)$ has m zeros in the open unit disk, and κ simple zeros on the unit circle at the κth roots of 1.*

Since in the above theorem the conditions concerning μ are exhaustive, also the reverse implication is valid in each of the three parts. In particular, if $a(z)$ has $m - \kappa$ zeros in the open unit disk, and κ simple zeros on the unit circle at the κth roots of 1, then $\mu < 0$.

If both $\boldsymbol{b}$ and $\sum_{i=-1}^{+\infty}(i+1)^2 A_i$ are finite then the three cases correspond, respectively, to the positive recurrent, null recurrent and transient cases identified in Theorem 4.7.

The following result, which characterizes the eigenvalues of $G_{\min}$ in terms of the zeros of $a(z)$, is a corollary of Proposition 7 of [40].

Theorem 4.10 *Under the hypotheses of Theorem 4.9 the eigenvalues of* $G_{\min}$ *are:*

- *the zeros of* $a(z)$ *in the closed unit disk if* $\mu < 0$;
- *the zeros of* $a(z)$ *in the open unit disk and the* κ*-th roots of 1 if* $\mu = 0$;
- *the zeros of* $a(z)$ *in the open unit disk if* $\mu > 0$.

From the above theorem it follows that the minimal nonnegative solution $G_{\min}$ of equation (4.4) is also the spectral minimal solution in the sense of Definition 3.14, i.e., $\rho(G_{\min}) \le \rho(X)$ for any other solution X of (4.4).

Turning to the roots in $|z| > 1$, we define the convergence radius r_a of $A(z)$ and observe that for r real in $(0, r_a)$, $A(r)$ is well defined, irreducible and nonnegative. Therefore, its Perron–Frobenius eigenvalue $\rho(A(r))$ is defined in that interval.

The following result, proved in [44], [43], shows that the decay of the steady state vector $\boldsymbol{\pi}$ is related to the zeros outside the unit disk of the function $a(z) = \det(zI - A(z))$. In particular, the decay is exponential when one of such zeros exists.

Theorem 4.11 *Let* $A(z)$ *and* $B(z)$ *be the matrix power series defining the first two block rows of an irreducible and positive recurrent M/G/1-type Markov chain. If* $A(z)$ *is analytic for* $|z| < r_a$*, where* $1 < r_a < +\infty$*, and if* $a(z)$ *has a zero of modulus greater than 1 and less than* r_a*, then there exists a zero* ξ *of smallest modulus such that* $1 < |\xi| < r_a$*, moreover,* ξ *is real, positive and simple and it coincides with the spectral radius of* $A(\xi)$*, that is* $\xi = \rho(A(\xi))$*. If in addition* $z = 1$ *is the only zero of* $a(z)$ *of modulus one, then* ξ *is the only zero of minimal modulus, moreover, if* $B(z)$ *is analytic for* $|z| < r_a$*, then the steady state vector* $\boldsymbol{\pi} = (\boldsymbol{\pi}_i)_{i \ge 0}$ *is such that* $\boldsymbol{\pi}_i = \frac{\boldsymbol{\gamma}}{\xi^{i+1}} + O(\xi^{-i})$ *for a suitable positive constant* $\boldsymbol{\gamma}$.

The following conditions which guarantee the existence of ξ are shown in [43]:

Theorem 4.12 *Let* $A(z)$ *be the generating function associated with an irreducible and positive recurrent M/G/1-type Markov chain. If the matrix function* $A(z)$ *is entire or rational, then there exists a real number* $\xi > 1$ *such that* $\det(\xi I - A(\xi)) = 0$*. Moreover,*

$$\xi = \min\{|z| : \ z \in \mathbb{C}, \ a(z) = 0, \ |z| > 1\}.$$

From the above results it follows that, if $A(z)$ is a matrix polynomial, then the vector $\boldsymbol{\pi}$ has exponential decay.

4.5 Canonical factorization

The crucial step in the results of Section 4.2 was the identification of $G_{\min}$ as the minimal nonnegative solution of (4.4). From that starting point, we might have followed a purely algebraic approach to Theorem 4.4 by using the results of Section 3.5.

In this section we obtain Ramaswami's formula (4.9) as an algebraic derivation from the canonical factorization introduced in Section 3.2.

The matrix $G_{\min}$ induces a weak canonical factorization of $I - S(z)$ as stated by the following.

Theorem 4.13 *Under the hypotheses of Theorem 4.9 the matrix Laurent power series $I - S(z)$ has the following weak canonical factorization*

$$\begin{aligned} &I - S(z) = U(z)L(z), \qquad |z| = 1, \\ &U(z) = I - \sum_{i=0}^{+\infty} z^i A_i^*, \quad L(z) = I - z^{-1}G_{\min}, \end{aligned} \tag{4.28}$$

where A_i^*, for $i \geq 0$, are defined in (4.8). Moreover,

- *if $\mu < 0$ then $U(z)$ is nonsingular for $|z| = 1$, $L(z)$ is singular for $z = \omega_\kappa^i$, $i = 0, \ldots, \kappa - 1$, and the coefficients of $L(z)^{-1}$ are uniformly bounded in norm by a constant;*
- *if $\mu = 0$ and $\sum_{i=-1}^{+\infty}(i+1)^2 A_i$ is finite, then both $L(z)$ and $U(z)$ are singular for $z = \omega_\kappa^i$, $i = 0, \ldots, \kappa - 1$ and the coefficients of $L(z)^{-1}$ are uniformly bounded in norm by a constant;*
- *if $\mu > 0$ then $L(z)$ is nonsingular for $|z| = 1$, $U(z)$ is singular for $z = \omega_\kappa^i$, $i = 0, \ldots, \kappa - 1$.*

Proof The factorization (4.28) can be proved by direct verification, using (4.11) and the fact that $A_i^* = A_i + A_{i+1}^* G_{\min}$ for $i \geq 0$ which follows from (4.8). The function $L(z)$ is in the Wiener algebra since it is a matrix polynomial. Since $\sum_{i=0}^{+\infty} |A_i^*| = \sum_{i=0}^{+\infty} A_i^*$ and since the infinity norm of the latter summation is $\| \sum_{i=0}^{+\infty} A_i^* \mathbf{1} \|_\infty \leq \| \sum_{i=0}^{+\infty} (i+1) A_i \mathbf{1} \|_\infty$, which is finite since $\boldsymbol{a}$ in (4.16) is finite by Theorem 4.7, we deduce that also $U(z)$ is in the Wiener algebra. Moreover, by construction, $L(z)$ is singular if z is an eigenvalue of $G_{\min}$. Therefore $L(z^{-1})$ is nonsingular for $|z| < 1$. For Theorem 4.9 we find that the zeros of $\det U(z)$ have modulus greater than or equal to 1. Therefore (4.28) is a weak canonical factorization. The spectral properties for $\mu > 0$, $\mu = 0$ and $\mu < 0$ follow from Theorems 4.9 and 4.10. If $\mu \leq 0$ then the coefficients of $L(z)^{-1}$ are uniformly bounded in norm by a constant since $G_{\min}^i \mathbf{1} \leq \mathbf{1}$ for any $i \geq 0$. □

An immediate consequence of the theorem above is the following result, which was proved in [92] by means of probabilistic arguments, and which will be used in Chapter 6 to show convergence properties of functional iterations:

Theorem 4.14 *Under the assumptions of Theorem 4.13 the matrix $I - \sum_{i=0}^{+\infty} A_i^*$ is:*

- *a nonsingular M-matrix if $\mu < 0$;*
- *a singular M-matrix if either $\mu > 0$ or $\mu = 0$ and $\sum_{i=-1}^{+\infty}(i+1)^2 A_i$ is finite.*

In particular, the following result holds:

Theorem 4.15 *Under the assumptions of Theorem 4.13 we have $\rho(A_0^*) < 1$, where $A_0^* = \sum_{i=0}^{+\infty} A_i G_{\min}^i$, therefore $I - A_0^*$ is a nonsingular M-matrix.*

Proof Since $G_{\min}\mathbf{1} \leq \mathbf{1}$, we have $0 \leq A_0^*\mathbf{1} = \sum_{i=0}^{+\infty} A_i G_{\min}^i \mathbf{1} \leq \sum_{i=0}^{+\infty} A_i \mathbf{1} = \mathbf{1} - A_{-1}\mathbf{1} \leq \mathbf{1}$. Therefore $\|A_0^*\|_\infty \leq 1$, which implies $\rho(A_0^*) \leq 1$ (see Theorem A.2 in the appendix). Since $A_0^* \geq 0$, then by the Perron–Frobenius theorem, its spectral radius is eigenvalue of A_0^*. Since by Theorem 4.13 $U(z)$ is the left factor of a weak canonical factorization, one has $\det U(z) \neq 0$ for $|z| < 1$ and, in particular, $\det(I - A_0^*) = \det U(0) \neq 0$. Therefore the spectral radius of A_0^* cannot be 1. □

In matrix form, we write (4.28) as $H = UL$, with

$$U = \begin{bmatrix} I - A_0^* & -A_1^* & -A_2^* & \dots \\ & I - A_0^* & -A_1^* & \ddots \\ & & \ddots & \ddots \\ 0 & & & \ddots \end{bmatrix},$$

$$L = \begin{bmatrix} I & & & 0 \\ -G_{\min} & I & & \\ & -G_{\min} & I & \\ 0 & & \ddots & \ddots \end{bmatrix},$$

and H defined in (4.6).

We transform $\boldsymbol{\pi}^{\mathrm{T}}(I - P) = 0$ into

$$[\boldsymbol{\pi}_1^{\mathrm{T}}, \boldsymbol{\pi}_2^{\mathrm{T}}, \dots]H = \boldsymbol{\pi}_0^{\mathrm{T}}[B_1, B_2, \dots],$$

which is of the form (3.38). Under the hypothesis of positive recurrence the assumptions of Theorem 3.29 are satisfied and we obtain equations (3.42) which coincide with the Ramaswami formula (4.9).

4.6 Solutions of the matrix equation

It is clear that (4.4) has only one substochastic solution when the Markov chain is recurrent: $G_{\min}$ is the minimal solution and is stochastic, and there cannot exist a stochastic matrix M such that $G_{\min} \leq M$, $G_{\min} \neq M$.

The situation is more involved in the transient case. On the one hand, $G_{\min}$ is not stochastic and, on the other hand, [92, page 87] shows that there is always at least one stochastic solution to (4.4). Therefore, there are at least two solutions and the question immediately arises as to how many there are and what are they.

The most detailed characterization is given in [40], which we now summarize.

Theorem 4.16 *The matrix G is a power-bounded solution of* (4.4) *if and only if it has the form*

$$G = G_{\min} + \sum_{j=1}^{k} (\sigma_j I - G_{\min}) \boldsymbol{x}_j \boldsymbol{y}_j^T \tag{4.29}$$

where

1. $\{\sigma_1, \ldots, \sigma_k\}$ *is a set of roots, not necessarily distinct, of* $a(z)$ *of modulus* 1;
2. $\{\boldsymbol{x}_1, \ldots, \boldsymbol{x}_k\}$ *is a set of linearly independent vectors such that* $A(\sigma_j)\boldsymbol{x}_j = \boldsymbol{x}_j$, $j = 1, \ldots, k$;
3. *the linear space* $\mathrm{span}\{\boldsymbol{y}_1, \ldots, \boldsymbol{y}_k\}$ *is invariant under multiplication by* $G_{\min}^{\mathrm{T}}$;
4. $\boldsymbol{y}_i^{\mathrm{T}}\boldsymbol{x}_j = \delta_{i,j}$

or, equivalently,

1′. $\{\sigma_1, \ldots, \sigma_k\}$ *are the eigenvalues of* G *with* $|\sigma_j| = 1$;
2′. $\boldsymbol{y}_j$ *is a left eigenvector of* G *for the eigenvalue* σ_j, $j = 1, \ldots, k$;
3′. $\{\boldsymbol{x}_1, \ldots, \boldsymbol{x}_k\}$ *is a basis of the right eigenspace corresponding to* $\{\sigma_1, \ldots, \sigma_k\}$ *such that* $\boldsymbol{y}_i^{\mathrm{T}}\boldsymbol{x}_j = \delta_{i,j}$.

The close connection between the eigencharacteristics of the solutions G and those of the fundamental matrix $A(z)$ is given in (4.22) and (4.23). Briefly stated, Theorem 4.16 tells us that we may obtain all the power-bounded solutions by starting from $G_{\min}$, ripping apart some of its eigenvalues and replacing them by other roots of $a(z)$ on $|z| = 1$.

In dealing with the nonnegative, and with the substochastic solutions, it is necessary to introduce the following constraint.

Condition 4.17 *The function* $a(z)$ *has no zero such that* $|z| = 1$, $z \neq 1$.

We shall return to this assumption in Section 4.7, meanwhile, we observe that if it holds, then the σ_i's in Theorem 4.16 are all equal to 1 and (4.29) reduces to $G = G_{\min} + \sum_{j=1}^{k}(I - G_{\min})\boldsymbol{x}_j\boldsymbol{y}_j^T$. Clearly, it suffices that $\boldsymbol{x}_j$ and $\boldsymbol{y}_j$ be nonnegative, and $\boldsymbol{x}_j \geq G_{\min}\boldsymbol{x}_j$ for all j, in order for G to be nonnegative. Remarkably, this is also necessary as the theorem below asserts.

Theorem 4.18 *[40] Let* $\boldsymbol{x}_j, \boldsymbol{y}_j \in \mathbb{R}^m$, $j = 1, \ldots, k$ *be such that:*

1. $A(1)\boldsymbol{x}_j = \boldsymbol{x}_j$, $j = 1, \ldots, k$;
2. *the vectors* $\boldsymbol{y}_j^{\mathrm{T}}$ *are left eigenvectors of* $G_{\min}$, $j = 1, \ldots, k$;
3. $\boldsymbol{y}_i^{\mathrm{T}}\boldsymbol{x}_j = \delta_{i,j}$
4. $\boldsymbol{x}_j \geq 0$, $\boldsymbol{y}_j \geq 0$, $G_{\min}\boldsymbol{x}_j \leq \boldsymbol{x}_j$, $j = 1, \ldots, k$.

Then

$$G = G_{\min} + \sum_{j=1}^{k}(I - G_{\min})\boldsymbol{x}_j\boldsymbol{y}_j^T \tag{4.30}$$

is a nonnegative power-bounded solution of (4.4).

Conversely, if Condition 4.17 holds, any nonnegative power bounded solution of (4.4) *has the form* (4.30) *where 1–4 are satisfied. Moreover, the vectors* $\boldsymbol{y}_j$, $j = 1, \ldots, k$ *can be chosen such that* $\boldsymbol{y}_i^{\mathrm{T}}\boldsymbol{y}_j = 0$ *if* $i \neq j$.

Finally, G *is substochastic if and only if* $\sum_{j=1}^{k}\boldsymbol{x}_j \leq \boldsymbol{1}$, *and* $\boldsymbol{y}_j^{\mathrm{T}}\boldsymbol{1} = 1$, *for* $j = 1, \ldots, k$, *and* G *is stochastic if and only if* $\sum_{j=1}^{k}\boldsymbol{x}_j = \boldsymbol{1}$.

Remark 4.19 In certain cases, the matrix equation (4.4) might have an infinite number of power-bounded solutions. This occurs, for instance, if

$$S(z) = \begin{bmatrix} p_1 z^{-1} + q_1 z & 0 & 0 \\ 0 & p_2 z^{-1} + q_2 z & 0 \\ 0 & 0 & p_3 z^{-1} + q_3 z \end{bmatrix}$$

with $p_3 < q_3$; in fact, any matrix of the form

$$G_\alpha = \begin{bmatrix} 1 & 0 & 0 \\ 0 & 1 & 0 \\ \alpha(1 - p_3/q_3) & (1-\alpha)(1 - p_3/q_3) & p_3/q_3 \end{bmatrix}$$

is a stochastic solution of (4.4), for any $0 \le \alpha \le 1$ [40, page 29].

It is important to point out that, if Condition 4.17 is satisfied, then we may apply the shifting technique introduced in Section 3.6 to the function $\varphi(z) = I - S(z)$. In fact, $z = 1$ is the only root of $\varphi(z)$ on the unit circle and $\varphi(1)\mathbf{1} = 0$. The shifted function is $\widehat{\varphi}(z) = \varphi(z)(I - z^{-1}Q)^{-1}$, where $Q = \mathbf{1}\boldsymbol{u}^{\mathrm{T}}$, and $\boldsymbol{u}$ is any vector such that $\boldsymbol{u}^{\mathrm{T}}\mathbf{1} = 1$. The new function $\widehat{\varphi}(z)$ is nonsingular for $|z| = 1$, and by denoting $\widehat{\varphi}(z) = I - \sum_{i=-1}^{+\infty} z^i \widehat{A}_i$, one has that $\widehat{G} = G_{\min} - Q$ is the spectral minimal solution of the matrix equation $\sum_{i=-1}^{+\infty} \widehat{A}_i X^{i+1} = X$ and $\rho(\widehat{G}) < 1$.

4.7 Irreducibility assumptions

In practice, the situation is much simpler than the results in the preceding sections make it appear. Consider again the Markov chain (4.24) defined by allowing the level to be negative. Assume that Condition 4.1 holds, but that P' is reducible. In such a situation, there are several classes of states in $\mathbb{N} \times \{1, 2, \ldots, m\}$ which communicate only through the level 0. To give one example, take $B_1 > 0$ and

$$S(z) = \begin{bmatrix} p_1 z^{-1} + q_1 z & 0 & 0 \\ 0 & p_2 z^{-1} + q_2 z & 0 \\ a r_3 & (1-a) r_3 & p_3 z^{-1} + q_3 z \end{bmatrix},$$

with parameter values such that $S(1)$ is a stochastic matrix.

The phases 1 and 2 are different from the phase 3: if the system is in a state $(n, 1)$ with $n > 0$, it will behave as if the other phases did not exist, until it returns to the level 0, and similarly for the phase 2. Phase 3 acts more like a transitory phase which the system occupies until it settles on either phase 1 or 2.

It is very unlikely that a model of physical interest would have the type of behavior exhibited by phases 1 and 2 and it is quite legitimate to impose the following constraint on Markov chains of M/G/1 type.

Condition 4.20 *The Markov chain on $\mathbb{Z} \times \{1, 2, \ldots, m\}$ with transition matrix (4.24) has only one final class $\mathbb{Z} \times S$, where $S \subseteq \{1, 2, \ldots, m\}$. Every other state is on a path to the final class.*

A stronger assumption (that of the irreducibility of P') has been introduced in [82] and has been very useful for the development of the theory.

The properties of the matrix $P' = T_{\pm\infty}[A(z)]$ of (4.24) are closely related to the properties of the finite matrix $A = A(1) = \sum_{i=-1}^{+\infty} A_i$. A first simple remark is that if $T_{\pm\infty}[A(z)]$ is irreducible then A is irreducible. In fact, for a contradiction, if there exist two disjoint subsets S_1 and S_2 of $\{1,\dots,m\}$ such that $S_1 \cup S_2 = \{1,\dots,m\}$ and $a_{i,j} = 0$ for $i \in S_1$ and $j \in S_2$, then it holds $a_{i,j}^{(n)} = 0$ for $i \in S_1$ and $j \in S_2$ and for any $n \geq -1$ where we set $A^{(n)} = (a_{i,j}^{(n)})$. Therefore looking at the structure of the bi-infinite matrix $T_{\pm\infty}[A(z)]$, we deduce that any state (n,i) for $i \in S_1$ cannot communicate with a state (k,j) with $j \in S_2$ since $a_{i,j}^{(k-n)} = 0$ for $i \in S_1$ and $j \in S_2$. The converse is not true as the example in Section 4.4 shows.

Another remark concerns irreducible classes. Let $S \subset \{1,\dots,m\}$ and $\mathcal{S} = \mathbb{Z}\times S$. Let $A(z)_S$ and $T_{\pm\infty}[A(z)]_{\mathcal{S}}$ be the submatrices of $A(z)$ and $T_{\pm\infty}[A(z)]$ with indices in S and $\mathcal{S}$, respectively. Then it holds that $T_{\pm\infty}[A(z)]_{\mathcal{S}} = T_{\pm\infty}[A(z)_S]$. Therefore, if $\mathcal{S}$ is an irreducible class of $T_{\pm\infty}[A(z)]$ then S is an irreducible class for A.

Similarly, if $\mathcal{S}$ is final, then S is final in A. Also this property can be easily verified by contradiction. In fact, if S is not final, there exist $j \notin S$, $k \in S$ and $n \in \mathbb{Z}$ such that the element of A_n in position (j,k) is nonzero. This implies that in $T_{\pm\infty}[A(z)]$ the node $(n,k) \notin \mathcal{S}$ can be reached from $(n,j) \in \mathcal{S}$, that is $\mathcal{S}$ would not be final.

The above arguments imply the following

Theorem 4.21 *Let the Markov chain* (4.24) *satisfy Condition 4.20, and let* $\mathcal{S} = \mathbb{Z}\times S$, $S \subset \{1,\dots,m\}$ *be the only final class of* $T_{\pm\infty}[A(z)]$, *then the matrix* $A = \sum_{i=-1}^{+\infty} A_i$ *has only one final class* S.

Proof According to the arguments introduced before the theorem, S is a final class of A. Assume that $S' \neq S$ is another final class for A, and that $k \in S'$. For Condition 4.20, for any $n \in \mathbb{Z}$ there exists a path connecting (n,k) to some $(p,h) \in \mathcal{S}$. To this path there corresponds a path in $\{1,\dots,m\}$ which connects $k \in S'$ to $h \in S$. Therefore, S' cannot be final. □

Finally, Condition 4.20 has the following simple consequence.

Remark 4.22 Observe that under the Condition 4.20, there exists a permutation matrix Π and an integer $K > 0$ such that

$$\Pi A(z)\Pi^{\mathrm{T}} = \begin{bmatrix} V_{1,1}(z) & 0 & \dots & 0 \\ V_{2,1}(z) & V_{2,2}(z) & \ddots & \vdots \\ \vdots & \vdots & \ddots & 0 \\ V_{K,1}(z) & V_{K,2}(z) & \dots & V_{K,K}(z) \end{bmatrix}$$

where $V_{1,1}(1)$ is stochastic and irreducible (it corresponds to the final class), $V_{i,i}(z)$ are irreducible and substochastic, for $i = 2,\dots,K$, moreover, $V_{i,1}(1) \neq 0$

for $i = 2, \ldots, K$. In this way, $\rho(V_{1,1}(1)) = 1$ since $V_{1,1}(1)$ is stochastic, moreover, $\rho(V_{i,i}(1) < 1$ for $i = 2, \ldots, K$. The latter property holds since $V_{i,i}(1) \leq \sum_{j=1}^{i} V_{i,j}(1) = B_i$ and $V_{i,i}(1) \neq B_i$ because of $V_{i,1} \neq 0$. Since $V_{i,i}(1)$ is irreducible, from the Perron–Frobenius theorem 1.27 one has $\rho(V_{i,i}(1) < \rho(B_i) = 1$.

Another consequence of Condition 4.20 is that the submatrix of A_{-1} with indices in S cannot be the null matrix. In fact, if the elements of A_{-1} with indices in S were zero then there would be no back transition from a state (k, i) to a state (h, j) for $h < k$ and $i, j \in S$ so that the class $\mathcal{S} = \mathbb{Z} \times S$ could not be irreducible. A similar, argument can be used to prove that the submatrix of $\sum_{i=1}^{+\infty} A_i = 0$ with indices in S cannot be the null matrix.

An important property which will be useful in Chapter 7 concerns certain Schur complements of $I - P$ where P is the bi-infinite matrix. (4.24)

Theorem 4.23 *Assume that the Markov chain* (4.24) *satisfies Condition 4.20, and let $\mathcal{S} = \mathbb{Z} \times S$, $S \subset \{1, \ldots, m\}$ be the only final class of $T_{\pm\infty}[A(z)]$. Consider a nonempty subset $U \subset \mathbb{Z}$ and partition the set of states $\mathbb{Z} \times \{1, \ldots m\}$ in the two disjoint subsets E_1 and E_2 such that $E_1 = U \times \{1, \ldots, m\}$. Assume that the sum $\sum_{i=0}^{+\infty} P_{E_1}^i$ is finite, where P_{E_1} is the submatrix of P with subscripts in the set E_1. Then, we may define the censored Markov chain $P' = P_{E_1} + P_{E_2}(\sum_{i=0}^{+\infty} P_{E_1}^i)P_{E_1}$, moreover, P' satisfies Condition 4.20 where the unique final class is $U \times S$.*

Proof The set of states $U \times S$ is an irreducible class for P'. In fact, for any pair of states $(n_1, i_1), (n_2, i_2) \in U \times S$ there exists a path in P connecting them. This implies that in the Markov chain P' there exists a path connecting (n_1, i_1) with (n_2, i_2). The set of states $U \times S$ is a final class, in fact if there exists a path from (n_1, i_1) to (n_2, i_2) where $(n_1, i_1) \in U \times S$ and $(n_2, i_2) \notin U \times S$, $(n_2, i_2) \in U \times \{1, \ldots, m\}$ then there will be a path out of $\mathbb{Z} \times S$ in P, which contraddicts the assumptions. Moreover every state $(n, i) \notin U \times S$, $(n, i) \in U \times \{1, \ldots, m\}$ is in a path to $U \times S$ in P'. In fact, there exists a path in P from (n, i) to $(n_1, i_1) \in \mathbb{Z} \times S$. If $n_1 \in U$ then the property holds, otherwise, since $\mathbb{Z} \times S$ is an irreducible class, there exists a path from (n_1, i_1) to (n_2, i_1) where $n_2 \in U$. This completes the proof. □

Another interesting consequence of Condition 4.20 is that the quantity κ defined in (4.27) is equal to one. In order to see this, take j in the set S. By Condition 4.20, all the states (n, j) are accessible to $(0, j)$, so that the set $\mathcal{D}_j$ defined on page 99 is equal to $\mathbb{Z}$. This implies that $\mathcal{D} = \mathbb{Z}$ and that $\kappa = 1$ by (4.26).

Thus, Condition 4.20 allows us to strengthen as follows Theorems 4.9 and 4.18.

Theorem 4.24 *Assume that m is finite, that Condition 4.20 holds, and that $\boldsymbol{a}$ is finite. The only zero of $a(z)$ on $|z| = 1$ is $z = 1$. It has multiplicity 1 if $\mu \neq 0$ and multiplicity 2 if $\mu = 0$ and $\sum_{i=-1}^{+\infty}(i+1)^2 A_i$ is finite where μ is the drift defined in* (4.17). *The only substochastic solutions of* (4.4) *are $G_{\min}$ and, if $\mu > 0$, $G_{\min} + (I - G_{\min})\mathbf{1}\boldsymbol{y}^{\mathrm{T}}$, where $\boldsymbol{y}$ is the left Perron–Frobenius eigenvector of $G_{\min}$.*

Theorem 4.7 remains valid in that case: the important point is that there should be a unique invariant probability vector for the matrix A.

Remark 4.25 In Lemma 15 of [43] it is shown that, if $A(z)$ satisfies the assumptions of Theorem 4.11, then any zero ζ of $a(z)$ with modulus $\xi = \min\{|z| : a(z) = 0, |z| > 1\}$ is such that ζ/ξ is zero of $a(z)$. Moreover Theorem 4.12 states that ξ is a zero of $a(z)$. If Condition 4.20 is satisfied, then $z = 1$ is the only zero of modulus 1 of $a(z)$, therefore ξ is the only zero of $a(z)$ of minimal modulus outside the unit circle.

An immediate consequence of Theorem 4.24 is that in the recurrent case $\lambda = 1$ is the only eigenvalue of modulus 1 of $G_{\min}$ and is simple. In general we have the following.

Theorem 4.26 *Let* $\eta = \max\{|z| : |z| < 1,\ a(z) = 0\}$. *If the drift* μ *is nonnegative then for any matrix norm and for any* ϵ *such that* $\eta + \epsilon < 1$, *there exists a positive* γ *such that*

- *if* $\mu > 0$ *then* $\|G_{\min}^n\| \leq \gamma(\eta + \epsilon)^n$
- *if* $\mu \leq 0$ *and Condition 4.20 is satisfied then* $\|G_{\min}^n - \mathbf{1}\boldsymbol{g}^{\mathrm{T}}\| \leq \gamma(\eta + \epsilon)^n$,

where $\boldsymbol{g} \geq 0$ *is the invariant probability vector of* $G_{\min}$, *i.e.,* $\boldsymbol{g}^{\mathrm{T}} G_{\min} = \boldsymbol{g}^{\mathrm{T}}$, $\boldsymbol{g}^{\mathrm{T}}\mathbf{1} = 1$.

Proof If $\mu > 0$ then $\rho(G_{\min}) = \eta$. Applying Corollary A.4 of the appendix to the matrix $G_{\min}$ we find that for any matrix norm $\|\cdot\|$ there exists $\gamma > 0$ such that $\|G_{\min}^n\| \leq \gamma(\eta + \epsilon)^n$. If $\mu \leq 0$ and Condition 4.20 is satisfied then $\rho(G_{\min}) = 1$ and 1 is the only eigenvalue of $G_{\min}$ of modulus equal to $\rho(G_{\min})$ and is simple, moreover η is the second largest modulus eigenvalue. Therefore, since $G_{\min}\mathbf{1}\boldsymbol{g}^{\mathrm{T}} = \mathbf{1}\boldsymbol{g}^{\mathrm{T}} G_{\min} = \mathbf{1}\boldsymbol{g}^{\mathrm{T}}$, then by using an inductive argument we find that $(G_{\min} - \mathbf{1}\boldsymbol{g}^{\mathrm{T}})^n = G_{\min}^n - \mathbf{1}\boldsymbol{g}^{\mathrm{T}}$. Applying Corollary A.4 of the appendix to the matrix $G_{\min} - \mathbf{1}\boldsymbol{g}^{\mathrm{T}}$ provides the bound $\|G_{\min}^n - \mathbf{1}\boldsymbol{g}^{\mathrm{T}}\| \leq \gamma(\eta + \epsilon)^n$. □

4.8 Bibliographic notes

One finds in [4], [58] and, above all, in [92], a more extensive analysis of M/G/1-type queues from the probabilistic point of view. The spectral analysis of the generating function (and its solutions) has been mainly conducted in [96] and in a series of papers by Gail, Hantler and Taylor, of which we have cited the most directly relevant ones. The relationship between Ramaswami's formula and canonical factorizations is pointed out in [87] and [24].

5

PHASE-TYPE QUEUES

5.1 Introduction

We consider here a series of processes on the two-dimensional state space $E = \mathbb{N} \times \{1, 2, \ldots, m\}$ for some finite m, with a variety of transition structures. We group these under the generic name of *phase-type queues (PH queues)* because they usually serve as models of queueing systems, and the queues defined in (4.3) are but the first family of PH queues which we have encountered.

We first deal with the so-called G/M/1-type Markov chains which are in some sense symmetrical to those analyzed in Chapter 4: starting from level n at some time t, the process may move to any of the levels $n-1$, $n-2$, ... down to the level zero at time $t+1$, but it may not move up beyond the level $n+1$; one says that it is skip-free to higher levels. Furthermore, the transition probabilities from $X_t = n$ to $X_{t+1} = k$ depend on $k - n$ only, for $k \geq 1$, and the transition matrix has the form

$$P = \begin{bmatrix} B_0 & A_1 & & & 0 \\ B_{-1} & A_0 & A_1 & & \\ B_{-2} & A_{-1} & A_0 & A_1 & \\ B_{-3} & A_{-2} & A_{-1} & A_0 & \ddots \\ \vdots & \vdots & \ddots & \ddots & \ddots \end{bmatrix}.$$

Observe that the transition matrix is block lower Hessenberg and, except for its first block column, is block Toeplitz.

The stationary distribution of such processes depends on a matrix equation similar to (4.4) and we show in Section 5.3 that there is an extremely tight connection between G/M/1 and M/G/1-type queues, so that one family may be considered as the dual of the other.

There are two ways of "combining" these two structures. The first is to allow the full generality of each, and to assume that the transition matrix has the form

$$P = \begin{bmatrix} B_0 & B_1 & B_2 & B_3 & \ldots \\ B_{-1} & A_0 & A_1 & A_2 & \ldots \\ B_{-2} & A_{-1} & A_0 & A_1 & \ddots \\ B_{-3} & A_{-2} & A_{-1} & A_0 & \ddots \\ \vdots & \vdots & \ddots & \ddots & \ddots \end{bmatrix}.$$

Here, the process is allowed to move from any level n to any level $n+j$, where $j = \ldots, -2, -1, 0, 1, 2, \ldots$ without any restriction save that the level may not become negative; furthermore, the transition probability is independent of n and depends on j only, as long as $n \neq 0$ and $n+j \neq 0$, that is, the transition matrix is block Toeplitz except for its first block row and its first block column. This leads to a rather unwieldy set of equations which becomes, as we show in Section 5.5, more manageable if one imposes some constraint on the jumps in one direction.

Alternately, one may impose both the restrictions of M/G/1 and G/M/1-type queues, and forbid transitions by more than one level at a time. The resulting processes are called *quasi-birth-and-death* (QBD) Markov chains and are said to be skip-free in both directions. Their transition matrix is

$$P = \begin{bmatrix} B_0 & A_1 & & & 0 \\ A_{-1} & A_0 & A_1 & & \\ & A_{-1} & A_0 & A_1 & \\ & & A_{-1} & A_0 & \ddots \\ 0 & & & \ddots & \ddots \end{bmatrix}.$$

QBD's are the most popular phase-type queues because their characteristic equations are very simple, as we show in Section 5.6.

We conclude with a short section on an interesting generalization of QBD's. We may think of a QBD as evolving on a simple linear list: each level is a node in the list and the process is allowed to move from one node to one of its two neighbors. For the processes in Section 5.8, the level evolves on a tree. This gives rise to a set of interconnected quadratic matrix equations.

5.2 G/M/1-type Markov chains

We analyze, in this and in the next section, Markov chains with transition matrix

$$P = \begin{bmatrix} B_0 & A_1 & & & 0 \\ B_{-1} & A_0 & A_1 & & \\ B_{-2} & A_{-1} & A_0 & A_1 & \\ B_{-3} & A_{-2} & A_{-1} & A_0 & \ddots \\ \vdots & \vdots & \ddots & \ddots & \ddots \end{bmatrix}. \tag{5.1}$$

Here, A_{-i}, $i \geq -1$, and B_{-i}, $i \geq 0$ are nonnegative matrices in $\mathbb{R}^{m \times m}$ such that $\sum_{i=-1}^{n-1} A_{-i} + B_{-n}$ is stochastic for all $n \geq 0$. Often, it so happens that $\lim_{n \to +\infty} B_{-n} = 0$ and then the matrix $A = \sum_{i=-1}^{+\infty} A_{-i}$ is stochastic, but this is not always the case.

We immediately assume that the Markov chains under study satisfy the following conditions.

Condition 5.1 *The Markov chain with transition matrix (5.1) is irreducible and nonperiodic.*

Condition 5.2 *The doubly infinite Markov chain on $\mathbb{Z} \times \{1, 2, \ldots, m\}$ with transition matrix*

$$P' = T_{\pm\infty}[A(z)] = \begin{bmatrix} \ddots & \ddots & & & & 0 \\ \ddots & A_0 & A_1 & & & \\ \ddots & A_{-1} & A_0 & A_1 & & \\ \ddots & A_{-2} & A_{-1} & A_0 & A_1 & \\ \ddots & A_{-3} & A_{-2} & A_{-1} & A_0 & \ddots \\ \ddots & \ddots & \ddots & \ddots & \ddots & \ddots \end{bmatrix} \tag{5.2}$$

has only one final class $\mathbb{Z} \times S$, where $S \subseteq \{1, 2, \ldots, m\}$. Every other state is on a path to the final class.

Our justification for making these assumptions is the same as for Conditions 4.1 and 4.20: the first one ensures the existence of limits for the state probabilities, the second removes from consideration examples which do not correspond to physical systems of practical interest.

An immediately useful consequence is that the matrix $A = \sum_{i=-1}^{+\infty} A_{-i}$ has only one irreducible class. If, in addition, it is stochastic, then, in view of the Perron–Frobenius Theorem 1.27, there exists a unique vector $\boldsymbol{\alpha}$ such that $\boldsymbol{\alpha}^{\mathrm{T}} A = \boldsymbol{\alpha}^{\mathrm{T}}$, $\boldsymbol{\alpha}^{\mathrm{T}}\mathbf{1} = 1$, with $\alpha_i > 0$ for i in the irreducible class and $\alpha_i = 0$ elsewhere.

Theorem 5.3 *Assume that $m < \infty$ and that Condition 5.1 holds.*

If $A = \sum_{i=-1}^{+\infty} A_{-i}$ is not stochastic then the Markov chain with transition matrix (5.1) is positive recurrent.

If A is stochastic and if Condition 5.2 holds, then the Markov chain is positive recurrent if $\delta < 0$, null recurrent if $\delta = 0$, and transient if $\delta > 0$, where $\delta = \boldsymbol{\alpha}^{\mathrm{T}}\boldsymbol{a}$, $\boldsymbol{\alpha}$ is such that $\boldsymbol{\alpha}^{\mathrm{T}} A = \boldsymbol{\alpha}^{\mathrm{T}}$, $\boldsymbol{\alpha}^{\mathrm{T}}\mathbf{1} = 1$, and $\boldsymbol{a} = \sum_{i=-1}^{+\infty} iA_{-i}\mathbf{1}$.

Detailed proofs may be found in [91, Section 1.3] in the case where A is irreducible and in [41] in all generality but, like in Chapter 4, the property is simple to understand. If A is not stochastic, then the sequence $\{B_{-n}\}_{n\geq -1}$ cannot converge to zero, and there is, at least for some phases, a positive probability of jumping straight to the level zero, no matter how far the process happens to be in the higher levels. Thus, returns to level zero are frequent and the process is positive recurrent. If A is stochastic, δ measures some average drift over unit time intervals. If $\delta < 0$, then the drift is toward level zero and the process is positive recurrent, the other two cases being similarly interpreted. Several equivalent conditions are given in [83] to characterize the stability of Markov chains of G/M/1 as well as M/G/1-types.

Similarly to the M/G/1 case, we partition the stationary probability vector $\boldsymbol{\pi}$, when it exists, into subvectors $\boldsymbol{\pi}_n \in \mathbb{R}^m$, $n \geq 0$ of length m, so that $\boldsymbol{\pi} = (\boldsymbol{\pi}_n)_{n=0,1,\ldots}$. It depends in a very simple manner on the matrix equation

$$X = \sum_{i=-1}^{+\infty} X^{i+1} A_{-i} \tag{5.3}$$

as we now explain.

Theorem 5.4 *If the Markov chain with transition matrix (5.1) is positive recurrent, then*

$$\boldsymbol{\pi}_n^{\mathrm{T}} = \boldsymbol{\pi}_0^{\mathrm{T}} R_{\min}^n \qquad \text{for } n \geq 1, \tag{5.4}$$

where $R_{\min}$ is the minimal nonnegative solution of (5.3).

That matrix is such that $(R_{\min})_{i,j}$ is the expected number of visits to $(1, j)$, starting from $(0, i)$, before the first return to level zero.

The vector $\boldsymbol{\pi}_0$ is characterized by the system

$$\boldsymbol{\pi}_0^{\mathrm{T}} = \boldsymbol{\pi}_0^{\mathrm{T}} \sum_{i=0}^{+\infty} R_{\min}^i B_{-i}, \qquad \boldsymbol{\pi}_0^{\mathrm{T}} (I - R_{\min})^{-1} \mathbf{1} = 1. \tag{5.5}$$

This holds even if $m = \infty$, in which case one needs to replace $(I - R_{\min})^{-1}$ in (5.5) by the series $\sum_{i=0}^{+\infty} R_{\min}^i$.

Proof The proof proceeds in several steps.

To begin with, we define the matrices $R^{(k)}$, for $k \geq 1$, with $(R^{(k)})_{i,j}$ being the expected number of visits to (k, j), starting from $(0, i)$, before the first return to level zero. We apply Theorem 1.23 with the set A being level zero, and the set B being the collection of all the other levels, and we obtain that

$$\boldsymbol{\pi}_n^{\mathrm{T}} = \boldsymbol{\pi}_0^{\mathrm{T}} R^{(n)} \qquad \text{for } n \geq 1. \tag{5.6}$$

Furthermore, in the notation of Section 1.4,

$$\begin{bmatrix} R^{(1)} & R^{(2)} & R^{(3)} & \ldots \end{bmatrix} = P_{A,B} S' = \begin{bmatrix} A_1 & 0 & 0 & \ldots \end{bmatrix} S',$$

where S' is the minimal nonnegative solution of $S'H = I$, with

$$H = \begin{bmatrix} I - A_0 & -A_1 & & & 0 \\ -A_{-1} & I - A_0 & -A_1 & & \\ -A_{-2} & -A_{-1} & I - A_0 & -A_1 & \\ -A_{-3} & -A_{-2} & -A_{-1} & I - A_0 & \ddots \\ \vdots & \ddots & \ddots & \ddots & \ddots \end{bmatrix}, \tag{5.7}$$

so that $\begin{bmatrix} R^{(1)} & R^{(2)} & R^{(3)} & \ldots \end{bmatrix}$ is the minimal nonnegative solution of the system

$$\begin{bmatrix} X_1 & X_2 & X_3 & \ldots \end{bmatrix} H = \begin{bmatrix} A_1 & 0 & 0 & \ldots \end{bmatrix}.$$

We show below that $R^{(n)} = (R^{(1)})^n$, for all n. This has two immediate consequences. The first is that $\begin{bmatrix} R^{(1)} & R^{(2)} & R^{(3)} & \ldots \end{bmatrix} = \begin{bmatrix} R_{\min}^1 & R_{\min}^2 & R_{\min}^3 & \cdots \end{bmatrix}$, where $R_{\min}$ is the minimal solution of the system

$$\begin{bmatrix} Z & Z^2 & Z^3 & \ldots \end{bmatrix} H = \begin{bmatrix} A_1 & 0 & 0 & \ldots \end{bmatrix}. \tag{5.8}$$

Since the equation obtained by equating the first block in both sides of (5.8) is equivalent to $Z = \sum_{i=-1}^{+\infty} Z^{i+1} A_{-i}$, we are led to the conclusion that $R_{\min}$ is the minimal nonnegative solution of (5.3) and, by (5.6), that the vector $\boldsymbol{\pi}$ has the matrix-geometric form (5.4).

The second consequence is that we obtain from the first set of equations in $\boldsymbol{\pi}^{\mathrm{T}} = \boldsymbol{\pi}^{\mathrm{T}} P$ that

$$\boldsymbol{\pi}_0^{\mathrm{T}} = \sum_{i=0}^{+\infty} \boldsymbol{\pi}_i^{\mathrm{T}} B_{-i} = \boldsymbol{\pi}_0^{\mathrm{T}} \sum_{i=0}^{+\infty} R_{\min}^i B_{-i}.$$

Since we assume that the Markov chain is positive recurrent, the series $\sum_{i=0}^{+\infty} \boldsymbol{\pi}_i^{\mathrm{T}}$ converges and, by (5.4), this implies that $\sum_{i=0}^{+\infty} R_{\min}^i$ converges to $(I - R_{\min})^{-1}$. Altogether, we see that the normalizing equation $\boldsymbol{\pi}^{\mathrm{T}}\mathbf{1} = 1$ becomes $\boldsymbol{\pi}_0^{\mathrm{T}}(I - R_{\min})^{-1}\mathbf{1} = 1$, which proves the last statement of the theorem

It remains for us to prove that $R^{(n)} = (R^{(1)})^n$. Since the process is skip-free to higher levels, and since we start from level zero, every visit to level n is preceded by a visit to level $n-1$. Conditioning on the visits to level $n-1$, we write that

$$R_{i,j}^{(n)} = \sum_{k=1}^{m} R_{i,k}^{(n-1)} R_{k,j}^{(n-1,n)}, \tag{5.9}$$

where $R_{k,j}^{(n-1,n)}$ is the expected number of visits to (n, j), starting from $(n-1, k)$, before the first return to level $n-1$. Now, as long as one avoids level zero, the transition probabilities depend on the difference between levels only and not on their specific values. This leads us to the conclusion that $R_{k,j}^{(n-1,n)}$ is independent of n and that we may rewrite (5.9) as $R^{(n)} = R^{(n-1)} R^{(1)} = (R^{(1)})^n$, which concludes the proof. □

Similarly to the case of M/G/1 Markov chains, here we associate with the transition matrix (5.1) the matrix power series

$$\begin{aligned} S(z) &= zA_1 + A_0 + z^{-1}A_{-1} + z^{-2}A_{-2} + \cdots = \sum_{i=-1}^{+\infty} z^{-i} A_{-i}, \\ A(z) &= z^{-1}S(z). \end{aligned} \tag{5.10}$$

Under our assumptions, $A(z)$ is a matrix power series in z^{-1}, and $A(z^{-1})$ is analytic in the open unit disk $\mathbb{D}$ and convergent in the closed unit disk. Similarly to the M/G/1 case, we call $A(z)$ the *generating function* of the Markov chain.

5.3 A duality property

The characteristic equations (4.4, 5.3) of M/G/1 and G/M/1-type Markov chains are very similar, and it comes as no surprise that they should have been the object of parallel developments in the literature, and that the results of Sections 4.4 and 4.6 apply here as well, mutatis mutandis. The theorem below corresponds to Theorems 4.9 and 4.24; details are to be found in [40].

Theorem 5.5 *Assume that m is finite, that Condition 5.2 holds, A is stochastic, and that $\boldsymbol{a}$ is finite. Define the drift δ as in Theorem 5.3 and $A(z)$ as in (5.10). Define $a(z) = \det(zI - A(z^{-1}))$.*

i If $\delta > 0$, then $a(z)$ has $m-1$ zeros in the open unit disk and one simple zero equal to 1;

ii if $\delta = 0$ and if $\sum_{-1}^{+\infty}(i+1)^2 A_{-i}$ is finite, then $a(z)$ has $m-1$ zeros in the open unit disk and $z = 1$ is a zero of multiplicity two;

iii If $\delta < 0$, then $a(z)$ has m zeros in the open unit disk and one simple zero equal to 1.

If R is a power-bounded solutions of (5.3) and if λ is one of its eigenvalues, then $|\lambda| \leq 1$ and $a(\lambda) = 0$. Thus, the only power bounded solution of (5.3) are $R_{\min}$ and, when $\delta < 0$, the matrix

$$R_1 = R_{\min} + \boldsymbol{z}\boldsymbol{\alpha}^{\mathrm{T}}(I - R_{\min}), \tag{5.11}$$

where $\boldsymbol{z}$ is the right Perron–Frobenius eigenvector of $R_{\min}$, normalized by $\boldsymbol{\alpha}^{\mathrm{T}}\boldsymbol{z} = 1$.

We had seen in the proof of Theorem 5.4 that, when the process is positive recurrent, the series $\sum_{i=0}^{+\infty} R_{\min}^i$ converges, which implies that $\rho(R_{\min}) < 1$. The converse is true, and the following property is proved in [91, Section 1.3].

Theorem 5.6 *The matrix $R_{\min}$ is power-bounded. It has spectral radius strictly less than one if and only if the G/M/1 Markov chain is positive recurrent.*

Like the case of the solution $G_{\min}$, the minimal nonnegative solution $R_{\min}$ of the equation (5.3) is also minimal in terms of spectral radius (see Definition 3.14). That is, $\rho(R_{\min}) \leq \rho(X)$ for any other solution X.

It is worth mentioning that, if the matrix $A = \sum_{i=-1}^{+\infty} A_{-i}$ is irreducible and stochastic, the connection between the two matrix equations is extremely simple. Define $D = \mathrm{Diag}(\boldsymbol{\alpha})$, where $\boldsymbol{\alpha}$ is the strictly positive stationary probability vector of A and define $\widetilde{A}_i = D^{-1}A_{-i}^{\mathrm{T}}D$, for $i = -1, 0, 1, \ldots$ Note that the matrix $\widetilde{A} = \sum_{i=-1}^{+\infty} \widetilde{A}_i$ is stochastic and has the same stationary probability vector $\boldsymbol{\alpha}$ as A.

Now, take any solution R of (5.3) and define $\widetilde{G} = D^{-1}R^{\mathrm{T}}D$. It is easy to verify that $\widetilde{G}$ is a solution of

$$X = \sum_{i=-1}^{+\infty} \widetilde{A}_i X^{i+1}, \tag{5.12}$$

which is the characteristic equation of the M/G/1-type Markov chain with transition matrix

$$P = \begin{bmatrix} \widetilde{B}_0 & \widetilde{B}_1 & \widetilde{B}_2 & \widetilde{B}_3 & \ldots \\ \widetilde{A}_{-1} & \widetilde{A}_0 & \widetilde{A}_1 & \widetilde{A}_2 & \ldots \\ & \widetilde{A}_{-1} & \widetilde{A}_0 & \widetilde{A}_1 & \ddots \\ & & \widetilde{A}_{-1} & \widetilde{A}_0 & \ddots \\ 0 & & & \ddots & \ddots \end{bmatrix}, \tag{5.13}$$

where $\widetilde{B}_i = D^{-1}B_{-i}^{\mathrm{T}}D$, for $i = 0, 1, 2, \ldots$. Furthermore, it is easy to see that the drift $\widetilde{\mu} = \boldsymbol{\alpha}^{\mathrm{T}} \sum_{i=-1}^{+\infty}(i+1)\widetilde{A}_i\boldsymbol{1}$ of the new Markov chain is equal to the drift δ of the Markov chain (5.1) and that $D^{-1}R_{\min}^{\mathrm{T}}D$ is the minimal nonnegative solution of (5.12).

These observations, combined with Theorems 4.7 and 5.3, lead us to the following conclusion.

Theorem 5.7 *To determine $R_{\min}$ for a positive recurrent G/M/1-type Markov chain is equivalent to determining $G_{\min}$ for a transient M/G/1-type Markov chain. Conversely, to determine $R_{\min}$ for a null recurrent or transient G/M/1-type Markov chain is equivalent to determining $G_{\min}$ for a recurrent M/G/1-type Markov chain.*

The construction above has a probabilistic interpretation, naturally, which was given in [5] and which we briefly present here. Consider the doubly infinite process $\{X_n, \varphi_n\}$ with transition matrix (5.2) and define the jumps $K_n = X_n - X_{n-1}$. Take the dual process $\{X_n^{\mathrm{d}}, \varphi_n^{\mathrm{d}}\}$ with transition matrix

$$P^{\mathrm{d}} = \begin{bmatrix} \ddots & \ddots & \ddots & \ddots & \ddots & \ddots \\ \ddots & \widetilde{A}_0 & \widetilde{A}_1 & \widetilde{A}_2 & \widetilde{A}_3 & \ddots \\ & \widetilde{A}_{-1} & \widetilde{A}_0 & \widetilde{A}_1 & \widetilde{A}_2 & \ddots \\ & & \widetilde{A}_{-1} & \widetilde{A}_0 & \widetilde{A}_1 & \ddots \\ & & & \widetilde{A}_{-1} & \widetilde{A}_0 & \ddots \\ 0 & & & & \ddots & \ddots \end{bmatrix}$$

and define the jumps $K_n^{\mathrm{d}} = X_n^{\mathrm{d}} - X_{n-1}^{\mathrm{d}}$. Finally, assume that the phase at time zero is chosen according to the stationary distribution $\boldsymbol{\alpha}$. Theorem 3.5 in [5] states that

$$\begin{aligned} &\mathrm{P}[\varphi_0 = j_0, K_1 = k_1, \ldots, K_n = k_n, \varphi_n = j_n] \\ &= \mathrm{P}[\varphi_0^{\mathrm{d}} = j_n, K_1^{\mathrm{d}} = k_n, \ldots, K_n^{\mathrm{d}} = k_1, \varphi_n^{\mathrm{d}} = j_0] \end{aligned}$$

and means that the dual process is obtained from the original process by reversing the flow of time and changing the sign of the jumps.

Based on this duality property, the canonical factorization associated with the M/G/1 Markov chain of Theorem 4.13 induces a canonical factorization of the function $I - S(z)$ in (5.10).

Theorem 5.8 *If A is irreducible, then the matrix Laurent power series $I - S(z)$ has the factorization*

$$I - S(z) = (I - zR_{\min})\left(I - \sum_{i=0}^{+\infty} z^{-i} A_{-i}^{*}\right), \tag{5.14}$$

where

$$A^*_{-i} = \sum_{j=i}^{+\infty} R_{\min}^{j-i} A_{-i}, \quad i = 0, 1, \ldots.$$

This factorization is canonical if A is substochastic, and is weak canonical if A is stochastic.

Proof If A is stochastic then, in view of Theorem 4.13, the function $I - \widetilde{S}(z) = I - \sum_{i=-1}^{+\infty} z^i \widetilde{A}_i$ has a weak canonical factorization

$$I - \widetilde{S}(z) = (I - \sum_{i=0}^{+\infty} z^i \widetilde{A}_i^*)(I - z^{-1} G_{\min}).$$

From the relations $G_{\min} = D^{-1} R_{\min}^{\mathrm{T}} D$, $\widetilde{A}_i^* = D^{-1}(A^*_{-i})^{\mathrm{T}} D$, and $\widetilde{S}(z^{-1}) = D^{-1} S(z)^{\mathrm{T}} D$ we arrive at (5.14). If A is substochastic then the function $I - \widetilde{S}(z)$ is nonsingular for $|z| = 1$ and the above weak canonical factorization turns into a canonical factorization. □

Rewriting (5.14) in matrix form yields the factorization $H = UL$ of the matrix H of (5.7), where

$$U = \begin{bmatrix} I & -R_{\min} & & 0 \\ & I & -R_{\min} & \\ & & \ddots & \ddots \\ 0 & & & \ddots \end{bmatrix} \quad \text{and} \quad L = \begin{bmatrix} I - A_0^* & & & 0 \\ -A^*_{-1} & I - A_0^* & & \\ \vdots & \ddots & \ddots & \\ \vdots & \ddots & \ddots & \ddots \end{bmatrix}.$$

From $\boldsymbol{\pi}^{\mathrm{T}}(I - P) = \mathbf{0}$, we deduce that

$$[\boldsymbol{\pi}_1^{\mathrm{T}}, \boldsymbol{\pi}_2^{\mathrm{T}}, \ldots] H = \boldsymbol{\pi}_0^{\mathrm{T}} [A_1, 0, 0, \ldots];$$

applying the factorization of H given above, we obtain that

$$[\boldsymbol{\pi}_1^{\mathrm{T}}, \boldsymbol{\pi}_2^{\mathrm{T}}, \ldots] U = \boldsymbol{\pi}_0^{\mathrm{T}} [A_1 (I - A_0^*)^{-1}, 0, 0, \ldots];$$

solving this block triangular system by forward substitution yields (5.4).

All of this explains why, in the rest of the book, we concentrate on solving the equation (4.4) to the exclusion of (5.3).

To conclude this section, we mention that there may exist a matrix-geometric invariant vector, that is, a vector $\boldsymbol{\pi}^{\mathrm{T}} = [\boldsymbol{\pi}_0^{\mathrm{T}}, \boldsymbol{\pi}_0^{\mathrm{T}} R, \boldsymbol{\pi}_0^{\mathrm{T}} R^2, \ldots]$ such that $\boldsymbol{\pi}^{\mathrm{T}} = \boldsymbol{\pi}^{\mathrm{T}} P$ even when the G/M/1-type queue is transient. Naturally, in that case, the vector is not summable. It is precluded on the existence of some $\xi > 1$ such that $A(\xi)$ converges and such that the spectral radius of $A(\xi)$ is ξ. The theorem below is proved in [43, Theorems 2 and 3].

Theorem 5.9 *Define $\hat{R}$ as follows:*

- *if $\delta = 0$, then $\hat{R} = R_{\min}$;*

- *if* $\delta > 0$ *and if there exists* $\xi > 1$ *and* $\boldsymbol{\ell} > 0$ *such that* $\boldsymbol{\ell}^{\mathrm{T}} A(\xi) = \xi \boldsymbol{\ell}$, *then* $\hat{R} = R_{\min} + \boldsymbol{z}\boldsymbol{\ell}^{\mathrm{T}}(\xi I - R_{\min})$, *where* $\boldsymbol{z}$ *is the right Perron–Frobenius eigenvector of* $R_{\min}$.

In both cases, there exists a vector $\boldsymbol{\pi}_0^{\mathrm{T}}$ *such that* $\boldsymbol{\pi}_0^{\mathrm{T}} = \boldsymbol{\pi}_0^{\mathrm{T}} \sum_{i=0}^{+\infty} \hat{R}^i B_{-i}$ *and* $\boldsymbol{\pi}_n^{\mathrm{T}} = \boldsymbol{\pi}_0^{\mathrm{T}} \hat{R}^n$ *yields a matrix-geometric invariant vector.*

5.4 Toeplitz-like transitions

We now turn our attention to Markov chains for which the transition matrix is

$$P = \begin{bmatrix} B_0 & B_1 & B_2 & B_3 & \dots \\ B_{-1} & A_0 & A_1 & A_2 & \dots \\ B_{-2} & A_{-1} & A_0 & A_1 & \ddots \\ B_{-3} & A_{-2} & A_{-1} & A_0 & \ddots \\ \vdots & \vdots & \ddots & \ddots & \ddots \end{bmatrix}, \tag{5.15}$$

where the blocks B_i and A_i (for all i) are nonnegative matrices in $\mathbb{R}^{m \times m}$. Since P is stochastic, the series $\sum_{i=-(n-1)}^{+\infty} A_i + B_{-n}$, for $n \geq 1$, and the series $\sum_{i=0}^{+\infty} B_i$ are all stochastic. Furthermore, the matrix $A = \sum_{i=-\infty}^{+\infty} A_i$ is stochastic if $\lim_{n \to +\infty} B_{-n} = 0$.

In order to determine the stationary probability vector of (5.15), we begin like we did in the proof of Theorem 4.4. We choose an arbitrary level n and we partition E into the subset A of levels less than or equal to n, and the subset B of levels strictly greater than n. Then we obtain the recurrence equation

$$\boldsymbol{\pi}_n^{\mathrm{T}} = \left(\boldsymbol{\pi}_0^{\mathrm{T}} C_n^* + \sum_{k=1}^{n-1} \boldsymbol{\pi}_k^{\mathrm{T}} A_{n-k}^* \right) (I - A_0^*)^{-1}, \qquad \text{for } n \geq 1, \tag{5.16}$$

where the matrices A_k^* and C_k^* have the following interpretations:

- for $k \geq 0$ and $i, j \in \{1, 2, \dots, m\}$, $(A_k^*)_{i,j}$ is the probability that, starting from $(n-k, i)$, the Markov chain visits (n, j) before any other state in the levels 0 to n, possibly after visiting the levels $n+1$ and higher, independently of n for $n \geq k+1$;
- for $n \geq 1$ and $i, j \in \{1, 2, \dots, m\}$, $(C_n^*)_{i,j}$ is the probability of the same event if the initial state is $(0, i)$; it does depend on n.

The next theorem is given without proof because it is an immediate consequence of the equation above.

Theorem 5.10 *If the Markov chain with transition matrix (5.15) is positive recurrent, then*

$$\boldsymbol{\pi}_n^{\mathrm{T}} = \boldsymbol{\pi}_0^{\mathrm{T}} R_{0,n} + \sum_{i=1}^{n-1} \boldsymbol{\pi}_i^{\mathrm{T}} R_{n-i},$$

for $n \geq 1$, *where* $R_k = A_k^*(I - A_0^*)^{-1}$ *gives the expected number of visits to the states in level* m, *starting from level* $m - k$, *before the first visit to any of the* *levels 0 to $m-1$, independently of $m \geq k$, and $R_{0m} = C^*_m(I - A^*_0)^{-1}$ gives the expected number of such visits starting from level 0.*

Observe that, in the case of G/M/1-type Markov chains, the only nonzero matrices are A^*_0, A^*_1 and C^*_1, with $A^*_1 = C^*_1 = A_1$, so that the equation above becomes $\boldsymbol{\pi}^{\mathrm{T}}_n = \boldsymbol{\pi}^{\mathrm{T}}_0 R_1$ for all n, which is identical to (5.4).

In order to characterize the first passage probability matrices A^*_k, [51] introduce the matrices G_k and $G_{k,0}$ defined as follows:

- for $k \geq 1$ and $i, j \in \{1, 2, \ldots, m\}$, $(G_k)_{i,j}$ is the probability that, starting from (n, i), the Markov chain visits $(n-k, j)$ before any other state in the levels 0 to $n-1$, possibly after spending some time in the levels n and higher, independently of $n \geq k+1$;
- similarly, for $n \geq 1$ and $i, j \in \{1, 2, \ldots, m\}$, $(G_{n,0})_{i,j}$ is the probability that, starting from (n, i), the Markov chain visits $(0, j)$ before any other state in the levels 0 to $n-1$.

Note that, for a Markov chain of M/G/1 type, the only such nonzero matrices are G_1 and $G_{1,0}$, which are then both equal to $G_{\min}$.

Theorem 5.11 *The matrices A^*_k, $(k \geq 0)$, C^*_k, G_k and $G_{k,0}$, $(k \geq 1)$, are such that*

$$A^*_k = A_k + \sum_{i=1}^{+\infty} A^*_{k+i} G_i, \tag{5.17}$$

$$C^*_k = C_k + \sum_{i=1}^{+\infty} C^*_{k+i} G_i, \tag{5.18}$$

$$G_k = A_{-k} + \sum_{i=0}^{+\infty} A^*_i G_{k+i}, \tag{5.19}$$

$$G_{k,0} = B_{-k} + \sum_{i=0}^{+\infty} A^*_i G_{k+i,0}. \tag{5.20}$$

Proof Assume that the initial state is in level $n-k$, for $n \geq k+1$. For the process to visit the level n before any of the levels 0 to $n-1$, two cases may occur: either the Markov chain immediately jumps to level n, or it visits some higher level. The probability of the first case is given by A_k, which justifies the first term in (5.17). In the second case, there will be some lowest level visited above n and the product $A^*_{k+i} G_i$ gives the probability that it should be $n+i$: A^*_{k+i} ensures that the level $n+i$ is actually visited before any other level between 0 and $n+i-1$, and G_i ensures that, afterward, the level n will be visited before any other level between 0 and $n+i-1$. This proves (5.17).

The same argument is used to prove the other equations. □

The coupled systems (5.17, 5.19) may also be written as

$$\left[A^*_0 \; A^*_1 \; A^*_2 \ldots\right] M_G = \left[A_0 \; A_1 \; A_2 \ldots\right] \tag{5.21}$$

and

$$M_A \begin{bmatrix} G_1 \\ G_2 \\ G_3 \\ \vdots \end{bmatrix} = \begin{bmatrix} A_{-1} \\ A_{-2} \\ A_{-3} \\ \vdots \end{bmatrix}, \tag{5.22}$$

where

$$M_G = \begin{bmatrix} I & & & 0 \\ -G_1 & I & & \\ -G_2 & -G_1 & I & \\ \vdots & \ddots & \ddots & \ddots \end{bmatrix}$$

and

$$M_A = \begin{bmatrix} I - A_0^* & -A_1^* & -A_2^* & \dots \\ & I - A_0^* & -A_1^* & \ddots \\ & & I - A_0^* & \ddots \\ 0 & & & \ddots \end{bmatrix};$$

furthermore, (5.18) may be written as $\left[C_1^* \; C_2^* \; C_3^* \dots\right] M_G = \left[C_1 \; C_2 \; C_3 \dots\right]$ and (5.20) as

$$M_A \begin{bmatrix} G_{1,0} \\ G_{2,0} \\ G_{3,0} \\ \vdots \end{bmatrix} = \begin{bmatrix} B_{-1} \\ B_{-2} \\ B_{-3} \\ \vdots \end{bmatrix}.$$

The next theorem generalizes Theorems 5.8 and 4.13

Theorem 5.12 *If $\sum_{i=0}^{+\infty}(I-M_A)^i$ and $\sum_{i=0}^{+\infty}(I-M_G)^i$ are finite then the matrix Laurent power series $I - S(z)$, $S(z) = \sum_{i=-\infty}^{+\infty} z^i A_i$, has the (weak) canonical factorization*

$$I - S(z) = (I - \sum_{i=0}^{+\infty} z^i A_i^*)(I - \sum_{i=1}^{+\infty} z^{-i} G_i)$$

where A_i^*, G_{i+1}, $i \geq 0$, are defined in Theorem 5.11.

Proof The factorization formally holds from (5.21) and (5.22). It is (weak) canonical since the matrix functions $I - \sum_{i=0}^{+\infty} z^i A_i^*$ and $I - \sum_{i=1}^{+\infty} z^i G_i$ are analytic and invertible for $|z| < 1$. □

Finally, one needs an equation to characterize $\boldsymbol{\pi}_0$; it is given in the next lemma.

Lemma 5.13 *If the Markov chain (5.15) is positive recurrent, then* $\boldsymbol{\pi}_0^{\mathrm{T}} = \boldsymbol{\pi}_0^{\mathrm{T}} B_0^*$, *where*

$$B_0^* = B_0 + \sum_{k=1}^{+\infty} C_k^* G_{k,0}. \tag{5.23}$$

and the normalizing equation is

$$\boldsymbol{\pi}_0^{\mathrm{T}} \left(I + \sum_{n=0}^{+\infty} R_{0,n}(I - \sum_{n=1}^{+\infty} R_n)^{-1} \right) \mathbf{1} = 1. \tag{5.24}$$

Proof We apply the censoring principle of Section 1.6 and obtain that $\boldsymbol{\pi}_0^{\mathrm{T}} = \boldsymbol{\pi}_0^{\mathrm{T}} B_0^*$ where B_0^* is defined as follows: its (i, j)th element is the probability that, starting from $(0, i)$, the Markov chain visits $(0, j)$ upon its first return to level 0.

To prove (5.23), we proceed as in Theorem 5.11 and note that either the process immediately returns to level 0, with probability given by B_0, or it moves to some higher level. In the second case, there will be some lowest level visited above 0 and $C_k^* G_{k,0}$ is the probability that it is level k. The normalizing equation is easily proved. □

It is interesting to point out that equations (5.16) and (5.24) can be proved by means of purely algebraic arguments by using the canonical factorization of $I - S(z)$ together with Theorem 3.30.

The characterization given in Theorem 5.10 and in Lemma 5.13, even if interesting from the theoretical point of view, is not always convenient for algorithmic purposes. In fact, unlike the case of M/G/1 and G/M/1 Markov chains, the computation of the stationary probability vector $\boldsymbol{\pi}$ is not reduced to solving a single matrix equation. However, it is interesting to observe that once the matrices G_i and A_i^*, $i = 0, 1, \ldots$, are known, then (5.16) provides a useful algorithmic tool for computing $\boldsymbol{\pi}$. On the other hand, in the light of Theorem 5.12, the matrices G_i and A_i^*, $i = 0, 1, \ldots$, are provided by the canonical factorization of $I - S(z)$. Therefore, an efficient algorithm for computing the latter factorization is a valid tool for the numerical solution of these Markov chains. In this regard we refer the reader to Section 3.3 for further details.

5.5 Limited displacements

Take the Toeplitz-like matrix (5.15) and assume that $A_i = 0$, $i \geq -N - 1$, for some positive N. That means that the displacements to lower levels are limited to a distance at most equal to N, with the possible exception of direct jumps to the level 0 if $\lim_{n \to -\infty} B_n \neq 0$. In that case, the first passage probability matrices $G_{N+1}, G_{N+2}, \ldots$ are all equal to zero since it is not possible to reach, from level n, any level between 1 and $n - N - 1$ without visiting either the level 0 or some intermediary level $n - N, n - N + 1, \ldots n - 1$.

Thus, the equations (5.17, 5.19) in Theorem 5.11 become

$$[A_0^* \; A_1^* \; A_2^* \ldots] \begin{bmatrix} I & & 0 \\ -G_1 & I & \\ \vdots & -G_1 & \ddots \\ -G_N & \vdots & \ddots \\ & -G_N & \\ 0 & & \ddots \end{bmatrix} = [A_0 \; A_1 \; A_2 \ldots] \tag{5.25}$$

and

$$\begin{bmatrix} I - A_0^* & -A_1^* & \ldots & -A_{N-1}^* \\ & I - A_0^* & \ddots & \vdots \\ & & \ddots & -A_1^* \\ 0 & & & I - A_0^* \end{bmatrix} \begin{bmatrix} G_1 \\ G_2 \\ \vdots \\ G_N \end{bmatrix} = \begin{bmatrix} A_{-1} \\ A_{-2} \\ \vdots \\ A_{-N} \end{bmatrix}. \tag{5.26}$$

Now, define the power series

$$S(z) = \sum_{i=-N}^{+\infty} z^i A_i, \qquad A(z) = z^N S(z).$$

The following result is a consequence of Theorem 5.12

Theorem 5.14 *The matrix Laurent power series $I - S(z)$ has a weak canonical factorization*

$$I - S(z) = (I - A^*(z))(I - G(z^{-1})), \tag{5.27}$$

where $A^*(z) = \sum_{i=0}^{+\infty} z^i A_i^*$ and $G(z) = \sum_{i=1}^{N} z^i G_i$. Furthermore, the matrix power series $A^*(z)$ and $G(z)$ are a solution of the equation

$$A(z) = G(z) + A^*(z)(z^N I - G(z)). \tag{5.28}$$

It is actually proved in [42] that $G(z)$ is the matrix polynomial with minimal, nonnegative coefficients that satisfies (5.28).

Another approach to the problem follows if one reorganizes the structure of the transition matrix into blocks of size mN, this is called "reblocking". In order to simplify this presentation, we assume that the matrix $A = \sum_{i=-N}^{+\infty} A_i$ is stochastic. Then the transition matrix is

$$P = \begin{bmatrix} B_0 & B_1 & B_2 & B_3 & \cdots \\ B_{-1} & A_0 & A_1 & A_2 & \cdots \\ \vdots & \vdots & \vdots & \vdots & \ddots \\ B_{-N+1} & A_{-N+2} & A_{-N+3} & A_{-N+4} & \ddots \\ A_{-N} & A_{-N+1} & A_{-N+2} & A_{-N+3} & \ddots \\ & A_{-N} & A_{-N+1} & A_{-N+2} & \ddots \\ & & A_{-N} & A_{-N+1} & \ddots \\ & & & A_{-N} & \ddots \\ 0 & & & & \ddots \end{bmatrix}. \tag{5.29}$$

We may view this matrix as having the structure (4.3) with blocks of size Nm, corresponding to what we might call "macro levels":

$$P = \begin{bmatrix} \mathcal{B}_0 & \mathcal{B}_1 & \mathcal{B}_2 & \mathcal{B}_3 & \cdots \\ \mathcal{A}_{-1} & \mathcal{A}_0 & \mathcal{A}_1 & \mathcal{A}_2 & \cdots \\ & \mathcal{A}_{-1} & \mathcal{A}_0 & \mathcal{A}_1 & \ddots \\ & & \mathcal{A}_{-1} & \mathcal{A}_0 & \ddots \\ 0 & & & \ddots & \ddots \end{bmatrix} \tag{5.30}$$

where

$$\mathcal{A}_i = \begin{bmatrix} A_{iN} & A_{iN+1} & \cdots & A_{iN+N-1} \\ A_{iN-1} & A_{iN} & \ddots & \vdots \\ \vdots & \ddots & \ddots & A_{iN+1} \\ A_{iN-N+1} & \cdots & A_{iN-1} & A_{iN} \end{bmatrix}, \tag{5.31}$$

for $i = -1, 0, 1, \ldots$, where we assume that $A_j = 0$ if $j < -N$, and

$$\mathcal{B}_0 = \begin{bmatrix} B_0 & B_1 & \cdots & B_{N-1} \\ B_{-1} & A_0 & \cdots & A_{N-2} \\ \vdots & \vdots & \ddots & \vdots \\ B_{-N+1} & A_{-N+2} & \cdots & A_0 \end{bmatrix},$$

$$\mathcal{B}_i = \begin{bmatrix} B_{Ni} & B_{Ni+1} & \cdots & B_{Ni+N-1} \\ A_{Ni-1} & A_{Ni} & \cdots & A_{Ni+N-2} \\ \vdots & \vdots & \ddots & \vdots \\ A_{Ni-N+1} & A_{Ni-N+2} & \cdots & A_{Ni} \end{bmatrix}, \quad i > 0.$$

The matrix $\mathcal{G}_{\min}$ of first passage probabilities from one macro level to the one below may be characterized as being the minimal nonnegative solution of the matrix equation

$$\mathcal{X} = \sum_{i=-1}^{+\infty} \mathcal{A}_i \mathcal{X}^{i+1} \tag{5.32}$$

but, as noted in [42], it may also be directly related to the first passage probability matrices $G_1, \ldots, G_N$ between ordinary levels. In the remainder of this section we shall write I_n for the identity matrix of size n.

Theorem 5.15 *The matrix $\mathcal{G}_{\min}$ is given by $\mathcal{G}_{\min} = \mathcal{C}^N$, where*

$$\mathcal{C} = \begin{bmatrix} 0 & I_m & 0 & \ldots & 0 \\ 0 & 0 & I_m & \ddots & \vdots \\ \vdots & \ddots & \ddots & \ddots & 0 \\ 0 & \ldots & 0 & 0 & I_m \\ G_N & G_{N-1} & \ldots & \ldots & G_1 \end{bmatrix}. \tag{5.33}$$

Proof A macro level is a set of N successive levels, say $\{L+1, L+2, \ldots, L+N-1\}$ for some ℓ. The matrix $\mathcal{G}_{\min}$ is partitioned into blocks $(\mathcal{G}_{i,j})_{1 \le i,j \le N}$ and $\mathcal{G}_{i,j}$ is the matrix of first passage probabilities from level $L+i$ to $L-N+j$ before any of the levels $L-N+1$, $L-N+2$, …, L; it is independent of L, provided that $L > N$. We shall use the short-hand notation

$$\mathcal{G}_{i,j} = \mathrm{P}[L+i \rightsquigarrow L-N+j \text{ before } L-N+1, \ldots, L].$$

With this notation, we have that

$$G_j = \mathrm{P}[L+1 \rightsquigarrow L-N+j \text{ before } L-N+1, \ldots, L]$$

and we immediately see that $\mathcal{G}_{1j} = G_{N+1-j}$ for all j.

For $i \ge 2$, we decompose the paths from $L+i$ to $L-N+j$ into two sets, according to whether they go through level $L+1$ or not. Thus,

$$\begin{aligned} \mathcal{G}_{i,j} = {} & \mathrm{P}[L+i \rightsquigarrow L-N+j \text{ before } L-N+2, \ldots, L+1] \\ & + \mathrm{P}[L+i \rightsquigarrow L+1 \text{ before } L-N+2, \ldots, L+1] \\ & \times \mathrm{P}[L+1 \rightsquigarrow L-N+j \text{ before } L-N+1, \ldots L]. \end{aligned}$$

The reason for this decomposition is that, starting from level $L+i \ge L+2$, there are certain lower levels, like $L-N+1$, which cannot be reached without passing through one of the levels between $L-N+2$ and $L+1$. Because of the homogeneity in the transition probabilities, this may also be written as

$$\begin{aligned} \mathcal{G}_{i,j} &= \mathrm{P}[L+i-1 \rightsquigarrow L-N+j-1 \text{ before } L-N+1, \ldots, L] \\ & \quad + \mathrm{P}[L+i-1 \rightsquigarrow L \text{ before } L-N+1, \ldots, L] \\ & \quad \times \mathrm{P}[L+1 \rightsquigarrow L-N+j \text{ before } L-N+1, \ldots, L] \\ &= \mathcal{G}_{i-1,j-1} + \mathcal{G}_{i-1,N} \mathcal{G}_{1j}. \end{aligned}$$

If we denote by $\mathcal{G}_{i\bullet}$ the row $\begin{bmatrix} \mathcal{G}_{i1} & \mathcal{G}_{i2} & \ldots & \mathcal{G}_{iN} \end{bmatrix}$, the equation above may be read as $\mathcal{G}_{i\bullet} = \mathcal{G}_{i-1,\bullet}\mathcal{C}$, so that

$$\mathcal{G}_{\min} = \begin{bmatrix} \mathcal{G}_{i\bullet} \\ \mathcal{G}_{i\bullet}\mathcal{C} \\ \vdots \\ \mathcal{G}_{i\bullet}\mathcal{C}^{N-1} \end{bmatrix}, \tag{5.34}$$

from which one obtains, after simple but tedious algebraic manipulations, that $\mathcal{G}_{\min} = \mathcal{C}^N$. □

This theorem makes it very clear that to solve the system (5.25, 5.26) is equivalent to solving the reblocked system. Theorem 5.15 can be proved by using purely algebraic arguments, as a direct consequence of Theorem 3.24. In fact, from the latter theorem and from Theorem 5.14 we deduce the following result:

Theorem 5.16 *Let*

$$\mathcal{S}(z) = z^{-1}\mathcal{A}_{-1} + \mathcal{A}_0 + z\mathcal{A}_1 + \cdots .$$

The matrix Laurent power series $I_{mN} - \mathcal{S}(z)$ has the following weak canonical factorization:

$$I_{mN} - \mathcal{S}(z) = \left(I_{mN} - \sum_{i=0}^{+\infty} z^i \mathcal{A}_i^* \right) (I_{mN} - z^{-1}\mathcal{G}_{\min}),$$

where $\mathcal{A}_n^* = \sum_{i=n}^{+\infty} \mathcal{A}_i \mathcal{G}_{\min}^{i-n}$. Moreover, $\mathcal{G}_{\min}$ can be factorized as

$$\mathcal{G}_{\min} = \begin{bmatrix} I_m & & & 0 \\ -G_1 & I_m & & \\ \vdots & \ddots & \ddots & \\ -G_{N-1} & \dots & -G_1 & I_m \end{bmatrix}^{-1} \begin{bmatrix} G_N & G_{N-1} & \dots & G_1 \\ & \ddots & \ddots & \vdots \\ & & G_N & G_{N-1} \\ 0 & & & G_N \end{bmatrix}. \tag{5.35}$$

In closing this section, we briefly discuss the roots of $\alpha(z) = \det(zI_{mN} - \mathcal{A}(z))$, where $\mathcal{A}(z) = z\mathcal{S}(z)$. Observe that as pointed out in Section 3.3.3 we have

$$\mathcal{A}(z) = \begin{bmatrix} \varphi_0(z) & \varphi_1(z) & \dots & \varphi_{N-1}(z) \\ z\varphi_{N-1}(z) & \varphi_0(z) & \ddots & \vdots \\ \vdots & \ddots & \ddots & \varphi_1(z) \\ z\varphi_1(z) & \dots & z\varphi_{N-1}(z) & \varphi_0(z) \end{bmatrix}$$

where $\varphi_j(z) = \sum_{i=-1}^{+\infty} z^{i+1} A_{iN+j}$, $j = 0, \dots, N-1$. That is, $\mathcal{A}(z)$ is a z-circulant matrix.

Observe that according to Theorem 3.23, the matrix $\mathcal{A}(z)$ can be block diagonalized by means of a scaled block DFT and we have

$$\det(z^N I_{mN} - \mathcal{A}(z^N)) = \prod_{i=0}^{N-1} \det(z^N I_m - A(z\omega_N^i)). \tag{5.36}$$

The following theorem is a direct consequence of the above equation.

Theorem 5.17 *If the matrix power series $A(z)$ is analytic for $|z| < R$, $R > 1$, then the matrix power series $\mathcal{A}(z)$ is analytic for $|z| < R^N$. Moreover, if λ is a zero of* $\det(z^N I - A(z))$ *then λ^N is a zero of* $\det(zI - \mathcal{A}(z))$, *if ν is a zero of* $\det \mathcal{A}(z)$ *then the N-th roots of ν are zeros of* $\det a(z)$.

Proof Indeed $\mathcal{A}(z)$ is analytic for $|z| < 1$ since $\mathcal{A}(1)$ is stochastic. Assuming $A(z)$ analytic for $|z| < R$, from Theorem 3.23 we deduce that $\mathcal{A}(z^N)$ is analytic for $0 < |z| < R$. Since $\mathcal{A}(z)$ is a matrix power series, then is analytic for $|z| < R^N$. The property relating the zeros of $\det(zI - \mathcal{A}(z))$ and $\det(z^N I - A(z))$ follows from (5.36). □

Therefore, if w is a zero of $\det(zI_m - A(z))$ then w^N is a zero of $\det(zI_{mN} - \mathcal{A}(z))$ and, if the function $a(z) = \det(zI_m - A(z))$ has $\kappa > 1$ zeros of modulus 1, then, by Theorem 4.9, these zeros are the κth roots of 1. Hence, by reblocking the matrix P into $\kappa m \times \kappa m$ blocks, we obtain a different representation of the same M/G/1 problem such that the generating function is $\mathcal{A}(z)$ and $\det(zI_{\kappa m} - \mathcal{A}(z))$ has only one zero of modulus 1. This, however, does not simplify the quest for all the solutions of the matrix equation since $\mathcal{A}$ is no longer irreducible and falls outside the scope of Theorem 4.9.

Observe that, from (5.34) as well as (5.35), it follows that the first block row of $\mathcal{G}_{\min}$ defines all the block elements. It has been shown in [42] that a nonminimal solution $\mathcal{G}$ of the matrix equation (5.32) does not have necessarily the structure (5.33).

As a last comment, we also like to point out that the factorization (5.35) with (5.33) provides the UL factorization of the Nth power of a block companion matrix. This result is a special instance of a more general result known as Barnett factorization (see Theorem 3.25).

If the displacements are constrained in the upward direction, that is, if $A_{N+1} = A_{N+2} = \ldots = 0$, then we may adapt equations (5.17, 5.19) to the new circumstances, and write that

$$[A_0^* \; A_1^* \; \ldots \; A_N^*] \begin{bmatrix} I & & & 0 \\ -G_1 & I & & \\ \vdots & \ddots & \ddots & \\ -G_N & \ldots & -G_1 & I \end{bmatrix} = [A_0 \; A_1 \; \ldots \; A_N] \tag{5.37}$$

and

$$\begin{bmatrix} I - A_0^* & -A_1^* & \ldots & -A_N^* & & 0 \\ & I - A_0^* & -A_1^* & \ldots & -A_N^* & \\ & & \ddots & \ddots & \ddots & \ddots \\ 0 & & & & & \end{bmatrix} \begin{bmatrix} G_1 \\ G_2 \\ G_3 \\ \vdots \end{bmatrix} = \begin{bmatrix} A_{-1} \\ A_{-2} \\ A_{-3} \\ \vdots \end{bmatrix}. \tag{5.38}$$

Alternately, we may restructure the transition matrix into blocks of size Nm and deal with a G/M/1-type queue. Then we must study its rate matrix $\mathcal{R}_{\min}$, either directly, like in [42], or by duality. In any event, it appears that

$$\mathcal{R}_{\min} = \begin{bmatrix} 0 & \dots & 0 & R_N \\ I & \ddots & \vdots & R_{N-1} \\ & \ddots & 0 & \vdots \\ 0 & & I & R_1 \end{bmatrix}^N .$$

If the displacements are bounded in both directions, that is, if $A_{N_u+1} = A_{N_u+2} = \dots = 0$ and $A_{-N_d-1} = A_{-N_d-2} = \dots = 0$ for some N_u and N_d, then we may specialize (5.37, 5.38) even further and, assuming that $N_u < N_d$, obtain

$$\begin{bmatrix} A_0^* \; A_1^* \; \dots \; A_{N_u}^* \end{bmatrix} \begin{bmatrix} I & & & 0 \\ -G_1 & I & & \\ \vdots & \ddots & \ddots & \\ -G_{N_u} & \dots & -G_1 & I \end{bmatrix} = \begin{bmatrix} A_0 \; A_1 \; \dots \; A_{N_u} \end{bmatrix}$$

and

$$\begin{bmatrix} I - A_0^* & -A_1^* & \dots & -A_{N_u}^* & & 0 \\ & I - A_0^* & -A_1^* & & \ddots & \\ & & \ddots & \ddots & & -A_{N_u}^* \\ & & & & & \vdots \\ 0 & & & & & I - A_0^* \end{bmatrix} \begin{bmatrix} G_1 \\ G_2 \\ \vdots \\ G_{N_d} \end{bmatrix} = \begin{bmatrix} A_{-1} \\ A_{-2} \\ \vdots \\ A_{-N_d} \end{bmatrix},$$

(with obvious modifications if $N_u \geq N_d$).

Of course, we may also restructure the transition matrix into blocks of size N^*m, where $N^* = \max(N_u, N_d)$; in that case, the process may be considered to be a QBD, one of the family of processes which we analyze in the next section.

5.6 Quasi-birth–death processes

Recall that we indicated two ways of combining the structures of M/G/1 and G/M/1-type Markov chains, one of which has been dealt with in Section 5.4. The other consists in imposing the constraints of each, so that the process may change by one level at a time, thereby becoming skip-free in both directions. Such Markov chains are called quasi-birth-and-death (QBD) processes and their transition matrix is

$$P = \begin{bmatrix} B_0 & A_1 & & & 0 \\ A_{-1} & A_0 & A_1 & & \\ & A_{-1} & A_0 & A_1 & \\ & & A_{-1} & A_0 & \ddots \\ 0 & & & \ddots & \ddots \end{bmatrix} \tag{5.39}$$

where $A_{-1}, A_0, A_1 \in \mathbb{R}^{m \times m}$ and B_0, $B_1 \in \mathbb{R}^{m \times m}$, are nonnegative matrices such that $A_{-1} + A_0 + A_1$ and $B_0 + B_1$ are stochastic.

Observe that a QBD can be viewed like an M/G/1-type Markov chain but also like a G/M/1-type Markov chain. Therefore, Theorem 5.4 applies and the stationary distribution, when it exists, is of the matrix-geometric form (5.4). This we state below as a theorem.

Theorem 5.18 *If the QBD is positive recurrent, then*

$$\begin{aligned}
&\boldsymbol{\pi}_n^{\mathrm{T}} = \boldsymbol{\pi}_0^{\mathrm{T}} R_{\min}^n, \qquad \text{for } n \geq 0 \\
&\boldsymbol{\pi}_0^{\mathrm{T}}(B_0 + B_1 G_{\min}) = \boldsymbol{\pi}_0^{\mathrm{T}} \\
&\boldsymbol{\pi}_0^{\mathrm{T}}(I - R_{\min})^{-1}\mathbf{1} = 1
\end{aligned}$$

where $R_{\min}$ is the minimal nonnegative solution of the matrix-quadratic equation

$$X = X^2 A_{-1} + X A_0 + A_1. \tag{5.40}$$

and $G_{\min}$ is the minimal nonnegative solution of

$$X = A_{-1} + A_0 X + A_1 X^2. \tag{5.41}$$

Proof As stated above, the matrix-geometric structure is a consequence of Theorem 5.4, and (5.40) is just a rewriting of (5.3) under the present circumstances.

Now, if we think of a QBD as being a simple case of an M/G/1 queue, then Theorem 4.4 applies as well and we see that $\boldsymbol{\pi}_0$ is given by (4.10) where $B_0^* = B_0 + B_1 G_{\min}$ where $G_{\min}$ is the minimal nonnegative solution of (4.4) which is written here as (5.41). □

Both Theorems 4.7 and 5.3 apply and we restate them here as a corollary.

Corollary 5.19 *Assume that the QBD is irreducible and m finite, assume that the stochastic matrix $A = A_{-1}+A_0+A_1$ is irreducible, and define $\mu = \boldsymbol{\alpha}^{\mathrm{T}}(-A_{-1}+A_1)\mathbf{1}$, where $\boldsymbol{\alpha}$ is the stationary probability vector of A. The QBD is*

- *positive recurrent if and only if $\mu < 0$*
- *null recurrent if and only if $\mu = 0$,*
- *transient if and only if $\mu > 0$.*

We have seen in Section 5.3 that there exists a close connection between the matrix $R_{\min}$ and the matrix $\widetilde{G}_{\min}$ of the dual process. In the case of a QBD, there is a direct connection between the matrix $R_{\min}$ and the matrix $G_{\min}$ of the *same* process. This connection is made through the matrix A_0^* which plays here an important role, and which we now denote by U.

As we saw in Theorem 5.10, and in (4.8, 4.11) we have

$$\begin{aligned}
G_{\min} &= (I - U)^{-1} A_{-1} \\
R_{\min} &= A_1 (I - U)^{-1}
\end{aligned} \tag{5.42}$$

where $U = A_0 + A_1 G_{\min}$. Thus, $R_{\min}$, $G_{\min}$ and U are three matrices which carry exactly the same information about the stationary distribution of the stochastic process. It is easy to see that U is a solution of

$$X = A_0 + A_1(I - X)^{-1}A_{-1} \tag{5.43}$$

and one shows ([79,]) that U is actually the minimal nonnegative solution of that equation.

The function $S(z)$ associated with a QBD is the matrix Laurent polynomial

$$S(z) = z^{-1}A_{-1} + A_0 + zA_1 \tag{5.44}$$

and the generating function is the quadratic polynomial $A(z) = A_{-1}+zA_0+z^2A_1$ if the QBD is viewed as an M/G/1-type queue, or $\widehat{A}(z) = A_1 + zA_0 + z^2A_{-1}$ if we take the G/M/1 view. We may switch between the two views by means of the transformation $A(z) \Rightarrow z^2A(z^{-1}) = \widehat{A}(z)$ which reverses the order of the matrix coefficients of the polynomial $A(z)$. The same relation occurs for the determinants

$$a(z) = \det(zI - A(z)) \tag{5.45}$$

and

$$\widehat{a}(z) = \det(zI - \widehat{A}(z)). \tag{5.46}$$

In fact, $\widehat{a}(z)$ is the reversed polynomial of $a(z)$, that is, $\widehat{a}(z) = z^{2m}a(z^{-1})$.

If A_1 is nonsingular, then $a(z) = \det(zI - A(z))$ has degree $2m$, otherwise, its degree is $r < 2m$. In the latter case we may look at $a(z)$ as a polynomial of degree $2m$ having $2m-r$ zeros at infinity. In this way, if we denote by ξ_i, $i = 1, \ldots, 2m$, the zeros of $a(z)$, then the zeros of $z^{2m}a(z^{-1})$ are $1/\xi_i$, $i = 1, \ldots, 2m$, where we assume that null zeros are transformed into infinity, and vice versa.

Since $A(z)$ is a polynomial, it is meromorphic in the complex plane and the results of Section 4.4 apply. In particular, by adding zeros equal to infinity if necessary, we may directly state the following property (it is the local version of Theorems 4.9 and 4.24, as well as Theorem 5.5).

Theorem 5.20 *Assume that m is finite and that Conditions 5.1 and 5.2 hold. The polynomial $a(z)$ defined in* (5.45) *has*

- *$m - 1$ zeros in the open unit disk, one simple zero equal to 1, and m zeros outside the closed unit disk, if $\mu < 0$;*
- *$m - 1$ zeros in the open unit disk, one zero of multiplicity two equal to 1, and $m - 1$ zeros outside the closed unit disk, if $\mu = 0$;*
- *m zeros in the open unit disk, one simple zero equal to 1, and $m - 1$ zeros outside the closed unit disk, if $\mu > 0$.*

If the QBD is positive recurrent, then $G_{\min}$ is the unique solution of (5.41) with spectral radius equal to one, in the set of nonnegative matrices and $R_{\min}$ is the unique solution of (5.40) spectral radius less than one, in the set of nonnegative matrices. In all cases, the eigenvalues of $G_{\min}$ are the zeros of smallest modulus of $a(z)$ in the closed unit disk and the eigenvalues of $R_{\min}$ are the zeros of smallest modulus of $z^{2m}a(z^{-1})$ in the closed unit disk. In particular, the spectral radius of $R_{\min}$ is $1/\xi$ where $\xi = \min\{|z| : a(z) = 0, |z| > 1\}$. In light of Theorems 4.11, 4.12 one has that $1/\xi$ is the only eigenvalue of maximum modulus of $R_{\min}$ and it is simple.

5.7 Reductions to QBDs

We have defined QBDs as a family at the intersection of M/G/1 and G/M/1-type Markov chains. It is interesting to note that QBDs may be seen as encompassing the two classes, so that the theory of QBDs may be thought of as a unifying umbrella.

Let us first take on M/G/1-type queues with transition matrix (4.3). We replace the simple transitions of the form $(n,i) \to (n+k,j)$ by a more elaborate, virtual procedure, whereby a new phase j is chosen first, as well as a number k of steps upward, and only then is the level increased, one unit at a time, until level $n+k$ is finally reached. Graphically, we have

$$(n,i) \to (n;k,j) \to (n+1;k-1,j) \cdots \to (n+k-1;1,j) \to (n+k,j).$$

At each transition of this virtual sequence, one keeps track of the number of steps which remain until the final level is reached. In order to have homogeneous notation, we shall write (n,i) as $(n;0,i)$.

We define in this way a three-dimensional process $\{X_t, K_t, \varphi_t\}_{t=0,1,\ldots}$ on the state space $\mathbb{N}\times\mathbb{N}\times\{1,2,\ldots,m\}$ for which the pair (K_t,φ_t) constitutes the phase. We also write that K_t is the *sublevel*. The transition matrix of the new process is

$$P = \begin{bmatrix} \mathcal{B} & \mathcal{A}_1 & & 0 \\ \mathcal{A}_{-1} & \mathcal{A}_0 & \mathcal{A}_1 & \\ & \mathcal{A}_{-1} & \mathcal{A}_0 & \ddots \\ 0 & & \ddots & \ddots \end{bmatrix}, \tag{5.47}$$

with

$$\mathcal{B} = \begin{bmatrix} B_0 & B_1 & B_2 & \cdots \\ 0 & 0 & 0 & \cdots \\ 0 & 0 & 0 & \cdots \\ \vdots & \vdots & \vdots & \ddots \end{bmatrix}, \tag{5.48}$$

$$\mathcal{A}_1 = \begin{bmatrix} 0 & & & 0 \\ I & 0 & & \\ & I & 0 & \\ & & \ddots & \ddots \\ 0 & & & \end{bmatrix}, \tag{5.49}$$

$$\mathcal{A}_0 = \begin{bmatrix} A_0 & A_1 & A_2 & \cdots \\ 0 & 0 & 0 & \cdots \\ 0 & 0 & 0 & \cdots \\ \vdots & \vdots & \vdots & \ddots \end{bmatrix}, \tag{5.50}$$

and

$$\mathcal{A}_{-1} = \begin{bmatrix} A_{-1} & 0 & 0 & \cdots \\ 0 & 0 & 0 & \cdots \\ 0 & 0 & 0 & \cdots \\ \vdots & \vdots & \vdots & \ddots \end{bmatrix}. \tag{5.51}$$

Theorem 5.21 *The matrix $\mathcal{G}_{\min}$ of the QBD with blocks (5.49–5.51) has the structure*

$$\mathcal{G}_{\min} = \begin{bmatrix} G_{\min} & 0 & 0 & \cdots \\ G_{\min}^2 & 0 & 0 & \cdots \\ G_{\min}^3 & 0 & 0 & \cdots \\ \vdots & \vdots & \vdots & \ddots \end{bmatrix}, \tag{5.52}$$

where $G_{\min}$ is the matrix of first passage probabilities from states of the form $(n, 0, i)$ to $(n-1, 0, j)$.

Proof Whenever the process moves to a lower level, the sublevel of the new state is necessarily 0, by (5.51). Thus, we have

$$\mathcal{G}_{\min} = \begin{bmatrix} G_0 & 0 & 0 & \cdots \\ G_1 & 0 & 0 & \cdots \\ G_2 & 0 & 0 & \cdots \\ \vdots & \vdots & \vdots & \ddots \end{bmatrix},$$

where the matrices $G_k, k \geq 0$, record the transition probability to level $n-1$, starting from the sublevel k in level n. For $k = 0$, this is $G_{\min}$ by definition. If $k \geq 1$, the process first reaches up from (n, k, i) to $(n+k, 0, i)$, then goes down k times by one level, so that $G_k = G_{\min}^k$. □

By Theorem 5.18, $\mathcal{G}_{\min}$ is the minimal nonnegative solution of the equation $G = \mathcal{A}_{-1} + \mathcal{A}_0 G + \mathcal{A}_1 G^2$. It is a simple matter to verify that it is equivalent to the property that $G_{\min}$ is the minimal nonnegative solution of (4.4). Furthermore, if we decompose each subvector $\boldsymbol{\pi}_n$ by the sublevel, so that $\boldsymbol{\pi}_n = \boldsymbol{\pi}_{n0}, \boldsymbol{\pi}_{n1}, \boldsymbol{\pi}_{n2}, \ldots$, one easily sees that $\boldsymbol{\pi}_{n0}$ satisfies the system (4.9): one applies the matrix-geometric property $\boldsymbol{\pi}_n = \boldsymbol{\pi}_{n-1} \mathcal{R}_{\min}$ with $\mathcal{R}_{\min} = \mathcal{A}_1 (I - \mathcal{A}_0 - \mathcal{A}_1 \mathcal{G}_{\min})^{-1}$, and uses the structure of $\mathcal{G}_{\min}$. For details, see [79, Theorem 13.1.5].

An important issue is to figure out if Condition 4.20 is satisfied by the QBD process (5.47) if the same condition is satisfied by the original M/G/1 Markov chain with transition matrix (4.3). This property is proved in the following

Theorem 5.22 *If the M/G/1 process* (4.3) *is such that Condition 4.20 is satisfied, then the QBD process* (5.47) *also satisfies that condition.*

Proof By assumption, the bi-infinite M/G/1-type Markov chain (4.24) on $\mathbb{Z} \times \{1, 2, \ldots, m\}$ has one final class and every other state is on a path to the final class; that final class is of the form $\mathbb{Z} \times S$, with $S \subseteq \{1, 2, \ldots, m\}$.

The state space of the bi-infinite QBD is $\mathbb{Z} \times \mathbb{N} \times \{1, 2, \ldots, m\}$. We show that there exists a set $\mathbb{Z} \times \mathcal{S}$ of states which communicate and that every other state is on a path to $\mathbb{Z} \otimes \mathcal{S}$. This will prove that there exists a unique final class which contains $\mathbb{Z} \times \mathcal{S}$ and, possibly, some other states as well.

We define

$$\mathcal{S} = \mathbb{N} \times S \setminus \{(k,j) : \quad (A_{k'})_{i,j} = 0, \forall i, \forall k' \geq k\}.$$

This means that we remove from consideration all the states (n,k,j) which cannot be reached either by a transition of the form $(n,0,i)$ to (n,k,j) or $(n-1,k+1,j)$ to (n,k,j). We now prove that the states in $\mathbb{Z} \times \mathcal{S}$ all communicate. We proceed in four steps.

Step 1. The set $\mathcal{S}$ obviously contains all the states $(0,i)$ with $i \in S$.

Step 2. For all $i,j \in S$, all n and n', there is a path from the state $(n,0,i)$ to the state $(n',0,j)$, which we write as $(n,0,i) \rightsquigarrow (n',0,j)$. This is because the state (n,i) of the M/G/1 Markov chain is identified with the state $(n,0,i)$ of the QBD and, by assumption, $(n,i) \rightsquigarrow (n',j)$ for all $i,j \in S$, all n and n'.

Step 3. Assume that (k,j) is in $\mathcal{S}$, with $k > 0$. For all $i \in S$, all n and n', $(n,0,i) \rightsquigarrow (n',k,j)$. To see this, we note that there exists some l and some $k' \geq k$ such that $(A_{k'})_{l,j} \neq 0$, which shows that there exists a path

$$(n,0,i) \rightsquigarrow (n'-k'+k,0,l) \rightarrow (n'-k'+k,k,j) \rightsquigarrow (n',k,j)$$

where the symbol $\rightarrow$ indicates a direct transition in one jump.

Step 4. For all $(k,j) \in \mathcal{S}$, all $i \in S$, all n and n', there is a path

$$(n,k,j) \rightsquigarrow (n+k,0,j) \rightsquigarrow (n',0,i)$$

where the path from (n,k,j) to $(n+k,0,j)$ is obvious and the one from $(n+k,0,j)$ to $(n',0,i)$ is justified in step 2 above.

At this stage we have proved that all the states in $\mathbb{Z} \times \mathcal{S}$ communicate. To conclude the proof, we need to show that for all $(k,j) \notin \mathcal{S}$, there exists a path from (n,k,j) to some state in $\mathbb{Z} \times \mathcal{S}$. This is done as follows:

$$(n,k,j) \rightsquigarrow (n+k,0,j) \rightsquigarrow (n',0,i),$$

for all n' and all $i \in S$. The first part of the path is obvious, and the existence of the second part is guaranteed by the fact that the states $(n+k,0,j)$ and $(n',0,i)$ are respectively identified with the states $(n+k,j)$ and (n',i) of the M/G/1 Markov chain and by Condition 4.20. □

In the case of a G/M/1-type queue with transition matrix (5.1), we proceed in a similar manner and we enlarge the state to (X, K, φ), where K is the number of steps which need to be taken in the downward direction. There is a difference with the previous situation, however: K must always be at most equal to X since the level cannot become negative. Thus, the transition matrix is

$$P = \begin{bmatrix} \mathcal{A}_0^0 & \mathcal{A}_1^0 & & 0 \\ \mathcal{A}_{-1}^1 & \mathcal{A}_0^1 & \mathcal{A}_1^1 & \\ & \mathcal{A}_{-1}^2 & \mathcal{A}_0^2 & \ddots \\ 0 & & \ddots & \ddots \end{bmatrix}, \tag{5.53}$$

where $\mathcal{A}_0^i$, for $i \geq 0$, is a matrix in $\mathbb{R}^{m(i+1)\times m(i+1)}$ and the other blocks are rectangular with appropriate dimensions. One has

$$\mathcal{A}_0^0 = B_0,$$

$$\mathcal{A}_0^i = \begin{bmatrix} A_{-1} & \dots & A_{-i+1} & B_{-i} \\ 0 & \dots & 0 & 0 \\ 0 & \dots & 0 & 0 \\ \vdots & \vdots & \vdots & \vdots \end{bmatrix}, \qquad \text{for } i \geq 1,$$

$$\mathcal{A}_1^i = \begin{bmatrix} A_1 & 0 & \dots & 0 \\ 0 & 0 & \dots & 0 \\ 0 & 0 & \dots & 0 \\ \vdots & \vdots & & \vdots \end{bmatrix}, \qquad \text{for } i \geq 0,$$

and

$$\mathcal{A}_{-1}^i = \begin{bmatrix} 0 & & & 0 \\ I & 0 & & \\ & I & \ddots & \\ & & \ddots & 0 \\ 0 & & & I \end{bmatrix}, \qquad \text{for } i \geq 1.$$

Observe that the transition probabilities in (5.53) are level-dependent, that is, the probability of moving from some state (n, k, i) to some other state is given by the matrices $\mathcal{A}_{-1}^n$, $\mathcal{A}_0^n$ and $\mathcal{A}_{-1}^n$, which depend on n. The theory of level-dependent QBDs is more involved that than presented in Section 5.6 and does not, in general, give rise to the same elegant equations.

We do not pursue the matter further and refer the reader to [79, Section 13.2], where the connection to Theorem 5.4 is given in full detail.

5.8 Tree-like processes

Let $\{(Y_n, \varphi_n) : n \geq 0\}$ be a discrete-time bivariate Markov process in which the values of the random variables Y_n are represented by the nodes of a d-ary tree, and the random variables φ_n take integer values in $\{1, \dots, m\}$. A d-ary tree is a tree for which each node has d children, where $d \geq 2$ is an integer.

Each node is represented by a string of integers taking values in $\{1, \dots, d\}$. If $J = (j_1, j_2, \dots, j_\ell)$ is one such string, its length ℓ is denoted by $|J|$ and its children by $J + k = (j_1, j_2, \dots, j_\ell, k)$, with $1 \leq k \leq d$. The root is represented by $\emptyset$, the empty string of length 0. Thus, the random variables (Y_n, φ_n) take their values in the state space $\mathcal{S}$ where

$$\mathcal{S} = \{(j_1, \dots, j_\ell; i) : \ell \geq 0, 1 \leq j_1, \dots, j_\ell \leq d, 1 \leq i \leq m\}.$$

The state space is partitioned into nodes

$$\mathcal{N}_J = \{(j_1, \dots, j_\ell; i) : 1 \leq i \leq m\}$$

where $J = (j_1, j_2, \ldots, j_\ell)$. We also partition $\mathcal{S}$ into levels: for fixed ℓ,

$$\mathcal{S}_\ell = \{(j_1, \ldots, j_\ell; i) : 1 \le j_1, \ldots, j_\ell \le d, 1 \le i \le m\}$$

is the reunion of all the nodes represented by strings of length ℓ. The first, or root level, comprises the unique node

$$\mathcal{N}_\emptyset = \{i : 1 \le i \le m\}.$$

The possible transitions are as follows:

1. within a node, from (J, i) to (J, i') with probability $(B_j)_{i,i'}$ where j is the rightmost integer in J;
2. within the root node, from i to i' with probability $(B_0)_{i,i'}$;
3. between a node and one of its children, from (J, i) to $(J + k, i')$ with probability $(A_k)_{i,i'}$, for $1 \le k \le d$;
4. between a node and its parent, from $(J + k, i)$ to (J, i') with probability $(D_k)_{i,i'}$, for $1 \le k \le d$.

Here, A_k, B_k and D_k are $m \times m$ matrices, moreover, $B_0 + A_1 + \cdots + A_d$ and $B_i + D_i + A_1 + \cdots + A_d$, $i = 1, \ldots, d$, are stochastic.

With these assumptions, it is clear that the transition graph between *nodes* is a tree, which is why we call these *tree-like processes*. In their full generality, tree-like processes allow a fifth type of transitions:

(5a) between a node at level ℓ and any node in the levels 0 to ℓ, for all ℓ

or

(5b) between a node and any of its descendants

but we do not consider this type of transitions here. Instead, we make the further simplifying assumption that

$$B_1 = B_2 = \cdots = B_d = B;$$

that is, we assume that, except at the root node, transition probabilities *within* a node do not depend on the specific node.

The structure of the matrix generator of this Markov process depends on the ordering of the states. There are two natural ways of ordering the nodes: level by level or lexicographically. If we enumerate the nodes level by level, then we recognize that the Markov process has the structure of a nonhomogeneous QBD.

It is, however, more fruitful to use the lexicographical order: $\mathcal{N}_\emptyset$ comes first, then $\mathcal{N}_1$ and all the nodes $\mathcal{N}_J$ for which the leftmost integer of J is 1, followed by $\mathcal{N}_2$ and all the nodes $\mathcal{N}_J$ for which the leftmost integer of J is 2, and so on up to $\mathcal{N}_d$ and all its descendants. With this ordering, the generator $Q = P - I$ (where P is the transition matrix of the Markov chain) is

$$Q = \begin{bmatrix} C_0 & \Lambda_1 & \Lambda_2 & \dots & \Lambda_d \\ V_1 & W & 0 & \dots & 0 \\ V_2 & 0 & W & \ddots & \vdots \\ \vdots & \vdots & \ddots & \ddots & 0 \\ V_d & 0 & \dots & 0 & W \end{bmatrix}, \tag{5.54}$$

where $\Lambda_1, \dots, \Lambda_d$ are matrices with m rows and infinitely many columns, given by

$$\Lambda_i = \left[A_i \; 0 \; 0 \; \dots \right], \qquad \text{for } 1 \le i \le d,$$

the matrices $V_1, \dots, V_d$ have m columns and infinitely many rows and are given by

$$V_i = \begin{bmatrix} D_i \\ 0 \\ 0 \\ \vdots \end{bmatrix}, \qquad \text{for } 1 \le i \le d,$$

the infinite matrix W is recursively defined by

$$W = \begin{bmatrix} C & \Lambda_1 & \Lambda_2 & \dots & \Lambda_d \\ V_1 & W & 0 & \dots & 0 \\ V_2 & 0 & W & \ddots & \vdots \\ \vdots & \vdots & \ddots & \ddots & 0 \\ V_d & 0 & \dots & 0 & W \end{bmatrix}, \tag{5.55}$$

and $C_0 = B_0 - I$, where B_0 is an $m \times m$ substochastic matrix.

The nature of this matrix can be better understood if we introduce the set of multi-indices

$$\mathcal{S} = \bigcup_{\ell=0}^{\infty} \mathcal{S}_\ell,$$
$$\mathcal{S}_0 = \{1, 2, \dots, m\},$$
$$\mathcal{S}_\ell = \{(j_1, \dots, j_\ell; i) : 1 \le j_1, \dots, j_\ell \le d, 1 \le i \le m\}, \; \ell = 1, 2, \dots.$$

In fact, W is a linear operator defined on the linear space $\ell^1(\mathcal{S})$ made up by all the multi-index sequences $\boldsymbol{x} = (x_k)_{k \in \mathcal{S}}$, $x_k \in \mathbb{R}$, having bounded 1-norm, i.e., $||\boldsymbol{x}|| = \sum_{k \in \mathcal{S}} |x_k| < \infty$. Here, the ordering of the components of $\boldsymbol{x}$, which provides the particular structure (5.55) to the matrix W, is lexicographic. That is, $(j_1, \dots, j_\ell; i) < (j'_1, \dots, j'_{\ell'}; i')$ if either $\ell < \ell'$ or there exists h, $1 \le h \le \ell$ such that $j_h < j'_h$, $j_q = j'_q$ for $q = 1, \dots, h-1$, or $j_q = j'_q$ for $q = 1, \dots, \ell$ and $i < i'$. In this way the multi-index sequence $\boldsymbol{x} = (x_k)_{k \in \mathcal{S}}$ can be naturally partitioned as

$$\boldsymbol{x} = (\boldsymbol{x}^{(0)}, \boldsymbol{x}^{(1)}, \dots, \boldsymbol{x}^{(d)}), \tag{5.56}$$

where $\boldsymbol{x}^{(0)}$ is indexed by the indices of $\mathcal{S}$ of length $\ell = 0$, i.e., $\boldsymbol{x}^{(0)} = (x_i)_{i \in \mathcal{S}_0}$; the sequence $\boldsymbol{x}^{(q)}$, for $q = 1, \dots, d$ is indexed by the indices $k = (j_1, \dots, j_\ell; i) \in \mathcal{S}$ of length $\ell \ge 1$ with the first component $j_1 = q$.

Observe that in each row and in each column of W there is a finite number of nonzero elements. From this property it follows that the 1-norm of W is finite. In order to prove this, consider a vector $\boldsymbol{x} = (x_k)_{k\in\mathcal{S}}$ such that $||\boldsymbol{x}||_1 = \sum_{k\in\mathcal{S}} |x_k| = 1$, and partition it as $\boldsymbol{x} = (\boldsymbol{x}^{(0)}, \boldsymbol{x}^{(1)}, \ldots, \boldsymbol{x}^{(d)})$, as described in (5.56). From the definition of 1-norm of $\boldsymbol{x}$ it follows that $1 = ||\boldsymbol{x}||_1 = \sum_{i=0}^{d} ||\boldsymbol{x}^{(i)}||_1$. Now let $\boldsymbol{y} = W\boldsymbol{x}$ and partition $\boldsymbol{y}$ as $\boldsymbol{y} = (\boldsymbol{y}^{(0)}, \boldsymbol{y}^{(1)}, \ldots, \boldsymbol{y}^{(d)})$, similarly to (5.56). Then

$$\begin{aligned} \boldsymbol{y}^{(0)} &= C\boldsymbol{x}^{(0)} + \textstyle\sum_{i=1}^{d} A_i \boldsymbol{x}^{(i,0)} \\ \boldsymbol{y}^{(i)} &= W\boldsymbol{x}^{(i)} + V_i \boldsymbol{x}^{(0)}, \quad i = 1, \ldots, d \end{aligned} \tag{5.57}$$

where $\boldsymbol{x}^{(i,0)}$ is the first block in the partitioning (5.56) of $\boldsymbol{x}^{(i)}$. Taking the norms in both sides of (5.57) we get

$$\begin{aligned} &||\boldsymbol{y}^{(0)}||_1 \le \gamma(||\boldsymbol{x}^{(0)}||_1 + \textstyle\sum_{i=1}^{d} ||\boldsymbol{x}^{(i,0)}||_1) \\ &||\boldsymbol{y}^{(i)}||_1 \le \gamma||\boldsymbol{x}^{(0)}||_1 + ||W\boldsymbol{x}^{(i)}||_1, \quad i = 1, \ldots, d, \end{aligned}$$

where γ is a positive constant depending on C, D_i and A_i, $i = 1, \ldots, d$. Therefore, we have

$$||W\boldsymbol{x}||_1 = ||\boldsymbol{y}^{(0)}||_1 + \cdots + ||\boldsymbol{y}^{(d)}||_1 \le (d+1)\gamma||\boldsymbol{x}^{(0)}||_1 + \gamma\sum_{i=1}^{d} ||\boldsymbol{x}^{(i,0)}||_1 + \sum_{i=1}^{d} ||W\boldsymbol{x}^{(i)}||_1$$

which recursively yields

$$||W\boldsymbol{x}||_1 \le (d+2)\gamma(||\boldsymbol{x}^{(0)}||_1 + \sum_{i=1}^{d} ||\boldsymbol{x}^{(i,0)}||_1 + \sum_{i,j=1}^{d} ||\boldsymbol{x}^{(i,j,0)}||_1 + \cdots) = \gamma(d+2)$$

where $\boldsymbol{x}^{(i,j,0)}$ is the first block component of the j-th block component of $\boldsymbol{x}^{(i)}$ in the recursive block decomposition (5.56). The latter inequality implies that W is a linear operator on $\ell^1(\mathcal{S})$ with bounded 1-norm.

The computation of the stationary distribution of the Markov process, i.e., the infinite nonnegative vector $\boldsymbol{\pi} = (\pi_i)_{i\in\mathcal{S}}$ such that $\boldsymbol{\pi}^{\mathrm{T}} Q = \boldsymbol{0}^{\mathrm{T}}$, $\sum_{i\in\mathcal{S}} \pi_i = 1$, is reduced to computing the UL factorization of (5.54) and this computation can be ultimately reduced to solving

$$X + \sum_{1\le i\le d} A_i X^{-1} D_i = C, \tag{5.58}$$

where we assume that

1. $C = B - I$, and B is substochastic;
2. A_i and D_i have nonnegative elements;
3. the matrices $I + C + D_i + A_1 + \cdots + A_d$, $i = 1, \ldots, d$, are stochastic.

The theorem below has been proved by different means in [122], [121], and [79]. In particular, the proof in [79] proceeds along purely probabilistic arguments. We provide here some remarks based on the recursive structure of W in (5.55).

Theorem 5.23 *Assume that the tree-like process with generator (5.54) is positive recurrent. Denote by $\boldsymbol{\pi}_J$ the subvector of stationary probabilities associated with the states in node $\mathcal{N}_J$.*

One has that

$$\boldsymbol{\pi}_J = \boldsymbol{\pi}_\emptyset R_{j_1} \cdots R_{j_\ell} \tag{5.59}$$

if $J = (j_1, \ldots, j_\ell)$, where

$$R_i = A_i(-S)^{-1}, \qquad \text{for } 1 \le i \le d,$$

and $S = T - I$, where $T_{k,k'}$ is the probability of moving from the state (J, k) to the state (J, k') at a later time, without visiting the node $\mathcal{N}_J$ or its parent in between, independently of $J \neq \emptyset$. The matrix S is the minimal solution of the nonlinear matrix equation

$$X + \sum_{i=1}^{d} A_i X^{-1} D_i = C, \tag{5.60}$$

with respect to the componentwise ordering. The vector $\boldsymbol{\pi}_\emptyset$ is the solution of

$$\boldsymbol{\pi}_\emptyset \left[C_0 + \sum_{i=1}^{m} A_i(-S)^{-1} D_i \right] = \mathbf{0} \tag{5.61}$$

normalized by

$$\boldsymbol{\pi}_\emptyset \sum_{n \ge 0} \left[\sum_{1 \le i \le d} R_i \right]^n \mathbf{1} = 1.$$

A first remark related to the previous theorem comes from the UL factorization of W as stated by the following

Theorem 5.24 *Let S be a nonsingular matrix. Then matrices L, U recursively defined by*

$$U = \begin{bmatrix} S & \Lambda_1 & \Lambda_2 & \ldots & \Lambda_d \\ 0 & U & 0 & \ldots & 0 \\ 0 & 0 & U & \ddots & \vdots \\ \vdots & \vdots & \ddots & \ddots & 0 \\ 0 & 0 & \ldots & 0 & U \end{bmatrix}, \qquad L = \begin{bmatrix} I & 0 & 0 & \ldots & 0 \\ Y_1 & L & 0 & \ldots & 0 \\ Y_2 & 0 & L & \ddots & \vdots \\ \vdots & \vdots & \ddots & \ddots & 0 \\ Y_d & 0 & \ldots & 0 & L \end{bmatrix},$$

with $Y_i = \begin{bmatrix} S^{-1}D_i \\ O \\ \vdots \end{bmatrix}$, for $i = 1, \ldots, d$, represent linear operators in $\ell^1(\mathcal{S})$ with bounded 1-norm. Moreover, S is a solution of (5.60) if and only if $W = UL$.

Proof The boundness of the operators L and U can be proved by following the same argument used for proving the boundness of W. The remaining part of the proposition can be proved by direct inspection. □

The matrices L and U have formal inverses recursively defined by

$$L^{-1} = \begin{bmatrix} I & 0 & 0 & \dots & 0 \\ -L^{-1}Y_1 & L^{-1} & 0 & \dots & 0 \\ -L^{-1}Y_2 & 0 & L^{-1} & \ddots & \vdots \\ \vdots & \vdots & \ddots & \ddots & 0 \\ -L^{-1}Y_d & 0 & \dots & 0 & L^{-1} \end{bmatrix}$$

and

$$U^{-1} = \begin{bmatrix} S^{-1} & -S^{-1}\Lambda_1 U^{-1} & -S^{-1}\Lambda_2 U^{-1} & \dots & -S^{-1}\Lambda_d U^{-1} \\ 0 & U^{-1} & 0 & \dots & 0 \\ 0 & 0 & U^{-1} & \ddots & \vdots \\ \vdots & \vdots & \ddots & \ddots & 0 \\ 0 & 0 & \dots & 0 & U^{-1} \end{bmatrix},$$

respectively. From the above relations it follows that L^{-1} and U^{-1} have finite elements, however, they may have an unbounded ℓ^1-norm even under the assumptions of stochasticity and positive recurrence of the tree-like process.

Once the matrix S is known, the stationary probability vector can be computed by using the UL factorization of W. In order to show this, we rewrite as follows the matrix Q given in (5.54):

$$Q = \begin{bmatrix} C_0 & \Lambda_0 \\ V_0 & W_0 \end{bmatrix},$$

and we decompose $\boldsymbol{\pi}$ as $\boldsymbol{\pi} = (\boldsymbol{\pi}_\emptyset, \boldsymbol{\pi}_\star)$, where $\boldsymbol{\pi}_*$ is the stationary probability vector for all the nodes below the root. The equation $\boldsymbol{\pi}Q = \mathbf{0}$ immediately leads to

$$\boldsymbol{\pi}_\emptyset \left[C_0 + \Lambda_0(-W_0)^{-1}V_0\right] = \mathbf{0} \tag{5.62}$$

$$\boldsymbol{\pi}_\star = \boldsymbol{\pi}_\emptyset \Lambda_0(-W_0)^{-1}. \tag{5.63}$$

Next we decompose $\boldsymbol{\pi}_\star$ as $\boldsymbol{\pi}_\star = (\boldsymbol{\pi}_{1\star}, \dots, \boldsymbol{\pi}_{d\star})$ where $\boldsymbol{\pi}_{i\star}$ is the stationary probability subvector of all the nodes $\mathcal{N}_J$ for which the string J begins with the integer i; we find from (5.63) that

$$\boldsymbol{\pi}_{i\star} = \boldsymbol{\pi}_\emptyset \Lambda_i(-W)^{-1} = \boldsymbol{\pi}_\emptyset \Lambda_i L^{-1}(-U)^{-1},$$

that is, $\boldsymbol{\pi}_{i\star} = \boldsymbol{\pi}_\emptyset \Lambda_i(-U)^{-1}$ and we readily find that

$$\left[\boldsymbol{\pi}_i\ \boldsymbol{\pi}_{i1\star} \dots \boldsymbol{\pi}_{id\star}\right] = \boldsymbol{\pi}_\emptyset R_i \left[I\ \Lambda_1(-U)^{-1} \dots \Lambda_d(-U)^{-1}\right]$$

where $\boldsymbol{\pi}_{ij\star}$ is the stationary probability vector for all the nodes $\mathcal{N}_J$ for which the string J begins with the integers i and j.

This shows that $\pi_i = \pi_\emptyset R_i$ for all i, and it suffices to iterate these calculations to prove (5.59) for all strings J. It is a simple matter to verify that (5.61) and (5.62) are identical equations.

We should emphasize that, although we deal with discrete-time Markov chains, nevertheless our results are directly adapted to continuous-time Markov processes; in that case, the matrices $B + D_i + A_1 + \cdots + A_d$ are conservative, stable matrices, and we set $C = B$ in (5.58).

5.9 Bibliographic notes

A systematic treatment of G/M/1-type Markov chains is made in [91]. The duality property between M/G/1 and G/M/1 queues has been analyzed in [5; 99]. A direct analysis of the spectral properties in both the M/G/1 and G/M/1 can be found in [40; 42]. The case of Toeplitz-like transition has been treated in [51; 52], while the case of limited displacements has been analyzed in [42]. Concerning QBD processes we refer the reader to the book [79]. The reduction of M/G/1 and G/M/1 Markov chains to the QBD process has been investigated in [100; 79]

Before concluding this chapter, we briefly discuss the case of semi-infinite blocks. We saw in Section 5.7 one example of a QBD for which the phase takes infinitely many values.

For such processes, it is known that the structure of the stationary distribution, assuming that it exists, is just as in the case of a finite set of phases. The article [112] deals with G/M/1-type queues on an arbitrary set of phases (continuous as well as discrete). Other authors have considered special cases; for instance, [101] shows that the theory applies to product-form stochastic networks.

It seems that conditions for ergodicity are very much dependent on each particular application and we have very few results equivalent to Theorems 4.7 or 5.3 when the number of phases is infinite. Even fewer results exist about the properties of the generating function of such processes, and the solutions of the matrix equations.

Typical applications of tree-like processes have been investigated by Yeung and Sengupta [122] (single server queues with LIFO service discipline), He and Alfa [60], [61] (an arriving customer interrupts the service in progress), and by Takine, Sengupta and Yeung [110] (an arriving customer is placed at the head of the queue but does not preempt the server). Van Houdt and Blondia [113] use tree-like processes to evaluate a medium access control protocol with an underlying stack structure. A brief general introduction is given in Latouche and Ramaswami [79, Chapter 14]. Further analysis is performed in [17] and [114].

Part III

Algorithms

6

FUNCTIONAL ITERATIONS

6.1 Basic concepts

In this chapter we analyze the algorithmic issues concerning the solution by means of functional iterations of matrix equations of the kind

$$X = \sum_{i=-1}^{+\infty} A_i X^{i+1} \tag{6.1}$$

where A_i, $i \geq -1$, and X are $m \times m$ matrices. By following the notation of Chapter 4 we set

$$A(z) = zS(z), \qquad S(z) = \sum_{i=-1}^{+\infty} z^i A_i.$$

For any solution G of (6.1) one has $G = A(G)$, where $A(G) = \sum_{i=-1}^{+\infty} A_i G^{i+1}$. Here and hereafter, given a matrix power series $V(z) = \sum_{i=0}^{+\infty} z^i V_i$, we denote by $V(X) = \sum_{i=0}^{+\infty} V_i X^i$ the extension of $V(z)$ to a function of the matrix variable X.

In general, given a matrix valued function $F(X)$, we call a *fixed point* of $F(X)$ any matrix W such that $W = F(W)$. According to this definition, any solution G of (6.1) is a fixed point of the function $A(X)$.

In this chapter, we assume that the matrix A_i is nonnegative for $i = -1, 0, \ldots$, and that $\sum_{i=-1}^{+\infty} A_i$ is stochastic. Here, the matrices A_i, $i = -1, 0, \ldots$, define the block Toeplitz part of the transition matrix (4.3) associated with an M/G/1-type Markov chain. We also assume that Condition 4.20 is satisfied by the Markov chain with transition matrix (4.3). This latter assumption ensures that the matrix $A = \sum_{i=-1}^{+\infty} A_i$ has only one final class and that $z = 1$ is the only zero of the function $a(z) = \det(zI - A(z))$ on the unit circle.

The computation of the minimal nonnegative solution $G_{\min}$ of (6.1) is fundamental in the numerical solution of M/G/1 type Markov chains, as explained in Chapter 4. We recall that if the drift μ of (4.17) is nonpositive then $G_{\min}$ is stochastic, while if $\mu > 0$ then $G_{\min}$ is substochastic.

The problem of approximating a fixed point W of a matrix function $F(X)$ is naturally solved by means of functional iteration methods.

Given a matrix valued function $F(X)$ the *functional iteration* or *fixed point iteration* defined by $F(X)$ is the set of all the matrix sequences $\{X_n\}_{n\in\mathbb{N}}$, defined by

$$X_n = F(X_{n-1}), \quad n = 1, 2, \ldots, \tag{6.2}$$

where X_0 is any matrix. It is immediate to verify that, if $F(X)$ is a continuous function and if, for a given X_0, there exists the matrix $W = \lim_n X_n$ then $F(W) = W$, that is, W is a fixed point of $F(X)$.

Our goal is to design and analyze suitable functional iterations which provide sequences converging to the *minimal* solution $G_{\min}$ or to the stochastic solution $G_{\rm sto}$ of the matrix equation (6.1). A natural choice is given by $F(X) = \sum_{i=-1}^{+\infty} A_i X^{i+1}$, whose fixed points satisfy the equation (6.1). Other functions can be obtained by formally manipulating the matrix equation (6.1).

Before introducing and analyzing specific functional iterations, we need to provide some concepts and definitions on which our subsequent analysis is based. Given the sequence $\{X_n\}_{n\geq 0}$ such that $\lim_n X_n = W$, it is useful to introduce the approximation error

$$E_n = W - X_n$$

and its infinity-norm

$$e_n = \|E_n\|_\infty$$

which represents the distance of the approximation X_n to W at the step n.

Based on the definition of e_n we may introduce a measure of the convergence speed of a sequence of matrices. Consider the limit, if it exists,

$$\gamma = \lim_n \frac{e_{n+1}}{e_n} \tag{6.3}$$

which expresses an asymptotic estimate of the error reduction at each step of the iteration. Observe that, since the sequence converges, then $\gamma \leq 1$. We call γ the *convergence rate* of the sequence. Indeed, the smaller is γ, the faster is the convergence speed of the sequence. In particular, if $\gamma = 1$, the convergence becomes extremely slow; if $\gamma = 0$ we have a very favorable situation where the convergence is very fast. We say that the convergence of the sequence $\{X_n\}_{n\geq 0}$ is *sublinear* if $\gamma = 1$, is *linear* if $0 < \gamma < 1$, and is *superlinear* if $\gamma = 0$. For sequences which have a superlinear convergence, if for a real $p > 1$ there exists finite and nonzero $\lim_n e_{n+1}/e_n^p$, we say that the *convergence has order of convergence* p. If $p = 2$ we say also that the convergence is *quadratic*.

The following example shows, in the scalar case, the different behavior of sublinear, linear and superlinear convergence.

Example 6.1 Consider the real valued function $f_1(x) = \log x + 1$, such that $f_1(1) = 1$. For the sequence generated by $x_{n+1} = f_1(x_n)$ with $x_0 = 1.5$ it holds $\lim_n x_n = 1$. Moreover, since $x_{n+1} - 1 = f_1(x_n) - f_1(1) = f_1'(\xi)(x_n - 1)$, where $|\xi - 1| < |x_n - 1|$, we have $\lim_n |x_{n+1} - 1|/|x_n - 1| = f_1'(1) = 1$. That is, the convergence is sublinear. By using similar arguments we deduce that the sequence generated by the function $f_2(x) = \frac{1}{2}\log x + 1$ with $x_0 = 1.5$ has linear convergence since $\lim_n |x_{n+1} - 1|/|x_n - 1| = f_2'(1) = 1/2$. Finally the sequence generated by the function $f_3(x) = \log^2 x + 1$ is such that $\lim_n |x_{n+1} - 1|/|x_n - 1| = f_3'(1) = 0$, and $\lim_n |x_{n+1} - 1|/|x_n - 1|^2 = 1$ so that the convergence is quadratic. In Figure

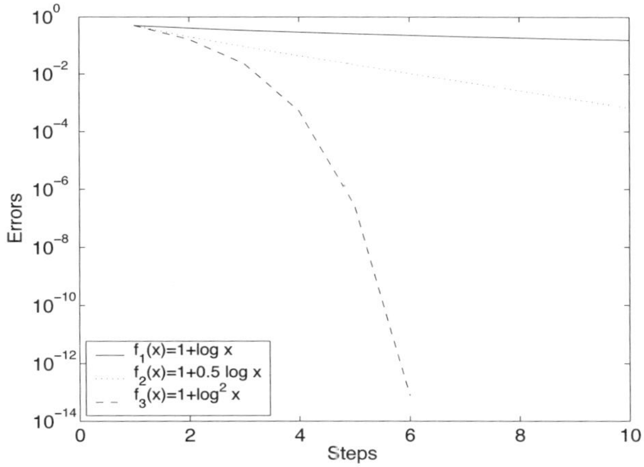

Fig. 6.1 Approximation errors of the three functional iterations of Example 6.1

6.1 we report the approximation errors of the three sequences generated in this way.

In order to better appreciate the difference between linear and superlinear convergence, consider the simple case where $e_{n+1} \le \gamma e_n$ for $0 < \gamma < 1$. In this situation it holds that $e_n \le \gamma^n e_0$. Here the error bound converges to zero exponentially with n. In the case of quadratic convergence, we find that $e_{n+1} \le \beta e_n^2$ for a suitable $\beta > 0$. By the latter bound we inductively deduce that $e_n \le \beta^{-1}(\beta e_0)^{2^n}$, so that, if $\beta e_0 < 1$ the convergence to zero of the error bound is *doubly exponential.* This further explains the rapid convergence which we encounter with superlinear convergence.

A functional iteration method is said to be *locally convergent* to W with sublinear, linear or superlinear convergence, if there exists a neighborhood $\mathcal{V}$ of W such that for any $X_0 \in \mathcal{V}$ the sequence $\{X_n\}_{n\ge 0}$ generated by (6.2) converges to W sublinearly, linearly or superlinearly, respectively.

A sufficient condition which ensures the local convergence of a functional iteration is expressed in the following general result.

Theorem 6.2 *Let $f : \mathcal{U} \subset \mathbb{R}^k \to \mathbb{R}^k$ be a continuous function having continuous partial derivatives, where $\mathcal{U}$ is an open subset of $\mathbb{R}^k$, $k \ge 1$. Denote $J(\boldsymbol{x}) = \left(\frac{\partial f_i(\boldsymbol{x})}{\partial x_j}\right)_{i,j=1,k}$ the Jacobian matrix of $f(\boldsymbol{x})$ at $\boldsymbol{x} = (x_i)_{i=1,k} \in \mathcal{U}$. Let $\boldsymbol{w} \in \mathcal{U}$ be a fixed point of $f(\boldsymbol{x})$, i.e., $f(\boldsymbol{w}) = \boldsymbol{w}$. If $\rho(J(\boldsymbol{w})) < 1$ then there exists a neighborhood $\mathcal{V} \subset \mathcal{U}$ of $\boldsymbol{w}$ such that for any $\boldsymbol{x}_0 \in \mathcal{V}$ the sequence generated by $\boldsymbol{x}_{n+1} = f(\boldsymbol{x}_n)$ converges to $\boldsymbol{w}$.*

We may express the convergence rate γ by means of a different formula which in certain circumstances is more convenient for our convergence analysis. Consider the geometric mean of the reduction of the error at each step, evaluated along the first n steps of the functional iteration,

$$\sigma(n) = \left(\frac{e_n}{e_{n-1}} \frac{e_{n-1}}{e_{n-2}} \cdots \frac{e_1}{e_0} \right)^{1/n}. \tag{6.4}$$

A simple calculation shows that

$$\sigma(n) = \left(\frac{e_n}{e_0} \right)^{1/n}.$$

The limit of $\sigma(n)$ for $n \to +\infty$ represents the asymptotic mean value of the reduction of the error per step. It is a simple matter to deduce from (6.4) that if there exists the limit (6.3) then

$$\lim_{n \to +\infty} \sigma(n) = \gamma; \tag{6.5}$$

moreover, this limit is independent of the norm with which $e_n = \|E_n\|$ is computed. The latter property holds true from the equivalence of norms on finite dimensional spaces (see Theorem A.1 in the appendix). Therefore, also the convergence rate, i.e., the limit (6.3) if it exists, is independent of the norm.

We observe that the existence of the limit (6.5) does not imply that the limit (6.3) exists as the following example shows. Therefore, (6.5) is a more general definition of the convergence rate γ.

Example 6.3 Define the sequence $\{e_n\}_{n \geq 0}$ in the following way

$$e_n = \begin{cases} 2\lambda^n & \text{if } n \text{ is even} \\ \lambda^n & \text{if } n \text{ is odd,} \end{cases}$$

where $\lambda > 0$ is a given real number. Then the limit (6.5) exists and is equal to λ, whereas $e_{n+1}/e_n = 2\lambda$ for n odd and $e_{n+1}/e_n = \lambda/2$ for n even so that the limit (6.3) does not exist.

6.2 Linearly convergent iterations

Functional iterations which generate sequences linearly convergent to the minimal solution $G_{\min}$ of (6.1) can be easily designed. The most natural way is provided by the function $F(X) = A(X)$ derived directly from the matrix equation (6.1), that is,

$$F(X) = \sum_{i=-1}^{+\infty} A_i X^{i+1}. \tag{6.6}$$

Different functional iterations are provided by the following functions obtained by means of simple formal manipulation of the matrix equation (6.1):

$$F(X) = (I - A_0)^{-1} \Big(A_{-1} + \sum_{i=1}^{+\infty} A_i X^{i+1} \Big), \tag{6.7}$$

$$F(X) = \Big(I - \sum_{i=0}^{+\infty} A_i X^i \Big)^{-1} A_{-1}. \tag{6.8}$$

It is immediate to verify that (6.6) and (6.7) are defined and continuous for any nonnegative matrix X such that $X\mathbf{1} \leq \mathbf{1}$. In fact, the power series

$\sum_{i=-1}^{+\infty} A_i X^{i+1}$ is convergent since it is an infinite summation of nonnegative matrices such that $\sum_{i=-1}^{+\infty} A_i X^{i+1}\mathbf{1} \leq \sum_{i=-1}^{+\infty} A_i \mathbf{1} = \mathbf{1}$; moreover $I - A_0$ is nonsingular. The function in (6.8) is defined for any X such that $0 \leq X \leq G_{\min}$; in fact, the power series $\sum_{i=0}^{+\infty} A_i X^i$ is obviously convergent, while $I - \sum_{i=0}^{+\infty} A_i X^i$ is nonsingular since $0 \leq \sum_{i=0}^{+\infty} A_i X^i \leq \sum_{i=0}^{+\infty} A_i G_{\min}^i$ and the spectral radius of the latter matrix is less than one, in light of Theorem 4.15.

6.2.1 Convergence properties: the case $X_0 = 0$

Let us denote by $\{X_n^{(N)}\}_{n\geq 0}$ the sequence defined by (6.6) (natural algorithm), by $\{X_n^{(T)}\}_{n\geq 0}$ the sequence defined by (6.7) (traditional algorithm) and by $\{X_n^{(U)}\}_{n\geq 0}$ the sequence defined by (6.8) (algorithm based on the matrix $U = \sum_{i=0}^{+\infty} A_i G_{\min}^i$), starting with a null initial approximation, i.e. $X_0^{(N)} = X_0^{(T)} = X_0^{(U)} = 0$.

The three sequences obtained in this way converge monotonically to $G_{\min}$ as stated by the following:

Theorem 6.4 *The sequences $\{X_n^{(N)}\}_{n\geq 0}$, $\{X_n^{(T)}\}_{n\geq 0}$, $\{X_n^{(U)}\}_{n\geq 0}$ converge monotonically to the matrix $G_{\min}$, that is $X_{n+1} - X_n \geq 0$ for X_n being any of $X_n^{(N)}$, $X_n^{(T)}$, $X_n^{(U)}$. Moreover, for any $n \geq 0$, it holds*

$$X_n^{(N)} \leq X_n^{(T)} \leq X_n^{(U)}.$$

Proof Consider the sequence $\{X_n^{(N)}\}_{n\geq 0}$. Let us prove its monotonicity by induction on n. For $n = 1$ the thesis is true, since $X_1^{(N)} = A_{-1} \geq 0 = X_0^{(N)}$. Suppose that $X_n^{(N)} \geq X_{n-1}^{(N)}$ and let us prove that $X_{n+1}^{(N)} \geq X_n^{(N)}$: from the inductive hypothesis, we obtain

$$X_{n+1}^{(N)} = \sum_{i=-1}^{+\infty} A_i {X_n^{(N)}}^{i+1} \geq \sum_{i=-1}^{+\infty} A_i {X_{n-1}^{(N)}}^{i+1} = X_n^{(N)}.$$

By using a similar argument we now prove that $X_n^{(N)} \leq G_{\min}$. Clearly $X_0^{(N)} = 0 \leq G_{\min}$. Assuming that $X_n^{(N)} \leq G_{\min}$, we have

$$X_{n+1}^{(N)} = \sum_{i=-1}^{+\infty} A_i {X_n^{(N)}}^{i+1} \leq \sum_{i=-1}^{+\infty} A_i G_{\min}^{i+1} = G_{\min}.$$

Since $0 \leq X_n^{(N)} \leq G_{\min}$, and since the sequence $\{X_n^{(N)}\}_{n\geq 0}$ is monotonic, there exists $G = \lim_n X_n^{(N)}$. Moreover G is a solution of the matrix equation (6.1) such that $0 \leq G \leq G_{\min}$. Since $G_{\min}$ is the minimal nonnegative solution, G must coincide with $G_{\min}$. By using similar arguments we may prove that the sequences $\{X_n^{(T)}\}_{n\geq 0}$, $\{X_n^{(U)}\}_{n\geq 0}$ converge mononically to $G_{\min}$.

Let us now show that $X_n^{(T)} \geq X_n^{(N)}$ for any n. For $n = 0$ equality holds. Assume that $X_n^{(T)} \geq X_n^{(N)}$ for fixed $n \geq 0$, and we show that $X_{n+1}^{(T)} \geq X_{n+1}^{(N)}$. We have, from the definition of $X_{n+1}^{(T)}$,

$$\begin{aligned} X_{n+1}^{(T)} &= (I - A_0)^{-1}\Big(A_{-1} + \sum_{i=1}^{+\infty} A_i X_n^{(T)^{i+1}}\Big) \\ &= A_{-1} + A_0(I - A_0)^{-1}A_{-1} + (I - A_0)^{-1}\sum_{i=1}^{+\infty} A_i X_n^{(T)^{i+1}} \\ &= A_{-1} + A_0 X_{n+1}^{(T)} + \sum_{i=1}^{+\infty} A_i X_n^{(T)^{i+1}}. \end{aligned}$$

From the latter equality, since $X_{n+1}^{(T)} \geq X_n^{(T)} \geq 0$, we deduce that

$$X_{n+1}^{(T)} \geq \sum_{i=-1}^{+\infty} A_i X_n^{(T)^{i+1}}.$$

For the inductive hypothesis $X_n^{(T)} \geq X_n^{(N)}$, therefore we obtain

$$X_{n+1}^{(T)} \geq \sum_{i=-1}^{+\infty} A_i X_n^{(N)^{i+1}} = X_{n+1}^{(N)}.$$

By using similar arguments we may prove that $X_n^{(U)} \geq X_n^{(T)}$ for any n. □

A first immediate consequence of the theorem above is that the three sequences are well defined when the initial approximation is the null matrix. In fact, every element X of the three subsequences is such that $0 \leq X \leq G_{\min}$; therefore, in light of the comments at the beginning of Section 6.2, the functions (6.6), (6.7) and (6.8) are well defined in X.

Theorem 6.4 allows one to deduce that the number of iterations I_N, I_T, I_U sufficient to obtain the approximations within the same error bound of the matrix $G_{\min}$ by means of the sequences $\{X_n^{(N)}\}_{n\geq 0}$, $\{X_n^{(T)}\}_{n\geq 0}$, $\{X_n^{(U)}\}_{n\geq 0}$, respectively, are such that

$$I_U \leq I_T \leq I_N.$$

Indeed, with the same number of iterations, the method based on (6.8) provides better approximations than the one based on (6.7), and the method based on (6.7) provides better approximations than the one based on (6.6). In the case where $\mu \leq 0$ we may give a more precise result by estimating also the rate of convergence of the three sequences.

Define, for each integer n, the matrices

$$\begin{aligned} E_n^{(N)} &= G_{\min} - X_n^{(N)}, \\ E_n^{(T)} &= G_{\min} - X_n^{(T)}, \\ E_n^{(U)} &= G_{\min} - X_n^{(U)}, \end{aligned}$$

which represent the error at step n for the sequences defined by (6.6), (6.7), (6.8), respectively, with $X_0 = 0$. Observe that, for the monotonicity stated by Theorem 6.4, the matrices $E_n^{(N)}$, $E_n^{(T)}$ and $E_n^{(U)}$ are nonnegative. Therefore, their infinity norms can be expressed in a simple way as

$$\|E_n\|_\infty = \|E_n\mathbf{1}\|_\infty$$

for E_n being any of the three matrices $E_n^{(N)}$, $E_n^{(T)}$ and $E_n^{(U)}$.

Moreover, for the three sequences, it is possible to give a bound on the error at step $n+1$ in terms of the error at step n as shown by the following

Theorem 6.5 *If $\mu \le 0$, then for every integer $n \ge 0$ one has*

$$\begin{aligned} E_{n+1}^{(N)}\mathbf{1} &= R_n^{(N)}E_n^{(N)}\mathbf{1},\\ E_{n+1}^{(T)}\mathbf{1} &= R_n^{(T)}E_n^{(T)}\mathbf{1},\\ E_{n+1}^{(U)}\mathbf{1} &= R_n^{(U)}E_n^{(U)}\mathbf{1} \end{aligned} \tag{6.9}$$

where, for $n \ge 0$,

$$\begin{aligned} R_n^{(N)} &= \sum_{i=0}^{+\infty}\sum_{j=i}^{+\infty} A_j X_n^{(N)^{j-i}},\\ R_n^{(T)} &= (I-A_0)^{-1}\Big(\sum_{i=0}^{+\infty}\sum_{j=i}^{+\infty} A_j X_n^{(T)^{j-i}} - A_0\Big),\\ R_n^{(U)} &= \Big(I - \sum_{i=0}^{+\infty} A_i X_n^{(U)^i}\Big)^{-1}\Big(\sum_{i=1}^{+\infty}\sum_{j=i}^{+\infty} A_j X_n^{(U)^{j-i}}\Big). \end{aligned} \tag{6.10}$$

Proof We rely on the following identity, valid for any $m \times m$ matrices X, Y, which can be proved by induction:

$$X^i - Y^i = \sum_{j=1}^{i} X^{j-1}(X-Y)Y^{i-j}, \quad i \ge 1 \tag{6.11}$$

Let us first analyze the sequence generated by the function (6.6). From (6.11), (6.1) and (6.2) it follows that

$$\begin{aligned} E_{n+1}^{(N)} =& G_{\min} - X_{n+1}^{(N)} = \sum_{i=1}^{+\infty} A_{i-1}\left(G_{\min}^i - X_n^{(N)^i}\right)\\ =& \sum_{i=1}^{+\infty}\sum_{j=1}^{i} A_{i-1} X_n^{(N)^{i-j}} E_n^{(N)} G_{\min}^{j-1}. \end{aligned}$$

Exchanging the order of the summations in the latter equality yields

$$E_{n+1}^{(N)} = \sum_{i=0}^{+\infty}\sum_{j=i}^{+\infty} A_j X_n^{(N)^{j-i}} E_n^{(N)} G_{\min}^i.$$

Whence, since $G_{\min}$ is stochastic, we have

$$E_{n+1}^{(N)}\mathbf{1} = \sum_{i=0}^{+\infty}\sum_{j=i}^{+\infty} A_j X_n^{(N)^{j-i}} E_n^{(N)}\mathbf{1} = R_n^{(N)} E_n^{(N)}\mathbf{1}.$$

Analogously, for the functional iteration based on (6.7), we obtain

$$\begin{aligned}E_{n+1}^{(T)} =& G_{\min} - X_{n+1}^{(T)} = (I - A_0)^{-1}\sum_{i=2}^{+\infty} A_{i-1}\left(G_{\min}^i - X_n^{(T)^i}\right)\\ =&(I-A_0)^{-1}\sum_{i=2}^{+\infty}\sum_{j=1}^{i} A_{i-1} X_n^{(T)^{i-j}} E_n^{(T)} G_{\min}^{j-1}.\end{aligned}$$

Once again, exchanging the order of the summations yields

$$E_{n+1}^{(T)} = (I - A_0)^{-1}\Big(\sum_{i=0}^{+\infty}\sum_{j=i}^{+\infty} A_j X_n^{(T)^{j-i}} E_n^{(T)} G_{\min}^i - A_0 E_n^{(T)}\Big).$$

Hence, from the stochasticity of $G_{\min}$, we obtain

$$E_{n+1}^{(T)}\mathbf{1} = R_n^{(T)} E_n^{(T)}\mathbf{1}.$$

Let us now analyze the functional iteration based on (6.8). Recalling that $A_0^* = \sum_{i=0}^{+\infty} A_i G_{\min}^i$, from (6.2) and (6.1) we have

$$\begin{aligned}E_{n+1}^{(U)} =& G_{\min} - X_{n+1}^{(U)}\\ =&\Big[(I - A_0^*)^{-1} - \Big(I - \sum_{i=0}^{+\infty} A_i X_n^{(U)^i}\Big)^{-1}\Big]A_{-1}\\ =&\Big(I - \sum_{i=0}^{+\infty} A_i X_n^{(U)^i}\Big)^{-1}\Big(A_0^* - \sum_{i=0}^{+\infty} A_i X_n^{(U)^i}\Big)(I - A_0^*)^{-1}A_{-1}\\ =&\Big(I - \sum_{i=0}^{+\infty} A_i X_n^{(U)^i}\Big)^{-1}\Big(A_0^* - \sum_{i=0}^{+\infty} A_i X_n^{(U)^i}\Big)G_{\min}.\end{aligned}$$

Whence, from (6.11), it follows that

$$E_{n+1}^{(U)} = \Big(I - \sum_{i=0}^{+\infty} A_i X_n^{(U)^i}\Big)^{-1}\Big(\sum_{i=1}^{+\infty}\sum_{j=1}^{i} A_i X_n^{(U)^{i-j}} E_n^{(U)} G_{\min}^{j-1}\Big)G_{\min}.$$

Interchanging the order of summations and using the stochasticity of $G_{\min}$ yields

$$E_{n+1}^{(U)}\mathbf{1} = R_n^{(U)} E_n^{(U)}\mathbf{1}.$$

□

Theorem 6.5 allows one to express the norm of the error at step n, by means of the matrices $R_n^{(N)}$, $R_n^{(T)}$, $R_n^{(U)}$ of (6.10):

Corollary 6.6 *If $\mu \le 0$, then for every integer $n \ge 1$ one has*

$$\begin{aligned} \|E_n^{(N)}\|_\infty &= \Big\| \prod_{i=0}^{n-1} R_i^{(N)} \Big\|_\infty, \\ \|E_n^{(T)}\|_\infty &= \Big\| \prod_{i=0}^{n-1} R_i^{(T)} \Big\|_\infty, \\ \|E_n^{(U)}\|_\infty &= \Big\| \prod_{i=0}^{n-1} R_i^{(U)} \Big\|_\infty, \end{aligned} \tag{6.12}$$

where the matrices $R_i^{(N)}$, $R_i^{(T)}$, $R_i^{(U)}$ are defined in (6.10) and $\prod_{i=0}^{n-1} H_i = H_{n-1}H_{n-2}\ldots H_0$, for $m\times m$ matrices H_i.

Proof For every integer n, since $E_n^{(N)} \ge 0$, we have $\|E_n^{(N)}\|_\infty = \|E_n^{(N)}\mathbf{1}\|_\infty$. Whence, for Theorem 6.5, $\|E_n^{(N)}\|_\infty = \|\prod_{i=0}^{n-1} R_i^{(N)} E_0^{(N)}\mathbf{1}\|_\infty$. On the other hand, since $X_0^{(N)} = 0$, we have $E_0^{(N)}\mathbf{1} = G_{\min}\mathbf{1} = \mathbf{1}$. Therefore, $\|E_n^{(N)}\|_\infty = \|\prod_{i=0}^{n-1} R_i^{(N)}\mathbf{1}\|_\infty = \|\prod_{i=0}^{n-1} R_i^{(N)}\|_\infty$. The analogous relations for $E_n^{(T)}$ and $E_n^{(U)}$ readily follow by using the same arguments. □

Now it is possible to estimate the asymptotic rate of convergence of the sequences $\{X_n^{(N)}\}_{n\ge0}$, $\{X_n^{(T)}\}_{n\ge0}$, $\{X_n^{(U)}\}_{n\ge0}$. An important role in the convergence analysis is played by the following matrices

$$\begin{aligned} R^{(N)} &= \sum_{i=0}^{+\infty} A_i^*, \\ R^{(T)} &= (I - A_0)^{-1}\Big(\sum_{i=0}^{+\infty} A_i^* - A_0\Big), \\ R^{(U)} &= (I - A_0^*)^{-1}\sum_{i=1}^{+\infty} A_i^* \end{aligned} \tag{6.13}$$

where A_i^* are defined in (4.8). Since the functions $X \to \sum_{i=0}^{+\infty}\sum_{j=i}^{+\infty} A_j X^{j-i}$ and $X \to \sum_{i=1}^{+\infty}\sum_{j=i}^{+\infty} A_j X^{j-i}$ are continuous at $X = G_{\min}$ we deduce that

$$R^{(N)} = \lim_n R_n^{(N)},\ R^{(T)} = \lim_n R_n^{(T)},\ R^{(U)} = \lim_n R_n^{(U)}. \tag{6.14}$$

Moreover, since the sequences $\{\|E_n^{(N)}\|\}_{n\ge0}$, $\{\|E_n^{(T)}\|\}_{n\ge0}$, $\{\|E_n^{(U)}\|\}_{n\ge0}$ are bounded from above by a constant, there exist the limits

$$r_N = \lim_n \sqrt[n]{\|E_n^{(N)}\|},\quad r_T = \lim_n \sqrt[n]{\|E_n^{(T)}\|},\quad r_U = \lim_n \sqrt[n]{\|E_n^{(U)}\|},$$

where $\|\cdot\|$ is any matrix norm. The above limits express the rates of convergence for the sequences $\{X_n^{(N)}\}_{n\geq 0}$, $\{X_n^{(T)}\}_{n\geq 0}$, $\{X_n^{(U)}\}_{n\geq 0}$, respectively. The following result relates these convergence rates:

Theorem 6.7 *If $\mu \leq 0$, one has*

$$r_N = \rho(R^{(N)}),\ r_T = \rho(R^{(T)}),\ r_U = \rho(R^{(U)}), \tag{6.15}$$

where the matrices $R^{(N)}$, $R^{(T)}$, $R^{(U)}$ are defined in (6.13).

Proof We prove the theorem only for the sequence $\{X_n^{(N)}\}_{n\geq 0}$; the same argument applies to the sequences $\{X_n^{(T)}\}_{n\geq 0}$ and $\{X_n^{(U)}\}_{n\geq 0}$. For Corollary 6.6 one has

$$r_N = \lim_n \sqrt[n]{\|E_n^{(N)}\|_\infty} = \lim_n \sqrt[n]{\Big\|\prod_{i=0}^{n-1} R_i^{(N)}\Big\|_\infty}. \tag{6.16}$$

For the monotonicity of the sequence $\{X_n^{(N)}\}_{n\geq 0}$ stated by Theorem 6.4 one has

$$0 \leq R_n^{(N)} \leq R_{n+1}^{(N)} \leq R^{(N)}.$$

Therefore, $\prod_{i=0}^{n-1} R_i^{(N)} \leq {R^{(N)}}^n$. Since $\|X\|_\infty \geq \|Y\|_\infty$ for $X \geq Y \geq 0$, we find that

$$\lim_n \sqrt[n]{\Big\|\prod_{i=0}^{n-1} R_i^{(N)}\Big\|_\infty} \leq \lim_n \sqrt[n]{\|{R^{(N)}}^n\|_\infty} = \rho(R^{(N)}),$$

where the last equality follows from Theorem A.3 in the appendix. Whence

$$r_N \leq \rho(R^{(N)}). \tag{6.17}$$

We now prove the opposite inequality. Suppose first that the nonnegative vector $R_0^{(N)}\mathbf{1}$ has no null components. For any integer $k < n$, for the monotonicity of the sequence $\{X_n^{(N)}\}_{n\geq 0}$ and for the monotonicity of $\|\cdot\|_\infty$, one has

$$\Big\|\prod_{i=0}^{n-1} R_i^{(N)}\Big\|_\infty \geq \Big\|{R_k^{(N)}}^{n-k}\prod_{i=0}^{k-1} R_i^{(N)}\Big\|_\infty \geq \|{R_k^{(N)}}^{n-k}{R_0^{(N)}}^k\|_\infty. \tag{6.18}$$

Hence, for any integer k, since the nonnegative vector ${R_0^{(N)}}^k\mathbf{1}$ also has no null components, it holds that

$$\begin{aligned}\|{R_k^{(N)}}^{n-k}{R_0^{(N)}}^k\|_\infty = &\|{R_k^{(N)}}^{n-k}{R_0^{(N)}}^k\mathbf{1}\|_\infty \geq \\ &c_k\|{R_k^{(N)}}^{n-k}\mathbf{1}\|_\infty = c_k\|{R_k^{(N)}}^{n-k}\|_\infty\end{aligned} \tag{6.19}$$

where $c_k > 0$ is the minimum value of the components of the positive vector ${R_0^{(N)}}^k\mathbf{1}$. From (6.14) it follows that, for any fixed $\epsilon > 0$, there exists an integer k_0 such that $\rho(R_{k_0}^{(N)}) \geq \rho(R^{(N)}) - \epsilon$. Whence, from (6.18) and (6.19), we obtain

$$r_N \geq \lim_n \sqrt[n]{\|R_{k_0}^{(N)^{n-k_0}} R_0^{(N)^{k_0}}\|_\infty} \geq \lim_n \sqrt[n]{c_{k_0}\|R_{k_0}^{(N)^{n-k_0}}\|_\infty}$$
$$= \lim_n \sqrt[n]{\|R_{k_0}^{(N)^{n-k_0}}\|_\infty} = \rho(R_{k_0}^{(N)}) \geq \rho(R^{(N)}) - \epsilon$$

where we have applied Theorem A.3 in the appendix.

Therefore, for the arbitrarity of ϵ, the above inequality, together with (6.17), leads to the thesis. Now, let us suppose that some components of the nonnegative vector $R_0^{(N)}\mathbf{1} = E_1^{(N)}\mathbf{1}$ are zero. Without loss of generality we may suppose that the first i components of the nonnegative vector $R_0^{(N)}\mathbf{1} = E_1^{(N)}\mathbf{1}$ are zero and that the remaining components are nonnull. Whence, for the monotonic convergence of the sequence $\{X_n^{(N)}\}_{n\geq 0}$, it follows that also the first i components of the nonnegative vector $E_n^{(N)}\mathbf{1}$ are zero for any $n \geq 1$. Moreover, it can be easily proved that, for any $n \geq 0$, the first i components of the vector $R_n^{(N)}\mathbf{1}$ are zero. Therefore, the matrices $R_n^{(N)}$ have the structure

$$R_n^{(N)} = \begin{bmatrix} 0 & 0 \\ T_n^{(N)} & U_n^{(N)} \end{bmatrix}$$

where $U_n^{(N)}$ is an $(m-i) \times (m-i)$ matrix and $T_n^{(N)}$ is an $(m-i) \times i$ matrix. Moreover, the matrix $R^{(N)}$ has the structure

$$R^{(N)} = \begin{bmatrix} 0 & 0 \\ T^{(N)} & U^{(N)} \end{bmatrix}$$

where $U^{(N)} = \lim_n U_n^{(N)}$, $T^{(N)} = \lim_n T_n^{(N)}$. If, except for the first i components, no other entry of the vector $E_n^{(N)}\mathbf{1}$ vanishes in a finite number of steps, then the matrices $U_n^{(N)}$ have no null columns for any $n \geq 1$. Let $\boldsymbol{v}$ be the $m-i$ dimensional vector defined by the nonnull components of the vector $E_1^{(N)}\mathbf{1}$. From the monotonicity of the sequence $\{X_n^{(N)}\}_{n\geq 0}$ and from the monotonicity of $\|\cdot\|_\infty$, it follows that, for any integer $k < n$

$$\Big\|\prod_{i=0}^{n-1} R_i^{(N)}\Big\|_\infty = \Big\|\prod_{i=1}^{n-1} U_i^{(N)}\boldsymbol{v}\Big\|_\infty \geq c\Big\|\prod_{i=1}^{n-1} U_i^{(N)}\Big\|_\infty \geq c\Big\|\prod_{i=1}^{k-1} U_k^{(N)^{n-k}} U_i^{(N)}\Big\|_\infty$$
$$\geq c\|U_k^{(N)^{n-k}} U_1^{(N)^{k-1}}\|_\infty \geq c_k\|U_k^{(N)^{n-k}}\|_\infty,$$

where $c, c_k > 0$ are suitable constants. Whence, for every $\epsilon > 0$, there exists k_0 such that

$$r_N \geq \lim_n \sqrt[n]{c_{k_0}\|U_{k_0}^{(N)^{n-k_0}}\|_\infty} = \rho(U_{k_0}^{(N)}) \geq \rho(U^{(N)}) - \epsilon = \rho(R^{(N)}) - \epsilon,$$

which leads to the thesis, for the arbitrarity of ϵ. If the components $i+1, \ldots, i+i_0$ of the vector $E_n^{(N)}\mathbf{1}$ vanish in a finite number of steps and no other component vanishes, then for $n \geq 1$, the matrices $U_n^{(N)}$ have the structure

$$U_n^{(N)} = \begin{bmatrix} \widehat{V}_n^{(N)} & 0 \\ \widehat{T}_n^{(N)} & \widehat{U}_n^{(N)} \end{bmatrix}$$

where $\widehat{U}_n^{(N)}$ is an $(m-i-i_0) \times (m-i-i_0)$ matrix having no null columns and $\widehat{V}_n^{(N)}$ is an upper triangular matrix with null diagonal elements. Therefore we may apply the same arguments to the matrix $\widehat{U}_n^{(N)}$, concluding the proof. □

Based on the properties of the regular splittings of M-matrices which are recalled in the appendix, we are able to compare the rate of convergence of the three sequences.

Theorem 6.8 *If $\mu < 0$ then the spectral radii of the matrices $R^{(N)}$, $R^{(T)}$ and $R^{(U)}$ are related by the following inequality*

$$\rho(R^{(U)}) \le \rho(R^{(T)}) \le \rho(R^{(N)}) < 1.$$

Moreover, if $\mu = 0$ then

$$\rho(R^{(U)}) = \rho(R^{(T)}) = \rho(R^{(N)}) = 1.$$

Proof Observe that

$$R^{(N)} = D^{(N)^{-1}} C^{(N)}, \ R^{(T)} = D^{(T)^{-1}} C^{(T)}, \ R^{(U)} = D^{(U)^{-1}} C^{(U)},$$

where

$$C^{(N)} = \sum_{i=0}^{+\infty} A_i^*, \qquad D^{(N)} = I,$$

$$C^{(T)} = \sum_{i=0}^{+\infty} A_i^* - A_0, \qquad D^{(T)} = I - A_0,$$

$$C^{(U)} = \sum_{i=1}^{+\infty} A_i^*, \qquad D^{(U)} = I - A_0^*,$$

are such that

$$D^{(N)} - C^{(N)} = D^{(T)} - C^{(T)} = D^{(U)} - C^{(U)} = I - \sum_{i=0}^{+\infty} A_i^*,$$

that is, the matrices $R^{(N)}$, $R^{(T)}$ and $R^{(U)}$ are obtained by means of regular splittings (compare with Definition A.14 in the appendix) of the M-matrix $I - \sum_{i=0}^{+\infty} A_i^*$. According to Theorem 4.14, if $\mu < 0$ then $I - \sum_{i=0}^{+\infty} A_i^*$ is a nonsingular M-matrix. For the properties of regular splittings (see Theorem A.15 in the appendix), since

$$C^{(U)} \le C^{(T)} \le C^{(N)},$$

it follows that

$$\rho(R^{(U)}) \le \rho(R^{(T)}) \le \rho(R^{(N)}) < 1.$$

If $\mu = 0$ then, according to Theorem 4.14, $I - \sum_{i=0}^{+\infty} A_i^*$ is a singular M-matrix, so that $\rho(R^{(U)}) = \rho(R^{(T)}) = \rho(R^{(N)}) = 1$ (see Theorem A.15). □

In light of Theorem 6.15, a direct consequence of the above theorem is that the convergence of the three sequences is linear if $\mu < 0$, and is sublinear if $\mu = 0$. Moreover, if $\mu < 0$, then the convergence rate of the sequence defined by (6.6) is less than or equal to the convergence rate of the sequence defined by (6.7). The latter rate is less than or equal to the convergence rate of the sequence defined by (6.8).

In the case where $\mu > 0$, and therefore $G_{\min}$ is substochastic, a similar convergence analysis can be performed, under the additional assumption that the matrices $G_{\min}$ and $\sum_{n=1}^{+\infty}\sum_{i=0}^{n-1}(I\otimes A_{i-1})((G_{\min}^{\mathrm{T}})^{n-1-i}\otimes G_{\min}^{i})$ are strictly positive. In this case Guo in [55] has shown that

$$\begin{aligned}
&\limsup_n \|G_{\min} - X_n^{(N)}\|^{1/n} \le \rho(M_1),\\
&\limsup_n \|G_{\min} - X_n^{(T)}\|^{1/n} \le \rho(M_2),\\
&\limsup_n \|G_{\min} - X_n^{(U)}\|^{1/n} \le \rho(M_3)
\end{aligned}$$

where

$$\begin{aligned}
M_1 &= \sum_{i=0}^{+\infty}(G_{\min}^{\mathrm{T}})^i A_i^*,\\
M_2 &= \sum_{i=0}^{+\infty}(G_{\min}^{\mathrm{T}})^i \otimes (I-A_0)^{-1}A_i^* - I\otimes (I-A_0)^{-1}A_0,\\
M_3 &= \sum_{i=1}^{+\infty}(G_{\min}^{\mathrm{T}})^i \otimes (I-A_0)^{-1}A_i^*.
\end{aligned}$$

Moreover, the spectral radii of the three matrices are such that

$$\rho(M_3) \le \rho(M_2) \le \rho(M_1) < 1.$$

As a consequence, the three sequences have a linear convergence, and one expects that the sequence $\{X_n^{(U)}\}_n$ converges faster than the sequence $\{X_n^{(T)}\}_n$, which converges faster than $\{X_n^{(N)}\}_n$.

6.2.2 A general class of functional iterations

Here we introduce a general class of functional iterations which contains the methods defined by (6.6), (6.7), (6.8). We show that the convergence properties of the previous section can be generalized to this more general class. Moreover, we prove that, in the case where the initial approximation is the null matrix, the functional iteration defined by (6.8) is the fastest of this class.

Consider the additive splitting $A(z) = zH(z) + K(z)$ of the matrix power series $A(z) = \sum_{i=-1}^{+\infty} z^{i+1}A_i$, where $H(z) = \sum_{i=0}^{+\infty} z^i H_i$ and $K(z) = \sum_{i=0}^{+\infty} z^i K_i$ are such that $H_i \ge 0$, $K_i \ge 0$, $i \ge 0$. In this way we have $A_{-1} = K_0$ and

$A_i = H_i + K_{i+1}$, for $i \geq 0$. Based on this splitting, we may introduce the matrix function

$$F(X) = (I - H(X))^{-1} K(X), \tag{6.20}$$

defined for each matrix X such that $I - H(X)$ is nonsingular. The function $F(X)$ defines the matrix equation

$$X = (I - H(X))^{-1} K(X). \tag{6.21}$$

Here, $K(X) = \sum_{i=0}^{+\infty} K_i X^i$, $H(X) = \sum_{i=0}^{+\infty} H_i X^i$.

It is immediate to observe that, if $H(z) = 0$ then (6.20) coincides with (6.6), if $H(z) = A_0$ then (6.20) coincides with (6.7), and finally if $H(z) = \sum_{i=0}^{+\infty} z^i A_i$ then (6.20) coincides with (6.8).

The matrix $\sum_{i=0}^{+\infty} A_i G_{\min}^i$ has spectral radius less than one (see Theorem 4.15). Therefore for Theorem 1.28, since $0 \leq H_i \leq A_i$ for $i \geq 0$, it follows that $\rho(H(G_{\min})) < 1$ so that $I - H(G_{\min})$ is a nonsingular M-matrix. In particular, the function (6.20) is defined for any matrix X such that $0 \leq X \leq G_{\min}$.

The solutions of the matrix equation (6.21) are related to the solution of (6.1), as stated by the following:

Theorem 6.9 *If G in an $m \times m$ matrix such that $I - H(G)$ is nonsingular, then G solves the matrix equation* (6.1) *if and only if it solves the matrix equation* (6.21). *Moreover, $I - H(G_{\min})$ is nonsingular and $G_{\min}$ is the minimal nonnegative solution of* (6.21).

Proof The matrix G solves the equation (6.1) if and only if it solves equation

$$(I - H(X))X = K(X).$$

Therefore, if $I - H(G)$ is nonsingular, G solves the matrix equation (6.1) if and only if it solves the matrix equation (6.21). Since $I - H(G_{\min})$ is nonsingular, $G_{\min}$ solves (6.21). If $G \geq G_{\min}$ were the minimal solution of (6.21), then, for the first part of the theorem, G would be also the minimal solution of (6.1), therefore $G = G_{\min}$. □

Equation (6.2) allows one to define the sequence of matrices

$$X_n = F(X_{n-1}), \quad n \geq 1$$

where $F(X)$ is defined in (6.20). The following result extends Theorem 6.4:

Theorem 6.10 *The sequence $X_n = F(X_{n-1})$, $X_0 = 0$, where $F(X)$ is given by* (6.20), *converges monotonically to the matrix $G_{\min}$.*

Proof We first prove that if $0 \leq Y \leq X \leq G_{\min}$ then $F(X) \geq F(Y)$. Observe that if $0 \leq Y \leq X \leq G_{\min}$ then $0 \leq H(Y) \leq H(X) \leq H(G_{\min})$ and $0 \leq K(Y) \leq K(X) \leq K(G_{\min})$. Moreover, since $\rho(H(G_{\min})) < 1$ from Theorem 1.28

it follows that $\rho(H(Y)) \leq \rho(H(X)) \leq \rho(H(G_{\min})) < 1$, therefore $I - H(X)$ and $I - H(Y)$ are nonsingular M-matrices. One has

$$(I - H(Y))^{-1} = \sum_{i=0}^{+\infty} H(Y)^i \leq \sum_{i=0}^{+\infty} H(X)^i = (I - H(X))^{-1}.$$

That is, $F(X) \geq F(Y) \geq 0$. Now we prove the monotonicity of the sequence $\{X_n\}_{n\geq 0}$ by induction on n. For $n = 1$ the thesis is true, since $X_1 = F(X_0) \geq 0 = X_0$. Suppose that $X_n \geq X_{n-1}$ and let us prove that $X_{n+1} \geq X_n$: from (6.20) and from the inductive hypothesis, we obtain

$$X_{n+1} = F(X_n) = (I - H(X_n))^{-1} K(X_n) \geq (I - H(X_{n-1}))^{-1} K(X_{n-1}) = X_n.$$

By using a similar argument we now prove that $X_n \leq G_{\min}$. Clearly $X_0 = 0 \leq G_{\min}$. Assuming that $X_n \leq G_{\min}$, we have

$$X_{n+1} = F(X_n) = (I - H(X_n))^{-1} K(X_n) \leq (I - H(G_{\min}))^{-1} K(G_{\min}) = G_{\min}.$$

Since $0 \leq X_n \leq G_{\min}$ and since the sequence is nondecreasing, there exists $G = \lim_n X_n$, with $G \leq G_{\min}$. Since $G_{\min}$ is the minimal nonnegative solution it follows that $G = G_{\min}$. □

Define, for each integer $n \geq 0$, the matrices $E_n = G - X_n$, which represent the error at step n of the sequence $\{X_n\}_{n\geq 0}$ converging to G. The matrices $\{E_n\}_{n\geq 0}$ satisfy the following general recursive relations which can be proved with the same arguments used in Section 6.2.1.

Theorem 6.11 *For any matrix X_0 consider the sequence $\{X_n\}_{n\geq 0}$ such that $X_n = F(X_{n-1})$, where $F(X)$ is defined by (6.20). Let $G = \lim_n X_n$ and $E_n = G - X_n$. Then for every integer $n \geq 0$ one has*

$$E_{n+1} = (I - H(X_n))^{-1} \Big(\sum_{i=1}^{+\infty} K_i X_n^{i-1} E_n + \sum_{j=1}^{+\infty} \sum_{i=j}^{+\infty} A_i X_n^{i-j} E_n G^j \Big). \tag{6.22}$$

From the above proposition we derive the generalization of Theorem 6.5:

Corollary 6.12 *Let $X_n = F(X_{n-1})$, $X_0 = 0$, where $F(X)$ is defined by (6.20). If $\mu \leq 0$, then for every integer $n \geq 0$ one has that*

$$E_{n+1}\mathbf{1} = R_n E_n \mathbf{1} \tag{6.23}$$

where, for $n \geq 0$,

$$R_n = (I - H(X_n))^{-1} \Big(\sum_{j=0}^{+\infty} \sum_{i=j}^{+\infty} A_i X_n^{i-j} - H(X_n) \Big). \tag{6.24}$$

Proof For the stochasticity of the matrix $G_{\min}$, from (6.22) and from the relation $A_i = H_i + K_{i-1}$ we obtain

$$E_{n+1}\mathbf{1} = (I - H(X_n))^{-1}\Big(\sum_{j=1}^{+\infty}\sum_{i=j}^{+\infty} A_{i-1}X_n^{i-j} - H(X_n)\Big)E_n\mathbf{1}.$$

□

The following result is a straightforward consequence of Corollary 6.12 and a generalization of Corollary 6.6.

Corollary 6.13 *Let $X_n = F(X_{n-1})$, $X_0 = 0$, where $F(X)$ is defined by (6.20). If $\mu \le 0$, then for every integer $n \ge 1$ one has*

$$\|E_n\|_\infty = \Big\|\prod_{i=0}^{n-1} R_i\Big\|_\infty \tag{6.25}$$

where the matrices R_n are defined in (6.24) and $\prod_{i=0}^{n-1} R_i = R_{n-1}R_{n-2}\ldots R_0$.

Define the matrix

$$R_F = (I - H(G_{\min}))^{-1}\Big(\sum_{i=0}^{+\infty} A_i^* - H(G_{\min})\Big) \tag{6.26}$$

such that $\lim_n R_n = R_F$. The matrix R_F allows one to express the asymptotic rate of convergence of the sequence $\{X_n\}_{n\ge 0}$, by means of the following result which can be easily proved by using the same arguments of Theorem 6.7.

Theorem 6.14 *Assume $\mu \le 0$, and let*

$$r = \lim_n \sqrt[n]{\|E_n\|}$$

be the convergence rate of the sequence $X_n = F(X_{n-1})$, $X_0 = 0$, where $F(X)$ is defined in (6.20) and $\|\cdot\|$ is any matrix norm. Then

$$r = \rho(R_F)$$

where R_F is given in (6.26).

The expression of the rate of convergence in terms of the spectral radius of the matrix R_F provided by the above theorem is very useful for comparing the convergence speed. In fact, observe that the matrix R_F can be written in the form

$$R_F = D^{-1}C,$$

where

$$C = \sum_{i=0}^{+\infty} A_i^* - H(G_{\min}), \quad D = I - H(G_{\min})$$

are such that

$$C \geq 0, \quad D^{-1} \geq 0, \quad D - C = I - \sum_{i=0}^{+\infty} A_i^*,$$

that is, R_F is obtained by means of a regular splitting (see Definition A.14 in the appendix) of the M-matrix $I - \sum_{i=0}^{+\infty} A_i^*$. Suppose that $R_1 = D_1^{-1}C_1$ and $R_2 = D_2^{-1}C_2$ are the matrices of (6.26) associated with the functions $F_1(X)$, $F_2(X)$ of (6.20), respectively, and suppose that $C_1 \leq C_2$. From Theorem A.15 in the appendix, it follows that

$$\rho(R_1) \leq \rho(R_2).$$

On the other hand observe that, for any choice of $F(X)$ defined by (6.20), since $0 \leq H_i \leq A_i$ for any i, one has that

$$H(G_{\min}) = \sum_{i=0}^{+\infty} H_i G_{\min}^i \leq \sum_{i=0}^{+\infty} A_i G_{\min}^i = A_0^*.$$

Whence

$$C = \sum_{i=0}^{+\infty} A_i^* - H(G_{\min}) \geq \sum_{i=1}^{+\infty} A_i^* = C^{(U)},$$

where $R^{(U)} = {D^{(U)}}^{-1} C^{(U)}$ is the matrix associated with the sequence $\{X_n^{(U)}\}_{n\geq 0}$ defined by (6.8). Therefore, if $\mu < 0$, the functional iteration method defined by (6.8) is the fastest one in the class of methods defined by (6.20), with $X_0 = 0$. If $\mu = 0$, all the methods of this class have sublinear convergence.

6.2.3 *Convergence properties: the case $X_0 = I$*

In the case where $\mu \leq 0$, since $G_{\min}$ is stochastic, it can be natural to start the functional iteration with a stochastic matrix. Consider sequences $\{X_n\}_{n\geq 0}$ generated by the recursion

$$X_n = F(X_{n-1}), \quad X_0 = I \tag{6.27}$$

where $F(X)$ is given by (6.20). It is immediate to verify that, if X_n is a stochastic matrix, and if $F(X_n)$ is defined, then X_{n+1} is stochastic. In fact, $H(X_n)\mathbf{1} + K(X_n)\mathbf{1} = \mathbf{1}$, so that $X_{n+1}\mathbf{1} = (I - H(X_n))^{-1}K(X_n)\mathbf{1} = \mathbf{1}$. Therefore, if the sequence defined by (6.27) can be generated without any breakdown, it is a sequence of stochastic matrices. The following theorem shows that, if $\mu \leq 0$, this sequence converges to $G_{\min}$.

Theorem 6.15 *Assume that $\mu \leq 0$. If the sequence $\{X_n\}_{n\geq 0}$ defined by (6.27) can be generated without any breakdown, then it converges to $G_{\min}$. Moreover the matrices obtained at each step of the iterative process are stochastic.*

Proof The proof is the extension of the proof of Theorem 3 of [76].

Let $\{Y_n\}_{n\geq 0}$ be the sequence defined by $Y_0 = 0$, $Y_n = F(Y_{n-1})$, for $n \geq 1$, where $F(X)$ is given by (6.20). In the previous section we have shown that the sequence $\{Y_n\}_{n\geq 0}$ converges monotonically to $G_{\min}$.

We now show by induction that $Y_n \leq X_n$ for any $n \geq 0$. The inequality holds for $n = 0$. Assume that $Y_n \leq X_n$ for a given $n \geq 0$. Then, as in the proof of Theorem 6.10, one has that $H(Y_n) \leq H(X_n)$ and $K(Y_n) \leq K(X_n)$; therefore

$$Y_{n+1} = (I - H(Y_n))^{-1}K(Y_n) \leq (I - H(X_n))^{-1}K(X_n) = X_{n+1}.$$

Since the sequence $\{X_n\}_{n\geq 0}$ is a sequence of stochastic matrices, it has at least one accumulation point, and it must be a stochastic matrix. Let X^* be one of these accumulation points. Since $Y_n \leq G_{\min}$, $\lim_n Y_n = G_{\min}$ and since $Y_n \leq X_n$ for any $n \geq 0$, we conclude that $G_{\min} \leq X^*$. Since both $G_{\min}$ and X^* are stochastic, they must coincide, therefore the sequence $\{X_n\}_{n\geq 0}$ converges to $G_{\min}$. □

It has been proved in [79] that for QBD processes no breakdowns are encountered in generating the sequence (6.27).

In this section we study the convergence of the sequence $\{X_n\}_{n\geq 0}$ to $G_{\min}$. In particular we show that, if $\mu \leq 0$, then the sequence of matrices obtained starting from any stochastic approximation converges faster than the sequence of matrices obtained by starting with the null matrix.

Consider the error $E_n = G_{\min} - X_n$ of the sequence $\{X_n\}_{n\geq 0}$ generated by (6.27). A recursive expression of the error E_n is provided by Theorem 6.11. We wish to evaluate the norm of E_n at each step n. Observe that, differently from the case analyzed in the previous section where $X_0 = 0$, the matrices E_n are not necessarily nonnegative. For this reason, the evaluation of the norm of E_n is not immediate as in the previous section. To this purpose we need to use Kronecker products (see Section 2.1) to rewrite (6.22) in a different way. Define $\boldsymbol{\epsilon}_n$, for $n \geq 0$, the m^2-dimensional vector associated with the matrix E_n, i.e., $\boldsymbol{\epsilon}_n = \mathrm{vec}(E_n)$. By using equation (2.2), the recursive formula (6.22) can be rewritten as

$$\boldsymbol{\epsilon}_{n+1} = \widehat{R}_n \boldsymbol{\epsilon}_n \tag{6.28}$$

where

$$\begin{aligned}\widehat{R}_n =& I \otimes (I - H(X_n))^{-1} \sum_{i=1}^{+\infty} K_i X_n^{i-1} \\ &+ \sum_{j=1}^{+\infty} (G_{\min}^{j-1})^{\mathrm{T}} \otimes (I - H(X_n))^{-1} \sum_{i=j}^{+\infty} A_i X_n^{i-j}.\end{aligned}$$

Observe that the matrix $\widehat{R}_n$, for $n \geq 0$, can be expressed in the form

$$\widehat{R}_n = \sum_{j=0}^{+\infty} (G_{\min}^{j})^{\mathrm{T}} \otimes Y_{j,n} \tag{6.29}$$

where

$$
\begin{aligned}
Y_{0,n} &= (I - H(X_n))^{-1} \sum_{i=1}^{+\infty} K_i X_n^{i-1}, \\
Y_{j,n} &= (I - H(X_n))^{-1} \sum_{i=j}^{+\infty} A_i X_n^{i-j}, \quad j \geq 1.
\end{aligned}
\tag{6.30}
$$

Let

$$
\begin{aligned}
Y_0 &= (I - H(G_{\min}))^{-1} \sum_{i=1}^{+\infty} K_i G_{\min}^{i-1}, \\
Y_j &= (I - H(G_{\min}))^{-1} \sum_{i=j}^{+\infty} A_i G_{\min}^{i-j}, \quad j \geq 1, \\
\widehat{R}_F &= \sum_{j=0}^{+\infty} (G_{\min}^j)^{\mathrm{T}} \otimes Y_j,
\end{aligned}
\tag{6.31}
$$

such that $Y_j = \lim_n Y_{j,n}$, $j \geq 0$, and $\widehat{R}_F = \lim_n \widehat{R}_n$. Observe that the matrix $Y_F = \sum_{j=0}^{+\infty} Y_j$ is given by

$$
Y_F = (I - H(G_{\min}))^{-1} \Big(\sum_{j=0}^{+\infty} \sum_{i=j}^{+\infty} A_i G_{\min}^{i-j} - H(G_{\min}) \Big). \tag{6.32}
$$

If $\mu \leq 0$, the matrix Y_F coincides with the matrix R_F of (6.26), which is related to the speed of convergence of the sequence $X_n = F(X_{n-1})$, $X_0 = 0$. This fact is the key to show that, when $\mu \leq 0$, the sequences obtained by starting with a stochastic matrix X_0 converge faster than the sequences obtained with $X_0 = 0$.

The following lemma which characterizes the eigenvalues of $\widehat{R}_F$ is fundamental to prove these convergence properties.

Lemma 6.16 *The eigenvalues of the matrix $\widehat{R}_F$ are given by the set*

$$
\bigcup_{\alpha \in \Lambda} \{ \eta | \eta \text{ is eigenvalue of } \sum_{j=0}^{+\infty} Y_j \alpha^j \},
$$

where Λ is the set of the eigenvalues of $G_{\min}$.

Proof Let S be the Schur canonical form [116] of the matrix $G_{\min}^{\mathrm{T}}$. Then, the matrix $\sum_{j=0}^{+\infty} (S^j \otimes Y_j)$ is similar to the matrix $\widehat{R}_F$. The lemma immediately follows since S is upper triangular with the diagonal elements equal to the eigenvalues of $G_{\min}$. □

Theorem 6.17 *If $\mu \le 0$ then $\rho(\widehat{R}_F) = \rho(R_F)$. Moreover, a nonnegative m^2-dimensional vector $\boldsymbol{w}$ such that*

$$\boldsymbol{w}^{\mathrm{T}}\widehat{R}_F = \lambda \boldsymbol{w}^{\mathrm{T}},$$

where $\lambda = \rho(R_F)$, is given by $\boldsymbol{w} = \mathbf{1} \otimes \boldsymbol{y}$, where $\boldsymbol{y}$ is a nonnegative m-dimensional vector such that $\boldsymbol{y}^{\mathrm{T}} R_F = \lambda \boldsymbol{y}^{\mathrm{T}}$.

Proof From Lemma 6.16, since $1 \in \Lambda$ and since $Y_F = R_F$, we deduce that the eigenvalues of the matrix R_F are also eigenvalues of the matrix $\widehat{R}_F$, therefore $\rho(\widehat{R}_F) \ge \rho(R_F)$. On the other hand, for Theorem 1.28, since for any $\alpha \in \Lambda$ one has that

$$\Big|\sum_{j=0}^{+\infty} Y_j \alpha^j\Big| \le \sum_{j=0}^{+\infty} Y_j |\alpha^j| \le \sum_{j=0}^{+\infty} Y_j = R_F,$$

we find that $\rho(\widehat{R}_F) \le \rho(R_F)$, whence $\rho(\widehat{R}_F) = \rho(R_F)$. From the properties of the tensor product and from the relation $G_{\min}\mathbf{1} = \mathbf{1}$, we obtain

$$\begin{aligned}(\mathbf{1}^{\mathrm{T}} \otimes \boldsymbol{y}^{\mathrm{T}})\widehat{R}_F &= (\mathbf{1}^{\mathrm{T}} \otimes \boldsymbol{y}^{\mathrm{T}}) \sum_{j=0}^{+\infty} (G_{\min}^j)^{\mathrm{T}} \otimes Y_j \\ &= \sum_{j=0}^{+\infty} (G_{\min}^j \mathbf{1})^{\mathrm{T}} \otimes (\boldsymbol{y}^{\mathrm{T}} Y_j) = \sum_{j=0}^{+\infty} \mathbf{1}^{\mathrm{T}} \otimes (\boldsymbol{y}^{\mathrm{T}} Y_j) \\ &= \mathbf{1}^{\mathrm{T}} \otimes \boldsymbol{y}^{\mathrm{T}} \sum_{i=0}^{+\infty} Y_j = \lambda(\mathbf{1}^{\mathrm{T}} \otimes \boldsymbol{y}^{\mathrm{T}}).\end{aligned}$$

Therefore, the vector $\boldsymbol{w}^{\mathrm{T}} = \mathbf{1}^{\mathrm{T}} \otimes \boldsymbol{y}^{\mathrm{T}}$ is a left nonnegative eigenvector of the matrix $\widehat{R}_F$. □

By following the same argument of the proof of Theorem 6.17, it is straightforward to prove the following.

Theorem 6.18 *If $\mu \le 0$ then for every $n \ge 0$ the spectral radius of the matrix $\widehat{R}_n$ is given by*

$$\rho(\widehat{R}_n) = \rho\Big(\sum_{j=0}^{+\infty} Y_{j,n}\Big).$$

Moreover, a nonnegative m^2-dimensional vector $\boldsymbol{w}_n$ such that

$$\boldsymbol{w}_n^{\mathrm{T}} \widehat{R}_n = \lambda_n \boldsymbol{w}_n^{\mathrm{T}},$$

where $\lambda_n = \rho(R_n)$, is given by $\boldsymbol{w}_n = \mathbf{1} \otimes \boldsymbol{y}_n$, where $\boldsymbol{y}_n$ is a nonnegative m-dimensional vector such that $\boldsymbol{y}_n^{\mathrm{T}} \sum_{j=0}^{+\infty} Y_{j,n} = \lambda_n \boldsymbol{y}_n^{\mathrm{T}}$.

From Thereom 6.18 it follows that, for any integer $n \geq 0$, the left nonnegative eigenvector $\boldsymbol{w}_n$ of the matrix $\widehat{R}_n$ associated with its spectral radius λ_n belongs to the linear space $\mathcal{S}$ generated by the orthogonal vectors

$$\boldsymbol{v}_1 = \frac{1}{\sqrt{m}}(\mathbf{1} \otimes \boldsymbol{e}_1), \quad \boldsymbol{v}_2 = \frac{1}{\sqrt{m}}(\mathbf{1} \otimes \boldsymbol{e}_2), \ldots, \boldsymbol{v}_m = \frac{1}{\sqrt{m}}(\mathbf{1} \otimes \boldsymbol{e}_m), \tag{6.33}$$

where $\boldsymbol{e}_i$ is the m-dimensional vector having the i-th component equal to 1 and the remaining components equal to zero. Let $\mathcal{T}$ be the linear space formed by the vectors orthogonal to $\boldsymbol{v}_1, \ldots, \boldsymbol{v}_n$. We may easily observe that, if A is any $m \times m$ matrix, then $\boldsymbol{v}_i^{\mathrm{T}} \operatorname{vec}(A) = \frac{1}{\sqrt{m}} \boldsymbol{e}_i^{\mathrm{T}} A \mathbf{1}$, $i = 1, \ldots, m$, therefore, since both X_n and $G_{\min}$ are stochastic, one has $\boldsymbol{v}_i^{\mathrm{T}} \boldsymbol{\epsilon}_n = \frac{1}{\sqrt{m}} \boldsymbol{e}_i^{\mathrm{T}} (G_{\min} - X_n)\mathbf{1} = 0$, for $i = 1, \ldots, m$, where $\boldsymbol{\epsilon}_n = \operatorname{vec}(G_{\min} - X_n)$. Therefore the error vector $\boldsymbol{\epsilon}_n$ belongs to the linear space $\mathcal{T}$, for any n.

In particular, if S and T are the matrices whose columns are an orthogonal basis of the linear spaces $\mathcal{S}$ and $\mathcal{T}$, respectively, then the $m^2 \times m^2$ matrix $\Gamma = [S|T]$ is such that

$$\Gamma^{\mathrm{T}} \Gamma = I, \quad \Gamma^{\mathrm{T}} \boldsymbol{\epsilon}_n = \begin{bmatrix} 0 \\ \hat{\boldsymbol{\epsilon}}_n \end{bmatrix},$$

where $\hat{\boldsymbol{\epsilon}}_n = T^{\mathrm{T}} \boldsymbol{\epsilon}_n$ is an $m(m-1)$-dimensional vector. Whence the recursive equations (6.28) can be rewritten in the form

$$\begin{bmatrix} 0 \\ \hat{\boldsymbol{\epsilon}}_{n+1} \end{bmatrix} = (\Gamma^{\mathrm{T}} \widehat{R}_n \Gamma)(\Gamma^{\mathrm{T}} \boldsymbol{\epsilon}_n) = (\Gamma^{\mathrm{T}} \widehat{R}_n \Gamma) \begin{bmatrix} 0 \\ \hat{\boldsymbol{\epsilon}}_n \end{bmatrix}. \tag{6.34}$$

On the other hand, observe that, if $\boldsymbol{w} \in \mathcal{S}$ then, from (6.29) and from the properties of the tensor product, it follows that $(\boldsymbol{w}^{\mathrm{T}} \widehat{R}_n)^{\mathrm{T}} \in \mathcal{S}$, whence $S^{\mathrm{T}} \widehat{R}_n T = 0$. Therefore the matrix $\Gamma^{\mathrm{T}} \widehat{R}_n \Gamma$ has the structure

$$\Gamma^{\mathrm{T}} \widehat{R}_n \Gamma = \begin{bmatrix} S^{\mathrm{T}} \\ - \\ T^{\mathrm{T}} \end{bmatrix} \widehat{R}_n \; (S|T) = \begin{bmatrix} V_n & 0 \\ T^{\mathrm{T}} \widehat{R}_n S & W_n \end{bmatrix}, \tag{6.35}$$

where $V_n = S^{\mathrm{T}} \widehat{R}_n S$, and

$$W_n = T^{\mathrm{T}} \widehat{R}_n T. \tag{6.36}$$

Moreover, from the representation (6.29) of $\widehat{R}_n$, we obtain that

$$V_n = S^{\mathrm{T}} \widehat{R}_n S = \frac{1}{m} (\mathbf{1}^{\mathrm{T}} \otimes I) \widehat{R}_n (\mathbf{1} \otimes I) = \sum_{j=0}^{+\infty} Y_{j,n}.$$

From the above equation and from (6.35) we deduce that

$$\Gamma^{\mathrm{T}} \widehat{R}_F \Gamma = \begin{bmatrix} Y_F & 0 \\ T^{\mathrm{T}} \widehat{R}_F S & W_F \end{bmatrix}, \tag{6.37}$$

where $W_F = T^{\mathrm{T}} \widehat{R}_F T$ and Y_F is the matrix of (6.32).

The recurrences (6.34) can be rewritten in the form

$$\hat{\boldsymbol{\epsilon}}_{n+1} = W_n \hat{\boldsymbol{\epsilon}}_n. \tag{6.38}$$

The above properties allow one to prove the following convergence results.

Theorem 6.19 *For the convergence rate*

$$r = \lim_n \sqrt[n]{\|\boldsymbol{\epsilon}_n\|}$$

of the sequence (6.27), where $F(X)$ is defined by (6.20) and $\|\cdot\|$ is any vector norm, we have

$$r \le \rho(W_F)$$

where $W_F = T^{\mathrm{T}} \widehat{R}_F T$ and $\widehat{R}_F$ is defined in (6.31). Moreover, if $\mu < 0$ and R_F is irreducible, then $\rho(W_F) < \rho(\widehat{R}_F) = \rho(R_F)$.

Proof From (6.34) and (6.38) it follows that

$$r = \lim_n \sqrt[n]{\|\boldsymbol{\epsilon}_n\|} = \lim_n \sqrt[n]{\|\hat{\boldsymbol{\epsilon}}_n\|}.$$

Therefore, we analyze the convergence of the sequence $\{\hat{\boldsymbol{\epsilon}}_n\}_{n\ge 0}$. Let $\epsilon > 0$ be fixed and let $\|\cdot\|_\epsilon$ be a matrix norm such that

$$\|W_F\|_\epsilon \le \rho(W_F) + \epsilon$$

(see Theorem A.2 in the appendix). Since the sequence $\{W_n\}_{n\ge 0}$ of (6.36) converges to W_F, there exists an integer i_0 such that

$$\|W_i\|_\epsilon \le \|W_F\|_\epsilon + \epsilon$$

for any $i \ge i_0$. Therefore

$$\begin{aligned}\|\hat{\boldsymbol{\epsilon}}_n\|_\epsilon = \|W_{n-1}W_{n-2}\dots W_0\hat{\boldsymbol{\epsilon}}_0\|_\epsilon &\le c_{i_0}\|W_{n-1}\|_\epsilon\|W_{n-2}\|_\epsilon\cdots\|W_{i_0}\|_\epsilon \\ &\le c_{i_0}(\|W_F\|_\epsilon+\epsilon)^{n-i_0} \le c_{i_0}(\rho(W_F)+2\epsilon)^{n-i_0}\end{aligned}$$

where c_{i_0} is a positive constant. Hence, by taking the n-th root, we have

$$\sqrt[n]{\|\hat{\boldsymbol{\epsilon}}_n\|_\epsilon} \le \sqrt[n]{c_{i_0}}(\rho(W_F)+2\epsilon)^{1-i_0/n}.$$

By taking the limit for $n \to +\infty$, we obtain

$$r = \lim_n \sqrt[n]{\|\hat{\boldsymbol{\epsilon}}_n\|_\epsilon} \le \rho(W_F) + 2\epsilon.$$

Therefore $r \le \rho(W_F)$ for the arbitrarity of ϵ.

To prove the second part of the theorem it is sufficient to show that $\rho(W_F) < \rho(\widehat{R}_F)$. For this purpose observe that from (6.37) the set of eigenvalues of $\widehat{R}_F$ is the union of the set of eigenvalues of Y_F and the set of eigenvalues of W_F.

Moreover, in light of Lemma 6.16 the set of eigenvalues of $\widehat{R}_F$ is the union of the set of eigenvalues of Y_F and the set $\bigcup_{\substack{\alpha\in\Lambda\\ \alpha\neq 1}}\{\eta|\eta \text{ is eigenvalue of } \sum_{j=0}^{+\infty} Y_j\alpha^j\}$. Therefore

$$\rho(W_F) = \max_{\substack{\alpha\in\Lambda\\ \alpha\neq 1}} \rho\Big(\sum_{j=0}^{+\infty} Y_j\alpha^j\Big).$$

Since $\mu < 0$, then $\sum_{j=0}^{+\infty} Y_j = R_F$, which is assumed irreducible. Therefore for the Perron–Frobenius Theorem 1.27 one has $\rho\big(\sum_{j=0}^{+\infty} Y_j\alpha^j\big) < \rho(R_F)$ for any α of modulus less than 1. □

In the cases where the matrix R_F is irreducible and $\mu < 0$, from the above theorem and from the results of the previous section, it follows that the rate of convergence of the sequence $X_n = F(X_{n-1})$ is:

– equal to the spectral radius of R_F, if $X_0 = 0$;

– strictly less than the spectral radius of R_F, if $X_0 = I$.

In many cases the modulus of the second largest eigenvalue of a matrix can be much smaller than its spectral radius. In the Figures 6.2, 6.3 and 6.4 we report, for the functional iterations defined by (6.6), (6.7), (6.8) the logarithm (to the base 10) of the residual $\|X_n - \sum_{i=0}^{+\infty} X_n^i A_i\|_\infty$, for the sequences obtained by starting with $X_0 = 0$ and with $X_0 = I$, for a problem arising from the modelling of a metropolitan network [3]. It is worth pointing out the increasing of the speed of convergence obtained by starting with $X_0 = I$ instead of $X_0 = 0$. Moreover it can be observed that the method defined by (6.8) (method based on the matrix U) converges more quickly than the methods based on (6.6) and (6.7) (natural and traditional algorithm, respectively). More precisely the spectral radius ρ_1 of the matrix $\widehat{R}_F$, which gives the mean asymptotic rate of convergence of the sequence obtained by starting with $X_0 = 0$, is given by $\rho_1 = 0.998448$ for the natural algorithm defined by (6.6), $\rho_1 = 0.998379$ for the traditional algorithm defined by (6.7) and $\rho_1 = 0.998176$ for the algorithm based on the matrix U defined by (6.8). On the other hand the second largest modulus eigenvalue of the matrix $\widehat{R}_F$ is $\rho_2 = 0.883677$ for the natural algorithm, $\rho_2 = 0.875044$ for the traditional algorithm and $\rho_2 = 0.858737$ for the algorithm based on the matrix U. Moreover, as can easily be observed from the figures 6.2, 6.3 and 6.4, for this particular example the rate of convergence of the sequence obtained by starting with $X_0 = I$ is equal to the second largest modulus eigenvalue of the matrix $\widehat{R}_F$.

An important role of the matrix $\widehat{R}_F$ is also played in the analysis of the local convergence of functional iterations $X_{n+1} = F(X_n)$.

Theorem 6.20 *Let $f_F(\boldsymbol{x}) : \mathbb{R}^{m^2} \to \mathbb{R}^{m^2}$ be the function defined by*

$$f_F(\boldsymbol{x}) = \operatorname{vec}\Big(F\big(\operatorname{vec}^{-1}(\boldsymbol{x})\big)\Big),$$

where $F(X)$ is given in (6.20), and denote by $J_F(\boldsymbol{x})$ the $m^2 \times m^2$ Jacobian matrix associated with the function $f_F(\boldsymbol{x})$. If $F(G) = G$, then $f_F(\boldsymbol{g}) = \boldsymbol{g}$ where $\boldsymbol{g} = \operatorname{vec}(G)$, and $J_F(\boldsymbol{g}) = \sum_{j=0}^{+\infty}(G^j)^{\mathrm{T}} \otimes Y_n(G)$, where

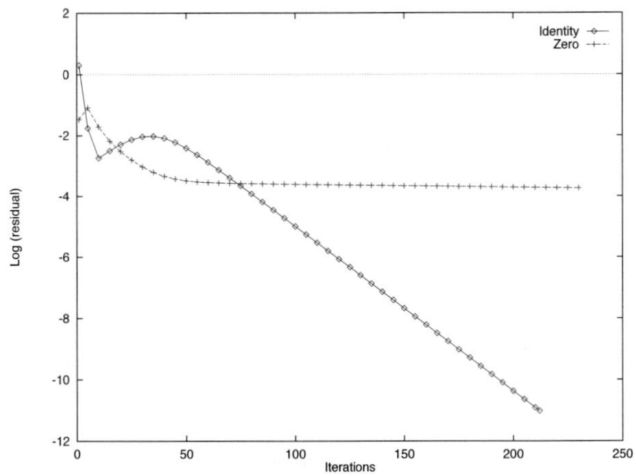

Fig. 6.2 Residual errors of the natural algorithm, for $X_0 = 0$ (graph marked with $+$), and for $X_0 = I$ (graph marked with $\diamond$).

$$Y_0(G) = (I - H(G))^{-1} \sum_{i=1}^{+\infty} K_i G^{i-1}$$

$$Y_j(G) = (I - H(G))^{-1} \sum_{i=j}^{+\infty} A_i G^{i-j}, \quad j \geq 1.$$

Proof The proof follows directly from the properties of the Kronecker product. □

Observe that, if $\mu \leq 0$, and hence $G = G_{\min}$, then $J_F(\boldsymbol{g}) = \widehat{R}_F$. If $\mu < 0$, one has $\rho(\widehat{R}_F) < 1$. Therefore, the above theorem implies the local convergence in light of Theorem 6.2.

A remarkable property, pointed out by Latouche and Taylor [81], is that if $\mu > 0$, so that $G_{\min} \neq G_{\text{sto}}$, there are at least two different nonnegative fixed points for the function $A(X)$. Moreover, while $G_{\min}$ is an attractive fixed point, the matrix G_{sto} is not attractive so that the sequence $\{X_n\}_{n \geq 0}$ generated by functional iterations with X_0 stochastic, or even belonging to a close neighborhood of G_{sto}, may fail to converge to G_{sto} and may converge to $G_{\min}$. The reason of this behaviour is that the spectral radius of the Jacobian of $J_F(\text{vec}(G_{\text{sto}}))$ is greater than 1 so that the conditions of Theorem 6.20 are not satisfied, while $\rho(J_F(\text{vec}(G_{\min}))) < 1$.

6.2.4 Computational issues

In this section we discuss some problems related to the computational aspects of functional iterations. A first approximation that we must do in order to numerically solve the matrix equation (6.1) is to truncate the matrix power series (6.1) to a matrix polynomial

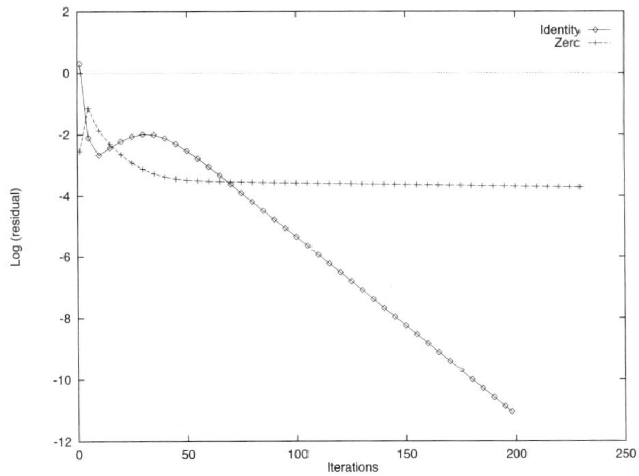

Fig. 6.3 Residual errors of the traditional algorithm, for $X_0 = 0$ (graph marked with $+$), and for $X_0 = I$ (graph marked with $\diamond$).

$$X = \sum_{i=-1}^{k} A_i X^{i+1}. \tag{6.39}$$

An important issue which must be taken into account is the analysis of the consequences of replacing the power series matrix equation (6.1) with the polynomial matrix equation (6.39). In fact, the truncated matrix equation has a minimal nonnegative solution $\widetilde{G}_{\min}$ such that $\widetilde{G}_{\min} \le G_{\min}$. The problem of giving bounds to the error $G_{\min} - \widetilde{G}_{\min}$ is not trivial at all as the following example of Gail, Hantler and Taylor [42] shows.

Example 6.21 Consider the matrix polynomial $A(z) = A_{-1} + zA_0 + z^2 A_1 + z^k A_{k-1}$, where $k > 2$ and

$$A_{-1} = \begin{bmatrix} p & 0 \\ 0 & 0 \end{bmatrix}, \quad A_0 = \begin{bmatrix} 0 & 1-p-q \\ 0 & 0 \end{bmatrix}, \quad A_1 = \begin{bmatrix} q & 0 \\ 0 & 1-\epsilon \end{bmatrix}, \quad A_{k-1} = \begin{bmatrix} 0 & 0 \\ \epsilon & 0 \end{bmatrix}$$

with $p+q \le 1$, $\epsilon > 0$. One has $\mu < 0$ if and only if $p > q + (1-p-q)(k-2+1/\epsilon)$. Consider also the approximation to $A(z)$ given by $\widetilde{A}(z) = A_{-1} + zA_0 + z^2 A_1$ and observe that $\widetilde{A}(1)$ is reducible whereas $A(1)$ is irreducible. The minimal nonnegative solutions of the equations $X = A(X)$ and $X = \widetilde{A}(X)$ are

$$G_{\min} = \begin{bmatrix} 1 & 0 \\ 1 & 0 \end{bmatrix}, \quad \widetilde{G}_{\min} = \begin{bmatrix} \alpha & 0 \\ 0 & 0 \end{bmatrix},$$

respectively, where $\alpha = [1 - (1 - 4pq)^{1/2}]/(2q)$ is independent of ϵ. Thus, even though the difference $A(z) - \widetilde{A}(z)$ becomes negligible as $\epsilon \to 0$, the difference $G_{\min} - \widetilde{G}_{\min}$ remains away from zero.

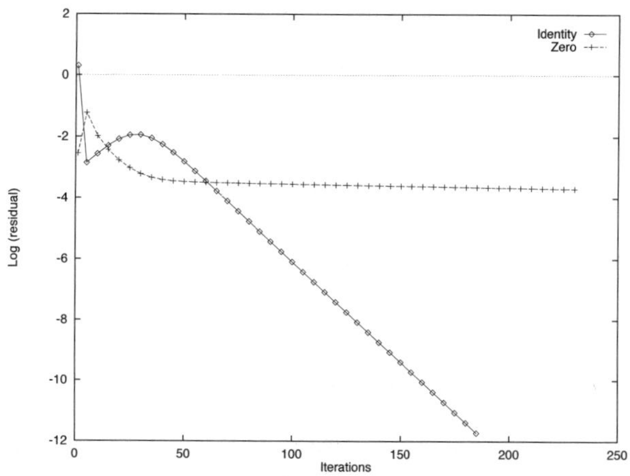

Fig. 6.4 Residual errors of the algorithm based on matrix U, for $X_0 = 0$ (graph marked with $+$), and for $X_0 = I$ (graph marked with $\diamond$).

The following result of Gail, Hantler and Taylor [42] provides bounds on the error $G_{\min} - \widetilde{G}_{\min}$.

Theorem 6.22 *Let $A(z) = \sum_{i=-1}^{+\infty} z^{i+1} A_i$ be the Laurent matrix power series associated with an M/G/1-type Markov chain such that $\mu < 0$. Let $G_{\min}$ the minimal nonnegative solution of the matrix equation (6.1). Consider the truncated polynomial $\widetilde{A}(z) = \sum_{i=-1}^{k} z^{i+1} A_i$ and let $\widetilde{G}_{\min}$ be the minimal nonnegative solution of the matrix equation $X = \widetilde{A}(X)$. Then $0 \le \widetilde{G}_{\min} \le G_{\min}$ and*

$$(G_{\min} - \widetilde{G}_{\min})\mathbf{1} \le \Big(I - \sum_{i\ge 0} A_i^*\Big)^{-1} (A(1) - \widetilde{A}(1))\mathbf{1}.$$

Observe that, since the term $(A(1)-\widetilde{A}(1))\mathbf{1}$ is the truncation error $\sum_{i>k} A_i\mathbf{1}$, the nonnegative matrix $(I-\sum_{i\ge 0} A_i^*)^{-1}$ expresses the amplification factor of the truncation error in the solution $\widetilde{G}_{\min}$.

Another computational issue related to the implementation of the functional iterations analyzed in this section concerns the evaluation of a matrix polynomial at a matrix value. More precisely, given the matrix polynomial $B(z) = \sum_{i=0}^{k} z^i B_i$ and the matrix X, compute the matrix $Y = B(X) = \sum_{i=0}^{k} B_i X^i$.

The Horner rule provides a means to perform this computation in a reliable and efficient way. It is based on the expression

$$B(X) = \Big(\dots((B_k X + B_{k-1})X + B_{k-2})X + \cdots\Big)X + B_0$$

which can be computed by means of Algorithm 6.1

Observe that the computational cost of Algorithm 6.1 is just k matrix multiplications and k matrix additions for an overall cost of $2km^3$ arithmetic operations.

Algorithm 6.1 Horner rule for matrix polynomials

INPUT: The degree k and the $m \times m$ matrix coefficients B_i, $i = 0, \ldots, k$, of the matrix polynomial $B(z) = \sum_{i=0}^{k} z^i B_i$, an $m \times m$ matrix X.

OUTPUT: The $m \times m$ matrix $Y = B(X)$.

COMPUTATION:

1. Set $Y = B_k$.
2. For $i = 1, \ldots, k$ compute $Y = YX + B_{k-i}$.

A different but not less relevant computational problem is to design criteria for stopping the functional iteration. The most natural way is to halt the iteration if $\|X_{n+1} - X_n\| \le \epsilon$ for a given positive ϵ and for a given norm $\|\cdot\|$. The following theorem relates the error $G_{\min} - X_n$ with ϵ when the iteration is halted with the above criterion. A similar result holds for general functional iterations generated by $F(X)$ of (6.20).

Theorem 6.23 *Let $\{X_n\}_{n\ge 0}$ be any sequence generated by the natural functional iteration $X_{n+1} = A(X_n)$ and assume that $\lim_n X_n = G$. Then we have*

$$\operatorname{vec}(G - X_n) = \Big(I - \sum_{i=0}^{+\infty}(G^{\mathrm{T}})^i \otimes \sum_{j=i}^{+\infty} A_j X_n^{j-i}\Big)^{-1} \operatorname{vec}(X_{n+1} - X_n).$$

Proof One has $X_{n+1} - X_n = -(G - X_{n+1}) + (G - X_n)$. Moreover, by applying the same arguments used in the proof of Theorem 6.5, we find that $G - X_{n+1} = \sum_{i=0}^{+\infty}\sum_{j=i}^{+\infty} A_j X_n^{j-i}(G - X_n)G^i$. Therefore, by using the properties of the Kronecker product we obtain the claim. □

Observe that, in the limit, the matrix in the above theorem turns into $(I - \sum_{i=0}^{+\infty}(G^{\mathrm{T}})^i \otimes \sum_{j=i}^{+\infty} A_j G^{j-i})$. In the case where $\mu < 0$ and $G = G_{\min}$ then the above matrix is a nonsingular M-matrix and coincides with $I - \widehat{R}_F$ of (6.31), therefore its inverse is nonnegative and has spectral radius $(1 - \rho(\widehat{R}_F))^{-1}$. In particular, if $\rho(\widehat{R}_F)$ is close to 1, besides a slow convergence rate of the sequence $\{X_n\}_{n\ge 0}$, we have the drawback that a small value of ϵ does not guarantee a small approximation error in $G_{\min}$.

In Algorithm 6.2 we report a scheme for the computation of $G_{\min}$ by using the general functional iteration generated by (6.20). Different specification of the polynomials $H(z)$ and $K(z)$ provide the different algorithms analyzed in this section.

In Table 6.1 we report the costs per step of the natural, traditional and U-based iterations applied to the polynomial equation (3.39), where, for the traditional algorithm the inversion of $I - A_0$ is performed once for all and is not included in the evaluation of the cost per step.

Natural choices for X_0 are $X_0 = 0$ and $X_0 = I$. If $\mu < 0$, in both cases the convergence to the minimal solution is guaranteed and all the matrices X_n

---

Algorithm 6.2 General functional iteration: M/G/1 case

INPUT: The integer $k > 0$ and the $m \times m$ matrix coefficients A_i, $i = -1, \ldots, k$, of the matrix polynomial $A(z) = \sum_{i=-1}^{k} z^{i+1} A_i$, such that $\mu < 0$; an $m \times m$ matrix X_0; a positive ϵ and the matrix polynomials $H(X)$ and $K(X)$ defining the matrix function $F(X)$ of (6.20).

OUTPUT: An approximation Y of the minimal nonnegative solution $G_{\min}$ of the equation $X = A(X)$ such that $\|Y - F(Y)\| \le \epsilon$ for a given norm $\|\cdot\|$.

COMPUTATION:

1. Set $X = X_0$
2. Apply Algorithm 6.1 to compute $H(X)$ and $K(X)$.
3. Set $Y = (I - H(X))^{-1} K(X)$.
4. If $\|X - Y\| \le \epsilon$ then output $Y = X$. Otherwise set $X = Y$ and repeat from step 2.

---

generated by the algorithm are nonnegative, moreover their computation involves addition of nonnegative numbers and inversion of an M-matrix. This fact guarantees that no numerical cancellation is encountered in the computation so that the algorithm is numerically stable. However, the choice $X_0 = I$ provides a faster convergence rate. If $\mu < 0$ and $X_0 = I$, the approximation Y to $G_{\min}$ is substochastic. Therefore, the condition $\|\mathbf{1} - Y\mathbf{1}\|_\infty \le \epsilon$ guarantees that $\|G_{\min} - Y\|_\infty \le \epsilon$. Thus the inequality $\|\mathbf{1} - Y\mathbf{1}\|_\infty \le \epsilon$ is a valid stop condition for the iteration.

If the Markov chain models a QBD problem, then $A_i = 0$ for $i > 1$ so that $A(z)$ is a polynomial of degree 2. In this case the functional iterations have a simpler formulation and their complexity per steps is very small. In Table 6.2 we report the cost of performing one step for the sequences $X_n^{(N)}$, $X_n^{(T)}$ and $X_n^{(U)}$, where the inversion of $I - A_0$ in the traditional algorithm is performed once for all and is not included in the evaluation of the cost per step.

6.2.5 Further developments

In Section 6.2.3 we have pointed out that, when $\mu \le 0$, we may improve the rate of convergence by choosing an appropriate initial approximation, which shares with $G_{\min}$ the dominant eigenvalue and the corresponding right eigenvector. In the case where $\mu > 0$ one could think of applying a similar strategy, if the dominant eigenvalue, and the corresponding right eigenvector of $G_{\min}$ are known. Unfortunately, in this case $G_{\min}$ is not stochastic, so the dominant eigenvalues is not readily available. An alternative strategy to improve the convergence, in the case where $\mu > 0$ and $A_i = 0$, for $i > k$, for a suitable $k > 1$, could be to apply the shifting technique of Section 3.6 to move the singular point $z = 1$ of $\det(zI - A(z))$ to infinity. Then, by means of functional iterations with null initial approximation, we compute the minimal (in the sense of spectral radius) solution of the new matrix equation, which provides the matrix $G_{\min}$. A precise analysis of the convergence rate obtained after the shifting has not been performed, but

Algorithm	Additions	Multiplications	Inversions
Natural	$k+1$	$k+1$	0
Traditional	k	$k+2$	0
Matrix U-based	k	$k+1$	1

Table 6.1 Costs per step of the functional iterations in terms of matrix operations.

several numerical experiments have shown that the speed of convergence is much improved in many cases.

6.3 Newton's iteration

Our purpose here is to show that Newton's method can be applied for solving (6.1), the iteration is well defined, and the convergence is monotonic and quadratic under very weak conditions which are extremely simple to verify, and which have a fundamental probabilistic interpretation.

Throughout the section we assume that $\mu < 0$ and that

$$\sum_{i=-1}^{+\infty} (i+1)^2 A_i < +\infty. \tag{6.40}$$

We define $\langle 0, G\rangle$ the set of matrices X such that $0 \leq X \leq G_{\min}$ (we use the natural partial order where $X \leq G_{\min}$ if $X_{i,j} \leq (G_{\min})_{i,j}$ for all i and j). We define the operator $\mathcal{M}$ on $\langle 0, G_{\min}\rangle$ as follows:

$$\mathcal{M}X = \sum_{i=0}^{+\infty} A_i X^i. \tag{6.41}$$

Observe that the series above converges for all X in $\langle 0, G_{\min}\rangle$. Moreover, we have that

$$\mathcal{M}X \leq \mathcal{M}Y, \qquad \text{for all } X \leq Y \text{ in } \langle 0, G_{\min}\rangle. \tag{6.42}$$

We also define the matrix $U = \mathcal{M}G_{\min}$, which is nonnegative, substochastic, and $\rho(U) < 1$.

The operator $\mathcal{L}$ on $\langle 0, U\rangle$ is defined as follows:

$$\mathcal{L}V = (I - V)^{-1} A_{-1}. \tag{6.43}$$

Since $\rho(U) < 1$, therefore $\rho(V) < 1$ for all V in $\langle 0, U\rangle$, so that $(I-V)^{-1}$ exists and is equal to the series $\sum_{\nu \geq 0} V^\nu$. As a consequence, we have that

$$\mathcal{L}V \leq \mathcal{L}W, \qquad \text{for all } V \leq W \text{ in} \langle 0, U\rangle. \tag{6.44}$$

We also have that $\mathcal{L}U = G_{\min}$ so that $\mathcal{M}$ applies $\langle 0, G\rangle$ on $\langle 0, U\rangle$, and $\mathcal{L}$ applies $\langle 0, U\rangle$ on $\langle 0, G\rangle$.

Algorithm	Additions	Multiplications	Inversions
Natural	2	2	0
Traditional	1	2	0
Matrix U – based	1	2	1

Table 6.2 Costs per step of the functional iterations for QBD problems in terms of matrix operations.

Finally, we define as follows the operator $\mathcal{F}$ on $\langle 0, G\rangle$:

$$\mathcal{F}X = \{\mathcal{I} - \mathcal{L}\mathcal{M}\}X, \tag{6.45}$$

where $\mathcal{I}$ is the identity operator such that $\mathcal{I}X = X$. In fact, the operator $\mathcal{F}$ is well defined for any (sub)stochastic matrix X. The matrix $G_{\min}$ is the minimal non-negative solution of the equation $\mathcal{F}X = 0$.

Our objective is to show that Newton's sequence

$$X_{n+1} = X_n - \mathcal{F}'(X_n)^{-1}\mathcal{F}X_n, \quad n \geq 0 \tag{6.46}$$

is well defined, converges monotonically to $G_{\min}$ if X_0 is in $\langle 0, G_{\min}\rangle$, and converges quadratically.

The Gateaux derivatives of $\mathcal{M}$ and $\mathcal{L}$ are operators such that

$$\mathcal{M}'(X)H = \sum_{i=1}^{+\infty} A_i \sum_{0\leq j\leq i-1} X^j H X^{i-1-j}, \tag{6.47}$$

$$\mathcal{L}'(V)H = (I - V)^{-1}H(I - V)^{-1}A_{-1}, \tag{6.48}$$

and for any $H \geq 0$ satisfy the inequalities

$$\mathcal{M}'(X)H \leq \mathcal{M}'(Y)H, \qquad \text{for all } X \leq Y \text{ in } \langle 0, G_{\min}\rangle, \tag{6.49}$$

$$\mathcal{L}'(V)H \leq \mathcal{L}'(W)H, \qquad \text{for all } V \leq W \text{ in } \langle 0, U\rangle. \tag{6.50}$$

We shall show that the chain rule applies, so that $\mathcal{F}'(X)$ is given by

$$\mathcal{F}'(X) = \mathcal{I} - \mathcal{L}'(\mathcal{M}X)\mathcal{M}'(X),$$

and we shall successively prove that $\mathcal{F}'$ is Lipschitz-continuous, order-concave, antitone and non-singular on $\langle 0, G_{\min}\rangle$.

Under condition (6.40) we may easily verify that the right-hand side of (6.47) is well defined for X in $\langle 0, G_{\min}\rangle$. For that purpose, it suffices to prove that $|\mathcal{M}'(X)H|\mathbf{1}$ is finite:

$$
\begin{aligned}
|\mathcal{M}'(X)H|\mathbf{1} &\le \sum_{i=1}^{+\infty} A_i \sum_{j=0}^{i-1} X^j|H|X^{i-1-j}\mathbf{1} \\
&\le \sum_{i=1}^{+\infty} A_i \sum_{j=0}^{i-1} X^j|H|\mathbf{1} \\
&\le c\sum_{i=1}^{+\infty} A_i \sum_{j=0}^{i-1} X^j\mathbf{1} \le c\sum_{i=1}^{+\infty} iA_i\mathbf{1},
\end{aligned}
$$

which is finite by assumption. The second and fourth inequalities result from $X\mathbf{1} \le G_{\min}\mathbf{1} \le \mathbf{1}$; for the third inequality, we choose c such that $|H|\mathbf{1} \le c\mathbf{1}$.

Here we use $\|\cdot\|$ to denote the infinity norm, however the results that we obtain are valid for any induced norm. We recall the following monotonicity property of the infinity norm:

$$
\|X\|_\infty \le \|Y\|_\infty \le 1 \qquad \text{for all } X \le Y \text{ in } \langle 0, G_{\min}\rangle \text{ or } \langle 0, U\rangle. \tag{6.51}
$$

Our first step is to analyze the operator $\mathcal{M}$ and its derivative.

Lemma 6.24 *The operators $\mathcal{M}$ and $\mathcal{M}'$ are Lipschitz-continuous on $\langle 0, G_{\min}\rangle$, and $\mathcal{M}'(X)$ is the Fréchet derivative of $\mathcal{M}$. The norm of $\mathcal{M}'$ is uniformly bounded on $\langle 0, G_{\min}\rangle$.*

Proof Assume that H is a matrix of norm 1, and that X, Y are two matrices in $\langle 0, G_{\min}\rangle$. We may write

$$
[\mathcal{M}'(X) - \mathcal{M}'(Y)]H = \sum_{i=1}^{+\infty} A_i \sum_{j=0}^{i-1}(X^jHX^{i-1-j} - Y^jHY^{i-1-j}),
$$

since we show below that the series in the right-hand side is absolutely convergent. Thus, we have that

$$
\begin{aligned}
\|[\mathcal{M}'(X) - \mathcal{M}'(Y)]H\| &\le \sum_{i=1}^{+\infty} \|A_i\| \sum_{j=0}^{i-1} \|X^jHX^{i-1-j} - Y^jHY^{i-1-j}\|, \\
&\le \|H\| \sum_{i=1}^{+\infty} \|A_i\| \sum_{j=0}^{i-1} \|X^j - Y^j\|(\|X^{i-1-j}\| + \|Y^{i-1-j}\|), \\
&\le 2\sum_{i=1}^{+\infty} \|A_i\| \sum_{j=0}^{i-1} \|X^j - Y^j\|,
\end{aligned}
$$

since $\|H\| = 1$, and $\|X\|, \|Y\| \le \|G_{\min}\| \le 1$ by (6.51). We also have that

$$
\begin{aligned}
\|X^j - Y^j\| &\le (\|X\|^{j-1} + \|X\|^{j-2}\|Y\| + \cdots + \|Y\|^{j-1})\|X - Y\| \\
&\le j\|X - Y\|.
\end{aligned}
$$

Hence, for the operator norm of $\mathcal{M}'(X) - \mathcal{M}'(Y)$ it holds

$$\|\mathcal{M}'(X) - \mathcal{M}'(Y)\| \leq \gamma_1 \|X - Y\|, \tag{6.52}$$

with $\gamma_1 = \sum_{i=1}^{+\infty} i(i-1)\|A_i\|$. This last series converges since the matrices A_i are nonnegative, and we assumed that the series $\sum_{i=1}^{+\infty}(i+1)^2 A_i$ converges due to the assumption (6.40).

Since the coefficient γ_1 in (6.52) is independent of X and Y, therefore $\mathcal{M}'$ is Lipschitz-continuous, it is the Fréchet derivative of $\mathcal{M}$ and $\mathcal{M}$ is Lipschitz-continuous (see Ortega Rheinboldt [95] sections 3.2.8 and 3.1.6).

Finally, the norm of $\mathcal{M}'(X)$ is bounded as follows:

$$\|\mathcal{M}'(X)\| \leq \|\mathcal{M}'(X) - \mathcal{M}'(0)\| + \|\mathcal{M}'(0)\|.$$

The right-hand side is independent of X since $\mathcal{M}'$ is Lipschitz-continuous and $\|X\| \leq \|G_{\min}\|$. □

Our second step is to prove similar properties for the operator $\mathcal{L}'$. We shall temporarily denote by $\mathcal{A}$ the operator $\mathcal{A}V = (I - V)^{-1}$, for V in $\langle 0, U\rangle$. It is easily seen that

$$\mathcal{A}V \leq \mathcal{A}W \leq \mathcal{A}U, \qquad \text{for all } V \leq W \text{ in } \langle 0, U\rangle, \tag{6.53}$$

and also that $\mathcal{A}$ is Fréchet-differentiable in $\langle 0, U\rangle$, with the derivative $\mathcal{A}'(V)$ being defined as follows:

$$\mathcal{A}'(V)H = (\mathcal{A}V)H(\mathcal{A}V). \tag{6.54}$$

Lemma 6.25 *The operator $\mathcal{L}'$ is Lipschitz-continuous and uniformly bounded in $\langle 0, U\rangle$.*

Proof Assume that H is a matrix of norm 1, and that V and W are in $\langle 0, G_{\min}\rangle$. Then

$$\begin{aligned}
&\|[\mathcal{L}'(V) - \mathcal{L}'(W)]H\| \\
&\leq \|(\mathcal{A}V)H(\mathcal{A}V)A_{-1} - (\mathcal{A}W)H(\mathcal{A}W)A_{-1}\| \\
&\leq (\|\mathcal{A}V\| + \|\mathcal{A}W\|)\|H\|\|\mathcal{A}V - \mathcal{A}W\|\|A_{-1}\| \\
&\leq 2\|A_{-1}\|\|\mathcal{A}U\|\|\mathcal{A}V - \mathcal{A}W\|, \qquad \text{by (6.53), (6.51)} \\
&\leq 2\|A_{-1}\|\|\mathcal{A}U\|(\sup_{0\leq t\leq 1} \|\mathcal{A}'(V_t)\|)\|V - W\|,
\end{aligned}$$

by [95], section 3.2.3, where $V_t = V + t(W - V)$ is in $\langle 0, U\rangle$ for $0 \leq t \leq 1$. From (6.53), (6.54), we obtain that

$$\sup_{0\leq t\leq 1} \|\mathcal{A}'(V_t)\| \leq (\sup_{0\leq t\leq 1} \|\mathcal{A}V_t\|)^2 \leq \|\mathcal{A}U\|^2.$$

and eventually that

$$\|\mathcal{L}'(V) - \mathcal{L}'(W)\| \leq \gamma_2 \|V - W\|,$$

where $\gamma_2 = 2\|A_{-1}\|\|\mathcal{A}U\|^2$ is independent of V and W. This proves that $\mathcal{L}'(V)$ is Lipschitz-continuous.

Finally, we have that

$$\|\mathcal{L}'(V)\| \leq \|\mathcal{L}'(V) - \mathcal{L}'(0)\| + \|\mathcal{L}'(0)\|,$$

where the right-hand side is independent of V, since $\mathcal{L}'$ is Lipschitz-continuous and $\|V\| \leq \|U\|$. □

Since both $\mathcal{M}'$ and $\mathcal{L}'$ are Fréchet derivatives, the chain-rule applies and $\mathcal{F}'(X) = \mathcal{I} - \mathcal{L}'(\mathcal{M}X)\mathcal{M}'(X)$ is the Fréchet derivative of $\mathcal{F}$ at X. The operator $\mathcal{F}$ and its derivative have the following properties.

Lemma 6.26 *The operator $\mathcal{F}$ is order-concave on $\langle 0, G_{\min}\rangle$; the operator $\mathcal{F}'$ is antitone and Lipschitz-continuous on $\langle 0, G_{\min}\rangle$.*

Proof Let us consider two matrices X and Y in $\langle 0, G_{\min}\rangle$ such that $X \leq Y$, and let us define $V = \mathcal{M}X, W = \mathcal{M}Y$. The following inequality immediately results from (6.42), (6.49), (6.50):

$$\mathcal{F}'(X) - \mathcal{F}'(Y) = \mathcal{L}'(W)\mathcal{M}'(Y) - \mathcal{L}'(V)\mathcal{M}'(X) \geq 0,$$

which proves that $\mathcal{F}'$ is antitone. Then, if $X \leq Y$,

$$[\mathcal{F}'(X) - \mathcal{F}'(Y)](X - Y) \leq 0,$$

which proves that $\mathcal{F}$ is order-concave (see [95] section 13.3.2).

The last statement is proved as follows. For any X, Y in $\langle 0, G_{\min}\rangle$, we have that

$$\begin{aligned}
&\|\mathcal{F}'(X) - \mathcal{F}'(Y)\| \\
&\quad = \|\mathcal{L}'(\mathcal{M}X)\mathcal{M}'(X) - \mathcal{L}'(\mathcal{M}Y)\mathcal{M}'(Y)\| \\
&\quad \leq \|\mathcal{L}'(\mathcal{M}X) - \mathcal{L}'(\mathcal{M}Y)\|\|\mathcal{M}'(X)\| + \|\mathcal{L}'(\mathcal{M}Y)\|\|\mathcal{M}'(X) - \mathcal{M}'(Y)\| \\
&\quad \leq c_1\|\mathcal{L}'(\mathcal{M}X) - \mathcal{L}'(\mathcal{M}Y)\| + c_2\|\mathcal{L}'(\mathcal{M}Y)\|\|X - Y\|,
\end{aligned}$$

for some constants c_1 an c_2 by Lemma 6.24. Since X and Y are in $\langle 0, G_{\min}\rangle$, therefore $\mathcal{M}X$ and $\mathcal{M}Y$ are in $\langle 0, U\rangle$ and we may apply Lemma 6.25, to obtain

$$\|\mathcal{F}'(X) - \mathcal{F}'(Y)\| \leq (c_3c_1 + c_4c_2)\|X - Y\|$$

for some constants c_3 an c_4. This proves that $\mathcal{F}'$ is Lipschitz-continuous in $\langle 0, G\rangle$. □

Finally, we prove that $\mathcal{F}'(X)$ is nonsingular. We shall write $\mathcal{F}'(X)$ as $\mathcal{I} - \mathcal{P}(X)$, with

$$\begin{aligned}
\mathcal{P}(X)H &= \mathcal{L}'(\mathcal{M}X)\mathcal{M}'(X)H \\
&= (I - \mathcal{M}X)^{-1}\sum_{i=1}^{+\infty} A_i \sum_{j=0}^{i-1} X^j H X^{i-1-j}(I - \mathcal{M}X)^{-1}A_{-1}.
\end{aligned}$$

The linear operator $\mathcal{P}(X)$ is nonnegative, and is increasing in $\langle 0, G_{\min}\rangle$: $\mathcal{P}(X) \leq \mathcal{P}(Y)$ for $X \leq Y$, by (6.42), (6.49), (6.50).

Lemma 6.27 *If the constant μ defined in Theorem 4.7 is such that $\mu < 0$, then the spectral radius of $\mathcal{P}(X)$ is strictly less than one for all X in $\langle 0, G_{\min}\rangle$. This implies that $\mathcal{F}'(X) = \mathcal{I} - \mathcal{P}(X)$ is nonsingular, and that $\mathcal{F}'(X)^{-1} = \sum_{\nu\geq 0}(\mathcal{P}(X))^\nu$ is nonnegative. If $\mu = 0$, then $\mathcal{F}'(G_{\min})$ is singular.*

Proof Since $\mathcal{P}(X)$ is increasing in $\langle 0, G_{\min}\rangle$, and since $\rho(A) \leq \rho(B)$ for $0 \leq A \leq B$, it is enough to prove the stated properties for $\mathcal{P}(G_{\min})$, which may be written as

$$\mathcal{P}(G_{\min})H = (I - U)^{-1}\sum_{i=1}^{+\infty} A_i \sum_{j=0}^{i-1} G^j H G^{i-j},$$

since $\mathcal{M}G_{\min} = U$, and $(I - U)^{-1}A_{-1} = \mathcal{L}U = G_{\min}$. After elementary manipulations, we find that

$$\mathcal{P}(G_{\min}) = (I - U)^{-1}[\mathcal{T}^*(G_{\min}) - U\mathcal{I}],$$

where the linear operator $\mathcal{T}^*(G_{\min})$ is defined as [90], Equation (46); its spectral radius is strictly less than one if $\mu < 0$, and equal to one if $\mu = 0$ ([90], Corollary 3 and Lemma 5).

Since we also have that

$$\mathcal{F}'(G_{\min}) = \mathcal{I} - \mathcal{P}(G_{\min}) = (I - U)^{-1}[\mathcal{I} - \mathcal{T}^*(G_{\min})],$$

we see that $\mathcal{F}'(G_{\min})$ is nonsingular if $\mu < 0$.

Define the matrix M which is the solution to the equation

$$(\mathcal{I} - \mathcal{P}(G_{\min}))M = (I - U)^{-1}E,$$

where E is the matrix with all elements equal to one. We have that

$$\begin{aligned} M &= (\mathcal{I} - \mathcal{P}(G_{\min}))^{-1}(I - U)^{-1}E \\ &= (\mathcal{I} - \mathcal{T}^*(G_{\min}))^{-1}E \\ &= \sum_{\nu\geq 0}(\mathcal{T}^*(G_{\min}))^\nu E \geq E > 0, \end{aligned}$$

since the spectral radius of $\mathcal{T}^*(G_{\min})$ is strictly less than one. As $(I - U)^{-1} \geq I$, we have that $(\mathcal{I} - \mathcal{P}(G_{\min}))M \geq E$, which implies that $\mathcal{P}(G_{\min})M \leq M - E < M$. This last inequality, together with the facts that $M > 0$ and that $\mathcal{P}(G_{\min})$ is nonnegative, implies that the spectral radius of $\mathcal{P}(G_{\min})$ is strictly less than one. □

The theorem below is stated without proof. In view of Lemmas 6.24–6.27, we merely need to paraphrase the proof of [95] 13.3.4. As indicated in [95] page 452, there are four sets of natural versions of that theorem. Ours correspond to the fourth column in [95] Table 13.1, page 444.

Theorem 6.28 *Let $\mathcal{F}$ be defined by* (6.41), (6.43), (6.45). *Assume that A is irreducible, that $\sum_{i=-1}^{+\infty}(i+1)^2 A_i$ is convergent, and that $\mu < 0$. Assume that X_0 is chosen such that $0 \le X_0 \le G_{\min}$, and $\mathcal{F}X_0 \le 0$. Then Newton's iterates*

$$X_{n+1} = X_n - \mathcal{F}'(X_n)^{-1}\mathcal{F}X_n, \qquad \text{for } n \ge 0,$$

are well defined and converge to $G_{\min}$ as n tends to infinity. They satisfy the inequality $X_n \le X_{n+1}$ for all $n \ge 0$. There exists a positive constant c such that

$$\|X_{n+1} - G_{\min}\| \le c\|X_n - G_{\min}\|^2,$$

for all $n \ge 0$.

6.3.1 Computational issues

The major problem is to evaluate expressions of the form $\mathcal{F}'(X)^{-1}W$ for certain substochastic matrices X and W, see equation (6.46). In other words, for given matrices X and W, we have to find the unique matrix Z such that $\mathcal{F}'(X)Z = W$, or

$$Z - (I - \mathcal{M}X)^{-1}\sum_{i=1}^{+\infty} A_i \sum_{j=0}^{i-1} X^j Z X^{i-1-j}(I - \mathcal{M}X)^{-1}A_{-1} = W.$$

One method proceeds by successive substitutions; another transforms the equation above into a more standard linear system equation. This is a well-known procedure, see Lancaster and Tismenetsky [74], which we have already introduced in Section 2.1. In fact, with the notation (2.1), in light of (2.2) one has $(I - D_X)\boldsymbol{z} = \mathrm{vec}(W)$, where $\boldsymbol{z} = \mathrm{vec}(\mathcal{F}'(X)^{-1}W)$,

$$D_X = \sum_{i=1}^{+\infty}\sum_{j=0}^{i-1} A_{-1}^{\mathrm{T}}(I - V^{\mathrm{T}})^{-1}(X^{\mathrm{T}})^{i-1-j} \otimes (I - V)^{-1}A_i X^j,$$

and $V = \mathcal{M}X$.

The resulting scheme is described in Algorithm 6.3 where we assumed that the matrices A_i are negligible for $i \ge k+1$. If $\mu < 0$, then it is known that $G_{\min}$ is stochastic, so that the stopping criterion $\|Y\mathbf{1} - \mathbf{1}\|_\infty \le \epsilon$ guarantees that $\|Y - G_{\min}\|_\infty \le \epsilon$.

A simple analysis shows that the number of multiplications and divisions needed by one step of Newton's iteration is given by

$$\frac{1}{3}m^6 + km^4 + \frac{1}{2}k^2m^3 + \frac{3}{2}km^3 - \frac{2}{3}m^3 + O(m). \tag{6.55}$$

This is much larger than the complexity of one step of the very simple algorithms which implements the classic iterations of Section 6.2 described in Algorithm 6.2. Newton's method, therefore, is not very attractive, except in cases where m and k have small values and where the number of iterations required by the classic iterations is very high. However, it may be used as a standard with which one may compare the speed of other algorithms.

Algorithm 6.3 Newton's iteration: M/G/1 case

INPUT: The integer $k > 0$ and the $m \times m$ matrix coefficients A_i, $i = -1, \dots, k$, of the matrix polynomial $A(z) = \sum_{i=-1}^{k} z^{i+1} A_i$, such that $\mu < 0$; a positive ϵ.

OUTPUT: An approximation Y to the minimal nonnegative solution $G_{\min}$ of the equation $X = A(X)$ such that $\|Y - G_{\min}\|_\infty \leq \epsilon$.

COMPUTATION

1. Set $Y = (I - A_0)^{-1} A_{-1}$.
2. Set $V = A_k$.
3. For $i = 1$ to $k+1$ compute $V = A_{k-i} + VY$.
4. Set $D_Y = 0$ and $T = A_{-1}^{\mathrm{T}} (I - V^{\mathrm{T}})^{-1}$.
5. For $i = 1$ to k
   (a) Set $S = A_k$.
   (b) For $j = k - i$ to 1 compute $S = A_{k-j} + SY$.
   (c) Compute $S = (I - V)^{-1} S$, $D_Y = D_Y + T \otimes S$ and $T = TY^{\mathrm{T}}$.
6. Compute $\boldsymbol{z} = (I - D_Y)^{-1} \operatorname{vec}((I - V)^{-1} A_{-1} - Y)$.
7. Compute $Y = Y + \operatorname{vec}_m^{-1}(\boldsymbol{z})$.
8. If $\|Y\mathbf{1} - \mathbf{1}\|_\infty \leq \epsilon$ then output Y, else repeat from step 2.

6.4 Bibliographic notes

A general theory on functional iterations can be found in the book [95]. Various iterative methods for solving matrix equations in Markov chains have been proposed by M. Neuts, V. Ramaswami and G. Latouche in [91], [92], [97], [75]. The monotonic convergence of natural and traditional iterations is investigated by M. Neuts in [92]. The iteration based on the matrix U has been introduced and analyzed in [76] where the different speeds of convergence of the sequences $\{X_n^{(N)}\}_{n \geq 0}$, $\{X_n^{(T)}\}_{n \geq 0}$ and $\{X_n^{(U)}\}_{n \geq 0}$ is proved and where the different behavior of the sequences obtained with $X_0 = 0$ and X_0 stochastic has been observed. A systematic analysis of the rate of convergence of these sequences is performed in [88]. The analysis of the stability of the convergence to $G_{\min}$ and to G_{sto} for transient QBD processes is performed in [81].

Newton's iteration has been introduced and studied in [77].

In the literature there exist other results concerning functional iterations for solving matrix equations in Markov chains, which are not reported in this book. In particular, acceleration techniques of classical functional iterations, based on the choice of the starting approximation X_0 and on relaxation techniques, are developed in [38] and [37]. These techniques, however, in order to be effective, require additional information on the spectral properties of $G_{\min}$. Different functional iterations which are inverse-free have been introduced in [6] and analyzed in [55]. The speed of convergence of these iterations is compared with the convergence of classical linearly convergent iterations only in the case where $X_0 = 0$. Functional iterations having linear convergence with an arbitrarily small convergence rate have been designed in [10] and are reported in Section 8.3.

Functional iterations for solving quadratic matrix equations encountered in different contexts are also investigated in [64], [65].

7

LOGARITHMIC REDUCTION AND CYCLIC REDUCTION

7.1 Introduction

In this chapter we continue the analysis of numerical methods for computing the minimal nonnegative solution $G_{\min}$ of the matrix equation (4.4), that is, $X = \sum_{i=-1}^{+\infty} A_i X^{i+1}$ associated with the M/G/1-type Markov chain with transition matrix (4.3). We recall that A_i, $i = -1, 0, \ldots$, are $m \times m$ matrices with nonnegative elements such that $\sum_{i=-1}^{+\infty} A_i$ is stochastic.

Here we follow an approach which is completely different from the one used in Chapter 6. The methods that we will design in this chapter, based on a divide-and-conquer strategy, have the nice feature of providing sequences of approximations which generally converge quadratically to $G_{\min}$ and do not require any initial approximation in order to be started.

Unlike Newton's iteration, which shares with the algorithms of this chapter the quadratic convergence speed, the cost of one step of the algorithms designed here is comparable with the cost of one step of the classic functional iterations of Chapter 6. This makes the methods particularly effective, especially in the cases where the Markov chain is close to being null recurrent. In fact in this situation the number of iterations needed by the functional iteration methods of Chapter 6 is generaly extremely large whereas the methods based on divide-and-conquer strategies still converge quickly. In particular, for null recurrent Markov chains, the convergence of classical functional iterations is sublinear whereas the convergence of the methods designed in this chapter is at least linear and it is possible to turn to quadratic convergence with a simple modification.

In the case of QBD processes, the methods based on the divide-and-conquer technique can be described and implemented in a very simple way. In the general case of M/G/1-type Markov chains the algorithms are more involved and they heavily rely on the structured matrix tools introduced in Chapters 2 and 3.

We start first by describing the algorithms in the case of QBD processes and then we treat the general M/G/1-type case by extending almost all the properties valid for QBDs.

Throughout this chapter we assume that Condition 4.20 is valid and that $\boldsymbol{a} = \sum_{i=0}^{+\infty}(i+1)A_i\mathbf{1}$ is finite, so that in light of Theorem 4.24, $z = 1$ is the only zero of modulus 1 of the function $a(z) = \det(zI - A(z))$ defined in (4.23), where $A(z) = \sum_{i=-1}^{+\infty} z^{i+1} A_i$.

7.2 Logarithmic reduction

The algorithm of *logarithmic reduction* applies to QBD processes that is computes the minimal nonnegative solution $G_{\min}$ of the quadratic matrix equation

$$X = A_{-1} + A_0 X + A_1 X^2. \tag{7.1}$$

The idea of logarithmic reduction relies on the possibility of expressing $G_{\min}$ in terms of $G_{\min}^2$, then expressing $G_{\min}^2$ in terms of $G_{\min}^4$ and so forth, so that one arrives at a formula expressing $G_{\min}$ in terms of $G_{\min}^{2^n}$. This formula provides an effective tool for approximating $G_{\min}$ since, in light of Theorem 4.26, $G_{\min}^{2^n}$ quadratically converges to 0 or to a matrix of rank 1 if the QBD is transient or positive recurrent, respectively.

Since $G_{\min}$ is a solution of (7.1), then we have

$$G_{\min} = B_{-1} + B_1 G_{\min}^2 \tag{7.2}$$

where $B_{-1} = (I - A_0)^{-1} A_{-1}$, $B_1 = (I - A_0)^{-1} A_1$. Post-multiplying (7.2) by $G_{\min}$ and by $G_{\min}^2$ yields

$$G_{\min}^2 = B_{-1} G_{\min} + B_1 G_{\min}^3, \tag{7.3}$$
$$G_{\min}^3 = B_{-1} G_{\min}^2 + B_1 G_{\min}^4. \tag{7.4}$$

By performing a simple formal manipulation of the two equations above and of (7.2) it is possible to arrive at a matrix equation in $G_{\min}^2$. More precisely, premultiply (7.2) by B_{-1}, premultiply (7.4) by B_1, sum the equations obtained in this way with (7.3) and get

$$G_{\min}^2 = B_{-1}^2 + (B_{-1} B_1 + B_1 B_{-1}) G_{\min}^2 + B_1^2 G_{\min}^4.$$

If the matrix $I - B_{-1} B_1 - B_1 B_{-1}$ is nonsingular, the latter equation allows one to express $G_{\min}^2$ as a function of $G_{\min}^4$ as

$$G_{\min}^2 = B_{-1}^{(1)} + B_1^{(1)} G_{\min}^4 \tag{7.5}$$

where

$$B_{-1}^{(1)} = (I - B_{-1} B_1 - B_1 B_{-1})^{-1} B_{-1}^2, \quad B_1^{(1)} = (I - B_{-1} B_1 - B_1 B_{-1})^{-1} B_1^2.$$

In this way, replacing (7.5) in (7.2) provides the following expression of $G_{\min}$ in terms of $G_{\min}^4$

$$G_{\min} = B_{-1} + B_1 B_{-1}^{(1)} + B_1 B_1^{(1)} G_{\min}^4.$$

Assuming that all the matrices which must be inverted are nonsingular, this process can be recursively repeated by generating successive expressions of $G_{\min}$ as functions of $G_{\min}^2, G_{\min}^4, G_{\min}^8, \ldots$. More formally, assuming that no singularity is encountered, we generate the sequences of matrices $\{B_{-1}^{(n)}\}_{n\geq 0}$ and $\{B_1^{(n)}\}_{n\geq 0}$ by means of the following recursions

$$\begin{aligned} B_{-1}^{(n+1)} &= (C^{(n)})^{-1}(B_{-1}^{(n)})^2, \\ B_1^{(n+1)} &= (C^{(n)})^{-1}(B_1^{(n)})^2, \quad n \geq 0, \end{aligned} \tag{7.6}$$

where

$$C^{(n)} = I - B_{-1}^{(n)}B_1^{(n)} - B_1^{(n)}B_{-1}^{(n)} \tag{7.7}$$

and

$$B_{-1}^{(0)} = (I - A_0)^{-1}A_{-1}, \quad B_1^{(0)} = (I - A_0)^{-1}A_1.$$

The sequences of matrices generated in this way constitute the *logarithmic reduction* algorithm.

We may easily verify that, for any $n \geq 0$,

$$G_{\min} = B_{-1}^{(0)} + \sum_{i=1}^{n}\Big(\prod_{j=0}^{i-1} B_1^{(j)}\Big)B_{-1}^{(i)} + \prod_{j=0}^{n} B_1^{(j)} G_{\min}^{2^{n+1}}, \tag{7.8}$$

where, given the matrices $R_0, \ldots, R_i$, the product $\prod_{j=0}^{i} R_i$ is equal to $R_0R_1\cdots R_i$.

In the case where the drift μ is nonzero (see equation (4.17)) so that the QBD is positive recurrent or transient, we may show that no breakdown is encountered and that the matrix $B_{-1}^{(n)} + B_1^{(n)}$ is stochastic for any $n \geq 0$ as stated by the following:

Theorem 7.1 *If $\mu \neq 0$ then the matrices $C^{(n)} = I - B_{-1}^{(n)}B_1^{(n)} - B_1^{(n)}B_{-1}^{(n)}$ are nonsingular M-matrices for any $n \geq 0$. Moreover, $B_{-1}^{(n)} \geq 0$, $B_1^{(n)} \geq 0$ and $B_{-1}^{(n)} + B_1^{(n)}$ is stochastic.*

Proof One shows by induction that $B_1^{(n)}$ contains the probability that, starting from level zero, the doubly infinite QBD with transition matrix

$$P' = \begin{bmatrix} \ddots & \ddots & & & & 0 \\ \ddots & A_0 & A_1 & & & \\ & A_{-1} & A_0 & A_1 & & \\ & & A_{-1} & A_0 & A_1 & \\ & & & A_{-1} & A_0 & \ddots \\ 0 & & & & \ddots & \ddots \end{bmatrix}$$

reaches the level 2^n before it reaches the level -2^n and $B_{-1}^{(n)}$ contains the probability that the QBD reaches the level -2^n before 2^n; the detailed argument is given in [79, page 191]. This leads us to the interpretation that $B_{-1}^{(n)}B_1^{(n)} + B_1^{(n)}B_{-1}^{(n)}$ is the probability that, starting from level zero, the QBD returns to level zero *after* having reached 2^n or -2^n but *before* having reached 2^{n+1} or -2^{n+1}, which shows that $C^{(n)}$ is an M-matrix. Under Condition 4.20, the QBD may not remain forever locked between these two levels, so that, by Theorem 1.14, $C^{(n)}$ is nonsingular.

We prove the second part of the theorem by induction on n. For $n = 0$ we have $B_{-1}^{(0)} = (I - A_0)^{-1}A_{-1} \geq 0$, $B_1^{(0)} = (I - A_0)^{-1}A_1 \geq 0$ and $(B_{-1}^{(0)} + B_1^{(0)})\mathbf{1} = (I - A_0)^{-1}(A_{-1} + A_1)\mathbf{1} = \mathbf{1}$. Now assume that the property holds for $n \geq 0$, and we prove it for $n + 1$. The matrices $B_{-1}^{(n+1)}$ and $B_1^{(n+1)}$ are clearly nonnegative. Since $(B_{-1}^{(n)} + B_1^{(n)})\mathbf{1} = \mathbf{1}$ by the inductive hypothesis, then $(B_{-1}^{(n)} + B_1^{(n)})^2\mathbf{1} = \mathbf{1}$, therefore

$$(I - B_{-1}^{(n)}B_1^{(n)} + B_{-1}^{(n)}B_1^{(n)})^{-1}((B_{-1}^{(n)})^2 + (B_1^{(n)})^2)\mathbf{1} = \mathbf{1}.$$

□

If μ is different from zero, nice convergence properties of logarithmic reduction allow to deduce from (7.8) that the minimal nonnegative solution $G_{\min}$ of (7.1) is

$$G_{\min} = B_{-1}^{(0)} + \sum_{i=1}^{+\infty}\Big(\prod_{j=0}^{i-1} B_1^{(j)}\Big)B_{-1}^{(i)}. \tag{7.9}$$

Before showing this important property, we need to introduce the matrix $R_{\min}$, that is the minimal nonegative solution of the quadratic matrix equation

$$X = A_1 + XA_0 + X^2A_{-1}. \tag{7.10}$$

A crucial role in the analysis of the convergence is played by the spectral properties of the matrices $R_{\min}$ and $G_{\min}$. We recall that (see Section 5.6), if $\mu < 0$ then $\rho(G_{\min}) = 1$ and $\rho(R_{\min}) < 1$, if $\mu > 0$ then $\rho(G_{\min}) < 1$ and $\rho(R_{\min}) = 1$. If $\mu = 0$ the situation is more delicate, since both $G_{\min}$ and $R_{\min}$ have spectral radius equal to 1, however, under additional assumptions, some weaker convergence properties hold.

It is also useful to provide a functional interpretation of logarithmic reduction. To this puropose let us introduce the matrix polynomial

$$B^{(n)}(z) = B_{-1}^{(n)} + z^2B_1^{(n)}, \quad n = 0, 1, \ldots.$$

From (7.6) and (7.7) we may easily deduce that

$$(zI - B^{(n)}(z))(-zI - B^{(n)}(-z)) = -C^{(n)}(z^2I - B^{(n+1)}(z^2)), \tag{7.11}$$

and since the right-hand side of (7.11) is a function of z^2, we also have

$$(zI - B^{(n)}(z))(-zI - B^{(n)}(-z)) = (-zI - B^{(n)}(-z))(zI - B^{(n)}(z)), \tag{7.12}$$

that is, the matrix polynomials $zI - B^{(n)}(z)$ and $zI + B^{(n)}(-z)$ commute. Equation (7.11) is important in the analysis of the convergence properties of logarithmic reduction. A simple observation which follows from (7.11) is that the roots of the matrix polynomial $zI - B^{(n+1)}(z)$ are the squares of the roots of the matrix polynomial $zI - B^{(n)}(z)$. Moreover, from (7.11) and (7.12) we find that the corresponding Jordan chains of $zI - B^{(n)}(z)$ and $zI - B^{(n+1)}(z)$ are the same. This implies that the number of roots of $zI - B^{(n)}(z)$ inside, outside the unit disk and on the unit circle, respectively, is the same for any $n \geq 0$. Observe that, since $zI - B^{(0)}(z) = (I - A_0)^{-1}(zI - A(z))$, the roots of $zI - B^{(0)}(z)$ coincide with the roots of $zI - A(z)$.

In order to better understand the role of (7.11) consider the case where the logarithmic reduction is convergent and denote by $B^{(\infty)}(z)$ the limit of $B^{(n)}(z)$. If $zI - B^{(0)}(z)$ has k roots of modulus less than 1 and h roots of modulus greater than 1, then $zI - B^{(\infty)}(z)$ has k roots equal to 0 and h roots at infinity. Moreover, the Jordan chains of $zI - B^{(\infty)}(z)$ corresponding to 0 and to infinity coincide with the union of the Jordan chains of $zI - B^{(0)}(z)$ corresponding to the roots inside the unit disk and outside the unit disk, respectively.

In the different cases where $\mu < 0$ or $\mu > 0$, we may provide more specific convergence properties of logarithmic reduction based on the different location of the roots of $zI - A(z)$ with respect to the unit circle.

Theorem 7.2 *If the drift μ is negative, equal to zero, or positive, respectively, then for each $n \geq 0$, the drift associated with the matrix polynomial $B^{(n)}(z)$ is negative, equal to zero, or positive, respectively. Moreover, the matrices $G_{\min}^{2^n}$ and $R_{\min}^{2^n}$ are the minimal nonnegative solutions of the matrix equations*

$$X = B_{-1}^{(n)} + B_1^{(n)} X^2,$$
$$X = B_1^{(n)} + X^2 B_{-1}^{(n)},$$

respectively.

Proof Assume that $\mu < 0$. Then the matrix polynomial $zI - B^{(0)}(z)$ has $m - 1$ roots of modulus less than 1, and one root equal to 1. Therefore, for the properties of logarithmic reduction, the matrix polynomial $zI - B^{(n)}(z)$ has $m - 1$ roots of modulus less than 1, and one root equal to 1, i.e. the drift associated with $B^{(n)}(z)$ is negative as well (compare with Theorem 5.20). A similar argument applies if $\mu = 0$ or $\mu > 0$.

We prove now the second part of the theorem by induction on n. For $n = 0$ the thesis is obviously true. Assume that $G_{\min}^{2^n}$ is the minimal nonnegative solution of $X = B_{-1}^{(n)} + B_1^{(n)} X^2$, for a fixed $n \geq 0$. Then we apply the same argument used for proving (7.5). That is, we consider the equation $G_{\min}^{2^n} = B_{-1}^{(n)} + B_1^{(n)}(G_{\min}^{2^n})^2$ and the two equations obtained by post-multiplying it by $G_{\min}^{2^n}$ and by $(G^{2^n})^2$, respectively. Manipulating these three equations as we already performed at the beginning of this section, yields

$$G_{\min}^{2^{n+1}} = (B_{-1}^{(n)})^2 + (B_{-1}^{(n)} B_1^{(n)} + B_1^{(n)} B_{-1}^{(n)}) G_{\min}^{2^{n+1}} + (B_1^{(n)})^2 (G_{\min}^{2^{n+1}})^2$$

which implies $G_{\min}^{2^{n+1}} = B_{-1}^{(n+1)} + B_1^{(n+1)}(G_{\min}^{2^{n+1}})^2$. That is, $G_{\min}^{2^{n+1}}$ is a solution of the matrix equation $X = B_{-1}^{(n+1)} + B_1^{(n+1)} X^2$. Moreover, $G_{\min}^{2^{n+1}}$ is the minimal solution since its eigenvalues are the m roots of the smallest moduli of $zI - B^{(n+1)}(z)$ (compare with Theorem 4.10).

The same argument can be applied for $R_{\min}$. □

It is useful to introduce special notation for the root with largest modulus of $zI - A(z)$ in the open unit disk and the root of smallest modulus out of the closed unit disk. More precisely we denote

$$\begin{aligned} \xi &= \min\{|z| : \ \det(zI - A(z)) = 0, |z| > 1\}, \\ \eta &= \max\{|z| : \ \det(zI - A(z)) = 0, |z| < 1\}. \end{aligned} \tag{7.13}$$

Recall that, if $\mu < 0$, then for Theorem 4.12 and Remark 4.25, ξ is a simple zero of $\det(zI - A(z))$ and there are no other zeros with the same modulus. Similarly, if $\mu > 0$, then η is a simple zero of $\det(zI - A(z))$ and there are no other zeros with the same modulus. This follows by applying Theorem 4.12 and Remark 4.25 to the reversed matrix polynomial $A'(z) = A_1 + zA_0 + z^2 A_{-1}$ which defines a QBD with the drift $\mu < 0$. In fact the roots of $zI - A'(z)$ are the reciprocal of the roots of $zI - A(z)$.

If $\mu \neq 0$, we may prove suitable convergence properties of logarithmic reduction which allow the approximation of $G_{\min}$.

Theorem 7.3 *If $\mu < 0$ then for any matrix norm and for any $\epsilon > 0$ such that $\epsilon + \eta < 1$, there exists $\gamma > 0$ such that*

$$\begin{aligned} &\|B_1^{(n)}\| \leq \gamma \xi^{-2^n}, \\ &\|B_{-1}^{(n)} - \mathbf{1}\boldsymbol{g}^{\mathrm{T}}\| \leq \gamma\big(\xi^{-2^n} + (\eta + \epsilon)^{2^n}\big), \\ &\|I - C^{(n)}\| \leq \gamma \xi^{-2^n} \end{aligned}$$

and

$$\|G_{\min} - B_{-1}^{(0)} - \sum_{i=1}^{n}\Big(\prod_{j=0}^{i-1} B_1^{(j)}\Big) B_{-1}^{(i)}\| \leq \gamma \xi^{-2^n},$$

where $\boldsymbol{g} \geq 0$ is such that $\boldsymbol{g}^{\mathrm{T}} G_{\min} = \boldsymbol{g}^{\mathrm{T}}$, $\boldsymbol{g}^{\mathrm{T}}\mathbf{1} = 1$.

If $\mu > 0$ then, for any matrix norm and for any $\epsilon > 0$ such that $\epsilon + \xi^{-1} < 1$, there exists $\gamma > 0$ such that

$$\begin{aligned} &\|B_{-1}^{(n)}\| \leq \gamma \eta^{2^n}, \\ &\|B_1^{(n)} - \mathbf{1}\boldsymbol{g}'^{\mathrm{T}}\| \leq \gamma\big((\xi^{-1} + \epsilon)^{2^n} + \eta^{2^n}\big), \\ &\|I - C^{(n)}\| \leq \gamma \eta^{2^n}, \end{aligned}$$

and

$$\|G_{\min} - B_{-1}^{(0)} - \sum_{i=1}^{n}\Big(\prod_{j=0}^{i-1} B_1^{(j)}\Big) B_{-1}^{(i)}\| \leq \gamma \eta^{2^n},$$

where $\boldsymbol{g}' \geq 0$ is such that $\boldsymbol{g}'^{\mathrm{T}} \widehat{G}_{\min} = \boldsymbol{g}'^{\mathrm{T}}$, $\boldsymbol{g}'^{\mathrm{T}}\mathbf{1} = 1$, and $G'_{\min}$ is the minimal nonnegative solution of the equation $X = A_1 + A_0 X + A_{-1} X^2$.

Proof Let us first suppose that $\mu < 0$. In this case the matrix equation (7.10) has a minimal nonnegative solution $R_{\min}$ such that $\rho(R_{\min}) = \xi^{-1}$; moreover ξ^{-1} is simple and there are no other eigenvalues with the same modulus (see Section 5.6). In light of Theorem 7.2 one has

$$0 \leq B_1^{(n)} = R_{\min}^{2^n} - R_{\min}^{2\cdot 2^n} B_{-1}^{(n)} \leq R_{\min}^{2^n}.$$

For Theorem A.2 in the appendix, there exists a matrix norm $\|\cdot\|$ such that $\|R_{\min}\| = \xi^{-1}$, therefore $\|R_{\min}^{2^n}\| \leq \xi^{-2^n}$. Now observe that $\|B_1^{(n)}\|_\infty \leq \|R_{\min}^{2^n}\|_\infty$, since $\|\cdot\|_\infty$ is monotonic when applied to nonnegative matrices. For the equivalence of matrix norms (see Theorem A.1), there exists $\alpha > 0$ such that $\|R_{\min}^{2^n}\|_\infty \leq \alpha\|R_{\min}^{2^n}\| \leq \alpha\xi^{-2^n}$, therefore $\|B_1^{(n)}\|_\infty \leq \alpha\xi^{-2^n}$. For the equivalence of matrix norms we deduce that for any matrix norm there exists $\gamma > 0$ such that $\|B_1^{(n)}\| \leq \gamma\xi^{-2^n}$. Concerning the convergence of the sequence $\{B_{-1}^{(n)}\}_{n\geq 0}$, observe that, for any matrix norm

$$\|B_{-1}^{(n)} - \mathbf{1}\boldsymbol{g}^{\mathrm{T}}\| \leq \|\mathbf{1}\boldsymbol{g}^{\mathrm{T}} - G_{\min}^{2^n}\| + \|B_{-1}^{(n)} - G_{\min}^{2^n}\|.$$

From Theorem 7.2 we deduce that

$$\|B_{-1}^{(n)} - \mathbf{1}\boldsymbol{g}^{\mathrm{T}}\| \leq \|\mathbf{1}\boldsymbol{g}^{\mathrm{T}} - G_{\min}^{2^n}\| + \|B_1^{(n)} G_{\min}^{2\cdot 2^n}\| \leq \|\mathbf{1}\boldsymbol{g}^{\mathrm{T}} - G_{\min}^{2^n}\| + \|B_1^{(n)}\| \cdot \|G_{\min}^{2\cdot 2^n}\|.$$

By Theorem 4.26 for any matrix matrix norm $\|\cdot\|$ there exists $\gamma' > 0$ such that $\|G_{\min}^n - \mathbf{1}\boldsymbol{g}^{\mathrm{T}}\| \leq \gamma'(\eta+\epsilon)^n$. Since the sequence $\{G_{\min}^{2^n}\}_{n\geq 0}$ is bounded we conclude that for any norm $\|\cdot\|$, we have $\|B_{-1}^{(n)} - \mathbf{1}\boldsymbol{g}^{\mathrm{T}}\| \leq \gamma\big((\eta+\epsilon)^{2^n} + \xi^{-2^n}\big)$ for a suitable positive γ. The convergence of the sequence $I - C^{(n)}$ follows directly from the convergence of the sequences $B_{-1}^{(n)}$ and $B_1^{(n)}$. Now, from (7.8) and from the inequality $0 \leq B_1^{(n)} \leq R_{\min}^{2^n}$, one has

$$\begin{aligned} 0 \leq G_{\min} - B_{-1}^{(0)} - \sum_{i=1}^{n}\Big(\prod_{j=0}^{i-1} B_1^{(j)}\Big) B_{-1}^{(i)} &= \prod_{j=0}^{n} B_1^{(j)} G_{\min}^{2^{n+1}} \\ \leq \prod_{j=0}^{n} R_{\min}^{2^j} G_{\min}^{2^{n+1}} &= R_{\min}^{2^{n+1}-1} G_{\min}^{2^{n+1}}. \end{aligned}$$

Taking the infinity norm of both sides of the latter equation, since $\|A\|_\infty \leq \|B\|_\infty$ for $0 \leq A \leq B$, we find that

$$\begin{aligned} \|G_{\min} - B_{-1}^{(0)} - \sum_{i=1}^{n}\Big(\prod_{j=0}^{i-1} B_1^{(j)}\Big) B_{-1}^{(i)}\|_\infty &\leq \|R_{\min}^{2^{n+1}-1} G_{\min}^{2^{n+1}}\|_\infty \\ &\leq \|R_{\min}^{2^{n+1}-1}\|_\infty \|G_{\min}\|_\infty^{2^{n+1}} \\ &\leq \|R_{\min}^{2^{n+1}-1}\|_\infty. \end{aligned}$$

The proof is completed by invoking Theorem A.2 in the appendix.

If $\mu > 0$, it is sufficient to consider the QBD defined by the matrices $A'_{-1} = A_1$, $A'_0 = A_0$, $A'_1 = A_{-1}$. For the latter QBD the drift is negative, and therefore the first part of the theorem holds. Whence we deduce the second part. □

Observe that from the above theorem we deduce that the sequences $\{B_{-1}^{(n)}\}_n$ and $\{B_1^{(n)}\}_n$ have quadratic convergence. At least one of the two sequences converges to zero in the case where μ is different from zero. It is also interesting to observe that the sequence $\{C^{(n)}\}$ generated by logarithmic reduction converges to the identity matrix.

It is worth pointing out that the matrices $X_n = B_{-1}^{(0)} + \sum_{i=1}^{n}\left(\prod_{j=0}^{i-1} B_1^{(j)}\right) B_{-1}^{(i)}$ provide a nondecreasing sequence of nonnegative matrices which quadratically converges to $G_{\min}$. This fact is very useful in order to design a stopping criterion for the logarithmic reduction in the positive recurrent case where $G_{\min}$ is stochastic. In fact, in this case we have

$$1 - \|X_n\|_\infty = \|G_{\min} - X_n\|_\infty.$$

The expression in the left-hand side provides a method for computing the approximation error $\|G_{\min} - X_n\|_\infty$.

If $\mu = 0$, the convergence of logarithmic reduction still holds under additional assumptions, however the convergence becomes linear. For instance, consider the case where $m = 1$ and $A_0 = 1/2$, $A_1 = A_{-1} = 1/4$, that is, $B_{-1} = B_1 = 1/2$. In this case we have $G_{\min} = 1$ and $C^{(n)} = B_{-1}^{(n)} = B_1^{(n)} = 1/2$ for any $n \geq 0$ so that $\lim_n B_1^{(n)} \neq 0$ and $\lim_n B_{-1}^{(n)} \neq 0$. From (7.8) we find that the approximation error

$$G_{\min} - B_{-1}^{(0)} - \sum_{i=1}^{n}\left(\prod_{j=0}^{i-1} B_1^{(j)}\right) B_{-1}^{(i)}$$

is equal to $1/2^{n+1}$. That is, the convergence of logarithmic reduction becomes linear with convergence rate 1/2.

Since the QBD defined by the blocks $B_{-1}, 0, B_1$ has a null drift, also the Markov chain obtained by interchanging B_{-1} and B_1 has null drift. Therefore the quadratic matrix equation

$$X = B_1 + B_{-1}X^2$$

has a minimal nonnegative solution $G'_{\min}$ which is stochastic. Let us denote by $\boldsymbol{g}'$ its probability vector. Since 1 is the only eigenvalue of modulus 1 for both $G_{\min}$ and $G'_{\min}$, one has $\lim_n G_{\min}^n = \mathbf{1}\boldsymbol{g}^{\mathrm{T}}$ and $\lim_n {G'_{\min}}^n = \mathbf{1}\boldsymbol{g}'^{\mathrm{T}}$. From Theorem 7.2 one has

$$\begin{aligned} G_{\min}^{2^n} &= B_{-1}^{(n)} + B_1^{(n)} G_{\min}^{2\cdot 2^n}, \\ {G'_{\min}}^{2^n} &= B_1^{(n)} + B_{-1}^{(n)} {G'_{\min}}^{2\cdot 2^n}. \end{aligned} \tag{7.14}$$

Since $B_{-1}^{(n)} + B_1^{(n)}$ is stochastic and $B_{-1}^{(n)}$, $B_1^{(n)}$ are nonnegative, the sequences $\{B_{-1}^{(n)}\}$ and $\{B_1^{(n)}\}$ are bounded. Therefore both of them have convergent subsequences. Let $B_{-1}^{(\infty)}$ and $B_1^{(\infty)}$ denote the limits of some subsequence of $\{B_{-1}^{(n)}\}$ and $\{B_1^{(n)}\}$, respectively. From (7.14) one has

$$\begin{aligned}\mathbf{1}\boldsymbol{g}^{\mathrm{T}} &= B_{-1}^{(\infty)} + B_1^{(\infty)}\mathbf{1}\boldsymbol{g}^{\mathrm{T}},\\ \mathbf{1}\boldsymbol{g}'^{\mathrm{T}} &= B_1^{(\infty)} + B_{-1}^{(\infty)}\mathbf{1}\boldsymbol{g}'^{\mathrm{T}}.\end{aligned}$$

From the above equations we deduce that

$$\begin{aligned}B_{-1}^{(\infty)} &= (\mathbf{1} - B_1^{(\infty)}\mathbf{1})\boldsymbol{g}^{\mathrm{T}} = B_{-1}^{(\infty)}\mathbf{1}\boldsymbol{g}^{\mathrm{T}},\\ B_1^{(\infty)} &= (\mathbf{1} - B_{-1}^{(\infty)}\mathbf{1})\boldsymbol{g}'^{\mathrm{T}} = B_1^{(\infty)}\mathbf{1}\boldsymbol{g}'^{\mathrm{T}}.\end{aligned}$$

In particular the limits of the converging subsequences of $B_{-1}^{(n)}$ and $B_1^{(n)}$ are matrices of rank 1 of the kind $\boldsymbol{a}\boldsymbol{g}^{\mathrm{T}}$ and $\boldsymbol{b}\boldsymbol{g}'^{\mathrm{T}}$ where $\boldsymbol{a} + \boldsymbol{b} = \mathbf{1}$.

We are ready to state the following convergence result whose proof can be found in [56].

Theorem 7.4 *If the QBD is null recurrent and if for any convergent subsequence of* $B_{-1}^{(n)}$ *its limit* $\boldsymbol{a}\boldsymbol{g}^{\mathrm{T}}$ *is such that* $0 < \boldsymbol{g}'^{\mathrm{T}}\boldsymbol{a} < 1$, *then the sequences* $\{B_{-1}^{(n)}\}$ *and* $\{B_1^{(n)}\}$ *are convergent and* $\lim_n B_{-1}^{(n)} = \frac{1}{2}\mathbf{1}\boldsymbol{g}^{\mathrm{T}}$, $\lim_n B_1^{(n)} = \frac{1}{2}\mathbf{1}\boldsymbol{g}'^{\mathrm{T}}$. *Moreover, for any matrix norm*

$$\lim_n \|G_{\min} - \sum_{i=0}^{n}\Big(\prod_{j=0}^{i-1} B_1^{(j)}\Big)B_{-1}^{(i)}\|^{1/n} = \frac{1}{2}.$$

According to [56], the assumptions of the above theorem concerning the properties of the limits of subsequences of $\{B_{-1}^{(n)}\}$ are satisfied if

$$S(\boldsymbol{g}') \subset S(\boldsymbol{g}), \quad \text{or} \quad S(\boldsymbol{g}) \subset S(\boldsymbol{g}'),$$

where for the vector $\boldsymbol{v} = (v_i)$, we define $S(\boldsymbol{v}) = \{i : 1 \le i \le m,\ v_i = 0\}$. The above condition is trivially satisfied if at least one of $G_{\min}$ and $G'_{\min}$ is irreducible, in fact, in this case one of the two vectors $\boldsymbol{g}$ and $\boldsymbol{g}'$ is positive by the Perron–Frobenius Theorem 1.27.

Observe that, according to Theorem 7.4, the convergence of logarithmic reduction applied to a null recurrent QBD is linear and the convergence rate is 1/2. Therefore logarithmic reduction is still more efficient than the functional iteration methods of Section 6.2 which for null recurrent QBDs have sublinear convergence. However, as we will show in Section 8.2, the application of suitable techniques enables one to transform the original null recurrent QBD into a new problem for which the convergence of logarithmic reduction still remains quadratic.

Algorithm 7.1 Logarithmic reduction for QBDs: the case $\mu < 0$

INPUT: The positive integer m and the $m \times m$ matrices A_{-1}, A_0, A_1, defining a positive recurrent QBD; a real $\epsilon > 0$.

OUTPUT: An approximation Y of the minimal nonnegative solution of the equation (7.1) such that $\|G_{\min} - Y\|_\infty \leq \epsilon$.

COMPUTATION:

1. Set $V_i = (I - A_0)^{-1}A_i$, $i = -1, 1$, $W = I - V_{-1}V_1 - V_1V_{-1}$, $Y = V_{-1}$, $U = I$.
2. Compute

$$\begin{aligned} V'_{-1} &= W^{-1}V_{-1}^2, \\ V'_1 &= W^{-1}V_1^2, \\ W &= I - V'_{-1}V'_1 - V'_1V'_{-1}, \\ U &= UV_1, \quad Y = Y + UV'_{-1} \end{aligned}$$

and set $V_i = V'_i$, $i = -1, 1$.

3. If $1 - \|Y\|_\infty \leq \epsilon$ then output Y and stop, otherwise repeat from step 2.

Algorithm 7.1 synthesizes logarithmic reduction in the positive recurrent case, where $\mu < 0$. Since $0 \leq Y \leq G_{\min}$ and $G_{\min}$ is stochastic, then the stop condition $1 - \|Y\|_\infty \leq \epsilon$ implies that $\|G_{\min} - Y\|_\infty \leq \epsilon$.

We observe that one step of logarithmic reduction can be performed with the cost of one matrix inversion and eight matrix multiplications. Observe that in the estimate of the computational cost we do not count matrix additions since they have a lower cost in terms of the matrix size. In fact $m \times m$ matrix addition costs m^2 ops while matrix multiplication costs $2m^3 - m^2$ ops and matrix inversion costs $2m^3 + O(m^2)$ ops if performed by means of LU factorization.

Concerning the numerical stability of logarithmic reduction, we observe that at each step we have to invert the M-matrix W and to compute some products of nonnegative matrices. Both computations are numerically stable. Possible numerical cancellation could be encountered in the computation of the diagonal elements of W. However, these subtractions of nonnegative numbers can be avoided by using the trick of Grassman, Taksar and Heyman [53] adjusted to this computation by Qiang Ye in [120].

It is interesting to observe that we may save one matrix product at each step of logarithmic reduction if the matrices V_{-1}^2, V_1^2 and $V_{-1}V_1 + V_1V_{-1}$ are computed with the following expressions:

$$\begin{aligned} T &= (V_{-1} + V_1)^2, \\ Y &= (V_{-1} - V_1)^2, \\ Z &= V_{-1}^2, \\ V_{-1}V_1 + V_1V_{-1} &= (T - Y)/2; \\ V_1^2 &= (T + Y)/2 - Z. \end{aligned}$$

In this way the cost per step is reduced to seven matrix multiplications and one matrix inversion. A drawback of these faster formulae is that the numerical stability properties of the classical implementation are not guaranteed. In fact numerical cancellation may occurr in the computation of Y, $(T - Y)/2$ and $(T + Y)/2 - Z$.

For transient QBDs, when $\mu > 0$, logarithmic reduction can still be applied but the stop condition must be modified since $G_{\min}$ is not stochastic. In order to do this, since the sequence $B_{-1}^{(n)}$ converges to zero we may halt the algorithm if $\|B_{-1}^{(n)}\| \le \epsilon$ for a suitable value of $\epsilon > 0$. This condition does not immediately provide an a-posteriori error bound on the approximation to $G_{\min}$ even though, in light of Theorem 7.3, the norm $\|B_{-1}^{(n)}\|$ and the error $\|G_{\min} - X_n\|$ have the same asymptotic decay.

The logarithmic reduction algorithm, modified in this way, is summarized in Algorithm 7.2.

Algorithm 7.2 Logarithmic reduction for QBDs: the case $\mu > 0$

INPUT: The positive integer m and the $m \times m$ matrices A_{-1}, A_0, A_1, defining a transient QBD; a real $\epsilon > 0$.

OUTPUT: An approximation Y of the minimal nonnegative solution of the equation (7.1).

COMPUTATION:

1. Set $V_i = (I - A_0)^{-1} A_i$, $i = -1, 1$, $W = I - V_{-1}V_1 - V_1V_{-1}$, $Y = V_{-1}$, $U = I$.
2. Compute

$$\begin{aligned} V_{-1}' &= W^{-1} V_{-1}^2, \\ V_1' &= W^{-1} V_1^2, \\ W &= I - V_{-1}'V_1' - V_1'V_{-1}', \\ U &= UV_1, \quad Y = Y + UV_{-1}' \end{aligned}$$

   and set $V_i = V_i'$, $i = -1, 1$.
3. If $\|V_{-1}\|_\infty \le \epsilon$ then output Y and stop, otherwise repeat from step 2.

For null recurrent QBDs, when $\mu = 0$, even though the logarithmic reduction algorithm still converges with a linear convergence, it is more convenient to apply the acceleration techniques described in Section 8.2 which guarantee the quadratic convergences even in this case.

7.3 Cyclic reduction for quasi-birth–death processes

The method of cyclic reduction, which we present in this section, was originally introduced in the late 1960s by B.L. Buzbee, G.H. Golub and C.W. Nielson, for solving certain block tridiagonal systems which arise in the numerical solution of elliptic equations. Cyclic reduction can also be viewed as a simple modification of logarithmic reduction. The advantage of this modification is that unlike logarithmic reduction, this technique can be extended to M/G/1-type Markov chains.

In Section 3.3 we have shown that the matrix equation (3.12) defined by the power series $A(z) = \sum_{i=-1}^{+\infty} z^{i+1} A_i$ can be transformed into the semi-infinite linear system (3.15). For the equation (7.1) this transformation leads to the linear system

$$\begin{bmatrix} I - A_0 & -A_1 & & 0 \\ -A_{-1} & I - A_0 & -A_1 & \\ & -A_{-1} & I - A_0 & \ddots \\ 0 & & \ddots & \ddots \end{bmatrix} \begin{bmatrix} G_{\min} \\ G_{\min}^2 \\ G_{\min}^3 \\ \vdots \end{bmatrix} = \begin{bmatrix} A_{-1} \\ 0 \\ 0 \\ \vdots \end{bmatrix}.$$

Applying an even–odd permutation to the block rows and to the block columns of the above system yields

$$\left[\begin{array}{ccc|ccc} I - A_0 & & 0 & -A_{-1} & -A_1 & 0 \\ & I - A_0 & & & -A_{-1} & \ddots \\ 0 & & \ddots & 0 & & \ddots \\ \hline -A_1 & & 0 & I - A_0 & & 0 \\ -A_{-1} & -A_1 & & & I - A_0 & \\ 0 & \ddots & \ddots & 0 & & \ddots \end{array}\right] \left[\begin{array}{c} G_{\min}^2 \\ G_{\min}^4 \\ \vdots \\ \hline G_{\min} \\ G_{\min}^3 \\ \vdots \end{array}\right] = \left[\begin{array}{c} 0 \\ 0 \\ \vdots \\ \hline A_{-1} \\ 0 \\ \vdots \end{array}\right], \tag{7.15}$$

which we rewrite in compact form as

$$\begin{bmatrix} I - U_{1,1} & -U_{1,2} \\ -U_{2,1} & I - U_{2,2} \end{bmatrix} \begin{bmatrix} \boldsymbol{x}_+ \\ \boldsymbol{x}_- \end{bmatrix} = \begin{bmatrix} 0 \\ \boldsymbol{b} \end{bmatrix},$$

where $\boldsymbol{x}_+$, $\boldsymbol{x}_-$ are the block vectors of components $G_{\min}^{2i}$ and $G_{\min}^{2i-1}$, respectively, $i = 1, 2, \ldots$, while $\boldsymbol{b}$ is the block vector with null components except the first one which is equal to A_{-1}.

Now we may apply one step of block Gaussian elimination to the 2×2 block system above and obtain

$$\begin{bmatrix} I - U_{1,1} & -U_{1,2} \\ 0 & H \end{bmatrix} \begin{bmatrix} \boldsymbol{x}_+ \\ \boldsymbol{x}_- \end{bmatrix} = \begin{bmatrix} 0 \\ \boldsymbol{b} \end{bmatrix},$$

where

$$H = I - U_{2,2} - U_{2,1}(I - U_{1,1})^{-1} U_{1,2}.$$

From the above system we deduce that

$$H\boldsymbol{x}_- = \boldsymbol{b}.$$

In this transformation there are two remarkable properties which it is worth pointing out. The matrix H, that is the Schur complement of $I - U_{2,2}$, is the block tridiagonal matrix

$$H = \begin{bmatrix} I - \widehat{A}_0^{(1)} & -A_1^{(1)} & & 0 \\ -A_{-1}^{(1)} & I - A_0^{(1)} & -A_1^{(1)} & \\ & -A_{-1}^{(1)} & I - A_0^{(1)} & \ddots \\ 0 & & \ddots & \ddots \end{bmatrix}$$

which is still block Toeplitz except for the block in position (1,1), where

$$\begin{aligned} A_{-1}^{(1)} &= A_{-1}(I - A_0)^{-1}A_{-1}, \\ A_0^{(1)} &= A_0 + A_{-1}(I - A_0)^{-1}A_1 + A_1(I - A_0)^{-1}A_{-1}, \\ A_1^{(1)} &= A_1(I - A_0)^{-1}A_1, \\ \widehat{A}_0^{(1)} &= A_0 + A_1(I - A_0)^{-1}A_{-1}. \end{aligned}$$

The other property is that the first block component of the unknown vector $\boldsymbol{x}_-$ in the system $H\boldsymbol{x}_- = \boldsymbol{b}$ has the first block component equal to $G_{\min}$, and that the right-hand side $\boldsymbol{b}$ is the same vector as in the original system.

In other words, the block even–odd permutation followed by one step of block Gaussian elimination has led to a new system having almost the same structure as the original system, namely

$$\begin{bmatrix} I - \widehat{A}_0^{(1)} & -A_1^{(1)} & & 0 \\ -A_{-1}^{(1)} & I - A_0^{(1)} & -A_1^{(1)} & \\ & -A_{-1}^{(1)} & I - A_0^{(1)} & \ddots \\ 0 & & \ddots & \ddots \end{bmatrix} \begin{bmatrix} G_{\min} \\ G_{\min}^3 \\ G_{\min}^5 \\ \vdots \end{bmatrix} = \begin{bmatrix} A_{-1} \\ 0 \\ 0 \\ \vdots \end{bmatrix}.$$

Now we may repeat this process by applying again the even–odd permutation followed by one step of block Gaussian elimination. After n steps of this process, which is called *cyclic reduction*, we obtain the system

$$\begin{bmatrix} I - \widehat{A}_0^{(n)} & -A_1^{(n)} & & 0 \\ -A_{-1}^{(n)} & I - A_0^{(n)} & -A_1^{(n)} & \\ & -A_{-1}^{(n)} & I - A_0^{(n)} & \ddots \\ 0 & & \ddots & \ddots \end{bmatrix} \begin{bmatrix} G_{\min} \\ G_{\min}^{2^n+1} \\ G_{\min}^{2\cdot 2^n+1} \\ G_{\min}^{3\cdot 2^n+1} \\ \vdots \end{bmatrix} = \begin{bmatrix} A_{-1} \\ 0 \\ 0 \\ \vdots \end{bmatrix} \tag{7.16}$$

where

$$\begin{aligned} A_{-1}^{(n+1)} &= A_{-1}^{(n)}(I - A_0^{(n)})^{-1}A_{-1}^{(n)}, \\ A_0^{(n+1)} &= A_0^{(n)} + A_{-1}^{(n)}(I - A_0^{(n)})^{-1}A_1^{(n)} + A_1^{(n)}(I - A_0^{(n)})^{-1}A_{-1}^{(n)}, \\ A_1^{(n+1)} &= A_1^{(n)}(I - A_0^{(n)})^{-1}A_1^{(n)}, \\ \widehat{A}_0^{(n+1)} &= \widehat{A}_0^{(n)} + A_1^{(n)}(I - A_0^{(n)})^{-1}A_{-1}^{(n)}, \end{aligned} \tag{7.17}$$

for $n = 0, 1, \ldots$, and $\widehat{A}_0^{(0)} = A_0$, $A_i^{(0)} = A_i$, $i = -1, 0, 1$.

If the entire process can be carried out, that is, if $\det(I - A_0^{(n)}) \neq 0$ for $n = 0, 1, 2, \ldots$, and if $\det(I - \widehat{A}_0^{(n)}) \neq 0$ then from the first equation of (7.16) we find that

$$G_{\min} = (I - \widehat{A}_0^{(n)})^{-1}(A_{-1} + A_1^{(n)} G_{\min}^{2^n+1}). \tag{7.18}$$

This formula is a valid tool for approximating the minimal solution $G_{\min}$ once we prove that the sequence $\{A_1^{(n)} G_{\min}^{2^n+1}\}$ converges quadratically to zero and that $\{I - \widehat{A}_0^{(n)}\}$ is quadratically convergent to a nonsingular matrix.

This property will be proven in the general case of M/G/1 Markov chains in the next section. Here, in the QBD case, we simply deduce this property from the relationships between cyclic reduction and logarithmic reduction.

In order to point out the similarities between logarithmic reduction and cyclic reduction, it is convenient to provide a functional formulation also of cyclic reduction. For this purpose let us introduce the quadratic matrix polynomial

$$A^{(n)}(z) = A_{-1}^{(n)} + z A_0^{(n)} + z^2 A_1^{(n)}.$$

From (7.17) we may easily deduce that

$$(zI - A^{(n)}(z))(I - A_0^{(n)})^{-1}(-zI - A^{(n)}(-z)) = -(z^2 I - A^{(n+1)}(z^2)). \tag{7.19}$$

Comparing (7.19) with (7.11) we deduce the following result which relates logarithmic reduction and cyclic reduction:

Theorem 7.5 *If $\mu \neq 0$ then cyclic reduction can be carried out and the following equations hold*

$$\begin{aligned} B_i^{(n)} &= (I - A_0^{(n)})^{-1} A_i^{(n)}, \quad i = -1, 1, \\ C^{(n)} &= (I - A_0^{(n)})^{-1}(I - A_0^{(n+1)}), \end{aligned} \tag{7.20}$$

for $n = 0, 1, \ldots$. Moreover the matrix $A_{-1}^{(n)} + A_0^{(n)} + A_1^{(n)}$ is stochastic for any $n \geq 0$.

Proof We proceed by induction. For $n = 0$ we have $B_i^{(0)} = (I - A_0)^{-1} A_i$ by definition, and

$$\begin{aligned} C^{(0)} =& I - B_{-1}^{(0)} B_1^{(0)} - B_1^{(0)} B_{-1}^{(0)} \\ =& (I - A_0)^{-1}(I - A_0 - A_{-1}(I - A_0)^{-1} A_1 - A_1 (I - A_0)^{-1} A_{-1}) \\ =& (I - A_0)^{-1}(I - A_0^{(1)}). \end{aligned}$$

Since $C^{(0)}$ is nonsingular for Theorem 7.1 we deduce that also $I - A_0^{(1)}$ is nonsingular so that the second step of cyclic reduction can be performed. For the inductive step, assume that $I - A_0^{(n)}$ is nonsingular and equations (7.20) hold for n. Then, since $C^{(n)}$ is nonsingular for Theorem 7.1, we deduce that also

$I - A_0^{(n+1)}$ is nonsingular. Moreover, from equation (7.6) and from the inductive hypothesis one has

$$\begin{aligned} B_i^{(n+1)} =& (C^{(n)})^{-1}(B_i^{(n)})^2 \\ =& (I - A_0^{(n+1)})^{-1}(I - A_0^{(n)}) \left[(I - A_0^{(n)})^{-1}A_i^{(n)}\right]^2 \\ =& (I - A_0^{(n+1)})^{-1} \left[A_i^{(n)}(I - A_0^{(n)})^{-1}A_i^{(n)}\right] \\ =& (I - A_0^{(n+1)})^{-1}A_i^{(n+1)}. \end{aligned}$$

Similarly, by using the expression $B_i^{(n+1)} = (I - A_0^{(n+1)})^{-1}A_i^{(n+1)}$ we deduce that

$$\begin{aligned} C^{(n+1)} =& I - B_{-1}^{(n+1)}B_1^{(n+1)} - B_1^{(n+1)}B_{-1}^{(n+1)} \\ =& I - (I - A_0^{(n+1)})^{-1}A_{-1}^{(n+1)}(I - A_0^{(n+1)})^{-1}A_1^{(n+1)} \\ & - (I - A_0^{(n+1)})^{-1}A_1^{(n+1)}(I - A_0^{(n+1)})^{-1}A_{-1}^{(n+1)} \\ =& (I - A_0^{(n+1)})^{-1}(I - A_0^{(n+2)}). \end{aligned}$$

In particular we find that $I - A_0^{(n+2)}$ is nonsingular, which completes the inductive proof. Finally, the stochasticity of $A_{-1}^{(n)} + A_0^{(n)} + A_1^{(n)}$ follows from the stochasticity of $B_{-1}^{(n)} + B_1^{(n)}$ which holds for Theorem 7.1. □

A simple consequence of the above result is obtained from the convergence properties of logarithmic reduction (see Theorem 7.3) and is summarized in the following theorem. The speed of convergence is related to the moduli of the roots of $A(z)$ closest to the unit circle, defined in equation (7.13).

Theorem 7.6 *If $\mu < 0$, then the following limits exist*

$$\begin{aligned} &\lim_n A_0^{(n)} = A_0^{(\infty)}, \quad \lim_n \widehat{A}_0^{(n)} = \widehat{A}_0^{(\infty)}, \\ &\lim_n A_{-1}^{(n)} = (I - A_0^{(\infty)})\mathbf{1}\boldsymbol{g}^{\mathrm{T}}, \quad \lim_n A_1^{(n)} = 0, \end{aligned}$$

where $\widehat{A}_0^{(\infty)}$ is the minimal nonnegative solution of

$$X = A_0 + A_1(I - X)^{-1}A_{-1}, \tag{7.21}$$

and $\boldsymbol{g} \geq 0$ is such that $\boldsymbol{g}^{\mathrm{T}}G_{\min} = \boldsymbol{g}^{\mathrm{T}}$, $\boldsymbol{g}^{\mathrm{T}}\mathbf{1} = 1$. Moreover, for any matrix norm and for any $\epsilon > 0$ such that $\epsilon + \eta < 1$, there exists $\gamma > 0$ such that

$$\begin{aligned} &\|A_0^{(n)} - A_0^{(\infty)}\| \leq \gamma\xi^{-2^n} \\ &\|A_1^{(n)}\| \leq \gamma\xi^{-2^n}, \\ &\|A_{-1}^{(n)} - (I - A_0^{(\infty)})\mathbf{1}\boldsymbol{g}^{\mathrm{T}}\| \leq \gamma\big(\xi^{-2^n} + (\eta + \epsilon)^{2^n}\big), \\ &\|\widehat{A}_0^{(n)} - \widehat{A}_0^{(\infty)}\| \leq \gamma\xi^{-2^n}, \end{aligned}$$

and

$$\begin{aligned}\|G_{\min} - (I - \widehat{A}_0^{(n)})^{-1}A_{-1}\| &\le \gamma\xi^{-2^n},\\ \|R_{\min} - A_1(I - \widehat{A}_0^{(n)})^{-1}\| &\le \gamma\xi^{-2^n},\end{aligned}$$

for any $n \ge 0$.

If $\mu > 0$, *then there exist the following limits*

$$\begin{aligned}&\lim_n A_0^{(n)} = A_0^{(\infty)}, \quad \lim_n \widehat{A}_0^{(n)} = \widehat{A}_0^{(\infty)},\\ &\lim_n A_1^{(n)} = (I - A_0^{(\infty)})\mathbf{1}\boldsymbol{g}'^{\mathrm{T}}, \quad \lim_n A_{-1}^{(n)} = 0,\end{aligned}$$

where $\widehat{A}_0^{(\infty)}$ *is the minimal nonnegative solution of*

$$X = A_0 + A_1(I - X)^{-1}A_{-1},$$

and $\boldsymbol{g}' \ge 0$ *is such that* $\boldsymbol{g}'^{\mathrm{T}}G'_{\min} = \boldsymbol{g}'^{\mathrm{T}}$, $\boldsymbol{g}'^{\mathrm{T}}\mathbf{1} = 1$, $G'_{\min}$ *is the minimal nonnegative solution of the equation* $X = A_1 + A_0X + A_{-1}X^2$. *Moreover, for any matrix norm and for any* $\epsilon > 0$ *such that* $\epsilon + \xi^{-1} < 1$, *there exists* $\gamma > 0$ *such that*

$$\begin{aligned}&\|A_0^{(n)} - A_0^{(\infty)}\| \le \gamma\eta^{2^n}\\ &\|A_{-1}^{(n)}\| \le \gamma\eta^{2^n},\\ &\|A_1^{(n)} - (I - A_0^{(\infty)})\mathbf{1}\boldsymbol{g}'^{\mathrm{T}}\| \le \gamma\big((\xi^{-1}+\epsilon)^{2^n} + \eta^{2^n}\big),\\ &\|\widehat{A}_0^{(n)} - \widehat{A}_0^{(\infty)}\| \le \gamma\eta^{2^n}\end{aligned}$$

and

$$\begin{aligned}\|G_{\min} - (I - \widehat{A}_0^{(n)})^{-1}A_{-1}\| &\le \gamma\eta^{2^n},\\ \|R_{\min} - A_1(I - \widehat{A}_0^{(n)})^{-1}\| &\le \gamma\eta^{2^n},\end{aligned}$$

for any $n \ge 0$.

Proof Let us first assume that $\mu < 0$. We may deduce from (7.17) that the sequence $\{A_0^{(n)}\}$ is nondecreasing. Moreover it is bounded since the matrices $A_0^{(n)}$, $n \ge 0$, are substochastic for Theorem 7.5. Therefore there exists $A_0^{(\infty)} = \lim_n A_0^{(n)}$ and from (7.20) one has that for any matrix norm

$$\|A_1^{(n)}\| \le \|I - A_0^{(n)}\| \cdot \|B_1^{(n)}\| \le c\|B_1^{(n)}\|,$$

for a suitable constant $c > 0$. Therefore, from Theorem 7.3 we deduce that for any matrix norm there exists $\gamma > 0$ such that

$$\|A_1^{(n)}\| \le \gamma\xi^{-2^n}.$$

From (7.20) and (7.17) we deduce that

$$I - C^{(n)} = (I - A_0^{(n)})^{-1} \left(A_{-1}^{(n)} (I - A_0^{(n)})^{-1} A_1^{(n)} + A_1^{(n)} (I - A_0^{(n)})^{-1} A_{-1}^{(n)} \right). \tag{7.22}$$

Observe that

$$A_0^{(n+1)} - A_0^{(n)} = A_{-1}^{(n)} (I - A_0^{(n)})^{-1} A_1^{(n)} + A_1^{(n)} (I - A_0^{(n)})^{-1} A_{-1}^{(n)}.$$

Therefore, since $(I - A_0^{(n)})^{-1} \geq I$ and $A_0^{(n+1)} - A_0^{(n)} \geq 0$, from (7.22) one has $A_0^{(n+1)} - A_0^{(n)} \leq I - C^{(n)}$. Since $\|X\|_\infty \leq \|Y\|_\infty$ if $0 \leq X \leq Y$, we obtain

$$\|A_0^{(n+1)} - A_0^{(n)}\|_\infty \leq \|I - C^{(n)}\|_\infty.$$

From Theorem 7.3 there exists a constant γ' such that $\|I - C^{(n)}\|_\infty \leq \gamma' \xi^{-2^n}$. Therefore $\|A_0^{(n+1)} - A_0^{(n)}\|_\infty \leq \gamma' \xi^{-2^n}$. From Lemma A.16 of the appendix we conclude that $\|A_0^{(n)} - A_0^{(\infty)}\|_\infty \leq \gamma \xi^{-2^n}$ for a suitable $\gamma > 0$. By using a similar argument we may prove that the sequence $\{\widehat{A}_0^{(n)}\}$ converges with the same rate to a limit $\widehat{A}_0^{(\infty)}$. Now, from the convergence of the sequences $\{A_0^{(n)}\}$ and $\{B_{-1}^{(n)}\}$, we may easily deduce the convergence of the sequence $\{A_{-1}^{(n)}\}$. In fact, from (7.20) one has that for any norm $\|\cdot\|$,

$$\begin{aligned}
\|A_{-1}^{(n)} - (I - A_0^{(\infty)})\mathbf{1}\boldsymbol{g}^{\mathrm{T}}\| =& \|(I - A_0^{(n)})B_{-1}^{(n)} - (I - A_0^{(\infty)})\mathbf{1}\boldsymbol{g}^{\mathrm{T}}\| \\
=& \|(A_0^{(\infty)} - A_0^{(n)})B_{-1}^{(n)} + (I - A_0^{(\infty)})(B_{-1}^{(n)} - \mathbf{1}\boldsymbol{g}^{\mathrm{T}})\| \\
\leq& c_1 \|A_0^{(\infty)} - A_0^{(n)}\| + c_2 \|B_{-1}^{(n)} - \mathbf{1}\boldsymbol{g}^{\mathrm{T}}\|,
\end{aligned}$$

where c_1 and c_2 are suitable positive numbers. From the latter inequality we conclude that for any matrix norm and for any $\epsilon > 0$ such that $\epsilon + \eta < 1$, there exists a positive constant γ such that

$$\|A_{-1}^{(n)} - (I - A_0^{(\infty)})\mathbf{1}\boldsymbol{g}^{\mathrm{T}}\| \leq \gamma(\xi^{-2^n} + (\eta + \epsilon)^{2^n}).$$

Let U be the minimal nonnegative solution of the matrix equation (7.21). We wish to show that $\lim_n \widehat{A}_0^{(n)} = U$. For this purpose, observe that by (5.42) the matrix $G_{\min}$ can be written as $G_{\min} = (I - U)^{-1} A_{-1}$. The equations $-A_{-1} + (I - A_0)G_{\min} - A_1 G_{\min}^2 = 0$ and $I - U = I - A_0 - A_1 G_{\min}$, which is derived from (7.21), can be rewritten in matrix form as

$$\begin{bmatrix} I - A_0 & -A_1 & & 0 \\ -A_{-1} & I - A_0 & -A_1 & \\ & -A_{-1} & I - A_0 & \ddots \\ 0 & & \ddots & \ddots \end{bmatrix} \begin{bmatrix} I \\ G_{\min} \\ G_{\min}^2 \\ \vdots \end{bmatrix} = \begin{bmatrix} I - U \\ 0 \\ 0 \\ \vdots \end{bmatrix}. \tag{7.23}$$

Applying cyclic reduction to the above system yields

$$\begin{bmatrix} I-\widehat{A}_0^{(n)} & -A_1^{(n)} & & 0 \\ -A_{-1}^{(n)} & I-A_0^{(n)} & -A_1^{(n)} & \\ & -A_{-1}^{(n)} & I-A_0^{(n)} & \ddots \\ 0 & & \ddots & \ddots \end{bmatrix} \begin{bmatrix} I \\ G_{\min}^{2^n} \\ G_{\min}^{2\cdot 2^n} \\ \vdots \end{bmatrix} = \begin{bmatrix} I-U \\ 0 \\ 0 \\ \vdots \end{bmatrix}.$$

From the first equation we deduce that

$$\widehat{A}_0^{(n)} + A_1^{(n)} G_{\min}^{2^n} = U. \tag{7.24}$$

This implies that $\lim_n \widehat{A}_0^{(n)} = U$. In particular, $I-U$ is nonsingular and from (7.18) one has $\|G_{\min} - (I-\widehat{A}_0^{(n)})^{-1}A_{-1}\| \le \gamma\xi^{-2^n}$.

If $\mu > 0$, we apply the above argument to the positive recurrent Markov chain obtained by interchanging A_{-1} with A_1 in order to show the concergence of the sequences $\{A_i^{(n)}\}_n$, $i = -1, 0, 1$. Concerning the convergence of the sequence $\{\widehat{A}_0^{(n)}\}$ we apply cyclic reduction to the system (7.23) and obtain (7.24). Since $\rho(G_{\min}) = \eta$ and $\{A_1^{(n)}\}$ is bounded, the convergence of $\{\widehat{A}_0^{(n)}\}$ to U holds for (7.24). □

The approximation of the minimal solution of the matrix equation (7.1) by means of cyclic reduction is synthesizes by Algorithm 7.3.

Algorithm 7.3 Cyclic reduction for QBDs: the case $\mu < 0$

INPUT: The positive integer m and the $m \times m$ matrices A_{-1}, A_0, A_1, defining a positive recurrent QBD; a real $\epsilon > 0$.

OUTPUT: An approximation Y of the minimal nonnegative solution of the equation (7.1) such that $\|Y - G_{\min}\|_\infty \le \epsilon\|(I-U)^{-1}\|_\infty$, where U is the minimal nonnegative solution of (7.21).

COMPUTATION:

1. Set $V_i = A_i$, $i = -1, 0, 1$, $\widehat{V} = A_0$.
2. Compute

$$\begin{aligned} &H = (I - V_0)^{-1}V_1, \quad K = (I - V_0)^{-1}V_{-1}, \\ &V_1' = V_1 H, \quad V_{-1}' = V_{-1}K, \\ &W = V_1 K, \quad \widehat{V}' = \widehat{V} + W, \\ &V_0' = V_0 + W + V_{-1}H, \end{aligned}$$

   and set $V_i = V_i'$, $i = -1, 0, 1$, $\widehat{V} = \widehat{V}'$.
3. If $\|V_1\|_\infty < \epsilon$ then output $Y = (I - \widehat{V})^{-1}A_{-1}$ and stop, otherwise repeat from step 2.

Observe that the stop condition guarantees that $\|Y - G_{\min}\|_\infty \le \epsilon\|(I - U)^{-1}\|_\infty$. In fact, since at the n-th step of cyclic reduction we have (7.18) we find that $G_{\min} - G^{(n)} = (I - \widehat{A}_0^{(n)})^{-1} A_1^{(n)} G_{\min}^{2^n+1}$, where $G^{(n)} = (I - \widehat{A}_0^{(n)})^{-1} A_{-1}$. Therefore,

$$\begin{aligned}\|G_{\min} - G^{(n)}\|_\infty =& \|(I - \widehat{A}_0^{(n)})^{-1} A_1^{(n)} G_{\min}^{2^n+1}\| \\ \le& \|(I - \widehat{A}_0^{(n)})^{-1}\|_\infty \|A_1^{(n)}\|_\infty \|G_{\min}^{2^n+1}\|_\infty \\ \le& \|(I - U)^{-1}\|_\infty \|A_1^{(n)}\|_\infty \|G_{\min}^{2^n+1}\|_\infty,\end{aligned} \tag{7.25}$$

since $(I - \widehat{A}_0^{(n)})^{-1} \le (I - U)^{-1}$ and the infinity norm is monotonic when applied to nonnegative matrices. From the property $\|G_{\min}\|_\infty = 1$ and $\|A_1^{(n)}\|_\infty \le \epsilon$, we find that $\|G_{\min} - G^{(n)}\|_\infty \le \epsilon \|(I - U)^{-1}\|_\infty$.

Concerning the computational cost we observe that the matrices V_i', $i = -1, 0, 1$ and $\widehat{V}'$ at step 2 of Algorithm 7.3 can be computed with one matrix inversion and six matrix multiplications. As in the case of logarithmnic reduction, in our complexity estimates we do not count matrix additions. This provides a slight improvement with respect to logarithmic reduction which requires one matrix inversion and eight matrix multiplications per step.

Likewise for logarithmic reduction, the cyclic reduction algorithm involves at each step the inversion of a nonsingular M-matrix and products and additions of nonnegative matrices. All these computation are numerically stable. The only possible source of cancellation is the computation of the diagonal elements of $I - V_0$ by subtraction. This subtraction can be avoided by using the Grassman, Taksar and Heyman technique of [53].

In the case of a transient QBD we may still apply cyclic reduction with the difference that in this case the sequence $A_{-1}^{(n)}$ converges to zero while the sequence $A_1^{(n)}$ is bounded from below by a nonnull matrix. Therefore, we have to replace the stop condition $\|A_1^{(n)}\| \le \epsilon$ with $\|A_{-1}^{(n)}\| \le \epsilon$, and at the same time we have to deduce an a-posteriori bound for the error $\|G_{\min} - G^{(n)}\|$ for $G^{(n)} = (I - \widehat{A}_0^{(n)})^{-1} A_{-1}$, given in terms of $\|A_{-1}^{(n)}\|$. For this purpose we rely on equation (7.25); since $\|A_1^{(n)}\|_\infty \le 1$ we have $\|G_{\min} - G^{(n)}\|_\infty \le \|(I - U)^{-1}\|_\infty \|G_{\min}^{2^n+1}\|_\infty$.

An estimate of $\|G_{\min}^{2^n+1}\|_\infty$ can be given in terms of $\|A_{-1}^{(n)}\|_\infty$ which converges to zero. In fact, since

$$G_{\min}^{2^n} = (I - A_0^{(n)})^{-1} A_{-1}^{(n)} + (I - A_0^{(n)})^{-1} A_1^{(n)} (G_{\min}^{2^n})^2,$$

we obtain that

$$\begin{aligned}\|G_{\min}^{2^n}\|_\infty \le& \|(I - A_0^{(n)})^{-1}\|_\infty \left(\|A_{-1}^{(n)}\|_\infty + \|A_1^{(n)}\|_\infty \|G_{\min}^{2^n}\|_\infty^2 \right) \\ \le& \|(I - A_0^{(n)})^{-1}\|_\infty \left(\|A_{-1}^{(n)}\|_\infty + \|G_{\min}^{2^n}\|_\infty^2 \right).\end{aligned}$$

Whence, if $\|A_{-1}^{(n)}\|_\infty \le \epsilon$ and $\sigma = \|(I - A_0^{(n)})^{-1}\|_\infty$, we deduce that

$$\|G_{\min}^{2^n}\|_\infty \le \sigma(\epsilon + \|G_{\min}^{2^n}\|_\infty^2),$$

which implies

$$\|G_{\min}^{2^n}\|_\infty \le \frac{1-\sqrt{1-4\sigma^2\epsilon}}{2\sigma} = \sigma\epsilon + O(\epsilon^2). \tag{7.26}$$

Observe that, since $(I-A_0^{(n)})^{-1}$ monotonically converges to $(I-A_0^{(\infty)})^{-1}$, we have $\sigma \le \|(I-A_0^{(\infty)})^{-1}\|_\infty$. Therefore, from (7.26) and (7.25) one has that

$$\|G_{\min} - G^{(n)}\|_\infty \le \|(I-U)^{-1}\|_\infty \frac{1-\sqrt{1-4\sigma^2\epsilon}}{2\sigma} = \sigma\epsilon\|(I-U)^{-1}\|_\infty + O(\epsilon^2).$$

The scheme that we obtain in this way is reported in Algorithm 7.4.

Algorithm 7.4 Cyclic reduction for QBDs: the case $\mu > 0$

INPUT: The positive integer m and the $m \times m$ matrices A_{-1}, A_0, A_1, defining a transient QBD; a real $\epsilon > 0$.

OUTPUT: An approximation Y of the minimal nonnegative solution of the equation (7.1) and an a-posteriori error bound $\delta \le \epsilon\|(I-A_0^{(\infty)})^{-1}\|_\infty + O(\epsilon^2)$, such that $\|G_{\min} - Y\|_\infty \le \delta\|(I-U)^{-1}\|_\infty$.

COMPUTATION:

1. Set $V_i = A_i$, $i=-1,0,1$, $\widehat{V} = A_0$.
2. Compute

$$\begin{aligned}
&H = (I-V_0)^{-1}V_1, \quad K = (I-V_0)^{-1}V_{-1},\\
&V_1' = V_1 H, \quad V_{-1}' = V_{-1}K,\\
&W = V_1 K, \quad \widehat{V}' = \widehat{V} + W,\\
&V_0' = V_0 + W + V_{-1}H,
\end{aligned}$$

   and set $V_i = V_i'$, $i=-1,0,1$, $\widehat{V} = \widehat{V}'$.
3. If $\|V_{-1}\|_\infty < \epsilon$ then output the approximation $Y = (I-\widehat{V})^{-1}A_{-1}$ and the error bound $\delta = \frac{1-\sqrt{1-4\sigma^2\epsilon}}{2\sigma} = \sigma\epsilon + O(\epsilon^2)$, where $\sigma = \|(I-V_0)^{-1}\|_\infty$, and stop; otherwise repeat from step 2.

As with logarithmic reduction, under some weak additional assumptions the cyclic reduction algorithm can still be applied in the case of a null recurrent QBD, where $\mu = 0$, but the quadratic convergence no longer holds. In this case it is much more convenient to apply the acceleration techniques shown in Section 8.2 which guarantee the quadratic convergence of cyclic reduction modified in this way.

It is interesting to point out that the cyclic reduction algorithm can be easily modified in order to approximate the minimal nonnegative solutions $G_{\min}$, $G'_{\min}$, $R_{\min}$ and $R'_{\min}$ of the following four matrix equations, respectively:

$$
\begin{aligned}
A_{-1} + A_0 X + A_1 X^2 &= X, \\
A_1 + A_0 X + A_{-1} X^2 &= X, \\
A_1 + X A_0 + X^2 A_{-1} &= X, \\
A_{-1} + X A_0 + X^2 A_1 &= X,
\end{aligned}
\tag{7.27}
$$

at almost the same cost of solving only one equation. In fact, besides the sequences $\{\widehat{A}_0^{(n)}\}$, $\{A_i^{(n)}\}$, $i = 0, 1, -1$, it is sufficient to compute the sequence $\{\widetilde{A}_0^{(n)}\}$ defined by

$$
\widetilde{A}_0^{(n+1)} = \widetilde{A}_0^{(n)} + A_{-1}^{(n)}(I - A_0^{(n)})^{-1} A_1^{(n)}, \quad n \geq 0,
$$

with $\widetilde{A}_0^{(0)} = A_0$. We may show that, if $\mu \neq 0$, the sequence $\{\widetilde{A}_0^{(n)}\}$ quadratically converges to the minimal nonnegative solution $\widetilde{A}_0^{(\infty)}$ of the matrix equation

$$
X = A_0 + A_{-1}(I - X)^{-1} A_1
$$

and, according to (5.42) applied to the first two equations of (7.28), we obtain that the solutions of the matrix equations (7.27) are given by

$$
\begin{aligned}
G_{\min} &= (I - \widehat{A}_0^{(\infty)})^{-1} A_{-1}, \\
G'_{\min} &= (I - \widetilde{A}_0^{(\infty)})^{-1} A_1, \\
R_{\min} &= A_1 (I - \widehat{A}_0^{(\infty)})^{-1}, \\
R'_{\min} &= A_{-1} (I - \widetilde{A}_0^{(\infty)})^{-1}.
\end{aligned}
\tag{7.28}
$$

Recall that these expressions have been proved in Theorem 3.20 under the additional condition that the polynomial $\det(A_{-1} + z(A_0 - I) + z^2 A_1)$ is nonzero for $|z| = 1$. This condition, which is not satisfied in our case, can be easily relaxed by means of a continuity argument under the assumptions of nonsingularity of the matrices $I - \widehat{A}_0^{(\infty)}$ and $I - \widetilde{A}_0^{(\infty)}$.

In Section 8.2, by means of a suitable shifting of the roots of the matrix polynomial $A_{-1} + z(A_0 - I) + z^2 A_1$ we will transform the QBD problem into a new problem where the hypotheses of Theorem 3.20 are satisfied. This transformation provides an acceleration of the convergence of cyclic reduction.

As a last remark of this section we provide a functional formulation of cyclic reduction which is slightly different from the one given in (7.19) and will be particularly useful in Section 9.2 to prove certain structural properties valid in the solution of non-skip-free M/G/1-type Markov chains. In fact, setting $\varphi^{(n)}(z) = zI - A^{(n)}(z)$, since $I - A_0^{(n)} = (\varphi^{(n)}(z) - \varphi^{(n)}(-z))/(2z)$, we deduce from (7.19) that

$$
\varphi^{(n+1)}(z^2) = \left(\frac{\varphi^{(n)}(z)^{-1} - \varphi^{(n)}(-z)^{-1}}{2z} \right)^{-1}, \quad n \geq 0,
$$

which holds for all the values of z such that $\det \varphi^{(n)}(z), \det \varphi^{(n)}(-z) \neq 0$ and $\det(\varphi^{(n)}(z)^{-1} - \varphi^{(n)}(-z)^{-1}) \neq 0$.

A nicer formulation of the above property can be given in terms of the functions $\psi^{(n)}(z) = (z^{-1}\varphi^{(n)}(z))^{-1}$. In fact, by means of a formal manipulation one can easily verify that

$$\psi^{(n+1)}(z^2) = \frac{1}{2}(\psi^{(n)}(z) + \psi^{(n)}(-z)), \quad n \geq 0.$$

In particular, if $\psi^{(0)}(z) = (I - z^{-1}A(z))^{-1} = \sum_{i=-\infty}^{+\infty} z^i H_i$, then

$$\psi^{(n)}(z) = \sum_{i=-\infty}^{+\infty} z^i H_{i \cdot 2^n}, \quad n \geq 0.$$

7.4 Cyclic reduction for M/G/1-type Markov chains

The cyclic reduction algorithm can be suitably extended to the case of M/G/1-type Markov chains where the minimal nonnegative solution $G_{\min}$ of the equation

$$X = \sum_{i=-1}^{+\infty} A_i X^{i+1} \tag{7.29}$$

can be computed, still maintaining the nice convergence features valid for QBD processes.

In order to show this, we follow the same line used in the QBD case, that is, we translate the matrix equation (7.29) into the semi-infinite linear system

$$\begin{bmatrix} I-A_0 & -A_1 & -A_2 & -A_3 & \dots \\ -A_{-1} & I-A_0 & -A_1 & -A_2 & \ddots \\ & -A_{-1} & I-A_0 & -A_1 & \ddots \\ & & -A_{-1} & I-A_0 & \ddots \\ 0 & & & \ddots & \ddots \end{bmatrix} \begin{bmatrix} G_{\min} \\ G_{\min}^2 \\ G_{\min}^3 \\ \vdots \end{bmatrix} = \begin{bmatrix} A_{-1} \\ 0 \\ \vdots \\ \vdots \end{bmatrix}. \tag{7.30}$$

Applying an even–odd permutation to the block rows and to the block columns of the above system yields

$$\left[\begin{array}{cccc|cccc} I-A_0 & -A_2 & -A_4 & \dots & -A_{-1} & -A_1 & -A_3 & \dots \\ & I-A_0 & -A_2 & \ddots & & -A_{-1} & -A_1 & \ddots \\ & & I-A_0 & \ddots & & & -A_{-1} & \ddots \\ 0 & & & \ddots & 0 & & & \ddots \\ \hline -A_1 & -A_3 & -A_5 & \dots & I-A_0 & -A_2 & -A_4 & \dots \\ -A_{-1} & -A_1 & -A_3 & \ddots & & I-A_0 & -A_2 & \ddots \\ & -A_{-1} & -A_1 & \ddots & & & I-A_0 & \ddots \\ 0 & & \ddots & \ddots & 0 & & & \ddots \end{array}\right] \left[\begin{array}{c} G_{\min}^2 \\ G_{\min}^4 \\ \vdots \\ \hline G_{\min} \\ G_{\min}^3 \\ \vdots \end{array}\right] = \left[\begin{array}{c} 0 \\ 0 \\ \vdots \\ \hline A_{-1} \\ 0 \\ \vdots \end{array}\right] \tag{7.31}$$

which we rewrite in compact form as

$$\begin{bmatrix} I - U_{1,1} & -U_{1,2} \\ -U_{2,1} & I - U_{2,2} \end{bmatrix} \begin{bmatrix} \boldsymbol{x}_+ \\ \boldsymbol{x}_- \end{bmatrix} = \begin{bmatrix} 0 \\ \boldsymbol{b} \end{bmatrix} \tag{7.32}$$

where $\boldsymbol{x}_+$, $\boldsymbol{x}_-$ are the block vectors of components $G_{\min}^{2i}$ and $G_{\min}^{2i-1}$, respectively, $i = 1, 2, \ldots$, while $\boldsymbol{b}$ is the block vector with null components except the first which is equal to A_{-1}.

In order to describe in compact form the entire process of cyclic reduction for M/G/1-type Markov chains it is convenient to apply the notation and the tools introduced in Section 3.1, where we denoted with $T_\infty[F(z)] = (F_{j-i})_{i,j\in\mathbb{N}}$ the semi-infinite block Toeplitz matrix associated with the matrix Laurent power series $F(z) = \sum_{i=-\infty}^{+\infty} z^i F_i$. Another useful tool which we need is the concept of even and odd part of a matrix Laurent power series $F(z)$. We define the even part $F_{\text{even}}(z)$ and the odd part $F_{\text{odd}}(z)$ of $F(z)$ as

$$\begin{aligned} F_{\text{even}}(z) &= \frac{1}{2}(F(\sqrt{z}) + F(-\sqrt{z})) = \sum_{i=-\infty}^{+\infty} z^i F_{2i}, \\ F_{\text{odd}}(z) &= \frac{1}{2\sqrt{z}}(F(\sqrt{z}) - F(-\sqrt{z})) = \sum_{i=-\infty}^{+\infty} z^i F_{2i+1}, \end{aligned}$$

so that $F(z) = F_{\text{even}}(z^2) + zF_{\text{odd}}(z^2)$. We will also use the notation $[F(z)]_{\text{even}} = F_{\text{even}}(z^2)$ and $[F(z)]_{\text{odd}} = F_{\text{odd}}(z^2)$.

Consider the generating function $A(z) = \sum_{i=-1}^{+\infty} z^{i+1} A_i$ associated with the M/G/1-type Markov chain and observe that the block triangular Toeplitz matrix $U_{1,1} = U_{2,2}$ is associated with the matrix power series $A_{\text{odd}}(z)$, while $U_{1,2}$ is associated with the matrix power series $A_{\text{even}}(z)$ and the block $U_{2,1}$ is associated with $z^{-1}A_{\text{even}}(z)$. That is, with formal notation we have

$$\begin{aligned} U_{1,1} &= U_{2,2} = T_\infty[A_{\text{odd}}(z)], \\ U_{1,2} &= T_\infty[A_{\text{even}}(z)], \\ U_{2,1} &= T_\infty[z^{-1}A_{\text{even}}(z)]. \end{aligned} \tag{7.33}$$

Observe also that, since $A_i \geq 0$ and $A(1) = \sum_{i=-1}^{+\infty} A_i$ is finite then the function $A(z)$ belongs to the Wiener algebra $\mathcal{W}_+$ (see Section 3.1.1), therefore also $A_{\text{odd}}(z)$ and $A_{\text{even}}(z)$ belong to $\mathcal{W}_+$. Moreover, we may show that $I - A_{\text{odd}}(z)$ is nonsingular for $|z| \leq 1$ so that its inverse exists and belongs to $\mathcal{W}_+$ in light of Theorem 3.2.

Theorem 7.7 *Let $A(z) = \sum_{i=-1}^{+\infty} z^{i+1} A_i$ be the generating function associated with an M/G/1-type Markov chain. Assume that Condition 4.20 is satisfied. Then $I - A_{\text{odd}}(z)$ is nonsingular for $|z| \leq 1$ and $\rho(A_0) < 1$.*

Proof Since the Markov chain (4.24) satisfies Condition 4.20, there exists only one final class $\mathcal{S} = \mathbb{Z} \times S$, $S \subset \{1, \ldots, m\}$ of $T_{\pm\infty}[A(z)]$. For simplicity we may assume $S = \{1, \ldots, k\}$ for some $k > 0$. Then for Theorem 4.21 and Remark 4.22 the matrix $A = \sum_{i=-1}^{+\infty} A_i$ has the only one final class S so that

$$A = \begin{bmatrix} B & 0 \\ C & D \end{bmatrix}$$

where $\rho(B) = 1$, $C \neq 0$, B is irreducible and $\rho(D) < 1$. Let us partition the matrix $A(z)$ according to the partition of A so that

$$A(z) = \begin{bmatrix} B(z) & 0 \\ C(z) & D(z) \end{bmatrix}$$

and

$$A = \begin{bmatrix} B & 0 \\ C & D \end{bmatrix} = \begin{bmatrix} B_{\text{even}} & 0 \\ C_{\text{even}} & D_{\text{even}} \end{bmatrix} + \begin{bmatrix} B_{\text{odd}} & 0 \\ C_{\text{odd}} & D_{\text{odd}} \end{bmatrix},$$

where $B_{\text{even}} = B_{\text{even}}(1)$, $B_{\text{odd}} = B_{\text{odd}}(1)$, $C_{\text{even}} = C_{\text{even}}(1)$, $C_{\text{odd}} = C_{\text{odd}}(1)$, $D_{\text{even}} = D_{\text{even}}(1)$, $D_{\text{odd}} = D_{\text{odd}}(1)$. Then we have $\rho(A_{\text{odd}}(1)) = \max\{\rho(B_{\text{odd}}), \rho(D_{\text{odd}})\}$. Since $D_{\text{odd}} \leq D_{\text{odd}} + D_{\text{even}} = D$, from Theorem 1.28 we have $\rho(D_{\text{odd}}) \leq \rho(D) < 1$. Observe that $B_{\text{even}} \geq (A_{-1})_S \neq 0$ in light of Remark 4.22, so that $B_{\text{odd}} \neq B$. Therefore, since $B_{\text{odd}} \leq B$ and B is irreducible, from the Perron–Frobenius Theorem 1.27 one has $\rho(B_{\text{odd}}) < \rho(B) = 1$. Whence we deduce that $\rho(A_{\text{odd}}(1)) < 1$. Since $|A_{\text{odd}}(z)| \leq A_{\text{odd}}(1)$ for $|z| \leq 1$, applying once again Theorem 1.28 we may conclude that $\rho(A_{\text{odd}}(z)) < 1$ for $|z| \leq 1$ so that $I - A_{\text{odd}}(z)$ is nonsingular for $|z| \leq 1$. Since $A_0 \leq A_{\text{odd}}(1)$, by Theorem 1.28 one has $\rho(A_0) \leq \rho(A_{\text{odd}}(1)) < 1$. □

Now, since the matrix functions $A_{\text{odd}}(z)$, $A_{\text{even}}(z)$, and $(I - A_{\text{odd}}(z))^{-1}$ belong to $\mathcal{W}_+$, they belong to $\mathcal{W}$. Therefore, for Theorem 3.1 and for the subsequent comments, the semi-infinite matrices $U_{1,2}$, $U_{2,1}$, $I - U_{2,2}$ and $(I - U_{1,1})^{-1}$ define bounded operators so that we may apply one step of block Gaussian elimination to the 2×2 block system (7.32) and obtain

$$\begin{aligned} &\begin{bmatrix} I - U_{1,1} & -U_{1,2} \\ 0 & H \end{bmatrix} \begin{bmatrix} \boldsymbol{x}_+ \\ \boldsymbol{x}_- \end{bmatrix} = \begin{bmatrix} 0 \\ \boldsymbol{b} \end{bmatrix}, \\ &H = I - U_{2,2} - U_{2,1}(I - U_{1,1})^{-1} U_{1,2}, \end{aligned} \tag{7.34}$$

where H is the Schur complement of $I - U_{2,2}$.

Now observe that since $(I - A_{\text{odd}}(z))^{-1} \in \mathcal{W}_+$ then $(I - U_{1,1})^{-1}$ is block upper triangular Toeplitz and also the product $W = (I - U_{1,1})^{-1} U_{1,2}$ is block upper triangular Toeplitz. Observe also that $U_{2,1} W$ is a block Hessenberg matrix, so that H is block Hessenberg.

Let us analyze more closely the structure of $U_{2,1} W$. Observe that the matrix $U'_{2,1}$ obtained by removing the first block row of $U_{2,1}$ is block upper triangular Toeplitz so that the product $U'_{2,1} W$ is block upper triangular Toeplitz. This fact implies that the matrix obtained by removing the first block row of H is block upper triangular Toeplitz. In other words, the Schur complement H is block

Hessenberg and block Toeplitz except for the first block row so that the system $H\boldsymbol{x}_- = \boldsymbol{b}$ turns into

$$\begin{bmatrix} I-\widehat{A}_0^{(1)} & -\widehat{A}_1^{(1)} & -\widehat{A}_2^{(1)} & -\widehat{A}_3^{(1)} & \dots \\ -A_{-1}^{(1)} & I-A_0^{(1)} & -A_1^{(1)} & -A_2^{(1)} & \dots \\ & -A_{-1}^{(1)} & I-A_0^{(1)} & -A_1^{(1)} & \ddots \\ & & -A_{-1}^{(1)} & I-A_0^{(1)} & \ddots \\ 0 & & & \ddots & \ddots \end{bmatrix} \begin{bmatrix} G_{\min} \\ G_{\min}^3 \\ G_{\min}^5 \\ \vdots \end{bmatrix} = \begin{bmatrix} A_{-1} \\ 0 \\ \vdots \end{bmatrix}. \tag{7.35}$$

The reduction obtained in this way can be described in a more simple way in functional form. Besides the matrix power series $A(z)$, $A_{\mathrm{odd}}(z)$ and $A_{\mathrm{even}}(z)$ we introduce the functions $\widehat{A}^{(1)}(z) = \sum_{i=0}^{+\infty} z^i \widehat{A}_i^{(1)}$ and $A^{(1)}(z) = \sum_{i=-1}^{+\infty} z^{i+1} A_i^{(1)}$ which completely define the block Hessenberg matrix of (7.35). In this way, by using the equivalence between block triangular Toeplitz matrices and matrix power series of Section 3.1.1, from (7.34) and (7.33) we obtain that

$$\begin{aligned} A^{(1)}(z) &= zA_{\mathrm{odd}}(z) + A_{\mathrm{even}}(z)(I - A_{\mathrm{odd}}(z))^{-1}A_{\mathrm{even}}(z), \\ \widehat{A}^{(1)}(z) &= \widehat{A}_{\mathrm{even}}(z) + \widehat{A}_{\mathrm{odd}}(z)(I - A_{\mathrm{odd}}(z))^{-1}A_{\mathrm{even}}(z), \end{aligned} \tag{7.36}$$

where we have introduced the function $\widehat{A}(z) = \sum_{i=0}^{+\infty} z^i A_i$. Observe that the functions $\widehat{A}^{(1)}(z)$ and $A^{(1)}(z)$ are in the Wiener algebra $\mathcal{W}_+$.

With an inductive argument we may deduce that this process can be recursively repeated by applying at each step the even–odd permutation followed by one step of block Gaussian elimination. After n steps of this cyclic reduction process we obtain the system

$$\begin{bmatrix} I-\widehat{A}_0^{(n)} & -\widehat{A}_1^{(n)} & -\widehat{A}_2^{(n)} & -\widehat{A}_3^{(n)} & \dots \\ -A_{-1}^{(n)} & I-A_0^{(n)} & -A_1^{(n)} & -A_2^{(n)} & \dots \\ & -A_{-1}^{(n)} & I-A_0^{(n)} & -A_1^{(n)} & \ddots \\ 0 & & \ddots & \ddots & \ddots \end{bmatrix} \begin{bmatrix} G_{\min} \\ G_{\min}^{2^n+1} \\ G_{\min}^{2\cdot 2^n+1} \\ G_{\min}^{3\cdot 2^n+1} \\ \vdots \end{bmatrix} = \begin{bmatrix} A_{-1} \\ 0 \\ 0 \\ \vdots \end{bmatrix}, \tag{7.37}$$

where the blocks $A_i^{(n)}$, $\widehat{A}_{i+1}^{(n)}$, $i = -1, 0, \dots$, are defined by means of the recursions

$$\begin{aligned} A^{(n+1)}(z) &= zA_{\mathrm{odd}}^{(n)}(z) + A_{\mathrm{even}}^{(n)}(z)(I - A_{\mathrm{odd}}^{(n)}(z))^{-1}A_{\mathrm{even}}^{(n)}(z), \\ \widehat{A}^{(n+1)}(z) &= \widehat{A}_{\mathrm{even}}^{(n)}(z) + \widehat{A}_{\mathrm{odd}}^{(n)}(z)(I - A_{\mathrm{odd}}^{(n)}(z))^{-1}A_{\mathrm{even}}^{(n)}(z), \end{aligned} \tag{7.38}$$

which generalize (7.36).

The applicability of cyclic reduction relies on the invertibility at each step of the matrix power series $I - A_{\mathrm{odd}}^{(n)}(z)$ for $|z| \le 1$.

Theorem 7.8 *In the hypotheses of Theorem 7.7, for any $n \ge 0$,*

1. *the functions* $A^{(n)}(z)$ *and* $\widehat{A}^{(n)}(z)$ *are in the Wiener algebra* $\mathcal{W}_+$*;*
2. $A_{i+1}^{(n)}, \widehat{A}_i^{(n)} \geq 0$*, for* $i \geq -1$*, and* $A^{(n)}(1)$*,* $A_{-1} + \widehat{A}^{(n)}(1)$ *are stochastic;*
3. *the bi-infinite Markov chain with transition matrix* $T_{\pm\infty}[z^{-1}A^{(n)}(z)]$ *satisfies Condition 4.20;*
4. *the function* $I - A_{\mathrm{odd}}^{(n)}(z)$ *is invertible for* $|z| \leq 1$ *and its inverse belongs to* $\mathcal{W}_+$*, so that cyclic reduction can be carried out, moreover,* $\rho(A_{\mathrm{odd}}^{(n)}(1)) < 1$*;*
5. *the matrices* $A_{\mathrm{even}}^{(n)}(1)(I - A_{\mathrm{odd}}^{(n)}(1))^{-1}$ *and* $(I - A_{\mathrm{odd}}^{(n)}(1))^{-1}A_{\mathrm{even}}^{(n)}(1)$ *have spectral radius 1, moreover the latter is stochastic.*

Proof We proceed by induction on n. For $n = 0$ the matrix $I - A_{\mathrm{odd}}^{(0)}(z)$ is invertible for $|z| \leq 1$ by Theorem 7.7, the remaining properties are satisfied by assumption. Assume that the thesis holds for n and let us prove it for $n+1$. Concerning part 1, since $I - A_{\mathrm{odd}}^{(n)}(z)$ is invertible and its inverse is in $\mathcal{W}_+$, and since $A^{(n)}(z), \widehat{A}^{(n)}(z) \in \mathcal{W}_+$, then the functions $A^{(n+1)}(z)$ and $\widehat{A}^{(n+1)}(z)$ exist and belong to $\mathcal{W}_+$ by (7.38). Concerning part 2, since $A_{\mathrm{odd}}^{(n)}(z)$ has nonnegative coefficients then the matrix power series $(I - A_{\mathrm{odd}}^{(n)}(z))^{-1} = \sum_{i=0}^{+\infty}(A_{\mathrm{odd}}^{(n)}(z))^i$ has nonnegative coefficients. Therefore, from (7.38) it follows that the matrices $A_i^{(n+1)}$ and $\widehat{A}_{i+1}^{(n+1)}$ for $i \geq -1$ are nonnegative since they are sums of products of nonnegative matrices. The matrix $A^{(n+1)}(1)$ is stochastic since from (7.38) it follows that $A^{(n+1)}(1)\mathbf{1} = A_{\mathrm{odd}}^{(n)}(1)\mathbf{1} + A_{\mathrm{even}}^{(n)}(1)(I - A_{\mathrm{odd}}^{(n)}(1))^{-1}A_{\mathrm{even}}^{(n)}(1)\mathbf{1} = A_{\mathrm{odd}}^{(n)}(1)\mathbf{1} + A_{\mathrm{even}}^{(n)}(1)\mathbf{1} = \mathbf{1}$. Here we used the property $(I - A_{\mathrm{odd}}^{(n)}(1))^{-1}A_{\mathrm{even}}^{(n)}(1)\mathbf{1} = \mathbf{1}$ which holds by induction. Similarly we may show that $(A_{-1} + \widehat{A}^{(n+1)}(1))\mathbf{1} = \mathbf{1}$. Concerning part 3, since the assumptions of Theorem 4.23 are satisfied then the bi-infinite Markov chain with transition matrix $T_{\pm\infty}[z^{-1}A^{(n+1)}(z)]$ satisfies Condition 4.20. Concerning part 4, the invertibility of $I - A_{\mathrm{odd}}^{(n+1)}(z)$ holds from Theorem 7.7 applied to the function $A^{(n+1)}(z)$. Since $A_{\mathrm{odd}}^{(n+1)}(1)$ is substochastic then $\rho(A_{\mathrm{odd}}^{(n+1)}(1)) < 1$. Finally, concerning part 5, since $A^{(n+1)}(1) = A_{\mathrm{odd}}^{(n+1)}(1) + A_{\mathrm{even}}^{(n+1)}(1)$ is stochastic, then $(A_{\mathrm{even}}^{(n+1)}(1) + A_{\mathrm{odd}}^{(n+1)}(1))\mathbf{1} = \mathbf{1}$, whence $(I - A_{\mathrm{odd}}^{(n+1)}(1))^{-1}A_{\mathrm{even}}^{(n+1)}(1)\mathbf{1} = \mathbf{1}$, that is, the matrix $(I - A_{\mathrm{odd}}^{(n+1)}(1))^{-1}A_{\mathrm{even}}^{(n+1)}(1)$ is stochastic and therefore has spectral radius 1. Moreover, since $(I - A_{\mathrm{odd}}^{(n+1)}(1))^{-1}A_{\mathrm{even}}^{(n+1)}(1)$ and $A_{\mathrm{even}}^{(n+1)}(1)(I - A_{\mathrm{odd}}^{(n+1)}(1))^{-1}$ are similar then $\rho(A_{\mathrm{even}}^{(n+1)}(1)(I - A_{\mathrm{odd}}^{(n+1)}(1))^{-1}) = 1$. □

From the first equation of (7.37) we find that

$$G_{\min} = (I - \widehat{A}_0^{(n)})^{-1}(A_{-1} + \sum_{i=1}^{+\infty} \widehat{A}_i^{(n)} G_{\min}^{i \cdot 2^n + 1}). \tag{7.39}$$

This formula is a valid tool for approximating the minimal solution $G_{\min}$ once we prove that the summation $\sum_{i=1}^{+\infty} \widehat{A}_i^{(n)} G_{\min}^{i \cdot 2^n + 1}$ converges quadratically to zero for $n \to +\infty$ and that $(I - \widehat{A}_0^{(n)})^{-1}$ is bounded and quadratically convergent. These properties will be proved in Section 7.4.1.

Similarly to the QBD case, defining $\varphi^{(n)}(z) = zI - A^{(n)}(z)$, we find that

$$I - A_{\mathrm{odd}}^{(n)}(z^2) = \frac{1}{2z}(\varphi^{(n)}(z) - \varphi^{(n)}(-z)) = \varphi_{\mathrm{odd}}^{(n)}(z^2)$$

and after some formal manipulations from (7.38) we obtain

$$\begin{aligned}\varphi^{(n+1)}(z^2) &= -\varphi^{(n)}(z)\left(\frac{\varphi^{(n)}(z) - \varphi^{(n)}(-z)}{2z}\right)^{-1}\varphi^{(n)}(-z) \\ &= -\varphi^{(n)}(z)\varphi_{\mathrm{odd}}^{(n)}(z^2)^{-1}\varphi^{(n)}(-z), \quad n \geq 0.\end{aligned} \tag{7.40}$$

The above formula can be rewritten in a slightly different form as

$$\varphi^{(n+1)}(z^2) = \left(\frac{\varphi^{(n)}(z)^{-1} - \varphi^{(n)}(-z)^{-1}}{2z}\right)^{-1}, \quad n \geq 0, \tag{7.41}$$

which holds for all the values of z such that $\det \varphi^{(n)}(z), \det \varphi^{(n)}(-z) \neq 0$ and $\det(\varphi^{(n)}(z)^{-1} - \varphi^{(n)}(-z)^{-1}) \neq 0$.

Moreover, setting $\psi^{(n)}(z) = (z^{-1}\varphi^{(n)}(z))^{-1}$, one has

$$\psi^{(n+1)}(z^2) = \frac{1}{2}(\psi^{(n)}(z) + \psi^{(n)}(-z)), \quad n \geq 0. \tag{7.42}$$

In particular, if $\psi^{(0)}(z) = (I - z^{-1}A(z))^{-1} = \sum_{i=-\infty}^{+\infty} z^i H_i$, then

$$\psi^{(n)}(z) = \sum_{i=-\infty}^{+\infty} z^i H_{i \cdot 2^n}. \tag{7.43}$$

We may provide similar expressions for the function $\widehat{A}^{(n)}(z)$ satisfying (7.38), which will be useful in Chapter 9, as shown in the following.

Theorem 7.9 *The function $\widehat{A}^{(n)}(z)$ defined by (7.38) can be represented in the following different ways*

$$\begin{aligned}\widehat{A}^{(n+1)}(z^2) &= \widehat{A}^{(n)}(z) - \widehat{A}_{\mathrm{odd}}^{(n)}(z^2)\varphi_{\mathrm{odd}}^{(n)}(z^2)^{-1}\varphi^{(n)}(z) \\ &= \widehat{A}^{(n)}(z) + \widehat{A}_{\mathrm{odd}}^{(n)}(z^2)\varphi^{(n)}(-z)^{-1}\varphi^{(n+1)}(z^2) \\ &= -\left[\widehat{A}^{(n)}(z)\varphi_{\mathrm{odd}}^{(n)}(z^2)^{-1}\varphi^{(n)}(-z)\right]_{\mathrm{odd}} \\ &= \left[\widehat{A}^{(n)}(z)\varphi^{(n)}(z)^{-1}\right]_{\mathrm{odd}}\varphi^{(n+1)}(z^2) \\ &= -\left[z^{-1}\widehat{A}^{(n)}(z)\varphi_{\mathrm{odd}}^{(n)}(z^2)^{-1}\varphi^{(n)}(-z)\right]_{\mathrm{even}} \\ &= \left[z^{-1}\widehat{A}^{(n)}(z)\varphi^{(n)}(z)^{-1}\right]_{\mathrm{even}}\varphi^{(n+1)}(z^2).\end{aligned} \tag{7.44}$$

Proof Rewrite the second equation in (7.38) with z replaced by z^2 and then replace $\widehat{A}^{(n)}_{\mathrm{even}}(z^2)$ with $\widehat{A}^{(n)}(z) - z\widehat{A}^{(n)}_{\mathrm{odd}}(z^2)$ and obtain

$$\widehat{A}^{(n+1)}(z^2) = \widehat{A}^{(n)}(z) - \widehat{A}^{(n)}_{\mathrm{odd}}(z^2)\left(zI - (I - A^{(n)}_{\mathrm{odd}}(z^2))^{-1}A^{(n)}_{\mathrm{even}}(z^2)\right).$$

By means of formal manipulations it turns out that the rightmost factor in the above expression can be written as $(I - A^{(n)}_{\mathrm{odd}}(z^2))^{-1}\varphi^{(n)}(z)$. Therefore, since $I - A^{(n)}_{\mathrm{odd}}(z) = \varphi^{(n)}_{\mathrm{odd}}(z)$, one obtains the first expression for $\widehat{A}^{(n+1)}(z^2)$ in (7.44).

The second expression can be simply deduced from the first one by observing that $\varphi^{(n)}_{\mathrm{odd}}(z^2)^{-1}\varphi^{(n)}(z) = -\varphi^{(n)}(-z)^{-1}\varphi^{(n+1)}(z^2)$ which holds from the equation obtained by replacing z with $-z$ in (7.40).

Concerning the third representation of $\widehat{A}^{(n+1)}(z^2)$, we rewrite the first formula of (7.44) by replacing $\widehat{A}^{(n)}_{\mathrm{odd}}(z^2)$ with $(\widehat{A}^{(n)}(z) - \widehat{A}^{(n)}(-z))/(2z)$ and obtain

$$\begin{aligned}&\widehat{A}^{(n+1)}(z^2)\\&= \frac{1}{2z}\widehat{A}^{(n)}(z)\left(2zI - \varphi^{(n)}_{\mathrm{odd}}(z^2)^{-1}\varphi^{(n)}(z)\right) + \frac{1}{2z}\widehat{A}^{(n)}(-z)\varphi^{(n)}_{\mathrm{odd}}(z^2)^{-1}\varphi^{(n)}(z).\end{aligned}$$

Now, since $2zI - \varphi^{(n)}_{\mathrm{odd}}(z^2)^{-1}\varphi^{(n)}(z) = \varphi_{\mathrm{odd}}(z^2)^{-1}(\varphi^{(n)}(z) - \varphi^{(n)}(-z) - \varphi^{(n)}(z)) = -\varphi^{(n)}_{\mathrm{odd}}(z^2)^{-1}\varphi^{(n)}(-z)$, from the latter displayed formula we get

$$\widehat{A}^{(n+1)}(z^2) = -\frac{1}{2z}\widehat{A}^{(n)}(z)\varphi^{(n)}_{\mathrm{odd}}(z^2)^{-1}\varphi^{(n)}(-z) + \frac{1}{2z}\widehat{A}^{(n)}(-z)\varphi^{(n)}_{\mathrm{odd}}(z^2)^{-1}\varphi^{(n)}(z),$$

that is $\widehat{A}^{(n+1)}(z^2) = -\left[\widehat{A}^{(n)}(z)\varphi^{(n)}_{\mathrm{odd}}(z^2)^{-1}\varphi^{(n)}(-z)\right]_{\mathrm{odd}}$.

The fourth representation is obtained from the third representation by replacing $\varphi^{(n)}_{\mathrm{odd}}(z^2)^{-1}\varphi^{(n)}(-z)$ with $-\varphi^{(n)}(z)^{-1}\varphi^{(n+1)}(z^2)$ so that

$$\begin{aligned}-\left[\widehat{A}^{(n)}(z)\varphi^{(n)}_{\mathrm{odd}}(z^2)^{-1}\varphi^{(n)}(-z)\right]_{\mathrm{odd}} &= \left[\widehat{A}^{(n)}(z)\varphi^{(n)}(z)^{-1}\varphi^{(n+1)}(z^2)\right]_{\mathrm{odd}}\\&= \left[\widehat{A}^{(n)}(z)\varphi^{(n)}(z)^{-1}\right]_{\mathrm{odd}}\varphi^{(n+1)}(z^2).\end{aligned}$$

The remaining representations can be proved similarly relying on the equation $[z^{-1}f(z)]_{\mathrm{even}} = [f(z)]_{\mathrm{odd}}$ which holds for any function $f(z)$. □

Observe that for the last equation in (7.44) one has

$$\widehat{A}^{(n+1)}(z^2)\varphi^{(n+1)}(z^2)^{-1} = \left[z^{-1}\widehat{A}^{(n)}(z)\varphi^{(n)}(z)^{-1}\right]_{\mathrm{even}}$$

and replacing $\varphi^{(n)}(z)^{-1}$ with $z^{-1}\psi(z)$ yields

$$\widehat{A}^{(n+1)}(z^2)\psi^{(n+1)}(z^2) = \left[\widehat{A}^{(n)}(z)\psi^{(n)}(z)\right]_{\mathrm{even}}. \tag{7.45}$$

A nice consequence of (7.40) and of Theorem 7.8 is that the roots of $\varphi^{(n+1)}(z)$ in the closed unit disk are the square of the roots of $\varphi^{(n)}(z)$ in the closed unit disk. This fact enables one to prove the following.

Theorem 7.10 *If the drift μ of the generating function $A(z)$ is negative, or positive, then the drift of the generating function $A^{(n)}(z)$ is negative, or positive, respectively.*

Proof The result follows from Theorem 4.9 and from the subsequent comments. □

Theorem 7.11 *Let $A(z)$ be the generating function associated with an M/G/1-type Markov chain. Assume that $A(z)$ is analytic for $|z| < r$, with $1 < r$, and that there exists ζ, such that $1 < |\zeta| < r$ and $\det(\zeta I - A(\zeta)) = 0$. If the drift μ is negative and if Condition 4.20 is satisfied, then:*

1. *There exists a root ξ of $\varphi(z) = zI - A(z)$ of smallest modulus among the roots of modulus greater than 1. Moreover, ξ is real and simple, $1 < \xi < r$, $\rho(A(\xi)) = \xi$, and there exists a positive vector $\boldsymbol{v}$ such that $A(\xi)\boldsymbol{v} = \xi\boldsymbol{v}$.*
2. *For $n \geq 0$ the matrix power series $\varphi^{(n)}(z) = zI - A^{(n)}(z)$ is convergent for $|z| = \xi^{2^n}$ and therefore is analytic for $|z| < \xi^{2^n}$.*
3. *For $n \geq 0$, ξ^{2^n} is the root of smallest modulus among the roots of modulus greater than 1 of $\varphi^{(n)}(z)$, moreover $\rho(A^{(n)}(\xi^{2^n})) = \xi^{2^n}$ and $A(\xi^{2^n})\boldsymbol{v} = \xi^{2^n}\boldsymbol{v}$.*

Proof The existence of ξ and its properties follow from Theorem 4.11. The positivity of $\boldsymbol{v}$ follows from Condition 4.20 in light of Theorem 1.29. Parts 2 and 3 are proved by induction. For $n = 0$, they are valid by assumption. Assume that parts 2 and 3 hold for n and prove them for $n+1$. In order to prove that $\varphi^{(n+1)}(z)$ is convergent for $|z| = \xi^{2^{n+1}}$, we first observe that from (7.40) one has $\varphi^{(n+1)}(z^2) = -\varphi^{(n)}(z)\varphi^{(n)}_{\text{odd}}(z^2)^{-1}\varphi^{(n)}(-z)$, so that it is sufficient to prove that $\varphi^{(n)}_{\text{odd}}(z)$ is nonsingular for $|z| = \xi^{2^{n+1}}$. Since $\varphi^{(n)}_{\text{odd}}(z) = I - A^{(n)}_{\text{odd}}(z)$, and the block coefficients of $A^{(n)}(z)$ are nonnegative, it is sufficient to prove that $\rho(A^{(n)}_{\text{odd}}(\xi^{2^{n+1}})) < 1$. Since $z^{-1}A^{(n)}(z) = A^{(n)}_{\text{odd}}(z^2) + z^{-1}A^{(n)}_{\text{even}}(z^2)$, then $A^{(n)}_{\text{odd}}(\xi^{2^{n+1}}) \leq \xi^{-2^n}A^{(n)}(\xi^{2^n})$ and from the Perron–Frobenius theorem we have $\rho(A^{(n)}_{\text{odd}}(\xi^{2^{n+1}})) \leq \rho(\xi^{-2^n}A^{(n)}(\xi^{2^n})) = 1$, where the latter equality holds by the inductive assumption. In order to prove the strict inequality $\rho(A^{(n)}_{\text{odd}}(\xi^{2^{n+1}})) < \rho(\xi^{-2^n}A^{(n)}(\xi^{2^n}))$ we rely on Condition 4.20 and follow the same argument used in the proof of Theorem 7.7. Moreover, from (7.40) we deduce that $\varphi^{(n+1)}(\xi^{2^{n+1}})\boldsymbol{v} = 0$, that is, $A^{(n+1)}(\xi^{2^{n+1}})\boldsymbol{v} = \xi^{2^{n+1}}\boldsymbol{v}$. The minimality of $\xi^{2^{n+1}}$ as root of $\varphi^{(n+1)}(z)$ follows from (7.40). Finally, since $\boldsymbol{v}$ is positive, then $\xi^{2^{n+1}}$ is the spectral radius of $A^{(n+1)}(\xi^{2^{n+1}})$. □

7.4.1 Convergence properties

The convergence of cyclic reduction can be proved under quite general assumptions which substantially reduce to Condition 4.20 which we have assumed at the beginning of the chapter. A first important property at the basis of convergence is expressed by the following.

Theorem 7.12 *Let $G_{\min}$ be the minimal nonnegative solution of the equation (7.29) and let . Then $G_{\min}^{2^n}$ is the minimal nonnegative solution of the equation*

$$X = \sum_{i=-1}^{+\infty} A_i^{(n)} X^{i+1}, \quad n \geq 0. \tag{7.46}$$

Proof First we show that $G_{\min}^{2^n}$ is a solution of (7.46). For this puropose, observe that applying n steps of cyclic reduction to the system

$$\begin{bmatrix} I - A_0 & -A_1 & -A_2 & -A_3 & \dots \\ -A_{-1} & I - A_0 & -A_1 & -A_2 & \ddots \\ & -A_{-1} & I - A_0 & -A_1 & \ddots \\ & & -A_{-1} & I - A_0 & \ddots \\ 0 & & & \ddots & \ddots \end{bmatrix} \begin{bmatrix} I \\ G_{\min} \\ G_{\min}^2 \\ \vdots \end{bmatrix} = \begin{bmatrix} I - A_0^* \\ 0 \\ \vdots \\ \vdots \end{bmatrix}, \tag{7.47}$$

where $A_0^* = \sum_{i=0}^{+\infty} A_i G_{\min}^i$, yields

$$\begin{bmatrix} I - \widehat{A}_0^{(n)} & -\widehat{A}_1^{(n)} & -\widehat{A}_2^{(n)} & -\widehat{A}_3^{(n)} & \dots \\ -A_{-1}^{(n)} & I - A_0^{(n)} & -A_1^{(n)} & -A_2^{(n)} & \ddots \\ & -A_{-1}^{(n)} & I - A_0^{(n)} & -A_1^{(n)} & \ddots \\ & & -A_{-1}^{(n)} & I - A_0^{(n)} & \ddots \\ 0 & & & \ddots & \ddots \end{bmatrix} \begin{bmatrix} I \\ G_{\min}^{2^n} \\ G_{\min}^{2\cdot 2^n} \\ \vdots \end{bmatrix} = \begin{bmatrix} I - A_0^* \\ 0 \\ \vdots \\ \vdots \end{bmatrix}. \tag{7.48}$$

From the second equation it follows that $G_{\min}^{2^n} = \sum_{i=-1}^{+\infty} A_i^{(n)} G_{\min}^{(i+1)\cdot 2^n}$. Concerning the minimality of $G_{\min}^{2^n}$ among the nonnegative solutions, we observe that the eigenvalues of $G_{\min}^{2^n}$ are the (2^n)-th power of the eigenvalues of $G_{\min}$ and therefore coincide with the minimum modulus roots of $\varphi^{(n)}(z)$ in light of (7.40). Therefore $G_{\min}^{2^n}$ is the spectral minimal solution of equation (7.46), and for Theorem 4.10, $G_{\min}^{2^n}$ is also the minimal nonnegative solution. □

For positive recurrent Markov chains we may prove the following convergence result.

Theorem 7.13 *Let $A(z)$ be the generating function associated with an M/G/1-type Markov chain with a negative drift μ satisfying Condition 4.20. Assume that $A(z)$ is analytic for $|z| < r$, with $1 < r$, and that there exists ζ, such that $1 < |\zeta| < r$ and $\det(\zeta I - A(\zeta)) = 0$. Denote $\eta = \max\{|z| : \ |z| < 1, \det \varphi(z) = 0\}$, $\xi = \min\{|z| : \ |z| > 1, \det \varphi(z) = 0\}$. Let ϵ be a positive number such that $\eta + \epsilon < 1$ and $\xi - \epsilon > 1$, and let $\|\cdot\|$ be any fixed matrix norm. Then the matrix power series $A^{(n)}(z) = \sum_{i=-1}^{+\infty} z^{i+1} A_i^{(n)}$, $\widehat{A}^{(n)}(z) = \sum_{i=0}^{+\infty} z^i \widehat{A}_i^{(n)}$, $n \geq 0$, generated by the recurrences (7.38) satisfy the following properties.*

1. *There exists a positive γ such that*

$$\begin{aligned} \|A_i^{(n)}\| &\leq \gamma \xi^{2^n} (\xi - \epsilon)^{-(i+1)2^n}, \quad i \geq 1, \\ \|\widehat{A}_i^{(n)}\| &\leq \gamma \xi^{2^n} (\xi - \epsilon)^{-(i+1)2^n}, \quad i \geq 1, \end{aligned} \tag{7.49}$$

for any $n \geq 0$.

2. *There exist* $A_0^{(\infty)} = \lim_n A_0^{(n)}$, $\widehat{A}_0^{(\infty)} = \lim_n \widehat{A}_0^{(n)}$, $A_{-1}^{(\infty)} = \lim_n A_{-1}^{(n)}$, *where* $A_{-1}^{(\infty)} = (I - A_0^{(\infty)})\mathbf{1}\boldsymbol{g}^{\mathrm{T}}$, $\widehat{A}_0^{(\infty)} = \sum_{i=0}^{+\infty} A_i G_{\min}^i$, *and* $\boldsymbol{g} \geq 0$ *is such that* $\boldsymbol{g}^{\mathrm{T}} G_{\min} = \boldsymbol{g}^{\mathrm{T}}$, $\boldsymbol{g}^{\mathrm{T}}\mathbf{1} = 1$; *moreover,*
$$\|\widehat{A}_0^{(n)} - \widehat{A}_0^{(\infty)}\| \leq \gamma \xi^{2^n} (\xi - \epsilon)^{-2^{n-1}}, \qquad n \geq 0.$$
3. *The sequence* $\{\varphi^{(n)}(z)\}_n$, *where* $\varphi^{(n)}(z) = zI - A^{(n)}(z)$ *for* $n \geq 0$, *uniformly converges to* $\varphi^{(\infty)}(z) = -A_{-1}^{(\infty)} - z(A_0^{(\infty)} - I)$ *over any compact subset of the disk* $\{z \in \mathbb{C} : \ |z| < \xi\}$, *moreover* $\rho(\widehat{A}_0^{(\infty)}) \leq \rho(A_0^{(\infty)}) < 1$.
4. *For any* $n \geq 0$
$$\|A_0^{(n)} - A_0^{(\infty)}\| \leq \gamma \xi^{2^n} (\xi - \epsilon)^{-2^{n+1}},$$
$$\|A_{-1}^{(n)} - A_{-1}^{(\infty)})\| \leq \gamma \left(\xi^{2^n} (\xi - \epsilon)^{-2^{n+1}} + (\eta + \epsilon)^{2^n} \right).$$
5. *For any* $n \geq 0$
$$\|G_{\min} - G^{(n)}\| \leq \gamma \xi^{2^n} (\xi - \epsilon)^{-2^{n+1}},$$
where $G^{(n)} = (I - \widehat{A}_0^{(n)})^{-1} A_{-1}$.

Proof By Theorem 7.11 the function $\varphi^{(n)}(z)$ is analytic for $|z| < \xi^{2^n}$. Therefore, from Theorem 3.7 one deduces that
$$A_{i-1}^{(n)} \leq M((\xi - \epsilon)^{2^n})(\xi - \epsilon)^{-i \cdot 2^n}, \quad i \geq 0, \tag{7.50}$$
for $0 < \epsilon < \xi - 1$, where the $m \times m$ matrix $M(\sigma)$ is such that $M(\sigma) = \max_{|z|=\sigma} |A^{(n)}(z)|$. Since $A^{(n)}(z)$ has nonnegative coefficients and since $1 < \xi - \epsilon < \xi$, one has $M((\xi - \epsilon)^{2^n}) = A^{(n)}((\xi - \epsilon)^{2^n}) \leq A^{(n)}(\xi^{2^n})$. By Theorem 7.11 one has $A^{(n)}(\xi^{2^n})\boldsymbol{v} = \xi^{2^n}\boldsymbol{v}$ where $\boldsymbol{v}$ is strictly positive. Whence, one deduces that $(A^{(n)}(\xi^{2^n}))_{i,j} \leq \xi^{2^n}\sigma$, $i, j = 1, \ldots, m$, where $\sigma = \max_i v_i / \min_i v_i$. Therefore from equation (7.50) one has $\|A_i^{(n)}\|_\infty \leq \gamma \xi^{2^n} (\xi - \epsilon)^{-(i+1) \cdot 2^n}$, $i \geq 1$ for a suitable $\gamma > 0$. Thus, the first equation of (7.49) follows from Theorem A.1 on the equivalence of matrix norms. Concerning the convergence of $\{\widehat{A}_i^{(n)}\}$, for $i \geq 1$, we observe that $0 \leq \widehat{A}_i^{(n)} \leq A_i^{(n)}$ for $i \geq 0$, which can be easily proved by induction on n relying on (7.38). Therefore, $\|\widehat{A}_i^{(n)}\|_\infty \leq \gamma \xi^{2^n} (\xi - \epsilon)^{-(i+1) \cdot 2^n}$, $i \geq 1$. This completes the proof of part 1.

Concerning part 2, from (7.38) one has that the sequences $\{A_0^{(n)}\}_n$, $\{\widehat{A}_0^{(n)}\}_n$ are nondecreasing. Since the matrices $A_0^{(n)}$ and $\widehat{A}_0^{(n)}$ are substochastic for any n, it follows that there exists $A_0^{(\infty)} = \lim_n A_0^{(n)}$ and $\widehat{A}_0^{(\infty)} = \lim_n \widehat{A}_0^{(n)}$. Now, from Theorem 7.12 one has
$$A_{-1}^{(n)} = (I - A_0^{(n)}) G_{\min}^{2^n} - \sum_{i=1}^{+\infty} A_i^{(n)} (G_{\min}^{2^n})^{i+1}. \tag{7.51}$$
Since by the properties stated in part 1 the summation in the above equation converges to zero for $n \to \infty$, and since $\lim_n G_{\min}^{2^n} = \mathbf{1}\boldsymbol{g}^{\mathrm{T}}$, $\lim_n A_0^{(n)} = A_0^{(\infty)}$,

then taking the limits in (7.51) as $n \to +\infty$ yields $\lim_n A_{-1}^{(n)} = (I - A_0^{(\infty)})\mathbf{1}\boldsymbol{g}^{\mathrm{T}}$. In order to prove that $\lim_n \widehat{A}_0^{(n)} = \sum_{i=0}^{+\infty} A_i G_{\min}^i$ we proceed as in the proof of Theorem 7.12. Namely, from (7.48) one has that

$$\widehat{A}_0^{(n)} = A_0^* - \sum_{i=1}^{+\infty} \widehat{A}_i^{(n)} G_{\min}^{i \cdot 2^n},$$

where $A_0^* = \sum_{i=0}^{+\infty} A_i G_{\min}^i$, therefore, by taking the infinity norm, since $G_{\min}$ is stochastic, we obtain

$$\|\widehat{A}_0^{(n)} - A_0^*\|_\infty = \left\|\sum_{i=1}^{+\infty} \widehat{A}_i^{(n)}\right\|_\infty \le \gamma \xi^{2^n} \sum_{i=1}^{+\infty} (\xi - \epsilon)^{-(i+1)2^n} \le \gamma' \xi^{2^n} (\xi - \epsilon)^{-2^{n+1}},$$

for suitable positive constants γ and γ'. This completes part 2.

Concerning part 3, in order to prove the uniform convergence we have to show that, for any compact subset $\mathcal{K}$ of the disk $\{z \in \mathbb{C} : \quad |z| < \xi\}$, one has $\lim_n \sup_{z\in\mathcal{K}} |\varphi^{(n)}(z) - \varphi^{(\infty)}(z)| = 0$. Let $\mathcal{K}$ be any compact subset and let $\sigma = \max_{z\in\mathcal{K}} |z|$. One has

$$|\varphi^{(n)}(z) - \varphi^{(\infty)}(z)| \le |A_{-1}^{(n)} - A_{-1}^{(\infty)}| + \sigma |A_0^{(n)} - A_0^{(\infty)}| + \sum_{i=1}^{+\infty} \sigma^{i+1} A_i^{(n)},$$

for any $z \in \mathcal{K}$. Therefore

$$\sup_{z\in\mathcal{K}} |\varphi^{(n)}(z) - \varphi^{(\infty)}(z)| \le |A_{-1}^{(n)} - A_{-1}^{(\infty)}| + \sigma |A_0^{(n)} - A_0^{(\infty)}| + \sum_{i=1}^{+\infty} \sigma^{i+1} A_i^{(n)}.$$

Since, for part 2, $\lim_n |A_{-1}^{(n)} - A_{-1}^{(\infty)}| = \lim_n |A_0^{(n)} - A_0^{(\infty)}| = 0$, it remains to prove that $\lim_n \sum_{i=1}^{+\infty} \sigma^{i+1} A_i^{(n)} = 0$. In order to show this, we apply part 1 and deduce that for any ϵ such that $0 < \epsilon < \xi - \sigma$ and for any norm there exists $\gamma > 0$ such that

$$\left\|\sum_{i=1}^{+\infty} \sigma^{i+1} A_i^{(n)}\right\| \le \gamma \xi^{2^n} \sum_{i=1}^{+\infty} \sigma^{i+1} (\xi - \epsilon)^{-(i+1)2^n} \le \gamma' \xi^{2^n} (\xi - \epsilon)^{-2^{n+1}}.$$

Whence $\lim_n \sum_{i=1}^{+\infty} \sigma^{i+1} A_i^{(n)} = 0$. Since the sequence $\{\varphi^{(n)}(z)\}_n$ uniformly converges over any compact subset $\mathcal{K}$ also the sequence $\{\det \varphi^{(n)}(z)\}_n$ uniformly converges over any compact subset $\mathcal{K}$ to $\det \varphi^{(\infty)}(z)$. Therefore we may apply Theorem A.19 to $\det \varphi^{(n)}(z)$ and deduce that for any θ, $1 < \theta < \xi$, the number of zeros in the disk $\mathcal{D} = \{z \in \mathbb{C} : \quad |z| < \theta\}$ of $\det \varphi^{(n)}(z)$ and of $\det \varphi^{(\infty)}(z)$ is the same for any n sufficiently large. Since $\mu < 0$, for any n the function $\det \varphi^{(n)}(z)$ has exactly m zeros in $\mathcal{D}$. Therefore also $\det \varphi^{(\infty)}(z)$ has exactly m zeros in $\mathcal{D}$. Now, since $A_{-1}^{(\infty)} = (I - A_0^{(\infty)})\mathbf{1}\boldsymbol{g}^{\mathrm{T}}$, we find that $\varphi^{(\infty)}(z) = (I - A_0^{(\infty)})(zI - \mathbf{1}\boldsymbol{g}^{\mathrm{T}})$.

Therefore, $I - A_0^{(\infty)}$ cannot be singular otherwise any $z \in \mathcal{D}$ would be a zero of $\det \varphi^{(\infty)}(z)$. Since from Theorem 7.8 $\rho(A_0^{(n)}) < 1$, then $\rho(A_0^{(\infty)}) \leq 1$. Now we prove that $\rho(A_0^{(\infty)}) \neq 1$. If by a condradiction $\rho(A_0^{(\infty)}) \neq 1$, then 1 would be eigenvalue of $A_0^{(\infty)}$ in light of Theorem 1.28, therefore the matrix $I - A_0^{(\infty)}$ would be singular.

In order to prove part 4 we provide an explicit espression of the block coefficients $A_0^{(n)}$ as follows. From (7.36) we deduce that

$$A_0^{(n+1)} = A_0^{(n)} + A_{-1}^{(n)} K_0^{(n)} A_1^{(n)} + A_{-1}^{(n)} K_1^{(n)} A_{-1}^{(n)} + A_1^{(n)} K_0^{(n)} A_{-1}^{(n)}$$

where $(I - A_{\mathrm{odd}}^{(n)}(z))^{-1} = \sum_{i=0}^{+\infty} z^i K_i^{(n)}$. Since $K_0^{(n)} = (I - A_0^{(n)})^{-1}$ and $K_1^{(n)} = (I - A_0^{(n)})^{-1} A_2^{(n)} (I - A_0^{(n)})^{-1}$, we find that

$$\begin{aligned} A_0^{(n+1)} - A_0^{(n)} =& A_{-1}^{(n)} (I - A_0^{(n)})^{-1} A_1^{(n)} \\ &+ A_{-1}^{(n)} (I - A_0^{(n)})^{-1} A_2^{(n)} (I - A_0^{(n)})^{-1} A_{-1}^{(n)} \\ &+ A_1^{(n)} (I - A_0^{(n)})^{-1} A_{-1}^{(n)}. \end{aligned}$$

From parts 2 and 3 the matrices $A_{-1}^{(n)}$ and $(I - A_0^{(n)})^{-1}$ have uniformly bounded norms; from part 1 we have that for any matrix norm and for any $\epsilon > 0$ such that $\xi - \epsilon > 1$, there exists a positive constant γ such that $\|A_1^{(n)}\| \leq \gamma \xi^{2^n} (\xi - \epsilon)^{-2^{n+1}}$ and $\|A_2^{(n)}\| \leq \gamma \xi^{2^n} (\xi - \epsilon)^{-3 \cdot 2^n}$. Whence we find that $\|A_0^{(n+1)} - A_0^{(n)}\| \leq \gamma' \xi^{2^n} (\xi - \epsilon)^{-2^{n+1}}$ for a suitable $\gamma' > 0$. Since there exists the limit of $\{A_0^{(n)}\}$, from the latter inequality and from Lemma A.16 in the appendix, we conclude that $\|A_0^{(n)} - A_0^{(\infty)}\| \leq \gamma'' \xi^{2^n} (\xi - \epsilon)^{-2^{n+1}}$ for a suitable $\gamma'' > 0$. Let us analyze the convergence speed of the sequence $\{A_{-1}^{(n)}\}$. From part 2 we have $A_{-1}^{(\infty)} = \mathbf{1}\boldsymbol{g}^{\mathrm{T}} - A_0^{(\infty)} \mathbf{1}\boldsymbol{g}^{\mathrm{T}}$, from (7.46) we have $A_{-1}^{(n)} = G_{\min}^{2^n} - A_0^{(n)} G_{\min}^{2^n} - \sum_{i=1}^{+\infty} A_i^{(n)} G_{\min}^{(i+1)2^n}$. Subtracting both sides of the latter two equations yields

$$\begin{aligned} A_{-1}^{(n)} - A_{-1}^{(\infty)} =& (G_{\min}^{2^n} - \mathbf{1}\boldsymbol{g}^{\mathrm{T}}) + A_0^{(\infty)} (\mathbf{1}\boldsymbol{g}^{\mathrm{T}} - G_{\min}^{2^n}) \\ &+ (A_0^{(\infty)} - A_0^{(n)}) G_{\min}^{2^n} - \sum_{i=1}^{+\infty} A_i^{(n)} G_{\min}^{(i+1)2^n}. \end{aligned} \tag{7.52}$$

Since from Theorem 4.26 $\|G_{\min}^{2^n} - \mathbf{1}\boldsymbol{g}^{\mathrm{T}}\| \leq \gamma' (\eta + \epsilon)^{2^n}$, for a suitable $\gamma' > 0$, from the convergence of the sequences $\{A_i^{(n)}\}_n$ for $i \geq 0$ and from (7.52) we deduce that

$$\|A_{-1}^{(n)} - A_{-1}^{(\infty)}\| \leq \gamma \left(\xi^{2^n} (\xi - \epsilon)^{-2^{n+1}} + (\eta + \epsilon)^{2^n} \right).$$

Part 5 is a direct consequence of (7.39) and of the quadratic convergence of the sequences $\{\widehat{A}_i^{(n)}\}_n$, $i \geq 1$, to the null matrix, and of the sequence $\{\widehat{A}_0^{(n)}\}_n$ to $\widehat{A}_0^{(\infty)}$, with $\rho(\widehat{A}_0^{(\infty)}) < 1$, stated in parts 1–4. □

If the Markov chain is transient then the convergence of cyclic reduction still holds even if with different features as shown in the following.

Theorem 7.14 *Let $A(z)$ be the generating function associated with an M/G/1-type Markov chain with positive drift μ. Let $A^{(n)}(z) = \sum_{i=-1}^{+\infty} z^{i+1} A_i^{(n)}$, $\widehat{A}^{(n)}(z) = \sum_{i=0}^{+\infty} z^i \widehat{A}_i^{(n)}$, $n \geq 0$, be the matrix power series generated by the recurrences (7.38), and let $\eta = \max\{|z| : \quad |z| < 1, \det \varphi(z) = 0\}$. Then:*

1. *For any $\epsilon > 0$ such that $\eta + \epsilon < 1$ and for any matrix norm $\|\cdot\|$ there exists a positive γ such that*
$$\|A_{-1}^{(n)}\| \leq \gamma(\eta + \epsilon)^{2^n}.$$
2. *There exist the limits $A_0^{(\infty)} = \lim_n A_0^{(n)}$ and $\widehat{A}_0^{(\infty)} = \lim_n \widehat{A}_0^{(n)}$, where $\widehat{A}_0^{(\infty)} = \sum_{i=0}^{+\infty} A_i G_{\min}^i$, moreover, $\rho(\widehat{A}_0^{(\infty)}) < 1$.*
3. *For any $\epsilon > 0$ such that $\eta + \epsilon < 1$ and for matrix norm $\|\cdot\|$ there exists a positive γ such that*
$$\|G_{\min} - G^{(n)}\| \leq \gamma(\eta + \epsilon)^{2^n}$$
where $G^{(n)} = (I - \widehat{A}_0^{(n)})^{-1} A_{-1}$.

Proof From Theorem 7.12 we find that
$$A_{-1}^{(n)} = (I - A_0^{(n)}) G_{\min}^{2^n} - \sum_{i=1}^{+\infty} A_i^{(n)} (G_{\min}^{2^n})^{i+1}.$$

By Theorem 4.26 there exists a positive γ' such that $\|G_{\min}^{2^n}\| \leq \gamma'(\eta + \epsilon)^{2^n}$. Since the matrices $A_i^{(n)}$, $i \geq 0$, are substochastic for any $n \geq 0$ by Theorem 7.8, we conclude that for any matrix norm $\|\cdot\|$ there exists a positive γ such that $\|\widehat{A}_{-1}^{(n)}\| \leq \gamma(\eta + \epsilon)^{2^n}$. Concerning part 2, from (7.38) the sequences $\{A_0^{(n)}\}$ and $\{\widehat{A}_0^{(n)}\}$ are monotonic and bounded, therefore they are convergent. Moreover, by proceeding as in the proof of Theorem 7.12, we find that $\widehat{A}_0^{(n)} = A_0^* - \sum_{i=1}^{+\infty} \widehat{A}_i^{(n)} G_{\min}^{i \cdot 2^n}$, where $A_0^* = \sum_{i=0}^{+\infty} A_i G_{\min}^i$. Therefore, from the convergence to zero of G^{2^n} one has $\lim_n \widehat{A}_0^{(n)} = A_0^*$. The inequality $\rho(\widehat{A}_0^{(\infty)}) < 1$ follows from Theorem 4.15. Part 3 follows from (7.39), from the substochasticity of the matrices $\widehat{A}_i^{(n)}$, $i \geq 1$, and from the quadratic convergence of $\{G_{\min}^{2^n}\}_n$ to the null matrix. □

For null recurrent Markov chains we no longer have the quadratic convergence of cyclic reduction. However, by using a suitable shift technique we may transform the Markov chain into a new problem for which cyclic reduction applies with quadratic convergence. This will be shown in Section 8.2.

7.5 Computational issues of cyclic reduction

The implementation of cyclic reduction for a M/G/1-type Markov chain is more delicate than for QBD problems. In fact in this case we have to deal with an infinite number of blocks A_i, $i \geq -1$. Let us examine closely the general step of cyclic reduction and consider the block Hessenberg matrix $H^{(n)}$ obtained at the nth step

$$H^{(n)} = \begin{bmatrix} I - \widehat{A}_0^{(n)} & -\widehat{A}_1^{(n)} & -\widehat{A}_2^{(n)} & -\widehat{A}_3^{(n)} & \dots \\ -A_{-1}^{(n)} & I - A_0^{(n)} & -A_1^{(n)} & -A_2^{(n)} & \dots \\ & -A_{-1}^{(n)} & I - A_0^{(n)} & -A_1^{(n)} & \ddots \\ 0 & & \ddots & \ddots & \ddots \end{bmatrix}. \quad (7.53)$$

Denote also by Π the permutation matrix associated with the block even–odd permutation so that

$$\Pi H^{(n)} \Pi^{\mathrm{T}} = \begin{bmatrix} I - U_{1,1}^{(n)} & -U_{1,2}^{(n)} \\ -U_{2,1}^{(n)} & I - U_{2,2}^{(n)} \end{bmatrix} \quad (7.54)$$

and

$$U_{1,1}^{(n)} = \begin{bmatrix} A_0^{(n)} & A_2^{(n)} & A_4^{(n)} & \dots \\ & A_0^{(n)} & A_2^{(n)} & \ddots \\ 0 & & \ddots & \ddots \end{bmatrix}, \qquad U_{1,2}^{(n)} = \begin{bmatrix} A_{-1}^{(n)} & A_1^{(n)} & A_3^{(n)} & \dots \\ & A_{-1}^{(n)} & A_1^{(n)} & \ddots \\ 0 & & \ddots & \ddots \end{bmatrix},$$

$$U_{2,1}^{(n)} = \begin{bmatrix} \widehat{A}_1^{(n)} & \widehat{A}_3^{(n)} & \widehat{A}_5^{(n)} & \dots \\ A_{-1}^{(n)} & A_1^{(n)} & A_3^{(n)} & \ddots \\ 0 & \ddots & \ddots & \ddots \end{bmatrix}, \qquad U_{2,2}^{(n)} = \begin{bmatrix} \widehat{A}_0^{(n)} & \widehat{A}_2^{(n)} & \widehat{A}_4^{(n)} & \dots \\ & A_0^{(n)} & A_2^{(n)} & \ddots \\ 0 & & \ddots & \ddots \end{bmatrix}.$$

The main computational task of cyclic reduction is computing the first two block rows of the matrix

$$H^{(n+1)} = I - U_{2,2}^{(n)} - U_{2,1}^{(n)}(I - U_{1,1}^{(n)})^{-1}U_{1,2}^{(n)}. \quad (7.55)$$

Here we follow two different approaches based on the matrix and on the functional interpretation of equation (7.55).

Before considering these two approaches we need to analyze the propagation of the errors at each step of cyclic reduction. In fact, in the computation of the cyclic reduction step we cannot avoid introducing errors. The source of these errors is twofold. On one hand, in dealing with infinite matrices and with power series we have to perform truncations and approximations: finite computations will necessarily introduce approximation errors. On the other hand, using floating point arithmetic introduces a roundoff error in each arithmetic operation.

In order to analyze the propagation of these errors it is useful to make a distinction between the local errors generated at a single step of cyclic reduction and the global error which results from the accumulations of the errors generated at the previous steps. We now formalize these two concepts.

Let us consider the pair of matrix functions $(A^{(n)}(z), \widehat{A}^{(n)}(z))$ which defines the matrix $H^{(n)}$ at the n-th step of cyclic reduction and let us call $\mathcal{F}$ the transformation which defines the single cyclic reduction step so that

$$(A^{(n+1)}(z), \widehat{A}^{(n+1)}(z)) = \mathcal{F}(A^{(n)}(z), \widehat{A}^{(n)}(z)).$$

Let us denote with $\widetilde{\mathcal{F}}$ the function which we actually implement in place of $\mathcal{F}$. In general $\widetilde{\mathcal{F}}$ is different from $\mathcal{F}$ due to rounding errors and to approximation errors generated by truncating power series to polynomials and infinite triangular matrices to banded matrices. Then we define the *local error* generated in one step of cyclic reduction at $(A^{(n)}, \widehat{A}^{(n)})$ as the expression

$$\mathcal{L}(A^{(n)}(z), \widehat{A}^{(n)}(z)) = \widetilde{\mathcal{F}}(A^{(n)}(z), \widehat{A}^{(n)}(z)) - \mathcal{F}(A^{(n)}(z), \widehat{A}^{(n)}(z)).$$

Let $(B^{(n)}(z), \widehat{B}^{(n)}(z))$ be the pairs actually generated by cyclic reduction when each single step is affected by the local error, so that we have

$$\begin{aligned} &(B^{(n+1)}(z), \widehat{B}^{(n+1)}(z)) = \widetilde{\mathcal{F}}(B^{(n)}(z), \widehat{B}^{(n)}(z)), \quad n \geq 0, \\ &B^{(0)}(z) = A^{(0)}(z), \quad \widehat{B}^{(0)}(z) = \widehat{A}^{(0)}(z). \end{aligned}$$

Define the *global error* at step n as

$$\mathcal{E}^{(n)}(z) = (E^{(n)}(z), \widehat{E}^{(n)}(z)) = (B^{(n)}(z), \widehat{B}^{(n)}(z)) - (A^{(n)}(z), \widehat{A}^{(n)}(z))$$

and deduce that

$$\mathcal{E}^{(n+1)}(z) = \widetilde{\mathcal{F}}(B^{(n)}(z), \widehat{B}^{(n)}(z)) - \mathcal{F}(A^{(n)}(z), \widehat{A}^{(n)}(z)).$$

Since $\widetilde{\mathcal{F}}(B^{(n)}(z), \widehat{B}^{(n)}(z)) = \mathcal{F}(B^{(n)}(z), \widehat{B}^{(n)}(z)) + \mathcal{L}(B^{(n)}(z), \widehat{B}^{(n)}(z))$, we find that

$$\mathcal{E}^{(n+1)}(z) = \mathcal{F}(B^{(n)}(z), \widehat{B}^{(n)}(z)) - \mathcal{F}(A^{(n)}(z), \widehat{A}^{(n)}(z)) + \mathcal{L}^{(n)}(z), \qquad (7.56)$$

where for simplicity we denote $\mathcal{L}^{(n)}(z) = \mathcal{L}(B^{(n)}(z), \widehat{B}^{(n)}(z)) = (L^{(n)}(z), \widehat{L}^{(n)}(z))$.

Therefore we may divide our analysis in two parts: estimating the error

$$\mathcal{F}(B^{(n)}(z), \widehat{B}^{(n)}(z)) - \mathcal{F}(A^{(n)}(z), \widehat{A}^{(n)}(z))$$

generated by the perturbation $\mathcal{E}^{(n)}(z)$ in the variables of the function $\mathcal{F}$, and analyzing the local error $\mathcal{L}^{(n)}(z) = (L^{(n)}(z), \widehat{L}^{(n)}(z))$.

While the latter issue depends on the way the cyclic reduction step is implemented, the former issue can be simply answered relying on the next lemma. Here and hereafter, we perform a first-order error analysis, that is, in the error expressions we consider only the terms which are linear in the errors and we neglect quadratic and higher-order terms. We use the symbol $\doteq$ to denote equality up to quadratic and higher-order terms in the errors, therefore we will use expression like $(1+\epsilon)^n \doteq 1 + n\epsilon$, and $1/(1-\epsilon) \doteq 1 + \epsilon$, where ϵ denotes the error. This formal analysis is realistic if the errors are sufficiently small so that their products or powers are actually negligible. Similarly we use the symbol $\dot{\leq}$ for denoting inequality up to quadratic and higher-order terms.

For a matrix power series $F(z) = \sum_{i=0}^{+\infty} z^i F_i$ in the Wiener algebra $\mathcal{W}$ we denote

$$\|F(z)\|_* = \| \sum_{i=0}^{+\infty} |F_i| \, \|_\infty .$$

It is a simple matter to show that $\| \cdot \|_*$ is a norm in $\mathcal{W}$.

The following lemma analyzes the propagation of the error at a general step of cyclic reduction. For notational simplicity we do not write the superscript (n) and we denote with $A(z)$ and $\widehat{A}(z)$ the matrix power series defining the first two block rows of $H^{(n)}$ in (7.53).

Lemma 7.15 *Let $A(z)$ and $\widehat{A}(z)$ be the matrix power series defining the first two block rows of $H^{(n)}$ in (7.53). Let $B(z)$ and $\widehat{B}(z)$ be approximations to $A(z)$ and $\widehat{A}(z)$, respectively, and denote with $R(z) = B(z) - A(z)$ and $\widehat{R}(z) = \widehat{B}(z) - \widehat{A}(z)$ the approximation errors, respectively. Then, for the matrix power series*

$$R^{(1)}(z) = B^{(1)}(z) - A^{(1)}(z), \qquad \widehat{R}^{(1)}(z) = \widehat{B}^{(1)}(z) - \widehat{A}^{(1)}(z),$$

where $(B^{(1)}(z), \widehat{B}^{(1)}(z)) = \mathcal{F}(B(z), \widehat{B}(z))$ and $(A^{(1)}(z), \widehat{A}^{(1)}(z)) = \mathcal{F}(A(z), \widehat{A}(z))$, one has

$$\begin{aligned} R^{(1)}(z) &\doteq z R_{\mathrm{odd}}(z) + R_{\mathrm{even}}(z) W(z) + V(z) R_{\mathrm{even}}(z) + V(z) R_{\mathrm{odd}}(z) W(z), \\ \widehat{R}^{(1)}(z) &\doteq \widehat{R}_{\mathrm{even}}(z) + \widehat{R}_{\mathrm{odd}}(z) W(z) + \widehat{V}(z) R_{\mathrm{even}}(z) + \widehat{V}(z) R_{\mathrm{odd}}(z) W(z), \end{aligned} \tag{7.57}$$

where $V(z) = A_{\mathrm{even}}(z)(I - A_{\mathrm{odd}}(z))^{-1}$, $W(z) = (I - A_{\mathrm{odd}}(z))^{-1} A_{\mathrm{even}}(z)$, $\widehat{V}(z) = \widehat{A}_{\mathrm{odd}}(z)(I - A_{\mathrm{even}}(z))^{-1}$. Moreover, one has

$$\begin{aligned} \|R^{(1)}(z)\|_* &\dot{\leq} 2\|R(z)\|_*(1 + \|V(1)\|_\infty), \\ \|\widehat{R}^{(1)}(z)\|_* &\dot{\leq} 2\|\widehat{R}(z)\|_* + 2\|R(z)\|_* \|\widehat{V}(1)\|_\infty . \end{aligned}$$

Proof Let us rewrite (7.36) replacing $A(z)$ with $B(z)$ and obtain

$$\begin{aligned} B^{(1)}(z) &= z B_{\mathrm{odd}}(z) + B_{\mathrm{even}}(z)(I - B_{\mathrm{odd}}(z))^{-1} B_{\mathrm{even}}(z), \\ \widehat{B}^{(1)}(z) &= \widehat{B}_{\mathrm{even}}(z) + \widehat{B}_{\mathrm{odd}}(z)(I - B_{\mathrm{odd}}(z))^{-1} B_{\mathrm{even}}(z). \end{aligned} \tag{7.58}$$

Now observe that

$$I - B_{\mathrm{odd}}(z) = I - A_{\mathrm{odd}}(z) - R_{\mathrm{odd}}(z) = (I - A_{\mathrm{odd}}(z))[I - (I - A_{\mathrm{odd}}(z))^{-1} R_{\mathrm{odd}}(z)]$$

so that

$$(I - B_{\mathrm{odd}}(z))^{-1} = [I - (I - A_{\mathrm{odd}}(z))^{-1} R_{\mathrm{odd}}(z)]^{-1} (I - A_{\mathrm{odd}}(z))^{-1}.$$

Moreover, since $[I-(I-A_{\text{odd}}(z))^{-1}R_{\text{odd}}(z)]^{-1}\doteq I+(I-A_{\text{odd}}(z))^{-1}R_{\text{odd}}(z)$, one has

$$(I-B_{\text{odd}}(z))^{-1}\doteq(I-A_{\text{odd}}(z))^{-1}+(I-A_{\text{odd}}(z))^{-1}R_{\text{odd}}(z)(I-A_{\text{odd}}(z))^{-1}.$$

Plugging the latter expression in both equations of (7.58) and subtracting equations (7.36) from the expressions obtained in this way, respectively, yields (7.57). For the second part, applying $\|\cdot\|_*$ at both sides of (7.57), and using the triangle inequality yields

$$\begin{aligned}\|R^{(1)}(z)\|_*\dot{\le}&\|R_{\text{odd}}(z)\|_*+\|R_{\text{even}}(z)\|_*\|W(z)\|_*+\|V(z)\|_*\|R_{\text{even}}(z)\|_*\\&+\|R_{\text{odd}}(z)\|_*\|V(z)\|_*\|W(z)\|_*.\end{aligned}$$

For the monotonicity of the infinity norm one has $\|R_{\text{odd}}(z)\|_*\le\|R(z)\|_*$ and $\|R_{\text{even}}(z)\|_*\le\|R(z)\|_*$. Moreover, since the block coeffcients of $W(z)$ are nonnegative and since by Theorem 7.8, $W(1)$ is a stochastic matrix then $\|W(z)\|_*=\|W(1)\|_\infty=1$. Whence we find that

$$\|R^{(1)}(z)\|_*\dot{\le}2\|R(z)\|_*(1+\|V(z)\|_*)=2\|R(z)\|_*(1+\|V(1)\|_\infty),$$

where the latter equality holds since the matrix power series $V(z)$ has nonnegative coefficients. Similarly we may prove the inequality for $\|\widehat{R}^{(1)}(z)\|_*$. □

It is interesting to point out that by Theorem 7.8 the matrices $V(1)$ and $W(1)$ have spectral radius 1.

From the above lemma and from equation (7.56) we immediately find that

$$\begin{aligned}E^{(n+1)}(z)\doteq&zE^{(n)}_{\text{odd}}(z)+V^{(n)}(z)E^{(n)}_{\text{even}}(z)\\&+(E^{(n)}_{\text{even}}(z)+V^{(n)}(z)E^{(n)}_{\text{odd}}(z))W^{(n)}(z)+L^{(n)}(z)\\\widehat{E}^{(n+1)}(z)\doteq&\widehat{E}^{(n)}_{\text{even}}(z)+\widehat{V}^{(n)}(z)E^{(n)}_{\text{even}}(z)\\&+(\widehat{E}^{(n)}_{\text{odd}}(z)+\widehat{V}^{(n)}(z)E^{(n)}_{\text{odd}}(z))W^{(n)}(z)+\widehat{L}^{(n)}(z)\end{aligned}\tag{7.59}$$

for $n=0,1,\ldots$, where $E^{(0)}(z)=\widehat{E}^{(0)}(z)=0$ and

$$\begin{aligned}V^{(n)}(z)&=A^{(n)}_{\text{even}}(z)(I-A^{(n)}_{\text{odd}}(z))^{-1},\\\widehat{V}^{(n)}(z)&=\widehat{A}^{(n)}_{\text{even}}(z)(I-A^{(n)}_{\text{odd}}(z))^{-1},\\W^{(n)}(z)&=(I-A^{(n)}_{\text{odd}}(z))^{-1}A^{(n)}_{\text{even}}(z).\end{aligned}$$

Taking norms $\|\cdot\|_*$ in (7.59), since $\|W^{(n)}(z)\|_*=1$ and $\|E^{(n)}_{\text{odd}}(z)\|_*,\|E^{(n)}_{\text{even}}(z)\|_*\le\|E^{(n)}(z)\|_*$, and $\|\widehat{E}^{(n)}_{\text{odd}}(z)\|_*,\|\widehat{E}^{(n)}_{\text{even}}(z)\|_*\le\|\widehat{E}^{(n)}(z)\|_*$, yields

$$\begin{aligned}&\|E^{(n+1)}(z)\|_*\dot{\le}2(1+\|V^{(n)}(1)\|_\infty)\|E^{(n)}(z)\|_*+\|L^{(n)}(z)\|_*\\&\|\widehat{E}^{(n+1)}(z)\|_*\dot{\le}2\|\widehat{E}^{(n)}(z)\|_*+2\|E^{(n)}(z)\|_*\|\widehat{V}^{(n)}(z)\|_*\\&\qquad+\|\widehat{L}^{(n)}(z)\|_*.\end{aligned}\tag{7.60}$$

In order to better understand the growth of $\|E^{(n)}(z)\|_*$ as a function of n, let us assume that $\|L^{(n)}(z)\|_*\le\nu$ and that $2(1+\|V^{(n)}(1)\|_\infty)=\gamma_n\le\gamma$ for some $\nu > 0$ and $\gamma > 1$. Then the upper bound provided by the first inequality in (7.60) takes the form

$$\|E^{(n+1)}(z)\|_* \le \nu(1 + \gamma_1 + \gamma_1\gamma_2 + \cdots + \gamma_1\gamma_2\cdots\gamma_n) \le n\nu\gamma^n, \tag{7.61}$$

which is exponential with n. A similar bound can be derived for $\|\widehat{E}^{(n)}(z)\|_*$.

In principle this exponential growth of the error bound might suggest that the algorithm is numerically unstable. On the other hand the results of the numerical experiments performed with different implementations of cyclic reduction show the reliability and the stability of the algorithm. This apparent contradiction can be explained with the following arguments. The upper bound (7.61) is obtained by applying several times the triangular inequality assuming each time the worst case error bound. This generally leads to a pessimistic error bound which rarely is reached in practice. Moreover, due to the quadratic convergence of cyclic reduction, the number of iterations generally required in practice is very small. This makes the exponential growth of the error bound less destructive as in principle it could seem.

A further observation relies on the fact that if $E^{(n)}(z)$ has null coefficients from degree 0 up to degree k and if $L^{(n)}(z)$ has null coefficients from degree 0 up to degree $k/2$, then from (7.59) one has that $E^{(n+1)}(z)$ has null coefficients from degree 0 up to degree $k/2$. Therefore, we deduce that if the ith step of cyclic reduction is performed so that $L^{(i)}(z)$ has null coefficients from degree 0 up to degree $k/2^i$, for $i = 0, \ldots, n-1$, then $E^{(n)}(z)$ has null coefficients from degree 0 up to degree $k/2^{n-1}$. In other words, the lower degree coefficients of $E^{(n)}(z)$ can be kept to zero by truncating the cyclic reduction step at a sufficiently large degree. A similar observation applies to $\widehat{E}^{(n)}(z)$.

Finally, another important observation concerning the error propagation is that, as we will see later on in Section 8.2, with a suitable shift technique we may transform the original problem into a different one where the quantities $\|W^{(n)}(z)\|_*$, $\|V^{(n)}(z)\|_*$, and $\|\widehat{V}^{(n)}(z)\|_*$, are bounded from above by $\theta\sigma^{2^n}$ for suitable $\theta > 0$ and $0 < \sigma < 1$. This property, together with (7.57) implies that $\|E^{(n+1)}(z)\|_* \dot{\le} (1+2\theta\sigma^{(2^n)})\|E^{(n)}(z)\|_* + \|L^{(n)}(z)\|_*$. Therefore, (7.61) holds with $\gamma_n = (1 + 2\theta\sigma^{2^n})$.

Since $\log(1+x) \le x$

$$\log \prod_{i=1}^{n} \gamma_i = \sum_{i=1}^{m} \log \gamma_i \le 2\theta \sum_{i=1}^{n} \sigma^{2^i} \le 2\theta\sigma^2/(1-\sigma^2),$$

so that from the first inequality in (7.61) we find that

$$\|E^{(n)}(z)\|_* \le \nu(n+1)e^{2\theta\frac{\sigma^2}{1-\sigma^2}} \tag{7.62}$$

which is a bound linear in n. A similar analysis can be performed for $\widehat{E}^{(n)}(z)$.

7.5.1 *Matrix implementation*

Concerning the matrix interpretation, the most expensive part in the implementation of (7.55) is the computation of the first two block rows of the matrix $U_{2,1}^{(n)}(I - U_{1,1}^{(n)})^{-1}U_{1,2}^{(n)}$. This computation can be split into the following parts:

1. compute $T_\infty[K(z)] = (I - U_{1,1}^{(n)})^{-1}$, where $K(z) = (I - A_{\text{odd}}^{(n)}(z))^{-1}$;
2. compute the block vectors

$$\boldsymbol{v}^{\mathrm{T}} = \boldsymbol{u}^{\mathrm{T}}(I - U_{1,1}^{(n)})^{-1}, \qquad \widehat{\boldsymbol{v}}^{\mathrm{T}} = \widehat{\boldsymbol{u}}^{\mathrm{T}}(I - U_{1,1}^{(n)})^{-1}, \tag{7.63}$$

   where $\widehat{\boldsymbol{u}}^{\mathrm{T}} = [\widehat{A}_1^{(n)}, \widehat{A}_3^{(n)}, \widehat{A}_5^{(n)}, \ldots]$ and $\boldsymbol{u}^{\mathrm{T}} = [A_{-1}^{(n)}, A_1^{(n)}, A_3^{(n)}, \ldots]$ are the first and the second block rows of $U_{2,1}^{(n)}$, respectively;
3. compute the products $\boldsymbol{y}^{\mathrm{T}} = \boldsymbol{v}^{\mathrm{T}}U_{1,2}^{(n)}$, $\widehat{\boldsymbol{y}}^{\mathrm{T}} = \widehat{\boldsymbol{v}}^{\mathrm{T}}U_{1,2}^{(n)}$.

In all three stages above we have to perform computations involving infinite block triangular Toeplitz matrices.

In order to deal with infinite matrices we could replace them with finite matrices obtained by means of truncation at a sufficiently large size d. However, this technique does not allow one reliable control of the truncation errors and a better approach can be devised by approximating the infinite triangular matrices with infinite banded matrices. In fact, despite their infinite size, banded Toeplitz matrices are defined by a finite number of parameters. This technique of band truncation relies on the decay properties of the block elements of the matrices involved.

In fact, according to Theorem 7.8 the functions $A^{(n)}(z)$, $\widehat{A}^{(n)}(z)$, are in the Wiener algebra $\mathcal{W}_+$ and $I - A_{\text{odd}}^{(n)}(z)$ is nonsingular and its inverse $K(z) = \sum_{i=0}^{+\infty} z^i K_i = (I - A_{\text{odd}}^{(n)}(z))^{-1}$ is in $\mathcal{W}_+$. In this way, the coefficients of $A^{(n)}(z)$, $\widehat{A}^{(n)}(z)$ and $K(z)$ decay to zero, i.e., for any $\epsilon > 0$ there exist positive integers $s, q > 0$ such that $\sum_{i \geq s} A_i^{(n)} < \epsilon E$, $\sum_{i \geq s+1} \widehat{A}_i^{(n)} < \epsilon E$, $\sum_{i \geq q} K_i < \epsilon E$, where E is the matrix with all elements equal to 1. Moreover, if in addition these functions are analytic in an open set containing the unit circle, then for Theorem 3.6 the decay is exponential.

Based on this property, we may replace the matrices $H^{(n)}$ with block banded matrices defined by the matrix polynomials $A_s^{(n)}(z) = \sum_{i=-1}^{s-1} z^{i+1} A_i^{(n)}$ and $\widehat{A}_s^{(n)}(z) = \sum_{i=0}^{s} z^i \widehat{A}_i^{(n)}$ where possibly s depends on n. For the sake of simplicity, without loss of generality we choose $s+1$ even. With this choice the number of block coefficients of the matrix polynomials $A_s^{(n)}(z)$ and $\widehat{A}_s^{(n)}(z)$ is the even number $d = s+1$. In this way the operations at stages 2 and 3 of the above scheme can be reduced to matrix polynomial multiplications and we may use the algorithms of Section 2.2 having cost $O(m^2 d \log d + m^3 d)$.

From the computational point of view, a first problem which we encounter is to estimate the value of d which guarantees a local error less than a given bound ϵ. Since the blocks $A_i^{(n)}$ and $\widehat{A}_i^{(n)}$ are nonnegative and $A^{(n)}(1)$, $A_{-1} + \widehat{A}^{(n)}(1)$ are stochastic (see Theorem 7.8), we may easily estimate the truncation error of replacing the matrix power series with the matrix polynomial. In fact, if $\|\mathbf{1} - A_{d-1}^{(n)}(1)\mathbf{1}\|_\infty \le \epsilon$ then $\sum_{i\ge d-1} A_i^{(n)} < \epsilon E$. A similar property holds for $\widehat{A}^{(n)}(z)$, namely, if $\|\mathbf{1} - (A_{-1} + \widehat{A}_{d-1}^{(n)}(1))\mathbf{1}\|_\infty \le \epsilon$ then $\sum_{i\ge d} \widehat{A}_i^{(n)} < \epsilon E$.

Concerning the computation of $(I - U_{1,1}^{(n)})^{-1}$, we observe that the inverse of a banded Toeplitz matrix is no longer banded. However, relying on the results of Chapter 3, we approximate $(I - U_{1,1}^{(n)})^{-1}$ with a block banded block Toeplitz matrix $T_\infty[K_q(z)]$, where $K_q(z) = K(z) \mod z^q$ is a matrix polynomial of degree at most $q-1$, where q is such that $\|T_\infty[K_q(z)] - T_\infty[K(z)]\|_\infty \le \epsilon$, for a given error bound ϵ. In other words, the matrix $T_\infty[K_q(z)]$ is obtained by truncating $T_\infty[K(z)]$ at a finite band q. In this way, $T_\infty[K_q(z)]$ and $(I - U_{1,1}^{(n)})^{-1}$ coincide in the nonzero part. This is obtained by applying Algorithm 3.1, where the input matrices $A_0, A_1, \ldots, A_{k-1}$ are replaced with the matrices $I - A_0^{(n)}, -A_2^{(n)}, \ldots, -A_{d-2}^{(n)}$, which outputs K_i, $i = 0, \ldots, q-1$, where K_i, $i = 0, \ldots, q-1$, are the first q block elements of the first block row of $(I - U_{1,1}^{(n)})^{-1}$ such that $\sum_{i=q}^{+\infty} K_i \le \epsilon E$.

In this way, the computation of $\boldsymbol{v}^{\mathrm{T}}$ and $\widehat{\boldsymbol{v}}^{\mathrm{T}}$ in (7.63) can be carried out in an approximate way by reducing it to multiplying an infinite vector with a finite number of nonnull elements and an infinite block upper triangular banded Toeplitz matrix.

For the equivalence of block banded Toeplitz matrices and matrix polynomials pointed out in Section 3.1, the nonnull block elements of $\boldsymbol{v} =^{\mathrm{T}} [V_0, V_1 \ldots]$ and $\widehat{\boldsymbol{v}}^{\mathrm{T}} = [\widehat{V}_0, \widehat{V}_1, \ldots]$ are given by the block coefficients of the matrix polynomials

$$\begin{aligned} \sum_{i=0}^{d/2+q-2} z^i V_i &= \left(\sum_{i=0}^{d/2-1} z^i A_{2i-1}^{(n)} \right) \left(\sum_{i=0}^{q-1} z^i K_i \right), \\ \sum_{i=0}^{d/2+q-2} z^i \widehat{V}_i &= \left(\sum_{i=0}^{d/2-1} z^i \widehat{A}_{2i+1}^{(n)} \right) \left(\sum_{i=0}^{q-1} z^i K_i \right). \end{aligned} \tag{7.64}$$

Similarly, the computation of $\boldsymbol{y}^{\mathrm{T}} = [Y_0, Y_1, \ldots]$ and $\widehat{\boldsymbol{y}} = [\widehat{Y}_0, \widehat{Y}_1, \ldots]$ is performed by means of

$$\begin{aligned} \sum_{i=0}^{d+q-3} z^i Y_i &= \left(\sum_{i=0}^{d/2+q-2} z^i V_i \right) \left(\sum_{i=0}^{d/2-1} z^i A_{2i-1}^{(n)} \right), \\ \sum_{i=0}^{d+q-3} z^i \widehat{Y}_i &= \left(\sum_{i=0}^{d/2+q-2} z^i \widehat{V}_i \right) \left(\sum_{i=0}^{d/2-1} z^i A_{2i-1}^{(n)} \right). \end{aligned} \tag{7.65}$$

The computations in (7.64) and (7.65) can be performed by means of Algorithm 2.1 in $O(m^2(d+q)\log(d+q) + m^3(d+q))$ ops.

Now we analyze the error generated in one step of cyclic reduction by truncating $K(z)$ to a polynomial of degree $q-1$. For the sake of notational simplicity we omit the superscript (n) from all the variables.

Let us denote with $A'(z)$ and $\widehat{A}'(z)$ the functions obtained by the cyclic reduction step applied to $A(z)$ and $\widehat{A}(z)$ and with $B'(z)$ and $\widehat{B}'(z)$ the functions obtained by performing the above step by truncating the power series $K(z) = (I - A_{\mathrm{odd}}^{(n)}(z))^{-1}$ to degree $q-1$, where q is such that the residual $\Gamma(z) = \sum_{i=q}^{+\infty} z^i K_i$ has norm at most ϵ, i.e., $\|\Gamma(z)\|_* = \|\Gamma(1)\|_\infty \le \epsilon$, where the latter equation holds since $K(z)$ has nonnegative block coefficients. More specifically, let

$$A'(z) = zA_{\mathrm{odd}}(z) - A_{\mathrm{even}}(z)(I - A_{\mathrm{odd}}(z))^{-1}A_{\mathrm{even}}(z),$$
$$\widehat{A}'(z) = \widehat{A}_{\mathrm{even}}(z) - \widehat{A}_{\mathrm{odd}}(z)(I - A_{\mathrm{odd}}(z))^{-1}A_{\mathrm{even}}(z),$$

and

$$B'(z) = zA_{\mathrm{odd}}(z) - A_{\mathrm{even}}(z)\left((I - A_{\mathrm{odd}}(z))^{-1} - \Gamma(z)\right)A_{\mathrm{even}}(z),$$
$$\widehat{B}'(z) = \widehat{A}_{\mathrm{even}}(z) - \widehat{A}_{\mathrm{odd}}(z)\left((I - A_{\mathrm{odd}}(z))^{-1} - \Gamma(z)\right)A_{\mathrm{even}}(z).$$

By subtracting the former equations from the latter ones, for the local error $\mathcal{L}(z) = (L(z), \widehat{L}(z)) = (B'(z), \widehat{B}'(z)) - (A'(z), \widehat{A}'(z))$ we obtain

$$L(z) = A_{\mathrm{even}}(z)\Gamma(z)A_{\mathrm{even}}(z),$$
$$\widehat{L}(z) = \widehat{A}_{\mathrm{odd}}(z)\Gamma(z)A_{\mathrm{even}}(z).$$

Moreover, since $L(z)$ has nonnegative block coefficients one has

$$\|L(z)\|_* = \|A_{\mathrm{even}}(1)\Gamma(1)A_{\mathrm{even}}(1)\|_\infty \le \|\Gamma(1)\|_\infty \le \epsilon$$

since $\|A_{\mathrm{even}}(1)\|_\infty \le 1$. Similarly we find that $\|L(z)\|_* \le \epsilon$. We also have $L(z) = \widehat{L}(z) = 0 \mod z^q$, that is the local errors in the first q coefficients are zero.

We may synthesize these properties in the following.

Theorem 7.16 *Let $d' = \max(d/2, q)$ and let $B'(z), \widehat{B}'(z)$ be the functions obtained by applying the cyclic reduction step modulo $z^{d'}$, where $A(1)$ is stochastic and $\widehat{A}(1)$ is substochastic. For the local error*

$$\mathcal{L}(z) = (L(z), \widehat{L}(z)) = (B'(z), \widehat{B}'(z)) - (A'(z), \widehat{A}'(z))$$

it holds

$$L(z) = A_{\mathrm{even}}(z)\Gamma(z)A_{\mathrm{even}}(z)$$
$$\widehat{L}(z) = \widehat{A}_{\mathrm{odd}}(z)\Gamma(z)A_{\mathrm{even}}(z)$$

where $\Gamma(z) = \sum_{i=q}^{+\infty} z^i K_i$, $K(z) = (I - A_{\mathrm{odd}}(z))^{-1}$, and $\|\Gamma(z)\|_* = \|\Gamma(1)\|_\infty \le \epsilon$. Moreover $\|L(z)\|_* = \|L(1)\|_\infty \le \epsilon$ and $\|\widehat{L}(z)\|_* = \|\widehat{L}(1)\|_\infty \le \epsilon$. Finally, $L(z) = 0$ mod z^q, $\widehat{L}(z) = 0 \mod z^q$, that is the local errors in the first q coefficients are zero, and $\|(\sum_{i=-1}^{d'-2} B_i')\mathbf{1} - \mathbf{1}\|_\infty \le \epsilon$, $\|(\sum_{i=0}^{d'-1} \widehat{B}_i' - \sum_{i=0}^{d-1} \widehat{A}_i)\mathbf{1}\|_\infty \le \epsilon$.

Algorithm 7.5 synthesizes a step of cyclic reduction applied in approximate way to an M/G/1-type Markov chain.

Algorithm 7.5 CR step: M/G/1 case, matrix version

INPUT: Positive even integer d and the $m \times m$ block elements $A_{i-1}, \widehat{A}_i$, $i = 0, \ldots, d-1$, defining the first two block rows of the matrix (7.53) at the general step of cyclic reduction; an error bound $\epsilon > 0$.

OUTPUT: Positive integers q, d' and the $m \times m$ block elements $B'_{i-1}, \widehat{B}'_i$, $i = 0, \ldots, d'-1$, approximating the block elements A'_{i-1}, $\widehat{A}'_i$, $i =, 0, \ldots, d'-1$ in the first two block rows of the matrix (7.53) after one step of cyclic reduction, such that $\|\mathbf{1} - \sum_{i=-1}^{d'-2} B'_i \mathbf{1}\|_\infty \le \epsilon$, $\|(\sum_{i=0}^{d'-1} \widehat{B}'_i - \sum_{i=0}^{d-1} \widehat{A}'_i)\mathbf{1}\|_\infty \le \epsilon$ and $B'_i = A'_i$, $i = -1, 0, \ldots, q-2$, $\widehat{B}'_i = \widehat{A}'_i$, $i = 0, \ldots, q-1$, $\|A'(z) - \sum_{i=-1}^{d'-1} z^{i+1} B'_i\|_* \le \epsilon$, $\|\widehat{A}'(z) - \sum_{i=0}^{d'-1} z^i \widehat{B}'_i\|_* \le \epsilon$.

COMPUTATION:

1. Apply Algorithm 3.1, with input: the integer $d/2$, the matrices $I - A_0, -A_2, \ldots, -A_{d-2}$ and the error bound ϵ. Obtain in output the integer q and matrices K_i, $i = 0, \ldots, q-1$, where K_i, $i = 0, \ldots, q-1$, are the first q block elements of the first block row of $(I - U_{1,1})^{-1}$ such that $\sum_{i=q}^{+\infty} K_i \le \epsilon E$.
2. Compute the block vectors $\boldsymbol{v}$ and $\widehat{\boldsymbol{v}}$ of (7.64) by applying Algorithm 2.1.
3. Compute the block vectors $\boldsymbol{y}$ and $\widehat{\boldsymbol{y}}$ of (7.65) by applying Algorithm 2.1.
4. Set $d' = \max(d/2, q)$ and

$$\begin{aligned} B'_i &= A_{2i} + Y_{i+1} \quad i = -1, \ldots, d'-1, \\ \widehat{B}'_i &= \widehat{A}_{2i} + \widehat{Y}_i \quad i = 0, \ldots, d'-1, \end{aligned}$$

where $A_i = 0$ if $i < -1$ or if $i \ge d-1$ and $\widehat{A}_i = 0$ if $i \ge d$.

The computational cost of this algorithm can be easily evaluated relying on the complexity estimates of Algorithms 2.1 and 3.1. In fact we have the following asymptotic complexity bound

$$O(m^3 d' + m^2 d' \log d').$$

Indeed, from Theorem 7.11, we have that $zI - A^{(n)}(z)$ as well as $I - A_{\text{odd}}^{(n)}(z)$ are analytic for $|z| < \xi^{2^n}$; moreover, the latter matrix power series is also nonsingular for $|z| < \xi^{2^n}$. Therefore $(I - A_{\text{odd}}^{(n)}(z))^{-1}$ is analytic for $|z| < \xi^{2^n}$ and for Theorem 3.7 the coefficients of the latter matrix power series decay esponentially to zero with exponential basis σ^{-2^n} for any $1 < \sigma < \xi$. This property guarantees that the numerical degree of $(I - A_{\text{odd}}^{(n)}(z))^{-1}$ is bounded from above by a constant. For positive recurrent Markov chains where the drift μ is negative, one should expect a rapid decrease of the value of d' after just a few iterations in view of Theorem 7.13.

Observe that if $A(1)$ is stochastic then the matrix $B'(1)$ is ϵ-stochastic. In order to preserve the stochasticity of the matrix provided by the cyclic reduction step we may modify Algorithm 7.5 by using a simple trick which uses the idea of the GTH trick of [53]. Indeed, at stage 4 we may replace the jth diagonal element of B'_0 with $1 - \boldsymbol{e}_j^{\mathrm{T}}(B'_{-1}\mathbf{1} + \sum_{i=1}^{d'-1} B'_i\mathbf{1} + B'_0(\mathbf{1} - \boldsymbol{e}_j))$, for $j = 1, \ldots, m$. Similarly we may do the same with $\widehat{B}'_0$ where we need to use the initial matrix $A_{-1}^{(0)}$ such that $A_{-1}^{(0)} + \sum_{i=0}^{+\infty} \widehat{A}_i^{(n)}$ is stochastic for any $n \geq 0$. In fact we replace the jth diagonal element of $\widehat{B}'_0$ with $1 - \boldsymbol{e}_j^{\mathrm{T}}(\boldsymbol{t} + \sum_{i=1}^{d'-1} \widehat{B}'_i\mathbf{1} + \widehat{B}'_0(\mathbf{1} - \boldsymbol{e}_j))$, for $j = 1, \ldots, m$, where $\boldsymbol{t} = A_{-1}^{(0)}\mathbf{1}$.

7.5.2 Functional implementation

A different way for implementing the cyclic reduction step relies on the fast power series arithmetic, based on the evaluation–interpolation technique, introduced in Chapter 3. More precisely, the evaluation of the expression $A_{\mathrm{even}}^{(n)}(z)(I - A_{\mathrm{odd}}^{(n)}(z))^{-1}A_{\mathrm{even}}^{(n)}(z)$ in (7.38) is not performed by applying the fast algorithms for multiplying and inverting block Toeplitz matrices, rather we evaluate the power series $zA_{\mathrm{odd}}^{(n)}(z) + A_{\mathrm{even}}^{(n)}(z)(I - A_{\mathrm{odd}}^{(n)}(z))^{-1}A_{\mathrm{even}}^{(n)}(z)$ at a sufficiently large number of points and then perform interpolation. Similarly we do the same for $\widehat{A}^{(n)}(z)$. This technique enables one to reduce the number of FFTs used still keeping full control on the local error generated by this computation.

In fact, we rely on Algorithm 3.5 applied twice for computing $A^{(n+1)}(z)$ and $\widehat{A}^{(n+1)}(z)$. In the former computation we have as input the matrix power series $X_1(z), X_2(z), X_3(z)$ and the rational function $F(X_1, X_2, X_3)$

$$\begin{aligned} &X_1(z) = A_{\mathrm{odd}}^{(n)}(z), \quad X_2(z) = A_{\mathrm{even}}^{(n)}(z), \quad X_3(z) = zA_{\mathrm{odd}}^{(n)}(z), \\ &F(X_1, X_2, X_3) = X_3 + X_2(I - X_1)^{-1}X_2, \end{aligned}$$

in the latter case we have

$$\begin{aligned} &X_1(z) = A_{\mathrm{odd}}^{(n)}(z), \quad X_2(z) = A_{\mathrm{even}}^{(n)}(z), \quad X_3(z) = \widehat{A}_{\mathrm{even}}^{(n)}(z), \quad X_4(z) = \widehat{A}_{\mathrm{odd}}^{(n)}(z), \\ &F(X_1, X_2, X_3, X_4) = X_3 + X_4(I - X_1)^{-1}X_2. \end{aligned}$$

As output we obtain the matrix polynomials $P^{(n+1)}(z)$ and $\widehat{P}^{(n+1)}(z)$ which interpolate $A^{(n+1)}(z)$ and $\widehat{A}^{(n+1)}(z)$ at the dth roots of 1 where d is dynamically determined so that

$$\begin{aligned} &\sum_{i=0}^{d-1} |A_{i-1}^{(n+1)} - P_i^{(n+1)}|\mathbf{1} + \sum_{i=d-1}^{+\infty} |A_i^{(n+1)}|\mathbf{1} \leq \epsilon\mathbf{1} \\ &\sum_{i=0}^{d-1} |\widehat{A}_i^{(n+1)} - \widehat{P}_i^{(n+1)}|\mathbf{1} + \sum_{i=d}^{+\infty} |\widehat{A}_i^{(n+1)}|\mathbf{1} \leq \epsilon\mathbf{1} \end{aligned} \tag{7.66}$$

for a given error bound $\epsilon > 0$. In this way the approximation error is bounded by ϵ.

In order to control the number d of interpolation points we have to describe the test used in Algoritm 3.5. For this purpose we introduce the following vectors

$$\boldsymbol{\alpha}^{(n)} = \sum_{i=0}^{+\infty}(i+1)A_i^{(n)}\mathbf{1}, \qquad \widehat{\boldsymbol{\alpha}}^{(n)} = \sum_{i=1}^{+\infty} i\widehat{A}_i^{(n)}\mathbf{1}, \tag{7.67}$$

which can be computed recursively, without computing the new blocks $A_i^{(n+1)}$ by means of the following.

Lemma 7.17 *For the vector sequences $\{\boldsymbol{\alpha}^{(n)}\}$, $\{\widehat{\boldsymbol{\alpha}}^{(n)}\}$ one has*

$$\begin{aligned}\boldsymbol{\alpha}^{(n+1)} &= \frac{1}{2}(\mathbf{1} + \boldsymbol{\alpha}^{(n)} - A_{\text{even}}^{(n)}(1)(I - A_{\text{odd}}^{(n)}(1))^{-1}(\mathbf{1} - \boldsymbol{\alpha}^{(n)})),\\ \widehat{\boldsymbol{\alpha}}^{(n+1)} &= \frac{1}{2}(\widehat{\boldsymbol{\alpha}}^{(n)} - \widehat{A}_{\text{odd}}^{(n)}(1)(I - A_{\text{odd}}^{(n)}(1))^{-1}(\mathbf{1} - \boldsymbol{\alpha}^{(n)})).\end{aligned}$$

Proof Rewrite (7.67) in matrix form as

$$H^{(n)}\begin{bmatrix} 0 \\ \mathbf{1} \\ 2\mathbf{1} \\ 3\mathbf{1} \\ \vdots \end{bmatrix} = \begin{bmatrix} -\widehat{\boldsymbol{\alpha}}^{(n)} \\ \mathbf{1} - \boldsymbol{\alpha}^{(n)} \\ \mathbf{1} - \boldsymbol{\alpha}^{(n)} \\ \mathbf{1} - \boldsymbol{\alpha}^{(n)} \\ \vdots \end{bmatrix},$$

where $H^{(n)}$ is the matrix in (7.53). Apply one step of cyclic reduction to the above system and obtain

$$2H^{(n+1)}\begin{bmatrix} 0 \\ \mathbf{1} \\ 2\mathbf{1} \\ 3\mathbf{1} \\ \vdots \end{bmatrix} = \begin{bmatrix} -\widehat{\boldsymbol{\alpha}}^{(n)} \\ \mathbf{1} - \boldsymbol{\alpha}^{(n)} \\ \mathbf{1} - \boldsymbol{\alpha}^{(n)} \\ \mathbf{1} - \boldsymbol{\alpha}^{(n)} \\ \vdots \end{bmatrix} + U_{2,1}^{(n)}(I - U_{1,1}^{(n)})^{-1}\begin{bmatrix} \mathbf{1} - \boldsymbol{\alpha}^{(n)} \\ \mathbf{1} - \boldsymbol{\alpha}^{(n)} \\ \mathbf{1} - \boldsymbol{\alpha}^{(n)} \\ \mathbf{1} - \boldsymbol{\alpha}^{(n)} \\ \vdots \end{bmatrix},$$

where $U_{2,1}^{(n)}$ and $U_{1,1}^{(n)}$ are the matrices defined in (7.54). By formal manipulations one can show that the vector in the right-hand side of the above equation has the first block component equal to

$$-\widehat{\boldsymbol{\alpha}}^{(n)} + \sum_{i=0}^{+\infty}\widehat{A}_{2i+1}^{(n)}(I - \sum_{i=0}^{+\infty}A_{2i}^{(n)})^{-1}(\mathbf{1} - \boldsymbol{\alpha}^{(n)})$$

and the remaining block components equal to

$$\mathbf{1} - \boldsymbol{\alpha}^{(n)} + \sum_{i=0}^{+\infty}A_{2i-1}^{(n)}(I - \sum_{i=0}^{+\infty}A_{2i}^{(n)})^{-1}(\mathbf{1} - \boldsymbol{\alpha}^{(n)}).$$

This completes the proof. □

The test used in Algorithm 3.5 for determining the number d of interpolation points which guarantee the error bound ϵ in the ouput relies on the following.

Theorem 7.18 *Let* $P^{(n)}(z) = \sum_{i=0}^{d-1} z^i P_i^{(n)}$, $\widehat{P}^{(n)}(z) = \sum_{i=0}^{d-1} z^i \widehat{P}_i^{(n)}$ *be the matrix polynomials of degree at most* $d-1$ *which interpolate the matrix power serie* $A^{(n)}(z)$ *and* $\widehat{A}^{(n)}(z)$*, respectively, at the dth roots of 1. Then one has*

$$\begin{aligned} d \sum_{i=d-1}^{+\infty} A_i^{(n)} \mathbf{1} &\le \boldsymbol{\alpha}^{(n)} - \sum_{i=1}^{d-1} i P_i^{(n)} \mathbf{1} \\ d \sum_{i=d}^{+\infty} \widehat{A}_i^{(n)} \mathbf{1} &\le \widehat{\boldsymbol{\alpha}}^{(n)} - \sum_{i=1}^{d-1} i \widehat{P}_i^{(n)} \mathbf{1}. \end{aligned} \tag{7.68}$$

Proof From Theorem 3.8 one has

$$\begin{aligned} P_i^{(n)} &= A_{i-1}^{(n)} + \sum_{h \ge 1} A_{i-1+hd}^{(n)}, \quad i = 0, 1, \ldots, d-1 \\ \widehat{P}_i^{(n)} &= \widehat{A}_i^{(n)} + \sum_{h \ge 1} \widehat{A}_{i+hd}^{(n)}, \quad i = 0, 1, \ldots, d-1. \end{aligned} \tag{7.69}$$

By using the first of (7.69) in the expression $\boldsymbol{\alpha}^{(n)} - \sum_{i=1}^{d-1} i P_i^{(n)} \mathbf{1}$ and by replacing $\boldsymbol{\alpha}^{(n)}$ with equation (7.67) one has

$$\boldsymbol{\alpha}^{(n)} - \sum_{i=1}^{d-1} i P_i^{(n)} \mathbf{1} = \sum_{i=d-1}^{+\infty} (i+1) A_i^{(n)} \mathbf{1} - \sum_{i=0}^{d-2} (i+1) \sum_{h \ge 1} A_{i+hd}^{(n)} \mathbf{1} \ge d \sum_{i=d-1}^{+\infty} A_i^{(n)} \mathbf{1},$$

which proves the first inequality. The second inequality can be similarly proved. □

From the above theorem it follows that if d is such that

$$\begin{aligned} \boldsymbol{\alpha}^{(n+1)} - \sum_{i=1}^{d-1} i P_i^{(n+1)} \mathbf{1} &\le \frac{1}{2} d\epsilon \mathbf{1} \\ \widehat{\boldsymbol{\alpha}}^{(n+1)} - \sum_{i=1}^{d-1} i \widehat{P}_i^{(n+1)} \mathbf{1} &\le \frac{1}{2} d\epsilon \mathbf{1} \end{aligned}$$

then condition (7.66) is satisfied. In fact, from (7.68) one has $\sum_{i=d-1}^{+\infty} A_i^{(n+1)} \mathbf{1} \le \frac{1}{2} \epsilon \mathbf{1}$ and $\sum_{i=d}^{+\infty} \widehat{A}_i^{(n+1)} \mathbf{1} \le \frac{1}{2} \epsilon \mathbf{1}$; combining the latter inequalities with equation (7.69) yields (7.66).

Algorithm 7.6 describes one step of cyclic reduction based on the fast power series arithmetic where we apply the technique shown in Algorithm 3.5.

The cost of the cyclic reduction step performed by means of Algorithm 7.6 is clearly

$$O(m^3 d' + m^2 d' \log d').$$

Observe also that, as pointed out in Remark 2.2, in the point-wise power series arithmetic technique, we do not have to recompute from scratch the matrices

$W_{1,j}$, $W_{2,j}$, $\widehat{W}_{1,j}$, $\widehat{W}_{2,j}$, F_j and $\widehat{F}_j$, for $j = 0, \ldots, M-1$, when we switch from M to $2M$ interpolation points. This allows one to reduce the complexity still keeping the same asymptotics.

Concerning the local error of the cyclic reduction step performed in functional form, we have the following.

Theorem 7.19 *For the local error* $\mathcal{L}(z) = (L(z), \widehat{L}(z)) = (B'(z), \widehat{B}'(z)) - (A'(z), \widehat{A}'(z))$ *it holds* $\|L(z)\|_* \le \epsilon$, $\|\widehat{L}(z)\|_* \le \epsilon$.

7.5.3 Cyclic reduction algorithm

The cyclic reduction scheme for M/G/1-type Markov chains is summarized in Algorithm 7.7.

Indeed, the overall cost of this algorithm depends on the maximum numerical degrees $d_{\max}$ of the matrix power series $A^{(n)}(z)$ and $\widehat{A}^{(n)}(z)$ generated by cyclic reduction. This cost amounts to

$$O(m^3 d_{\max} + m^2 d_{\max} \log d_{\max})$$

ops.

It is important to observe that, under the assumptions of the convergence theorem 7.13, the coefficients of z^i in the matrix power series $A^{(n)}(z)$ and $\widehat{A}^{(n)}(z)$ have norm bounded from above by $\gamma\sigma^{-i\cdot 2^n}$ for a constant γ, where $1 < \sigma < \xi$ and ξ is the root of $zI - A(z)$ of minimum modulus among the roots of modulus greater than 1. That is, the decay to zero of the coefficients of $A^{(n)}(z)$ holds with the exponential basis σ^{-2^n}. This means that the numerical degree of the matrix power series $A^{(n)}(z)$ and $\widehat{A}^{(n)}(z)$ is bounded from above by a constant and that one should expect a rapid decrease of the numerical degrees of the power series involved after just a few iterations.

For transient Markov chains the above algorithm can still be applied provided that we replace the stop condition which relies on the propertie of positive recurrence of the Markov chain. A possible stop condition is $\|G^{(n+1)} - G^{(n)}\|_\infty \le \epsilon$, where $G^{(n)} = (I - \widehat{B}_0^{(n)})^{-1} A_{-1}$ is the approximation to $G_{\min}$ provided at the n-th step of the algorithm.

Algorithm 7.6 CR step: M/G/1 case, functional version

INPUT: Positive integer $d = 2^h$ and the $m \times m$ block elements $A_{-1}, A_0, \dots, A_{d-2}$, $\widehat{A}_0, \dots, \widehat{A}_{d-1}$, defining the first two block rows of the matrix (7.53) at the general step of cyclic reduction; the vectors $\boldsymbol{\alpha}$ and $\widehat{\boldsymbol{\alpha}}$ of (7.67); an error bound $\epsilon > 0$.

OUTPUT: Positive integer d' and the $m \times m$ matrices $B'_{-1}, B'_0, \dots, B'_{d'-2}$, $\widehat{B}'_0, \dots, \widehat{B}'_{d'-1}$, which approximate the block elements $A'_{-1}, A'_0, \dots, A'_{d'-2}$, $\widehat{A}'_0, \dots, \widehat{A}'_{d'-1}$, defining the first two block rows of the matrix (7.53) after one step of cyclic reduction such that (7.66) is satisfied; the vectors $\boldsymbol{\alpha}' = \sum_{i=0}^{d'-2}(i+1)A'_i\mathbf{1}$ and $\widehat{\boldsymbol{\alpha}}' = \sum_{i=1}^{d'-1} i\widehat{A}'_i\mathbf{1}$.

COMPUTATION:

1. Compute $\boldsymbol{\alpha}' = \frac{1}{2}(\mathbf{1} + \boldsymbol{\alpha} - A_{\text{even}}(1)(I - A_{\text{odd}}(1))^{-1}(\mathbf{1} - \boldsymbol{\alpha}))$, $\widehat{\boldsymbol{\alpha}}' = \frac{1}{2}(\widehat{\boldsymbol{\alpha}} - \widehat{A}_{\text{odd}}(1)(I - A_{\text{odd}}(1))^{-1}(\mathbf{1} - \boldsymbol{\alpha}))$.
2. Set $M = d/2$.
   (a) Compute $W_{1,j} = A_{\text{odd}}(\omega_M^j)$, $W_{2,j} = A_{\text{even}}(\omega_M^j)$, $\widehat{W}_{1,j} = \widehat{A}_{\text{odd}}(\omega_M^j)$, $\widehat{W}_{2,j} = \widehat{A}_{\text{even}}(\omega_M^j)$, for $j = 0, \dots, M-1$ in the following way:
      i. Set $\boldsymbol{x} = (X_i)_{i=1,M}$ where $X_i = A_{2i-2}$, for $i = 1, \dots, d/2$, $X_i = 0$ for $i > d/2$; compute $\boldsymbol{y} = \text{IDFT}_M(\boldsymbol{x})$, set $W_{1,j-1} = Y_j$, $j = 1, \dots, M$.
      ii. Set $\boldsymbol{x} = (X_i)_{i=1,M}$ where $X_i = A_{2i-3}$, for $i = 1, \dots, d/2$, $X_i = 0$ for $i > d/2$; compute $\boldsymbol{y} = \text{IDFT}_M(\boldsymbol{x})$, set $W_{2,j-1} = Y_j$, $j = 1, \dots, M$.
      iii. Set $\boldsymbol{x} = (X_i)_{i=1,M}$ where $X_i = \widehat{A}_{2i-1}$, for $i = 1, \dots, d/2$, $X_i = 0$ for $i > d/2$; compute $\boldsymbol{y} = \text{IDFT}_M(\boldsymbol{x})$, set $\widehat{W}_{1,j-1} = Y_j$, $j = 1, \dots, M$.
      iv. Set $\boldsymbol{x} = (X_i)_{i=1,M}$ where $X_i = \widehat{A}_{2i-2}$, for $i = 1, \dots, d/2$, $X_i = 0$ for $i > d/2$; compute $\boldsymbol{y} = \text{IDFT}_M(\boldsymbol{x})$, set $\widehat{W}_{2,j-1} = Y_j$, $j = 1, \dots, M$.

   (b) Compute $F_j = \omega_M^j W_{1,j} + W_{2,j}(I - W_{1,j})^{-1}W_{2,j}$, $\widehat{F}_j = \widehat{W}_{2,j} + \widehat{W}_{1,j}(I - W_{1,j})^{-1}W_{2,j}$, $j = 0, \dots, M-1$.

   (c) Compute the coefficients P_i, $\widehat{P}_i$, $i = 0, \dots, M-1$, of the matrix polynomials $P(z)$ and $\widehat{P}(z)$, respectively, such that $P(\omega_M^j) = F_j$ and $\widehat{P}(\omega_M^j) = \widehat{F}_j$, $j = 0, \dots, M-1$, in the following way
      i. Set $\boldsymbol{y} = (Y_i)_{i=1,M}$ where $Y_i = F_{i-1}$, for $i = 1, \dots, M$; compute $\boldsymbol{x} = \text{DFT}_M(\boldsymbol{y})$, set $P_{j-1} = X_j$, $j = 1, \dots, M$.
      ii. Set $\boldsymbol{y} = (Y_i)_{i=1,M}$ where $Y_i = \widehat{F}_{i-1}$, for $i = 1, \dots, M$; compute $\boldsymbol{x} = \text{DFT}_M(\boldsymbol{y})$, set $\widehat{P}_{j-1} = X_j$, $j = 1, \dots, M$.

   (d) If $\boldsymbol{\alpha}' - \sum_{i=1}^{M-1} iP_i\mathbf{1} > \frac{1}{2}M\epsilon\mathbf{1}$ or $\widehat{\boldsymbol{\alpha}}' - \sum_{i=1}^{M-1} i\widehat{P}_i\mathbf{1} > \frac{1}{2}M\epsilon\mathbf{1}$ set $M = 2M$ and continue from step 2b.
3. Output $d' = M$ and $B'_{j-1} = P_j$, $\widehat{B}'_j = \widehat{P}_j$, $j = 0, \dots, d'-1$.

Algorithm 7.7 CR for M/G/1 type Markov chains: the case $\mu < 0$

INPUT: Positive integer k and the $m \times m$ block elements $A_{-1}, A_0, \ldots, A_{k-1}$, defining the block Toeplitz block Hessenberg matrix in (7.50) such that $\mu < 0$; an error bound $\epsilon > 0$.

OUTPUT: An approximation Y to the matrix $G_{\min}$.

COMPUTATION:

1. Set $n = 0$, $A^{(n)}(z) = \sum_{i=-1}^{k-1} z^{i+1} A_i$, $\widehat{A}^{(n)}(z) = \sum_{i=0}^{k-1} z^i A_i$.
   (a) Compute approximations $B^{(n+1)}(z)$ and $\widehat{B}^{(n+1)}(z)$ to $A^{(n+1)}(z)$ and $\widehat{A}^{(n+1)}(z)$, respectively, by means either of Algorithm 7.5 or of Algorithm 7.6.
   (b) If $\|\mathbf{1} - (A_{-1} + \widehat{B}_0^{(n)})\mathbf{1}\|_\infty \le \epsilon$, set $n = n+1$, $A^{(n)}(z) = B^{(n)}(z)$, $\widehat{A}^{(n)}(z) = \widehat{B}^{(n)}(z)$, and repeat from step 1a.
2. Output $Y = (I - \widehat{B}_0^{(n)})^{-1} A_{-1}$.

7.6 Bibliographic notes

Logarithmic reduction was introduced by Latouche and Ramaswami in [78]. Convergence properties in the null recurrent case are shown by Guo [56]. An error analysis is performed by Ye in [120].

The cyclic reduction technique was first introduced in the late 1960s by Buzbee, Golub and Nielson [28] for the solution of certain block tridiagonal systems which discretize elliptic differential equations. Actually, the method was originally invented by Gene H. Golub when he was a PhD student at the Urbana-Champaign University. There is a wide literature concerning the analysis and the implementation of cyclic reduction. Among the stability and convergence properties we refer the reader to the papers [2], [118], [117].

Cyclic reduction is applied to solving matrix equations in QBD problems in [119], [84] by San-Qi Li *et al.*, even though they call it folding algorithm. Cyclic reduction is extended to M/G/1-type Markov chains in [18], where this technique is used for computing $\boldsymbol{\pi}$. Its application to solve power series matrix equations is in [11] for the polynomial version and in [19] for the evaluation–interpolation technique; the non-skip-free case is investigated in [22], [21] where the properties of displacement operators are used for decreasing the computational cost. Cyclic reduction for QBD problems derived from M/G/1-type Markov chains is analyzed in [25]; the case of PH/PH/1 queues is studied in [13]. Cyclic reduction, logarithmic reduction, and matrix sign iteration are compared in [16]. A survey on displacement structure, cyclic reduction, and the divide-and-conquer method can be found in [23]. General properties of cyclic reduction, its relation with Graeffe's iteration and applications to infinite Toeplitz systems and to nonlinear matrix equations are presented in [14].

8

ALTERNATIVE APPROACHES

8.1 Introduction

In this chapter we describe diverse methods for the solution of the matrix equation

$$X = \sum_{i=-1}^{+\infty} A_{i+1} X^{i+1} \tag{8.1}$$

and for the computation of the invariant probability vector $\boldsymbol{\pi}$ such that $\boldsymbol{\pi}^{\mathrm{T}} P = \boldsymbol{\pi}^{\mathrm{T}}$ where the matrix P is defined in (4.3). These methods rely in part on the tools and properties presented in the previous chapters and in part on different techniques.

We first deal, in Section 8.2, with the application of the tools introduced in Section 3.6, that is, the shifting technique, where the matrix equation (8.1) is transformed to a new one having better computational features and show that the convergence of cyclic reduction is much faster for the new equation than for the original one. The algorithm obtained in this way also has better stability properties, that is, the local errors obtained in the cyclic reduction steps are not much amplified during the iteration.

Then, in Section 8.3, we consider a method which is obtained by combining together the divide-and-conquer strategy of cyclic reduction and the technique of functional iteration. This method shares the self-correcting features of functional iterations and its convergence is linear but with an arbitrarily high convergence rate. This method relies on the fact that at each step of cyclic reduction a functional iteration formula is generated. This formula may be used to continue the approximation in place of cyclic reduction. The method can be useful in the cases where cyclic reduction, combined with the shifting technique, encounters some breakdown or near-breakdown problem. In fact in this case, keeping the cyclic reduction iteration might lead to large amplification of the errors whereas continuing with the functional iteration still improves the approximation and avoids the breakdown.

In Section 8.4 we describe a different doubling technique which is still based on a divide-and-conquer strategy, but where the recursive doubling is performed on the size of the finite matrices obtained by truncating $T_\infty[I - z^{-1}A(z)]$ to a finite size.

The subsequent section deals with methods based on the evaluation–interpolation technique. Here we rely on the tools introduced in Section 3.5 where the vector $\boldsymbol{\pi}$ is expressed in terms of a canonical factorization of the function $I -$

$z^{-1}A(z)$, and this factorization is expressed as a function of the block coefficients of $(I - z^{-1}A(z))^{-1}$. Since the function $I - z^{-1}A(z)$ is singular for $z = 1$, in order to approximate its coefficients, we have to apply the shift technique of Section 3.6 and deal with the modified function $I - z^{-1}\widetilde{A}(z)$ which is nonsingular for $|z| = 1$. Thus, the block coefficients of the inverse of the latter function are approximated by means of evaluation interpolation at the roots of 1 and from these coefficients we obtain the coefficients of $(I - z^{-1}A(z))^{-1}$ and then the weak canonical factorization of $I - z^{-1}A(z)$. Also the case of QBD stochastic processes is considered in this section where we rely on Theorem 3.20 which relates the coefficients of $(I - z^{-1}\widetilde{A}(z))^{-1}$ with the minimal solution $\widetilde{G}_{\min}$ of the matrix equation $X = \widetilde{A}(X)$. The coefficients of the inverse matrix function are computed by means of evaluation–interpolation at the roots of 1. This approach is particularly effective for Markov chains with limited displacement (NSF Markov chains) of Section 5.5.

Invariant subspace methods are the subject of Section 8.6.

Throughout this chapter we assume that Condition 4.20 is valid and that $\boldsymbol{a} = \sum_{i=0}^{+\infty}(i+1)A_i\mathbf{1}$ is finite, so that in light of Theorem 4.24, $z = 1$ is the only zero of modulus 1 of the function $a(z) = \det(zI - A(z))$ defined in (4.23), where $A(z) = \sum_{i=-1}^{+\infty} z^{i+1}A_i$.

8.2 Convergence acceleration

The shifting technique, introduced in Section 3.6, can be applied to the function $\varphi(z) = zI - A(z)$ in order to remove the root $\lambda = 1$ of $\varphi(z)$ corresponding to the eigenvector $\mathbf{1}$ such that $\varphi(1)\mathbf{1} = 0$. More precisely, a new function $\widetilde{\varphi}(z) = \varphi(z)(I - z^{-1}Q)^{-1}$ is constructed so that $\widetilde{\varphi}(z)$ has the same roots of $\varphi(z)$ except for $z = 1$ which is replaced by the root $z = 0$. Here we set $Q = \mathbf{1}\boldsymbol{u}^{\mathrm{T}}$ where $\boldsymbol{u}$ is any vector such that $\boldsymbol{u}^{\mathrm{T}}\mathbf{1} = 1$. Moreover, if $\varphi(z) = zI - \sum_{i=-1}^{+\infty} z^{i+1}A_i$, then $\widetilde{\varphi}(z) = zI - \sum_{i=-1}^{+\infty} z^{i+1}\widetilde{A}_i$ where

$$\begin{aligned} \widetilde{A}_{-1} &= A_{-1}(I - Q) \\ \widetilde{A}_i &= A_i - (\sum_{j=-1}^{i} A_j - I)Q, \quad i = 0, 1, \ldots. \end{aligned} \tag{8.2}$$

For more details of this technique we refer the reader to Section 3.6.

To be more specific, consider the case where the drift μ is negative, and assume that the function $\varphi(z)$ is analytic for $|z| < r$, for a given $r > 1$ and that there exists at least a root of $\varphi(z)$ of modulus greater than 1[1]. Moreover, without loss of generality let us arrange the roots ξ_i, $i = 1, 2 \ldots$, of $\varphi(z)$ so that

$$|\xi_1| \le |\xi_2| \le \cdots \le |\xi_{m-1}| < \xi_m = 1 < |\xi_{m+1}| \le \cdots < r, \tag{8.3}$$

where $\xi_m = 1$ is simple. Then, in light of Theorem 3.32, the roots $\widetilde{\xi}_i$, $i = 1, 2, \ldots$, of the new function $\widetilde{\varphi}(z)$ are such that $\widetilde{\xi}_1 = 0$, $\widetilde{\xi}_{i+1} = \xi_i$, $i = 1, \ldots, m-1$, $\widetilde{\xi}_i = \xi_i$, $i = m+1, \ldots$, that is,

[1]Conditions under which this assumption is verified are given in Theorem 4.12.

$$\widetilde{\xi}_1 = 0 \leq |\widetilde{\xi}_2| \leq \cdots \leq |\widetilde{\xi}_m| = |\xi_{m-1}| < 1 < |\widetilde{\xi}_{m+1}| = |\xi_{m+1}| \leq \cdots < r,$$

in fact, by construction, the root $\xi_m = 1$ is moved into the root $\widetilde{\xi}_1 = 0$. Therefore the function $\widetilde{\varphi}(z)$ as well as the matrix Laurent power series $z^{-1}\widetilde{\varphi}(z)$ is analytic and invertible in the annulus $\mathbb{A} = \{z \in \mathbb{C} : \quad |\xi_{m-1}| < |z| < |\xi_{m+1}|\}$. In this way the coefficients of the matrix Laurent power series $z^{-1}\widetilde{\varphi}(z)$ as well as the coefficients of the matrix Laurent power series $\widetilde{\psi}(z) = (z^{-1}\widetilde{\varphi}(z))^{-1}$ have an exponential decay (compare with Theorem 3.6). This fact allows one to prove faster convergence properties of cyclic reduction applied to the shifted function $\widetilde{\varphi}(z)$.

If $\widetilde{G}_{\min}$ is the solution with minimal spectral radius of the matrix equation

$$\widetilde{A}_{-1} + \widetilde{A}_0 X + \widetilde{A}_1 X^2 + \cdots = X \tag{8.4}$$

associated with $\widetilde{\varphi}(z)$, which exists in light of Theorem 3.33, then the minimal solution $G_{\min}$ of the original matrix equation (7.29) can be immediately recovered by means of the simple expression

$$G_{\min} = \widetilde{G}_{\min} + Q.$$

According to (3.51), there exists a canonical factorization of the function $z^{-1}\widetilde{\varphi}(z)$, namely,

$$z^{-1}\widetilde{\varphi}(z) = U(z)(I - z^{-1}\widetilde{G}_{\min})$$

where $U(z)$ is a suitable matrix power series analytic for $|z| < 1$ and invertible for $|z| \leq 1$. A similar factorization can be proved for the function $z\widetilde{\varphi}(z^{-1})$ as shown by the following.

Theorem 8.1 *Let $A(z)$ be the generating function associated with an M/G/1-type Markov chain with negative drift μ which verifies Condition 4.20. Assume that $A(z)$ is analytic for $|z| < r$ with $1 < r < +\infty$. Let $Q = \mathbf{1}\boldsymbol{u}^{\mathrm{T}}$, $\boldsymbol{u} > 0$ be such that $\boldsymbol{u}^{\mathrm{T}}\mathbf{1} = 1$ and let $\widetilde{A}(z) = \sum_{i=-1}^{+\infty} z^{i+1}\widetilde{A}_i$ be the matrix power series whose coefficients $\widetilde{A}_i$, $i \geq -1$ are defined by (8.2). Then there exists the spectral minimal solution $\widetilde{R}_{\min}$ of the matrix equation*

$$X = \sum_{i=-1}^{+\infty} X^{i+1}\widetilde{A}_i \tag{8.5}$$

and the function $I - z\widetilde{A}(z^{-1})$ has a canonical factorization

$$I - z\widetilde{A}(z^{-1}) = (I - z\widetilde{R}_{\min})\widetilde{L}(z), \tag{8.6}$$

where $\widetilde{L}(z) = \sum_{i=0}^{+\infty} z^{-i}\widetilde{L}_{-i}$.

Proof Since the drift μ is negative, then there exists the minimal nonnegative solution $R_{\min}$ of the equation $X = \sum_{i=-1}^{+\infty} X^{i+1}A_i$, and the matrix $R_{\min}$ has spectral radius $\rho(R_{\min}) = 1$. Moreover $\rho(R_{\min})$ is a simple eigenvalue and is the unique eigenvalue of modulus 1 of $R_{\min}$. In light of Theorem 5.8 there exists a weak canonical factorization

$$I - zA(z^{-1}) = (I - zR_{\min})S(z) \tag{8.7}$$

where $S(z) = \sum_{i=0}^{+\infty} z^{-i}S_i$, S_0 is nonsingular and $S(z)$ is analytic for $|z| > r^{-1}$. Observe that since $0 = (I - A(1))\mathbf{1} = (I - R_{\min})S(1)\mathbf{1}$, then $\boldsymbol{v} = S(1)\mathbf{1}$ is a right eigenvector of $R_{\min}$ corresponding to the eigenvalue $\rho(R_{\min}) = 1$. Moreover, $\boldsymbol{v}$ is nonnegative since $R_{\min}$ is nonnegative. Let us denote by $\boldsymbol{w}$ a nonnegative vector such that $\boldsymbol{w}^{\mathrm{T}}\boldsymbol{v} = 1$. Rewrite (8.7) as

$$I - zA(z^{-1}) = (I - zR_{\min})(I - z\boldsymbol{v}\boldsymbol{w}^{\mathrm{T}})^{-1}(I - z\boldsymbol{v}\boldsymbol{w}^{\mathrm{T}})S(z).$$

A simple calculation shows that $(I - zR_{\min})(I - z\boldsymbol{v}\boldsymbol{w}^{\mathrm{T}})^{-1} = I - z\widetilde{R}_{\min}$ for $\widetilde{R}_{\min} = R_{\min} - \boldsymbol{v}\boldsymbol{w}^{\mathrm{T}}$. Therefore, since $I - z\widetilde{A}(z^{-1}) = (I - zA(z^{-1}))(I - zQ)^{-1}$, for $Q = \mathbf{1}\boldsymbol{u}^{\mathrm{T}}$, we have

$$I - z\widetilde{A}(z^{-1}) = (I - z\widetilde{R}_{\min})(I - z\boldsymbol{v}\boldsymbol{w}^{\mathrm{T}})S(z)(I - zQ)^{-1}. \tag{8.8}$$

Now, taking determinants of $I - z\widetilde{R}_{\min} = (I - zR_{\min})(I - z\boldsymbol{v}\boldsymbol{w}^{\mathrm{T}})^{-1}$ yields $\det(tI - \widetilde{R}_{\min}) = \det(tI - R_{\min})\frac{t}{t-1}$, where we set $t = z^{-1}$. This shows that 0 is an eigenvalue of $\widetilde{R}_{\min}$ together with all the eigenvalues of $R_{\min}$ different from 1 and that, since 1 is a simple eigenvalue of $R_{\min}$ it cannot be eigenvalue of $\widetilde{R}_{\min}$. This implies that $\rho(\widetilde{R}_{\min}) < 1$ and that the matrix function $I - z\widetilde{R}_{\min}$ is invertible for $|z| \le 1$. Now we prove that the rightmost factor $\widetilde{L}(z) = (I - z\boldsymbol{v}\boldsymbol{w}^{\mathrm{T}})S(z)(I - zQ)^{-1}$ in (8.8) is a matrix power series in z^{-1} in the Wiener algebra $\mathcal{W}_-$ and invertible for $|z| \ge 1$. We first analyze the product $B(z) = S(z)(I - zQ)^{-1} = \sum_{i=-\infty}^{+\infty} z^i B_i$. It holds

$$B_i = \begin{cases} S_{-i} + \sum_{j=-i+1}^{+\infty} S_j Q & \text{if } i \le 0 \\ \boldsymbol{v}\boldsymbol{u}^{\mathrm{T}} & \text{if } i > 0 \end{cases} \tag{8.9}$$

in fact, if $i > 0$ then $B_i = \sum_{j=0}^{+\infty} S_j Q = S(1)Q = S(1)\mathbf{1}\boldsymbol{u}^{\mathrm{T}} = \boldsymbol{v}\boldsymbol{u}^{\mathrm{T}}$. Now we are ready to show that for a suitable $\boldsymbol{w}$ the product

$$\widetilde{L}(z) = (I - z\boldsymbol{v}\boldsymbol{w}^{\mathrm{T}})B(z)$$

is a matrix power series in $\mathcal{W}_-$. Indeed, since

$$\widetilde{L}(z) = \sum_{i=-\infty}^{+\infty} z^i B_i - \sum_{i=-\infty}^{+\infty} z^i(\boldsymbol{v}\boldsymbol{w}^{\mathrm{T}}B_{i-1}),$$

we obtain

$$\widetilde{L}_i = B_i - \boldsymbol{v}\boldsymbol{w}^{\mathrm{T}}B_{i-1}, \quad i = 0, \pm 1, \pm 2, \ldots, \tag{8.10}$$

and from (8.9) we deduce that $\widetilde{L}_1 = B_1 - \boldsymbol{v}\boldsymbol{w}^{\mathrm{T}}B_0 = \boldsymbol{v}(\boldsymbol{u}^{\mathrm{T}} - \boldsymbol{w}^{\mathrm{T}}B_0)$. Therefore, since $S_0^{-1} \ge 0$ (see Theorem 5.8), and $\boldsymbol{u} > 0$, then $\boldsymbol{u}^{\mathrm{T}}S_0^{-1}\boldsymbol{v} > 0$, and choosing

$$\boldsymbol{w}^{\mathrm{T}} = \frac{1}{\boldsymbol{u}^{\mathrm{T}} S_0^{-1} \boldsymbol{v}} \boldsymbol{u}^{\mathrm{T}} S_0^{-1}$$

yields $\boldsymbol{w}^{\mathrm{T}}\boldsymbol{v} = 1$ and

$$\begin{aligned}
\boldsymbol{u}^{\mathrm{T}} - \boldsymbol{w}^{\mathrm{T}} B_0 =& \boldsymbol{u}^{\mathrm{T}} - \frac{1}{\boldsymbol{u}^{\mathrm{T}} S_0^{-1} \boldsymbol{v}} \boldsymbol{u}^{\mathrm{T}} S_0^{-1} (S_0 + \sum_{j=1}^{+\infty} S_j Q) \\
=& \left(1 - \frac{1}{\boldsymbol{u}^{\mathrm{T}} S_0^{-1} \boldsymbol{v}}\right) \boldsymbol{u}^{\mathrm{T}} - \frac{1}{\boldsymbol{u}^{\mathrm{T}} S_0^{-1} \boldsymbol{v}} \boldsymbol{u}^{\mathrm{T}} S_0^{-1} (\boldsymbol{v} - S_0 \mathbf{1}) \boldsymbol{u}^{\mathrm{T}} = 0,
\end{aligned}$$

where we used the property $\sum_{j=0}^{+\infty} S_j \mathbf{1} = \boldsymbol{v}$. This implies that $\widetilde{L}_1 = 0$. From (8.9) and (8.10) it follows that $\widetilde{L}_i = 0$ for $i > 1$, that is, $\widetilde{L}(z)$ is a matrix power series in z^{-1}. Moreover, $\sum_{i=0}^{+\infty} |\widetilde{L}_{-i}| \le \sum_{i=0}^{+\infty} |B_{-i}| + \sum_{i=0}^{+\infty} |\boldsymbol{v}\boldsymbol{w}^{\mathrm{T}} B_{-i-1}|$. Therefore, in order to prove that $\widetilde{L}(z) \in \mathcal{W}_-$ it is sufficient to show that $\sum_{i=0}^{+\infty} |B_{-i}|$ is finite. To prove this property it is sufficient to show that $B(z) = S(z)(I - zQ)^{-1}$ is analytic for $r^{-1} < |z| < 1$, in fact the convergence of $\sum_{i=0}^{+\infty} z^{-i} B_i$ for $r^{-1} < |z| < 1$ implies the convergence of $\sum_{i=0}^{+\infty} B_{-i}$ since $B_i \ge 0$ for $i \le -1$. The analiticity of $B(z)$ for $r^{-1} < |z| < 1$ holds since $S(z)$ is analytic for $|z| > r^{-1}$ and $(I - zQ)^{-1}$ is analytic for $|z| < 1$. Finally, in order to complete the proof we have to show that $\widetilde{L}(z)$ is invertible for $|z| \ge 1$. This property holds since $\det(I - z\widetilde{A}(z^{-1}))$ has exactly m zeros outside the closed unit disk, and no zeros of modulus 1. Therefore, the only values of z of modulus greater than 1 which make $I - z\widetilde{A}(z^{-1})$ noninvertible are the zeros of $\det(I - z\widetilde{R}_{\min})$, that is, the reciprocal of the eigenvalues of $\widetilde{R}_{\min}$, and the factorization $I - z\widetilde{A}(z^{-1}) = (I - z\widetilde{R}_{\min})\widetilde{L}(z)$ implies the nonsingularity of $\widetilde{L}(z)$ for $|z| \ge 1$. □

Since $\widetilde{\varphi}(z)$ is analytic and invertible for $|\xi_{m-1}| < |z| < |\xi_{m+1}|$, then the function $\widetilde{\psi}(z) = (z^{-1}\widetilde{\varphi}(z))^{-1}$ is analytic for $|\xi_{m-1}| < |z| < |\xi_{m+1}|$. Let us denote by $\widetilde{\psi}(z) = \sum_{i=-\infty}^{+\infty} z^i \widetilde{H}_i$ its Laurent series. Denote also $\widetilde{\psi}^{(n)}(z) = \sum_{i=-\infty}^{+\infty} z^i \widetilde{H}_{i \cdot 2^n}$, $n = 0, 1, \ldots$, so that

$$\widetilde{\psi}^{(n+1)}(z^2) = \frac{1}{2}(\widetilde{\psi}^{(n)}(z) + \widetilde{\psi}^{(n)}(-z)). \tag{8.11}$$

Let us apply cyclic reduction to the function $\widetilde{\varphi}(z)$ generating in this way the sequence of matrix power series $\widetilde{\varphi}^{(n)}(z)$ and $\widehat{\widetilde{\varphi}}^{(n)}(z)$ by means of (7.40) suitably adjusted by replacing any ocurrence of $\varphi^{(n)}(z)$ with $\widetilde{\varphi}^{(n)}(z)$ and any occurrence of $\widehat{\varphi}^{(n)}(z)$ with $\widehat{\widetilde{\varphi}}^{(n)}(z)$. For the sake of simplicity we use the same notation of Chapter 7 and set

$$\eta = |\xi_{m-1}|, \qquad \xi = |\xi_{m+1}|. \tag{8.12}$$

We have the following convergence result concerning cyclic reduction applied to the function $\widetilde{\varphi}(z)$.

Theorem 8.2 *Let $A(z)$ be the generating function associated with an M/G/1-type Markov chain with negative drift μ which satisfies Condition 4.20. Assume that $A(z)$ is analytic for $|z| < r$ with $1 < r < +\infty$, and that there exists ζ such that $1 < |\zeta| < r$ and $\det(\zeta I - A(\zeta)) = 0$. Let ξ_i, $i \geq 1$ be the roots of $\varphi(z) = zI - A(z)$ such that (8.3) and (8.12) hold. Let $\boldsymbol{u} > 0$ be such that $^{\mathrm{T}}\mathbf{1} = 1$, set $Q = \mathbf{1}\boldsymbol{u}^{\mathrm{T}}$, $\widetilde{\varphi}(z) = \varphi(z)(I - z^{-1}Q)^{-1}$. Then for any $\epsilon > 0$ such that $\eta + \epsilon < 1 < \xi - \epsilon$, there exist a real $0 < \theta < 1$ and a positive integer n_0 such that for any $n \geq n_0$ the function $\widetilde{\psi}^{(n)}(z)$ defined in (8.11) is analytic and invertible in the annulus $\mathbb{A}_n(\theta, \epsilon) = \{z \in \mathbb{C} : \quad \theta^{-1}(\eta + \epsilon)^{2^n} < |z| < \theta(\xi - \epsilon)^{2^n}\}$, and has a uniformly bounded inverse for $z \in \mathbb{A}_n(\theta, \epsilon)$. Moreover, for any $n \geq n_0$, the function*

$$\widetilde{\varphi}^{(n)}(z) = z(\widetilde{\psi}^{(n)}(z))^{-1} = -\widetilde{A}_{-1}^{(n)} + z(I - \widetilde{A}_0^{(n)}) - z^2 \widetilde{A}_1^{(n)} - \cdots,$$

is analytic and invertible in $\mathbb{A}_n(\theta, \epsilon)$, and for any operator norm $\|\cdot\|$ there exist positive constants c_i, $i = -1, 1, 2, \ldots$, such that

$$||\widetilde{A}_i^{(n)}|| \leq \begin{cases} c_i(\xi - \epsilon)^{-i2^n}, & \text{for } i > 0, \\ c_i(\eta + \epsilon)^{-i2^n} & \text{for } i = -1. \end{cases} \tag{8.13}$$

for any $n \geq n_0$. Therefore, cyclic reduction applied with the shift technique is convergent.

Proof Let $\epsilon > 0$ be such that $\eta + \epsilon < 1 < \xi - \epsilon$. By using an inductive argument on n, from (8.11) it follows that $\widetilde{\psi}^{(n)}(z)$ is analytic in the annulus $\{z \in \mathbb{C} : \quad (\eta + \epsilon)^{2^n} < |z| < (\xi - \epsilon)^{2^n}\}$ which contains $\mathbb{A}_n(\theta, \epsilon)$ for any $0 < \theta < 1$.

Now we prove that there exist $n_0 > 0$ and $0 < \theta < 1$ such that, for any $n \geq n_0$, the matrix $\widetilde{\psi}^{(n)}(z)$ is nonsingular for $z \in \mathbb{A}_n(\theta, \epsilon)$ and has a uniformly bounded inverse. Recall that $z^{-1}\widetilde{\varphi}(z)$ has a canonical factorization and, by Theorem 8.1, also the function $z\widetilde{\varphi}(z^{-1})$ has a canonical factorization. Therefore we may apply Theorem 3.22 and deduce that the matrix $\widetilde{H}_0$ is nonsingular. Thus, we may write $\widetilde{\psi}^{(n)}(z) = \widetilde{H}_0(I + W^{(n)}(z))$ where

$$W^{(n)}(z) = \widetilde{H}_0^{-1} \sum_{i=1}^{+\infty} (z^i \widetilde{H}_{i2^n} + z^{-i} \widetilde{H}_{-i2^n}). \tag{8.14}$$

In order to prove the nonsingularity of $\widetilde{\psi}^{(n)}(z)$ and the boundedness of its inverse for $z \in \mathbb{A}_n(\theta, \epsilon)$ it is sufficient to prove that there exists an operator norm such that $\|W^{(n)}(z)\| \leq 1/2$ for $z \in \mathbb{A}_n(\theta, \epsilon)$. In fact, the latter bound implies that $\rho(W^{(n)}(z)) \leq 1/2$ for $z \in \mathbb{A}_n(\theta, \epsilon)$ (see Lemma A.2 in the appendix) so that $\det(I + W^{(n)}(z)) \neq 0$. Moreover, since $\|\widetilde{\psi}^{(n)}(z)^{-1}\| \leq \|\widetilde{H}_0^{-1}\| \cdot \|(I + W^{(n)}(z))^{-1}\|$ and since $\|(I + W^{(n)}(z))^{-1}\| \leq 1/(1 - \|W^{(n)}(z)\|)$ (see Lemma A.5 in the appendix), then the condition $\|W^{(n)}(z)\| \leq 1/2$ for $z \in \mathbb{A}_n(\theta, \epsilon)$, implies $\|(I + W^{(n)}(z))^{-1}\| \leq 1/2$ for $z \in \mathbb{A}_n(\theta, \epsilon)$, that is, the uniform boundedness of $\widetilde{\psi}^{(n)}(z)^{-1}$.

Let us prove that there exists an operator norm $\|\cdot\|$ and there exists $n_0 > 0$ such that for any $n \geq n_0$, $\|W^{(n)}(z)\| < 1/2$ for $z \in \mathbb{A}_n(\theta, \epsilon)$. Since $\widetilde{\psi}(z)$ is analytic in $\mathbb{A}$ then, by Theorem 3.6, for any operator norm $\|\cdot\|$, there exists a constant c such that

$$||\widetilde{H}_i|| \leq \begin{cases} c(\xi-\epsilon)^{-i}, & \text{for} \quad i > 0, \\ c(\eta+\epsilon)^{-i} & \text{for } i < 0. \end{cases} \tag{8.15}$$

Let n_0 be such that $2(\eta+\epsilon)^{2^n} < 1 < \frac{1}{2}(\xi-\epsilon)^{2^n}$ for any $n \geq n_0$. Then from (8.15) and (8.14) we obtain that

$$\begin{aligned}
\|W^{(n)}(z)\| \leq & \|\widetilde{H}_0^{-1}\| \cdot \|\sum_{i=1}^{+\infty}(z^i \widetilde{H}_{i2^n} + z^{-i}\widetilde{H}_{-i2^n})\| \\
\leq & c\|\widetilde{H}_0^{-1}\| \sum_{i=1}^{+\infty}(|z|^i(\xi-\epsilon)^{-i\cdot 2^n} + |z|^{-i}(\eta+\epsilon)^{i\cdot 2^n}) \\
= & c\|\widetilde{H}_0^{-1}\| \left(\frac{|z|(\xi-\epsilon)^{-2^n}}{1-|z|(\xi-\epsilon)^{-2^n}} + \frac{|z|^{-1}(\eta+\epsilon)^{2^n}}{1-|z|^{-1}(\eta+\epsilon)^{2^n}}\right) \\
\leq & \gamma(|z|(\xi-\epsilon)^{-2^n} + |z|^{-1}(\eta+\epsilon)^{2^n}),
\end{aligned}$$

where the latter inequality is valid for any $n \geq n_0$ provided that $2(\eta+\epsilon)^{2^n} < |z| < \frac{1}{2}(\xi-\epsilon)^{2^n}$, with $\gamma = 4c\|\widetilde{H}_0^{-1}\|$. Therefore it is sufficient to prove that there exists a constant $0 < \theta < 1$ such that

$$\gamma(|z|(\xi-\epsilon)^{-2^n} + |z|^{-1}(\eta+\epsilon)^{2^n}) \leq 1/2, \quad \text{for} \quad z \in \mathbb{A}_n(\theta, \epsilon), \tag{8.16}$$

and

$$\theta^{-1}(\eta+\epsilon)^{2^n} < \theta(\xi-\epsilon)^{2^n}. \tag{8.17}$$

Since $(\eta+\epsilon)/(\xi-\epsilon) < 1$, equation (8.17) is satisfied for $\left(\frac{\eta+\epsilon}{\xi-\epsilon}\right)^{2^{n-1}} < \theta < 1$. Moreover, (8.16) is satisfied for any $\theta \leq 1/(4\gamma)$. Choosing $\theta = 1/(4\gamma)$ and n_0 such that $\left(\frac{\eta+\epsilon}{\xi-\epsilon}\right)^{2^{n_0-1}} < 1/(4\gamma)$ then (8.17) and (8.16) are satisfied for any $n \geq n_0$.

Concerning the last part of the theorem, recall that $\widetilde{\varphi}^{(n)}(z) = z(\widetilde{\psi}^{(n)}(z))^{-1}$ and $\widetilde{\psi}^{(n)}(z)^{-1}$ is analytic and bounded in $\mathbb{A}_n(\theta, \epsilon)$ for any $n \geq n_0$. Therefore, from Theorem 3.6 applied to $\widetilde{\psi}^{(n)}(z)^{-1}$ one deduces that there exists a constant γ such that $\|\widetilde{A}_i^{(n)}\| \leq \gamma\theta^i(\xi-\epsilon)^{-i2^n}$ for $i > 0$ and $\|\widetilde{A}_i^{(n)}\| \leq \gamma\theta^i(\eta+\epsilon)^{-i2^n}$ for $i = -1$. Therefore, choosing $c_i = \gamma\theta^i$ yields (8.13). □

From the above theorem we deduce the following result concerning the convergence to $G_{\min}$ of the sequence $G^{(n)} = (I - \widehat{\widetilde{A}}_0^{(n)})^{-1}A_{-1}$, $n = 0, 1, 2, \ldots$.

Theorem 8.3 *In the assumption of Theorem 8.2, if* $\det(I - \widehat{\widetilde{A}}_0^{(n)}) \neq 0$ *and if* $\|(I - \widehat{\widetilde{A}}_0^{(n)})^{-1}\|$ *is uniformly bounded from above, then* $\lim_n G^{(n)} = G_{\min}$, *where* $G^{(n)} = (I - \widehat{\widetilde{A}}_0^{(n)})^{-1} A_{-1}$. *Moreover,*

$$\widetilde{A}_{-1}^{(n)} \mathbf{1} = 0, \quad (I - \widehat{\widetilde{A}}_0^{(n)})\mathbf{1} = A_{-1}\mathbf{1}, \quad n \geq 0 \tag{8.18}$$

and for any matrix norm and for any $\epsilon > 0$ *such that* $\eta + \epsilon < 1 < \xi - \epsilon$, *there exist* $\gamma > 0$ *and* $\sigma_i > 0$, $i \geq 1$, *such that*

$$\|G_{\min} - G^{(n)}\| \leq \gamma \left(\frac{\eta + \epsilon}{\xi - \epsilon} \right)^{2^n}, \quad n \geq 0,$$

and

$$\|\widehat{\widetilde{A}}_i^{(n)}\| \leq \gamma_i (\xi - \epsilon)^{-i \cdot 2^n}, \quad i \geq 1. \tag{8.19}$$

Proof Let us prove (8.19). From (7.45), denoting $V(z) = \widehat{\widetilde{A}}(z)\widetilde{\psi}(z)$, one has $\widehat{\widetilde{A}}^{(n)}(z)\widetilde{\psi}^{(n)}(z) = V^{(n)}(z)$, where $V^{(n)}(z) = \sum_{i=-\infty}^{+\infty} z^i V_{i2^n}$. Therefore,

$$\widehat{\widetilde{A}}^{(n)}(z) = V^{(n)}(z)\widetilde{\psi}^{(n)}(z)^{-1} = z^{-1} V^{(n)}(z) \widetilde{\varphi}^{(n)}(z).$$

Comparing the coefficients in the above expression yields

$$\widehat{\widetilde{A}}_k^{(n)} = V_{k \cdot 2^n} - \sum_{i=0}^{+\infty} V_{(k-i+1)2^n} \widetilde{A}_{i-1}^{(n)}, \quad k \geq 0.$$

Taking norms of both sides yields

$$\|\widehat{\widetilde{A}}_k^{(n)}\| \leq \|V_{k \cdot 2^n}\| + \sum_{i=0}^{+\infty} \|V_{(k-i+1)2^n}\| \cdot \|\widetilde{A}_{i-1}^{(n)}\|.$$

Since $V(z)$ is analytic for $\eta < |z| < \xi$, there exists $\theta > 0$ such that $\|V_{k \cdot 2^n}\| \leq \theta(\xi - \epsilon)^{-k \cdot 2^n}$ for any $n, k > 0$ and $\|V_{k \cdot 2^n}\| \leq \theta(\eta + \epsilon)^{-k \cdot 2^n}$ for any $k < 0$, $n > 0$. Moreover, from (8.13) we have $\|\widetilde{A}_i^{(n)}\| \leq \gamma_i (\xi - \epsilon)^{-i \cdot 2^n}$. Therefore we find that there exists $\sigma_k > 0$ such that

$$\|\widehat{\widetilde{A}}_k^{(n)}\| \leq \sigma_k (\xi - \epsilon)^{-k 2^n}$$

for any $n, k > 0$. Equations (8.18) can be proved by induction. Concerning the convergence of $G^{(n)}$ to $G_{\min}$, since

$$(I - \widehat{\widetilde{A}}_0^{(n)})\widetilde{G}_{\min} = \widetilde{A}_{-1} + \sum_{i=1}^{+\infty} \widehat{\widetilde{A}}_i^{(n)} \widetilde{G}_{\min}^{i \cdot 2^n},$$

from equation (8.18) one has $(I-\widehat{\widetilde{A}}_0^{(n)})G_{\min} = A_{-1}+\sum_{i=1}^{+\infty}\widehat{\widetilde{A}}_i^{(n)}\widetilde{G}_{\min}^{i\cdot 2^n}$. Therefore we obtain

$$G_{\min} - G^{(n)} = (I-\widehat{\widetilde{A}}_0^{(n)})^{-1}\sum_{i=1}^{+\infty}\widehat{\widetilde{A}}_i^{(n)}\widetilde{G}_{\min}^{i\cdot 2^n}. \tag{8.20}$$

Since $\rho(\widetilde{G}_{\min}) = \eta$, for Theorem A.2 in the appendix we have that for any $\epsilon > 0$ there exists a matrix norm $\|\cdot\|$ such that $\|\widetilde{G}_{\min}\| = \eta+\epsilon$. Therefore, for the equivalence of matrix norms, for any matrix norm there exists a constant $\theta > 0$ such that $\|\widetilde{G}_{\min}^{i2^n}\| \le \theta(\eta+\epsilon)^{i2^n}$, so that

$$\|G_{\min} - G^{(n)}\| \le \|(I-\widehat{\widetilde{A}}_0^{(n)})^{-1}\|\sum_{i=1}^{+\infty}\theta(\eta+\epsilon)^{i\cdot 2^n}\|\widehat{\widetilde{A}}_i^{(n)}\|.$$

Since $\|(I-\widehat{\widetilde{A}}_0^{(n)})^{-1}\|$ is bounded from above by a constant, from (8.19) we have $\|G_{\min} - G^{(n)}\| \le \gamma(\frac{\eta+\epsilon}{\xi-\epsilon})^{2^n}$ for a suitable γ. □

Algorithm 8.1 summarizes the entire process of cyclic reduction with shift applied to an M/G/1-type Markov chain where $\mu < 0$.

Algorithm 8.1 Shifted cyclic reduction for M/G/1-type problems with $\mu < 0$

INPUT: The $m\times m$ matrices A_i, $i = -1,0,\ldots,d$, which define the block Toeplitz block Hessenberg matrix in (7.30) and the matrix polynomial $A(z) = \sum_{i=-1}^{d} z^{i+1}A_i$ where $\mu < 0$; a real $\epsilon > 0$.

OUTPUT: An approximation Y to the minimal solution $G_{\min}$ of the equation $A_{-1}+A_0X+\cdots+A_dX^{d+1} = X$ and a real $\sigma > 0$ such that $\|G_{\min}-Y\|_\infty \le \epsilon\sigma$.

COMPUTATION:

1. Choose any vector $\boldsymbol{u} > 0$ such that $\boldsymbol{u}^{\mathrm{T}}\mathbf{1} = 1$ and set $Q = \mathbf{1}\boldsymbol{u}^{\mathrm{T}}$.
2. Set $\widetilde{A}_i = A_i + (\sum_{j=i+1}^{d} A_j)Q$ for $i = 0,\ldots,d$, $\widetilde{A}_{-1} = A_{-1} - A_{-1}Q$, and $\widetilde{\varphi}(z) = zI - \sum_{i=-1}^{d} z^{i+1}\widetilde{A}_i$, $\widehat{\widetilde{\varphi}}(z) = I - \sum_{i=0}^{d} z^i\widetilde{A}_i$.
3. Apply one step of cyclic reduction to the functions $\widetilde{\varphi}(z)$ and $\widehat{\widetilde{\varphi}}(z)$, that is, compute

$$\widetilde{\varphi}'(z^2) = -\widetilde{\varphi}(z)\widetilde{\varphi}_{\mathrm{odd}}(z^2)^{-1}\widetilde{\varphi}(-z),$$
$$\widehat{\widetilde{\varphi}}'(z^2) = \widehat{\widetilde{\varphi}}(z) + \widehat{\widetilde{\varphi}}_{\mathrm{odd}}(z^2)\widetilde{\varphi}_{\mathrm{odd}}(z^2)^{-1}\widetilde{\varphi}(z).$$

4. Set $\widetilde{\varphi}(z) = \widetilde{\varphi}'(z)$, $\widehat{\widetilde{\varphi}}(z) = \widehat{\widetilde{\varphi}}'(z)$.
5. If $\sum_{i=1}^{\widehat{d}}\|\widetilde{A}_i\|_\infty \le \epsilon$, where $\widehat{d}$ is the numerical degree of $\widetilde{\varphi}(z)$, then output $Y = (I-\widehat{\widetilde{A}}_0)^{-1}A_{-1}$ and $\sigma = 2\|(I-\widehat{\widetilde{A}}_0)^{-1}\|_\infty$; otherwise, repeat from stage 3.

The error bound $\|G_{\min}-Y\|_\infty \le \epsilon\sigma$ holds since $\widetilde{G}^i_{\min} = G^i_{\min} - Q$ for any $i \ge 1$ so that $\|\widetilde{G}^i_{\min}\|_\infty \le \|G^i_{\min}\|_\infty + \|Q\|_\infty = 2$. Therefore from the stop condition we deduce that $\sum_{i=1}^{+\infty} \|\widehat{\widetilde{A}}_i^{(n)}\|_\infty \doteq \sum_{i=1}^{\widehat{d}} \|\widehat{\widetilde{A}}_i^{(n)}\|_\infty \le \epsilon$ whence the bound follows from (8.20).

The cyclic reduction step at stage 3 of this algorithm can be implemented with the functional approach as in Algorithm 7.6 or with the matrix approach as in Algorithm 7.5.

We recall that cyclic reduction applied to the new functions $\widetilde{\varphi}(z)$ and $\widehat{\widetilde{\varphi}}(z)$ is more stable than cyclic reduction applied to the functions $\varphi(z)$ and $\widehat{\varphi}(z)$ since $|\xi_{m-1}| < 1 < |\xi_{m+1}|$ in light of (7.62).

From the above result, it follows that the shifted cyclic reduction can be much faster than the original one. Indeed, if the second largest modulus eigenvalue ξ_{m-1} of $G_{\min}$ is far from the unit circle, then the rate of convergence is much improved.

Thus, the deflating technique leads to a better speed of convergence, but destroys the nonnegativity and the M-matrix properties of the blocks generated at each step. Indeed, in general the matrices $\widetilde{A}_i^{(n)}$ and $\widehat{\widetilde{A}}_{i+1}^{(n)}$ generated by the shifted cyclic reduction are not nonnegative. In principle this fact could lead to a loss of accuracy of the results obtained with the shifting technique. In practice, no differences in terms of accuracy have been pointed out between the results obtained with the two algorithms [59]. Furthermore, as has been proved in [59] for the QBD case, the shifted equation is numerically better conditioned than the original one. We recall also that, according to the analysis performed in Section 7.5, the amplification of the errors generated at each step of cyclic reduction is much more limited than in the unshifted case.

8.2.1 The QBD case

If $\varphi(z) = -A_{-1} + z(I - A_0) - z^2 A_1 z$ and $\widetilde{\varphi}(z) = -\widetilde{A}_{-1} + z(I - \widetilde{A}_0) - z^2 \widetilde{A}_1 z$, then we may show that cyclic reduction applied to $\widetilde{\varphi}(z)$ can be carried out with no break-down. We first recall that in principle cyclic reduction can encounter a break-down if the matrix $\left[\varphi^{(n)}(z)\right]_{\text{odd}} = (I - A_0^{(n)})$ is singular. It is shown in [14] that $\det(I - A_0^{(n)})$ is equal to zero if and only if the block tridiagonal matrix $T_{2^{n+1}-1}$ is singular, where we denote by T_k the $mk \times mk$ leading principal submatrix of the semi-infinite block tridiagonal block Toeplitz matrix

$$T = \begin{bmatrix} I - A_0 & -A_1 & & \\ -A_{-1} & I - A_0 & -A_1 & \\ & \ddots & \ddots & \ddots \end{bmatrix}.$$

Moreover, by Theorem 7.5, under Condition 4.20 cyclic reduction can be carried out with no breakdown so that $\det T_{2^{n+1}-1} \neq 0$ for any $n \ge 0$.

In order to show that shifted cyclic reduction can be carried out with no break-down it is sufficient to prove that $\det \widetilde{T}_{2^{n+1}-1} \neq 0$ for any $n \ge 0$ where $\widetilde{T}_k$

is the $mk \times mk$ leading principal submatrix of the semi-infinite block tridiagonal block Toeplitz matrix

$$\widetilde{T} = \begin{bmatrix} I - \widetilde{A}_0 & -\widetilde{A}_1 & & \\ -\widetilde{A}_{-1} & I - \widetilde{A}_0 & -\widetilde{A}_1 & \\ & \ddots & \ddots & \ddots \end{bmatrix}.$$

In order to prove this property we need some preliminary results.

First of all we prove that, if cyclic reduction can be applied to $\varphi(z)$, then $\det T_k \neq 0$ for any $k > 1$. Assume by a contradiction that there exists some k_0 such that $\det T_{k_0} = 0$ then $\rho(I - T_{k_0}) = 1$. Let k be any integer such that $k > k_0$. Denoting with U_k the $mk \times mk$ matrix having $I - T_{k_0}$ as leading principal submatrix and null elements elsewhere, it holds $0 \leq U_k \leq I - T_k$. From Theorem 1.28 we get $1 = \rho(U_k) \leq \rho(I - T_k) \leq 1$. Thus we obtain that $\rho(I - T_k) = 1$ and applying once again Theorem 1.28 we find that there exists an eigenvalue of $I - T_k$ equal to 1, therefore the matrix T_k is singular for any $k > k_0$. This would imply that cyclic reduction cannot be carried out in contradiction with our assumptions.

Theorem 8.4 *Let T_k be nonsingular, set $W_k = T_k^{-1}$ and denote with P_k the $m \times m$ block of W_k in the lower rightmost corner. Then the matrix $\widetilde{T}_k$ is singular if and only if $\boldsymbol{u}^{\mathrm{T}} P_k A_1 \mathbf{1} = 1$.*

Proof The equation $\widetilde{\varphi}(z) = \varphi(z)(I - z^{-1}Q)^{-1}$ can be equivalently rewritten in the matrix form as

$$\widetilde{T} = T \begin{bmatrix} I & & & \\ Q & I & & \\ Q & Q & I & \\ \vdots & \ddots & \ddots & \ddots \end{bmatrix},$$

whence we deduce that

$$\widetilde{T}_k = T_k \begin{bmatrix} I & & & \\ Q & I & & \\ \vdots & \ddots & \ddots & \\ Q & \dots & Q & I \end{bmatrix} + \begin{bmatrix} 0 \\ \vdots \\ 0 \\ -A_1 \end{bmatrix} \begin{bmatrix} Q \, Q \dots Q \end{bmatrix}.$$

This implies that $\widetilde{T}_k$ is singular if and only if the matrix in the right hand side of the above expression is singular, that is, if and only if

$$\det(I - W_k \begin{bmatrix} 0 \\ \vdots \\ 0 \\ A_1 \mathbf{1} \end{bmatrix} \begin{bmatrix} 0 \dots 0 \; \boldsymbol{u}^{\mathrm{T}} \end{bmatrix}) = 0.$$

The latter determinant is given by $1 - \boldsymbol{u}^{\mathrm{T}} P_k A_1 \mathbf{1}$. □

Theorem 8.5 *Let $\boldsymbol{u} > 0$ and $\det T_k \neq 0$, then $\boldsymbol{u}^{\mathrm{T}} P_k A_1 \mathbf{1} = 1$ if and only if $P_k A_1 \mathbf{1} = \mathbf{1}$.*

Proof It is clear that if $P_k A_1 \mathbf{1} = \mathbf{1}$ then $\boldsymbol{u}^{\mathrm{T}} P_k A_1 \mathbf{1} = \boldsymbol{u}^{\mathrm{T}} \mathbf{1} = 1$. In order to prove the reverse implication, observe that, since $(A_{-1} + A_0 + A_1)\mathbf{1} = \mathbf{1}$, then

$$T_k \begin{bmatrix} \mathbf{1} \\ \vdots \\ \boldsymbol{e} \end{bmatrix} = \begin{bmatrix} A_{-1}\mathbf{1} \\ 0 \\ \vdots \\ 0 \\ A_1 \mathbf{1} \end{bmatrix}.$$

Whence we obtain

$$\begin{bmatrix} \mathbf{1} \\ \vdots \\ \boldsymbol{e} \end{bmatrix} = W_k \begin{bmatrix} A_{-1}\mathbf{1} \\ 0 \\ \vdots \\ 0 \\ A_1 \mathbf{1} \end{bmatrix}.$$

Therefore from the last m rows in the above expression we get

$$Q_k A_{-1} \mathbf{1} + P_k A_1 \mathbf{1} = \mathbf{1}, \tag{8.21}$$

where Q_k is the $m \times m$ submatrix in the lower leftmost corner of W_k. Since W_k is the inverse of a nonsingular M-matrix, it is nonnegative, so that $Q_k A_{-1} \mathbf{1} \geq 0$ and $0 \leq P_k A_1 \mathbf{1} \leq \mathbf{1}$. This implies that if $P_k A_1 \mathbf{1} \neq \mathbf{1}$ then $\boldsymbol{u}^{\mathrm{T}} P_k A_1 \mathbf{1} - \mathbf{1} \neq 0$, which completes the proof. □

Now we are ready to prove the main applicability result:

Theorem 8.6 *Let A_{-1}, A_0 and A_1 be $m \times m$ matrices defining a QBD process. Let us assume that cyclic reduction can be applied to $\varphi(z)$. If $\boldsymbol{u} > 0$ and if $I - A_0 - A_1 Q$ is nonsingular, then cyclic reduction can be applied to the shifted problem associated with $\widetilde{\varphi}(z)$ with no break-down.*

Proof We prove that $\det \widetilde{T}_k \neq 0$ for $k \geq 1$. If $k = 1$ then $\widetilde{T}_1 = I - \widehat{A}_0 = I - A_0 - A_1 Q$. Therefore the nonsingularity of $\widetilde{T}_1$ follows by assumption.

Let us consider the case $k > 1$. Since cyclic reduction can be applied, then $\det T_k \neq 0$ for any k. Apply Theorems 8.4, 8.5 and deduce that $\det \widetilde{T}_k = 0$ if and only if $P_k A_1 \mathbf{1} = \mathbf{1}$. Assume by a contradiction that $\det \widetilde{T}_{k_0} = 0$ for some k_0, so that $P_{k_0} A_1 \mathbf{1} = \mathbf{1}$. Then we prove that $P_k A_1 \mathbf{1} = \mathbf{1}$ for any $k \geq k_0$ so that, in view of Theorems 8.4 and 8.5, the matrix $\widetilde{T}_k$ would be singular for any $k \geq k_0$. This property contradicts the fact that, by Theorem 8.2, there exists n_0 such that for any $n \geq n_0$ the matrices $\widetilde{\psi}^{(n)}(z)$ and $\widetilde{\varphi}^{(n)}(z)$ are invertible for $|z| = 1$. In order to prove that $P_k A_1 \mathbf{1} = \mathbf{1}$, consider the Schur complement S_k of T_{k-1} in T_k and recall that since T_k and T_{k-1} are nonsingular then S_k exists, is nonsingular and $S_k^{-1} = P_k$. Moreover, from the definition of the Schur complement and from the block tridiagonal structure of T_k and T_{k+1} it holds that $S_{k+1} = (I - A_0) - A_{-1}S_k^{-1}A_1 = (I - A_0) - A_{-1}P_kA_1$. Therefore, if $P_kA_1\mathbf{1} = \mathbf{1}$ then $S_{k+1}\mathbf{1} = (I - A_0)\mathbf{1} - A_{-1}S_k^{-1}A_1\mathbf{1} = \mathbf{1} - A_0\mathbf{1} - A_{-1}\mathbf{1} = A_1\mathbf{1}$, where the latter equation holds since $A_{-1} + A_0 + A_1$ is stochastic. Thus we get $\mathbf{1} = P_{k+1}A_1\mathbf{1}$ which completes the proof. □

For QBD processes the shifted cyclic reduction technique can be much simplified. Algorithm 8.2 synthesizes cyclic reduction for solving the matrix equations associated with a positive recurrent QBD.

Algorithm 8.2 Shifted yclic reduction for QBDs: the case $\mu < 0$

INPUT: The $m \times m$ matrices A_{-1}, A_0 and A_1 which define the matrix polynomial $A(z) = A_{-1} + zA_0 + z^2A_1$ associated with a QBD process with a negative drift μ; a real $\epsilon > 0$.

OUTPUT: An approximation Y to the minimal solution $G_{\min}$ of the equation $A_{-1} + A_0X + A_1X^2 = X$ and a real $\sigma > 0$ such that $\|G_{\min} - Y\|_\infty \le \epsilon\sigma$.

COMPUTATION:

1. Choose any vector $\boldsymbol{u} > 0$ such that $\boldsymbol{u}^{\mathrm{T}}\mathbf{1} = 1$ and set $Q = \mathbf{1}\boldsymbol{u}^{\mathrm{T}}$.
2. Set $\widetilde{A}_{-1} = A_{-1} - A_{-1}Q$, $\widetilde{A}_0 = A_0 + A_1Q$, $\widetilde{A}_1 = A_1$, $\widehat{\widetilde{A}}_0 = \widetilde{A}_0$.
3. Compute the matrices

$$\begin{aligned}
\widetilde{A}'_{-1} &= \widetilde{A}_{-1}(I - \widetilde{A}_0)^{-1}\widetilde{A}_{-1},\\
\widetilde{A}'_0 &= \widetilde{A}_0 + \widetilde{A}_1(I - \widetilde{A}_0)^{-1}\widetilde{A}_{-1} + \widetilde{A}_{-1}(I - \widetilde{A}_0)^{-1}\widetilde{A}_1,\\
\widetilde{A}'_1 &= \widetilde{A}_1(I - \widetilde{A}_0)^{-1}\widetilde{A}_1,\\
\widehat{\widetilde{A}}{}'_0 &= \widehat{\widetilde{A}}_0 + \widetilde{A}_1(I - \widetilde{A}_0)^{-1}\widetilde{A}_{-1}.
\end{aligned}$$

4. Set $\widetilde{A}_i = \widetilde{A}'_i$, for $i = -1, 0, 1$ and $\widehat{\widetilde{A}}_0 = \widehat{\widetilde{A}}{}'_0$.
5. If $\|\widetilde{A}_1\|_\infty \le \epsilon$ then output $Y = (I - \widehat{\widetilde{A}}_0)^{-1}A_{-1}$, $\sigma = 2\|(I - \widehat{\widetilde{A}}_0)^{-1}\|_\infty$; otherwise repeat from stage 3.

In the case $\mu > 0$ the shift technique can be similarly applied. Here, the root $z = 1$ is mapped to infinity. If $\mu = 0$, then $\det(zI - A(z))$ has two unit zeros $z_1 = z_2 = 1$. In this case we still can apply this technique by shifting $z_1 = 1$ into 0. The convergence speed can be increased if, in addition, $z_2 = 1$ is shifted into ∞.

8.3 Cyclic reduction and functional iterations

In this section we present a class of functional iterations for the solution of the matrix equation

$$-A_{-1} + (I - A_0)X - A_1X^2 = 0. \tag{8.22}$$

which exploit the strategy of divide-and-conquer on the basis of the methods of logarithmic reduction and of cyclic reduction.

We recall that the algorithms based on functional iterations which we have analyzed in Chapter 6 have a linear convergence and are self-correcting. The latter property means that possible errors introduced at any step, say by floating point arithmetic, are corrected in the subsequent steps due to the local convergence of the iteration. On the other hand, logarithmic reduction and cyclic reduction provide iterations which generally have a quadratic convergence but they are not self-correcting.

In this section we try to combine the nice features of functional iterations like the self-correction, and the high speed of the divide-and-conquer techniques in order to design a class of functional iterations which share the following features: each iteration in this class is globally convergent, self-correcting, and the local convergence is linear with an arbitrarily large rate of convergence. More precisely, the kth element of this one-parameter family of functional iterations generates a sequence of approximations such that the error $e_n^{(k)}$ at the generic n-th step converges to zero as $e_n^{(k)} = O(\sigma^{n2^k})$ and the cost of a single step is *independent* of k, where $0 < \sigma < 1$. This places the algorithms in this class in between the functional iterations of Chapter 6 and the cyclic reduction process of Section 7.3, obtained in the limit as $k \to \infty$.

Let us assume that the roots $\xi_1, \ldots, \xi_{2m}$ of $a(z) = \det(-A_{-1} + (I - A_0)z - A_1 z^2)$ satisfy the *splitting property*

$$0 \leq |\xi_1| \leq \cdots \leq |\xi_m| < 1 < |\xi_{m+1}| \leq \cdots \leq |\xi_{2m}| \leq \infty, \tag{8.23}$$

where we assume roots at infinity if the polynomial $a(z)$ has degree less than $2m$. In this way the minimal solution of (8.22) has eigenvalues $\xi_1, \ldots, \xi_m$.

This assumption is no loss of generality since, under the general Condition 4.20, we may reduce to this case by applying the shift technique of Section 8.2.

Let us consider the following functional iteration

$$W^{(n+1)} = S_1(W^{(n)}), \qquad n = 0, 1, 2, \ldots, \tag{8.24}$$

which is started with an initial matrix $W^{(0)}$, where

$$S_1(W) = I - A_0 - A_1 W^{-1} A_{-1};$$

our goal is to relate the fixed points of (8.24) and the solutions of (8.22). To this end we have the following.

Remark 8.7 Observe that if there exists a nonsingular solution R of the equation $S_1(R) = R$ then $X = R^{-1}A_{-1}$ is a solution of the equation $-A_{-1} + (I - A_0)X - A_1 X^2 = 0$. In fact, for $X = R^{-1}A_{-1}$ it follows that

$$-A_{-1} + (I - A_0)X - A_1 X^2 = (-R + (I - A_0) - A_1 R^{-1} A_{-1}) R^{-1} A_{-1} = 0$$

since $-R + (I - A_0) - A_1 R^{-1} A_{-1} = -R + S_1(R) = 0$. Similarly we find that the matrix $Y = A_1 R^{-1}$ solves $-Y^2 A_{-1} + Y(I - A_0) - A_1 = 0$.

Consider the function $\boldsymbol{s}(\boldsymbol{w}) = \mathrm{vec}(S_1(W)) : \mathbb{R}^{m^2} \to \mathbb{R}^{m^2}$, where $\boldsymbol{w} = \mathrm{vec}(W)$. We indicate with $\mathcal{J}(W)$ the $m^2 \times m^2$ Jacobian matrix of the function $\boldsymbol{s}(\boldsymbol{w})$ and with $\rho(\mathcal{J}(w))$ its spectral radius. Concerning the convergence of the sequence $\{W^{(n)}\}$ we have the following.

Theorem 8.8 *Let ξ_i, $i = 1, \ldots, 2m$, be the zeros of $\det(-A_{-1}+z(I-A_0)-z^2A_1)$ satisfying (8.23) and let V be a solution of $S_1(V) = V$, i.e., a fixed point of the function $S_1(W)$. Then $\rho(\mathcal{J}(V)) < 1$ if and only if $X = V^{-1}A_{-1}$ and $Y = A_1V^{-1}$ are minimal solutions of the equations $-A_{-1} + (I - A_0)X - A_1X^2 = 0$ and $-A_1+Y(I-A_0)-Y^2A_{-1} = 0$, respectively, that is, X has eigenvalues $\xi_1, \ldots, \xi_m$, and Y has eigenvalues $\xi_{m+1}^{-1}, \ldots, \xi_{2m}^{-1}$. Moreover, if the fixed point V satisfies $\rho(\mathcal{J}(V)) < 1$ then V is unique and, given an operator norm $\|\cdot\|$, there exists a neighborhood $\mathcal{S} = \{W \in \mathbb{R}^{m\times m} : \ \|W - V\| \le \epsilon\}$ of V such that for any $W_0 \in \mathcal{S}$ the sequence (8.24) is well defined and converges to V. More specifically, for any σ, $|\xi_m/\xi_{m+1}| < \sigma < 1$, there exists a positive constant γ such that $\|V - W^{(n)}\| < \gamma\sigma^n$.*

Proof It can be easily verified that the Jacobian matrix of $\boldsymbol{s}(\boldsymbol{w})$ is $\mathcal{J}(W) = (A_1W^{-1})^{\mathrm{T}} \otimes (W^{-1}A_{-1})$ and that $\rho(\mathcal{J}(W)) = \rho(A_1W^{-1})\rho(W^{-1}A_{-1})$. If $V = S_1(V)$ then from Remark 8.7 the matrices $X = V^{-1}A_{-1}$ and $Y = A_1V^{-1}$ are solutions of the equations $-A_{-1} + (I - A_0)X - A_1X^2 = 0$, $-A_1 + Y(I - A_0) - Y^2A_{-1} = 0$, respectively. Since the eigenvalues of X are a subset of m entries (counting multiplicities) of $\{\xi_1, \ldots, \xi_{2m}\}$, and the eigenvalues of Y are a subset of m entries of $\{\xi_1^{-1}, \ldots, \xi_{2m}^{-1}\}$ (compare with Section 3.3), the condition $\rho(\mathcal{J}(V)) = \rho(X)\rho(Y) < 1$ implies that the eigenvalues of X are $\xi_1, \ldots, \xi_m$, and the eigenvalues of Y are $\xi_{m+1}^{-1}, \ldots, \xi_{2m}^{-1}$, so that $\rho(\mathcal{J}(V)) = |\xi_m/\xi_{m+1}|$. Conversely, if X and Y are minimal solutions then $\rho(X) = |\xi_m|$, $\rho(Y) = |\xi_{m+1}|^{-1}$, and this implies $\rho(\mathcal{J}(V)) = |\xi_m/\xi_{m+1}| < 1$ so that V is an attractive fixed point in light of Theorem 6.2. Concerning uniqueness, if V_1 and V_2 were fixed points then $V_1 - V_2 = S_1(V_1) - S_1(V_2) = -A_1(V_1^{-1} - V_2^{-1})A_{-1} = A_1V_1^{-1}(V_1 - V_2)V_2^{-1}A_{-1}$. That is, $\mathrm{vec}(V_2-V_1)$ would be an eigenvector of the matrix $(V_2^{-1}A_{-1})^{\mathrm{T}}\otimes(A_1V_1^{-1})$ corresponding to the eigenvalue 1, that is, $\rho((V_2^{-1}A_{-1})^{\mathrm{T}} \otimes (A_1V_1^{-1})) \ge 1$. This contradicts the fact that $\rho((V_2^{-1}A_{-1})^{\mathrm{T}} \otimes (A_1V_1^{-1})) = \rho(V_2^{-1}A_{-1})\rho(A_1V_1^{-1}) < 1$ since $\rho(V_2^{-1}A_{-1}) < 1$ and $\rho(A_1V_1^{-1}) < 1$. Whence $V_1 - V_2 = 0$ and the uniqueness is proved. Finally, the local convergence properties follow from customary arguments on the convergence theory of iterative methods (see Theorem 6.2). □

Under the assumptions of the above theorem the iteration (8.24) provides a means to solve the matrix equation (8.22) if complemented with the expression $X^{(n)} = (W^{(n)})^{-1}A_{-1}$. According to Theorem 8.8 the local convergence of $\{X^{(n)}\}$ to the minimal solution $G_{\min}$ of (8.22) is linear with rate of convergence $|\xi_m/\xi_{m+1}| < \sigma < 1$.

Let us inductively define $S_n(W) = S_{n-1}(S_1(W))$ so that

$$S_n(W) = S_1(\cdots(S_1(S_1(W))\cdots), \quad n \ \text{times}.$$

In this way the function $S_h(W)$ defines a functional iteration

$$Z^{(n+1)} = S_h(Z^{(n)}), \quad n = 0, 1, \ldots,$$

such that, for $Z^{(0)} = W^{(0)}$ then $Z^{(n)} = W^{(n \cdot h)}$, $n = 0, 1, \ldots$. In other words, the iteration defined by the function $S_h(Z)$ generates a subsequence of $\{W^{(n)}\}$ obtained by picking up elements at the distance of h steps from each other. Indeed the sequence $\{Z^{(n)}\}$ still has a linear convergence but its convergence rate is σ^h.

Now, our goal is to show that we may perform one step of the functional iteration associated with the function $S_{2^k}(W)$ at a cost *independent* of k. In order to do this we will prove the following expression

$$S_{2^k}(W) = \widehat{A}_0^{(k)} + A_1^{(k)}(W - \widetilde{A}_0^{(k)})^{-1} A_{-1}^{(k)}.$$

for suitable matrices $\widehat{A}_0^{(k)}$, $\widetilde{A}_0^{(k)}$, $A_i^{(k)}$, $i = -1, 1$ which are computed once for all at a low cost by applying a slight modification of cyclic reduction.

For this purpose, consider the $n \times n$ block tridiagonal matrix

$$\mathcal{T}_n = \mathcal{T}_n(W) = \begin{bmatrix} W & -A_{-1} & & 0 \\ -A_1 & I - A_0 & \ddots & \\ & \ddots & \ddots & -A_{-1} \\ 0 & & -A_1 & I - A_0 \end{bmatrix},$$

and assume that it is nonsingular for any $n > 0$. From $\mathcal{T}_2$ we deduce that $S_1(W)$ is nothing else but the Schur complement of $\mathcal{T}_1$ in $\mathcal{T}_2$. In general, from the block LU decomposition of $\mathcal{T}_n$

$$\mathcal{T}_n = \begin{bmatrix} I & & & 0 \\ * & I & & \\ & \ddots & \ddots & \\ 0 & & * & I \end{bmatrix} \begin{bmatrix} W & -A_{-1} & & 0 \\ & S_1(W) & \ddots & \\ & & \ddots & -A_{-1} \\ 0 & & & S_{n-1}(W) \end{bmatrix}$$

where "$*$" denotes generally nonzero blocks, and from the properties of the Schur complement (see Definition A.6 and Theorem A.7) we may inductively deduce that $S_{n-1}(W)$ is the Schur complement of $\mathcal{T}_{n-1}$ in $\mathcal{T}_n$. In fact, from the block Gaussian elimination we deduce that

$$\begin{aligned} S_1(W) &= I - A_0 - A_1 W^{-1} A_{-1}, \\ S_2(W) &= I - A_0 - A_1 S_1(W)^{-1} A_{-1} = S_1(S_1(W)), \\ S_{i+j}(W) &= S_i(S_j(W)) \quad \forall i, j > 0. \end{aligned}$$

Now let P_{n-1} be any permutation matrix and define $\mathcal{T}'_{n-1} = P_{n-1}^{\mathrm{T}} \mathcal{T}_{n-1} P_{n-1}$, $\mathcal{T}'_n = P_n^{\mathrm{T}} \mathcal{T}_n P_n$, where $P'_n = \begin{bmatrix} P'_{n-1} & 0 \\ 0 & 1 \end{bmatrix}$. Then the Schur complement of of $\mathcal{T}_{n-1}$ in $\mathcal{T}_n$ coincides with the Schur complement of of $\mathcal{T}'_{n-1}$ in $\mathcal{T}'_n$.

In order to provide a suitable expression for $S_{2^k}(W)$, we choose $n = 2^k + 1$, for a positive integer k, and apply a suitable permutation P_n to the rows and columns of $\mathcal{T}_n$ which leaves unchanged the last block row and the last block column and then apply the Schur complementation to the matrices $\mathcal{T}'_n$ and $\mathcal{T}'_{n-1}$ obtained in this way. The permutation that we choose is the one obtained by applying cyclic reduction. By following the same strategy used in Section 7.3, let us apply the even–odd permutation to block rows and to block columns of $\mathcal{T}_h$ followed by a step of block Gaussian elimination. For simplicity we recall this transformation in the case $n = 5$ so that

$$\mathcal{T}_5 = \begin{bmatrix} W & -A_{-1} & & & 0 \\ -A_1 & I - A_0 & -A_{-1} & & \\ & -A_1 & I - A_0 & -A_{-1} & \\ & & -A_1 & I - A_0 & -A_{-1} \\ 0 & & & -A_1 & I - A_0 \end{bmatrix}$$

and after the even–odd permutation we get

$$\left[\begin{array}{cc|ccc} I - A_0 & 0 & -A_1 & -A_{-1} & 0 \\ 0 & I - A_0 & 0 & -A_1 & -A_{-1} \\ \hline -A_{-1} & 0 & W & 0 & 0 \\ -A_1 & -A_{-1} & 0 & I - A_0 & 0 \\ 0 & -A_1 & 0 & 0 & I - A_0 \end{array}\right].$$

The Schur complementation of the 2×2 block matrix yields

$$\begin{bmatrix} \widetilde{W}^{(1)} & -A_{-1}^{(1)} & 0 \\ -A_1^{(1)} & I - A_0^{(1)} & -A_{-1}^{(1)} \\ 0 & -A_1^{(1)} & I - \widehat{A}_0^{(1)} \end{bmatrix} \tag{8.25}$$

with

$$\begin{aligned} A_0^{(1)} &= A_0 + A_{-1}(I - A_0)^{-1}A_1 + A_1(I - A_0)^{-1}A_{-1}, \\ A_{-1}^{(1)} &= A_{-1}(I - A_0)^{-1}A_{-1}, \quad A_1^{(1)} = A_1(I - A_0)^{-1}A_1, \\ \widehat{A}_0^{(1)} &= A_0 + A_{-1}(I - A_0)^{-1}A_1, \\ \widetilde{W}^{(1)} &= W - A_1(I - A_0)^{-1}A_{-1}. \end{aligned}$$

It is evident that in the Schur complement the block tridiagonal structure is kept. Also the block Toeplitz structure is kept except for the blocks $\widetilde{W}^{(1)}$ and $I - \widehat{A}_0^{(1)}$ in the north-west and south-east corners.

Another step of cyclic reduction applied to the 3×3 block matrix (8.25) generates the 2×2 block matrix

$$\begin{bmatrix} \widetilde{W}^{(2)} & -A_{-1}^{(2)} \\ -A_1^{(2)} & I - \widehat{A}_0^{(2)} \end{bmatrix}$$

where $\widetilde{W}^{(2)} = W - A_1^{(1)}(I - A_0^{(1)})^{-1}A_{-1}^{(1)}$. The overall permutation that we obtain in this way is the composition of the even–odd permutations of blocks:

$(2,4,1,3,5)$, $(2,4,3,1,5)$ which leave the last block unchanged. Therefore, the last step of Schur complement $I-\widehat{A}_0^{(2)}-A_{-1}^{(2)}(\widetilde{W}^{(2)})^{-1}A_1^{(2)}$, provides $S_4(W)$, that is,

$$S_4(W) = \widehat{A}_0^{(2)} - A_1^{(2)}(\widetilde{W}^{(2)})^{-1}A_{-1}^{(2)}.$$

In general, for any $n = 2^k + 1$, we may recursively apply this transformation provided that the matrices $I - A_0^{(i)}$ are nonsingular for $i = 1, 2, \ldots$, until we arrive at a 2×2 block

$$\begin{bmatrix} \widetilde{W}^{(k)} & -A_{-1}^{(k)} \\ -A_1^{(k)} & I - \widehat{A}_0^{(k)} \end{bmatrix},$$

and then compute the Schur complement

$$S_{2^k}(W) = I - \widehat{A}_0^{(k)} - A_1^{(k)}(\widetilde{W}^{(k)})^{-1}A_{-1}^{(k)}$$

provided that $\widetilde{W}^{(k)}$ is nonsingular. In fact, the composition of the even–odd permutations leave the last block unchanged. In this way we obtain the following sequence of matrices where we have introduced $\widetilde{A}_0^{(k)} = W - \widetilde{W}^{(k)}$:

$$\begin{aligned}
A_0^{(n+1)} &= A_0^{(n)} + A_{-1}^{(n)}(I - A_0^{(n)})^{-1}A_1^{(n)} + A_1^{(n)}(I - A_0^{(n)})^{-1}A_{-1}^{(n)}, \\
A_{-1}^{(n+1)} &= A_{-1}^{(n)}(I - A_0^{(n)})^{-1}A_{-1}^{(n)}, \\
A_1^{(n+1)} &= A_1^{(n)}(I - A_0^{(n)})^{-1}A_1^{(n)}, \\
\widehat{A}_0^{(n+1)} &= \widehat{A}_0^{(n)} + A_{-1}^{(n)}(I - A_0^{(n)})^{-1}A_1^{(n)}, \\
\widetilde{A}_0^{(n+1)} &= \widetilde{A}_0^{(n)} + A_1^{(n)}(I - A_0^{(n)})^{-1}A_{-1}^{(n)},
\end{aligned} \tag{8.26}$$

with the initial conditions $A_0^{(0)} = A_0$, $A_{-1}^{(0)} = A_{-1}$, $A_1^{(0)} = A_1$, $\widehat{A}_0^{(0)} = A_0$, $\widetilde{A}_0^{(0)} = 0$, and so

$$S_{2^k}(W) = \widehat{A}_0^{(k)} + A_{-1}^{(k)}(W - \widetilde{A}_0^{(k)})^{-1}A_1^{(k)}. \tag{8.27}$$

The above formula provides a means to compute the sequence $\{W^{(i\cdot 2^k)}\}_{i=0,1,\ldots}$ by performing only a finite number of matrix operations per step, once the precomputation of the blocks $A_i^{(k)}$, $i = -1, 0, 1$, $\widetilde{A}_0^{(i)}$ and $\widehat{A}_0^{(i)}$ has been completed. Moreover the sequence $\{Z^{(n)}\}_n$, $Z^{(n)} = W^{(n\cdot 2^k)}$ converges linearly to the fixed point V with the convergence rate σ^{2^k} for $\sigma > |\xi_m/\xi_{m+1}|$.

We may synthesize the results of this section with Algorithm 8.3 for the solution of the matrix equation (8.1).

The local convergence of the sequence generated with this algorithm has a linear speed such that $\|W^{(i\cdot 2^k)} - V\| = O(\sigma^{i\cdot 2^k})$ for $|\xi_m/\xi_{m+1}| < \sigma < 1$, moreover the algorithm is self-correcting.

Algorithm 8.3 Functional iteration with cyclic reduction

INPUT: The $m \times m$ block elements A_{-1}, A_0, A_1, such that $\det(-A_{-1} + z(I - A_0) - z^2 A_1)$ has zeros ξ_i such that (8.23) holds; an error bound $\epsilon > 0$, a positive integer k; either an initial approximation $W^{(0)}$ of the fixed point V of $S(W)$, or the matrix $W^{(0)} = I$.

OUTPUT: An approximation Y to the matrix $G_{\min}$.

COMPUTATION:

1. (initialization) Compute $\widehat{A}^{(k)}$, $\widetilde{A}^{(k)}$, $A_{-1}^{(k)}$ and $A_{-1}^{(k)}$ by applying (8.26).
2. Compute

$$V^{(n+1)} = \widehat{A}_0^{(k)} + A_{-1}^{(k)}(V^{(n)} - \widetilde{A}_0^{(k)})^{-1} A_1^{(k)}, \quad n = 0, 1, \ldots,$$

until the residual $\|V^{(n+1)} - V^{(n)}\|_\infty \le \epsilon$.

3. Set $Y = (V^{(n+1)})^{-1} A_1$.

8.4 Doubling methods

The even–odd permutation at the basis of logarithmic reduction and of cyclic reduction, relies on a divide and conquer strategy applied to the coefficients of the power series $A(z)$ or, equivalently, to the block diagonals of the block Toeplitz matrix (7.30). A different idea relies on applying the divide and conquer strategy to the size of the matrices obtained by truncating (7.30) to a finite size. More precisely, consider (7.30) which we rewrite below for the sake of clarity

$$\begin{bmatrix} I - A_0 & -A_1 & -A_2 & -A_3 & \ldots \\ -A_{-1} & I - A_0 & -A_1 & -A_2 & \ddots \\ & -A_{-1} & I - A_0 & -A_1 & \ddots \\ & & -A_{-1} & I - A_0 & \ddots \\ 0 & & & \ddots & \ddots \end{bmatrix} \begin{bmatrix} G_{\min} \\ G_{\min}^2 \\ G_{\min}^3 \\ \vdots \end{bmatrix} = \begin{bmatrix} A_{-1} \\ 0 \\ \vdots \\ \vdots \end{bmatrix}, \tag{8.28}$$

and consider the linear system obtained by truncating (8.28) to finite block size n, that is

$$\begin{bmatrix} I - A_0 & -A_1 & -A_2 & \ldots & -A_{n-1} \\ -A_{-1} & I - A_0 & -A_1 & \ddots & \vdots \\ & -A_{-1} & I - A_0 & \ddots & -A_2 \\ & & \ddots & \ddots & -A_1 \\ 0 & & & -A_{-1} & I - A_0 \end{bmatrix} \begin{bmatrix} X_1^{(n)} \\ X_2^{(n)} \\ \vdots \\ X_n^{(n)} \end{bmatrix} = \begin{bmatrix} A_{-1} \\ 0 \\ \vdots \\ 0 \end{bmatrix}. \tag{8.29}$$

Here, the idea is to approximate $G_{\min}$ with $X_1^{(n)}$ for n sufficiently large with the hope that the components of the solution of the finite system (8.29) converge to the corresponding components of the solution of the infinite system (8.28). In this context we assume that the drift μ is negative, that $A(z)$ is analytic for $|z| < r$, where $r > 1$, and that $\det \varphi(z)$, with $\varphi(z) = zI - A(z)$, has at least a zero of modulus greater than 1. The latter property is satisfied under the assumptions of Theorem 4.12. In this way, the zeros $z = \xi_i$, $i = 1, 2, \ldots$, of $\det \varphi(z)$ can be ordered so that $|\xi_1| \leq \cdots \leq \xi_m = 1 < |\xi_{m+1}| \leq \cdots$. As in Chapter 7 we denote

$$\eta = \max\{|z| : \quad |z| < 1, \det \varphi(z) = 0\},$$
$$\xi = \min\{|z| : \quad |z| > 1, \det \varphi(z) = 0\}.$$

The convergence property of the doubling technique can be proved relying on the weak canonical factorization (see Theorem 4.13)

$$\begin{aligned} &I - z^{-1}A(z) = U(z)(I - z^{-1}G_{\min}), \\ &U(z) = \sum_{i=0}^{+\infty} z^i U_i, \quad \det U(z) \neq 0 \text{ for } |z| \leq 1, \\ &\rho(G_{\min}) = 1, \end{aligned} \tag{8.30}$$

of the function $I - z^{-1}A(z)$ and on the nonnegativity of the coefficients A_i, $i \geq -1$.

Theorem 8.9 *Let $A(z)$ be the generating function associated with an M/G/1-type Markov chain. Assume that $A(z)$ is analytic for $|z| < r$, with $1 < r < +\infty$, and that there exists ζ, such that $1 < |\zeta| < r$ and $\det(\zeta I - A(\zeta)) = 0$. If the drift μ is negative and if Condition 4.20 is satisfied, then for the matrix $X_1^{(n)}$ defined in (8.29) one has $0 \leq X_1^{(n)} \leq X_1^{(n+1)} \leq G_{\min}$ for $n \geq 1$, $X_i^{(n)} \leq G_{\min}^i$ for $i = 1, \ldots, n$. Moreover $G_{\min} - X_1^{(n)} = U_n^{(-1)} A_{-1} X_n^{(n)}$, where $U_n^{(-1)}$ is the coefficient of the term of degree n in the matrix power series $U(z)^{-1}$ of (8.30), and for any $\epsilon > 0$ there exist a positive constant γ such that*

$$\|G_{\min} - X_1^{(n)}\|_\infty \leq \gamma(\xi - \epsilon)^{-n},$$

for any $n > 0$.

Proof Denote by $H_\infty = T_\infty[I - z^{-1}A(z)]$ the block Hessenberg block Toeplitz matrix (8.29) and $H_n = T_n[I - z^{-1}A(z)]$ its section to block size n and observe that $H_n = I - T_n[z^{-1}A(z)]$. By Theorem 1.14 and Corollary 1.15 the series $\sum_{i=0}^{+\infty} T_\infty[z^{-1}A(z)]^i$ is convergent, and is the minimal nonnegative inverse of H_∞. Since $0 \leq K_\infty \leq T_\infty[z^{-1}A(z)]$ where

$$K_\infty = \begin{bmatrix} T_n[z^{-1}A(z)] & 0 \\ 0 & 0_\infty \end{bmatrix}$$

and 0_∞ is the semi-infinite matrix with all the elements equal to zero, then $K_\infty^i \leq T_\infty[z^{-1}A(z)]^i$ for any $i \geq 0$. Therefore $\sum_{i=0}^{+\infty} K_\infty^i \leq \sum_{i=0}^{+\infty} T_\infty[z^{-1}A(z)]^i < +\infty$, so that the series $\sum_{i=0}^{+\infty} T_n[z^{-1}A(z)]^i$ is convergent, is a nonnegative matrix and

$H_n^{-1} = \sum_{i=0}^{+\infty} T_n[z^{-1}A(z)]^i$. This implies that $X_i^{(n)} \geq 0$, for $i = 1, \ldots, n$. Moreover observe that the $(n+1) \times (n+1)$ block matrix

$$K_{n+1} = \begin{bmatrix} T_n[z^{-1}A(z)] & 0 \\ 0 & 0_m \end{bmatrix},$$

where 0_m is the $m \times m$ matrix with all the elements equal to zero, is such that $0 \leq K_{n+1} \leq T_{n+1}[z^{-1}A(z)]$, where the inequality holds element-wise. Therefore, $0 \leq \sum_{i=0}^{+\infty} K_{n+1}^i \leq \sum_{i=0}^{+\infty} T_{n+1}[z^{-1}A(z)]^i$ and $0 \leq (I - K_{n+1})^{-1} \leq H_{n+1}^{-1}$. Since

$$(I - K_{n+1})^{-1} = \begin{bmatrix} H_n^{-1} & 0 \\ 0 & I \end{bmatrix},$$

one has $(H_n^{-1})_{1,1} \leq (H_{n+1}^{-1})_{1,1}$, where $(H_i^{-1})_{1,1}$ denotes the $m \times m$ block of H_i^{-1} in position $(1,1)$, which implies that $0 \leq X_1^{(n)} \leq X_1^{(n+1)}$.

In order to prove that $G_{\min}^i - X_i^{(n)} \geq 0$ for $i = 1, \ldots, n$, observe that

$$H_\infty \begin{bmatrix} G_{\min} - X_1^{(n)} \\ \vdots \\ G_{\min}^n - X_n^{(n)} \\ G_{\min}^{n+1} \\ G_{\min}^{n+2} \\ \vdots \end{bmatrix} = \begin{bmatrix} 0 \\ \vdots \\ 0 \\ A_{-1}X_n^{(n)} \\ 0 \\ \vdots \end{bmatrix}, \tag{8.31}$$

where the nonzero block in the right-hand side is the $(n+1)$st block component. Recall that $I - z^{-1}A(z)$ admits the weak canonical factorization (8.30) where $U_0 = I - A_0^*$, $U_i = -A_i^*$ for $i \geq 1$, where A_i^* are defined in (4.8), and U_0 is an M-matrix in light of Theorem 4.15 so that we have $U_0^{-1} \geq 0$. In matrix form we have

$$H_\infty = \begin{bmatrix} U_0 & U_1 & U_2 & \ldots \\ & U_0 & U_1 & \ddots \\ 0 & & \ddots & \ddots \end{bmatrix} \begin{bmatrix} I & & & 0 \\ -G_{\min} & I & & \\ & -G_{\min} & I & \\ 0 & & \ddots & \ddots \end{bmatrix}. \tag{8.32}$$

Since U_0 is an M-matrix and U_i are nonpositive for $i \geq 1$, then any finite section of the block upper triangular Toeplitz matrix in the above expression are M-matrices, so that their inverses are nonnegative. Since the minimum modulus zero of $\det U(z)$ is ξ, then $U(z)$ is analytic and invertible for $|z| < \xi$ so that we may define the matrix power series $U(z)^{-1} = \sum_{i=0}^{+\infty} z^i U_i^{(-1)}$, such that $T_\infty[U(z)]^{-1} = T_\infty[U(z)^{-1}]$ for $|z| < \xi$. Then from (8.31) and (8.32) we find that

$$\begin{bmatrix} I & & & 0 \\ -G_{\min} & I & & \\ & -G_{\min} & I & \\ 0 & & \ddots & \ddots \end{bmatrix} \begin{bmatrix} G_{\min} - X_1^{(n)} \\ \vdots \\ G_{\min}^n - X_n^{(n)} \\ G_{\min}^{n+1} \\ G_{\min}^{n+2} \\ \vdots \end{bmatrix} = \begin{bmatrix} U_n^{(-1)} \\ \vdots \\ U_0^{(-1)} \\ 0 \\ \vdots \end{bmatrix} A_{-1} X_n^{(n)}. \tag{8.33}$$

The nonnegativity of the matrices $U_i^{(-1)}$, $i \geq 0$, of A_{-1} and of $X_n^{(n)}$ implies that $G_{\min}^i - X_i^{(n)} \geq 0$ for $i = 1, \ldots, n$. Moreover, from the first equation of (8.33) one has $G_{\min} - X_1^{(n)} = U_n^{(-1)} A_{-1} X_n^{(n)}$.

Since the function $U(z)$ is analytic and invertible for $|z| < \xi$, therefore, applying Theorem 3.6 to $U(z)^{-1}$ we find that for any ϵ there exists $\gamma > 0$ such that $\|U_i^{(-1)}\|_\infty \leq \gamma(\xi - \epsilon)^i$ for any $i > 0$. This implies that $\|G_{\min} - X_1^{(n)}\|_\infty \leq \gamma(\xi - \epsilon)^n \|A_{-1}\|_\infty \|X_n^{(n)}\|_\infty \leq \gamma(\xi - \epsilon)^n$. □

If the drift μ is positive we may prove a similar convergence result. Indeed, the weak canonical factorization (8.30) still holds, where $U(z)$ is nonsingular for $|z| < 1$, $U(z)^{-1}$ has uniformly bounded block coefficients, and $\rho(G_{\min}) = \eta < 1$. Therefore, from the relation $G_{\min} - X_1^{(n)} = U_n^{(-1)} A_{-1} X_n^{(n)}$ we deduce that for any matrix norm $\|\cdot\|$ and for any $\epsilon > 0$ such that $\eta + \epsilon < 1$ there exists a constant σ such that $\|G_{\min} - X_1^{(n)}\|_\infty \leq \sigma(\eta + \epsilon)^n$, for $n \geq 1$.

It is natural to design an algorithm which allows one to compute $X_1^{(2n)}$ given $X_1^{(n)}$. This can be done relying on the Shermann–Morrison–Woodbury formula (A.2). Denoting by $H_n = T_n[I - z^{-1}A(z)]$ the block Hessenberg block Toeplitz matrix (8.29) we may partition H_{2n} into a 2×2 block matrix obtaining

$$H_{2n} = \begin{bmatrix} H_n & T_n \\ -\boldsymbol{f}_n A_{-1} \boldsymbol{l}_n^{\mathrm{T}} & H_n \end{bmatrix},$$

where we denoted by $\boldsymbol{f}_n$ and $\boldsymbol{l}_n$ the first and the last block columns of size $mn \times m$ of an $mn \times mn$ identity matrix, and where T_n is the block Toeplitz matrix

$$T_n = - \begin{bmatrix} A_n & A_{n+1} & \ldots & A_{2n-1} \\ A_{n-1} & A_n & \ddots & \vdots \\ \vdots & \ddots & \ddots & A_{n+1} \\ A_1 & \ldots & A_{n-1} & A_n \end{bmatrix}.$$

In particular we may decompose H_{2n} as

$$H_{2n} = \begin{bmatrix} H_n & T_n \\ 0 & H_n \end{bmatrix} + \begin{bmatrix} 0 & 0 \\ -\boldsymbol{f}_n A_{-1} \boldsymbol{l}_n^{\mathrm{T}} & 0 \end{bmatrix} = S_{2n} - \boldsymbol{u}_{2n} A_{-1} \boldsymbol{v}_{2n}^{\mathrm{T}}, \tag{8.34}$$

where

$$S_{2n} = \begin{bmatrix} H_n & T_n \\ 0 & H_n \end{bmatrix}, \quad \boldsymbol{u}_{2n} = \begin{bmatrix} 0 \\ \boldsymbol{f}_n \end{bmatrix}, \quad \boldsymbol{v}_{2n} = \begin{bmatrix} \boldsymbol{l}_n \\ 0 \end{bmatrix}. \tag{8.35}$$

Assume that both H_n and H_{2n} are nonsingular, then by applying the Sherman–Morrison–Woodbury formula (A.2) to the decomposition (8.34), we immediately find that $I - \boldsymbol{v}^{\mathrm{T}} S_{2n}^{-1} \boldsymbol{u}_{2n} A_{-1}$ is nonsingular and

$$H_{2n}^{-1} = S_{2n}^{-1} + S_{2n}^{-1}\boldsymbol{u}_{2n}A_{-1}(I - \boldsymbol{v}_{2n}^{\mathrm{T}} S_{2n}^{-1}\boldsymbol{u}_{2n}A_{-1})^{-1}\boldsymbol{v}_{2n}^{\mathrm{T}} S_{2n}^{-1}, \tag{8.36}$$

where

$$S_{2n}^{-1} = \begin{bmatrix} H_n^{-1} & -H_n^{-1}T_nH_n^{-1} \\ 0 & H_n^{-1} \end{bmatrix}. \tag{8.37}$$

In this way the matrix inverse of H_{2n} is expressed by means of the matrix inverse of H_n. By using the concept of displacement rank of Section 2.4, we may find even more useful relations between H_{2n}^{-1} and H_n^{-1}. Due to the block upper Hessenberg structure of H_n it is suitable to consider the displacement operator $\Delta_2(A) = A\mathcal{Z}^{\mathrm{T}} - \mathcal{Z}^{\mathrm{T}}A$, for $\mathcal{Z} = I \otimes Z$, introduced in Section 2.4. For the sake of notational simplicity, we will denote by $\Delta(\cdot)$ the operator $\Delta_2(\cdot)$. It is a simple matter to verify that

$$\Delta(H_n) = \boldsymbol{f}_n A_{-1} \boldsymbol{f}_n^{\mathrm{T}} - \boldsymbol{l}_n A_{-1} \boldsymbol{l}_n^{\mathrm{T}}. \tag{8.38}$$

Therefore, in light of Theorem 2.15,

$$\Delta(H_n^{-1}) = -H_n^{-1}\boldsymbol{f}_n A_{-1}\boldsymbol{f}_n^{\mathrm{T}} H_n^{-1} + H_n^{-1}\boldsymbol{l}_n A_{-1}\boldsymbol{l}_n^{\mathrm{T}} H_n^{-1},$$

and

$$H_n^{-1} = \Big(I + L(\mathcal{Z}\boldsymbol{c}_n^{(1)}A_{-1})\Big)U((\boldsymbol{r}_n^{(1)})^{\mathrm{T}}) - L(\mathcal{Z}\boldsymbol{c}_n^{(2)}A_{-1})U((\boldsymbol{r}_n^{(2)})^{\mathrm{T}}), \tag{8.39}$$

where

$$\begin{aligned} \boldsymbol{c}_n^{(1)} &= H_n^{-1}\boldsymbol{f}_n, \quad \boldsymbol{c}_n^{(2)} = H_n^{-1}\boldsymbol{l}_n, \\ (\boldsymbol{r}_n^{(1)})^{\mathrm{T}} &= \boldsymbol{f}_n^{\mathrm{T}} H_n^{-1}, \quad (\boldsymbol{r}_n^{(2)})^{\mathrm{T}} = \boldsymbol{l}_n^{\mathrm{T}} H_n^{-1}. \end{aligned} \tag{8.40}$$

From (8.39) we deduce that the first and the last block columns and rows of H_n^{-1} are sufficient to fully define H_n^{-1}. By using the decomposition (8.36), we may provide recursive relations among the four vectors of (8.40), which allow the efficient computation of H_n^{-1}, when n is a power of 2:

Theorem 8.10 *The first and the last block columns $\boldsymbol{c}_{2n}^{(1)}$, $\boldsymbol{c}_{2n}^{(2)}$, and the first and the last block rows $(\boldsymbol{r}_{2n}^{(1)})^{\mathrm{T}}$, $(\boldsymbol{r}_{2n}^{(2)})^{\mathrm{T}}$ of H_{2n}^{-1} are related to the corresponding block vectors of H_n^{-1} by means of the following equations:*

$$
\begin{aligned}
\boldsymbol{c}_{2n}^{(1)} &= \begin{bmatrix} \boldsymbol{c}_n^{(1)} \\ 0 \end{bmatrix} + \begin{bmatrix} -H_n^{-1}T_n\boldsymbol{c}_n^{(1)} \\ \boldsymbol{c}_n^{(1)} \end{bmatrix} W_n(\boldsymbol{l}_n^{\mathrm{T}}\boldsymbol{c}_n^{(1)}), \\
\boldsymbol{c}_{2n}^{(2)} &= \begin{bmatrix} -H_n^{-1}T_n\boldsymbol{c}_n^{(2)} \\ \boldsymbol{c}_n^{(2)} \end{bmatrix} - \begin{bmatrix} -H_n^{-1}T_n\boldsymbol{c}_n^{(1)} \\ \boldsymbol{c}_n^{(1)} \end{bmatrix} W_n((\boldsymbol{r}_n^{(2)})^{\mathrm{T}}T_n\boldsymbol{c}_n^{(2)}), \\
(\boldsymbol{r}_{2n}^{(1)})^{\mathrm{T}} &= [(\boldsymbol{r}_n^{(1)})^{\mathrm{T}}, -(\boldsymbol{r}_n^{(1)})^{\mathrm{T}}T_nH_n^{-1}] \\
&\quad - \Big((\boldsymbol{r}_n^{(1)})^{\mathrm{T}}T_n\boldsymbol{c}_n^{(1)}\Big)W_n[(\boldsymbol{r}_n^{(2)})^{\mathrm{T}}, -(\boldsymbol{r}_n^{(2)})^{\mathrm{T}}T_nH_n^{-1}], \\
(\boldsymbol{r}_{2n}^{(2)})^{\mathrm{T}} &= [0, (\boldsymbol{r}_n^{(2)})^{\mathrm{T}}] + (\boldsymbol{l}_n^{\mathrm{T}}\boldsymbol{c}_n^{(1)})W_n[(\boldsymbol{r}_n^{(2)})^{\mathrm{T}}, -(\boldsymbol{r}_n^{(2)})^{\mathrm{T}}T_nH_n^{-1}],
\end{aligned}
\tag{8.41}
$$

where $W_n = A_{-1}\Big(I + (\boldsymbol{r}_n^{(2)})^{\mathrm{T}}T_n\boldsymbol{c}_n^{(1)}A_{-1}\Big)^{-1}$.

Proof From the definition (8.35) of $\boldsymbol{u}_n$ and $\boldsymbol{v}_n$, and from (8.37) one has that

$$
\begin{aligned}
S_{2n}^{-1}\boldsymbol{u}_{2n} &= \begin{bmatrix} -H_n^{-1}T_n\boldsymbol{c}_n^{(1)} \\ \boldsymbol{c}_n^{(1)} \end{bmatrix}, \\
\boldsymbol{v}_{2n}^{\mathrm{T}}S_{2n}^{-1} &= [(\boldsymbol{r}_n^{(2)})^{\mathrm{T}}, -(\boldsymbol{r}_n^{(2)})^{\mathrm{T}}T_nH_n^{-1}], \\
\boldsymbol{v}_{2n}^{\mathrm{T}}S_{2n}^{-1}\boldsymbol{u}_{2n} &= -(\boldsymbol{r}_n^{(2)})^{\mathrm{T}}T_n\boldsymbol{c}_n^{(1)}.
\end{aligned}
$$

By using the above relations, and by multiplying (8.36) by $\boldsymbol{f}_{2n}$, $\boldsymbol{l}_{2n}$ on the right, and by $\boldsymbol{f}_{2n}^{\mathrm{T}}$, $\boldsymbol{l}_{2n}^{\mathrm{T}}$ on left, respectively, we obtain (8.41). □

The theorem above can be used to generate the sequence of vectors $\{\boldsymbol{c}_{2^n}^{(1)}\}_n$, $\{\boldsymbol{c}_{2^n}^{(2)}\}_n$, $\{\boldsymbol{r}_{2^n}^{(1)}\}_n$, $\{\boldsymbol{r}_{2^n}^{(2)}\}_n$, which define the sequence of inverses $\{H_{2^n}^{-1}\}_n$. The computation of $\boldsymbol{c}_{2n}^{(1)}$, $\boldsymbol{c}_{2n}^{(2)}$, $(\boldsymbol{r}_{2n}^{(1)})^{\mathrm{T}}$, $(\boldsymbol{r}_{2n}^{(2)})^{\mathrm{T}}$, starting from $\boldsymbol{c}_n^{(1)}$, $\boldsymbol{c}_n^{(2)}$, $(\boldsymbol{r}_n^{(1)})^{\mathrm{T}}$, $(\boldsymbol{r}_n^{(2)})^{\mathrm{T}}$, requires the computation of block Toeplitz matrix-block vector products, which can be performed by means of the algorithms of Section 2.3. More specifically, one step of the doubling method can be summarized in Algorithm 8.4.

The most expensive parts of this algorithm are the stages where the products of $n \times n$ block Toeplitz matrices and block vectors must be computed. The cost of this part amounts to $O(m^3n + m^2n\log n)$ ops.

The doubling method for approximating the matrix $G_{\min}$ in the case $\mu < 0$ is described in Algorithm 8.5.

Observe that, since $X_1^{(n)}$ converges monotonically to $G_{\min}$ then $\|G_{\min} - X_1^{(n)}\|_\infty = \|(G_{\min} - X_1^{(n)})\boldsymbol{1}\|_\infty = \|\boldsymbol{1} - X_1^{(n)}\boldsymbol{1}\|_\infty$. Therefore, the stop condition $\|X_1^{(n)}\boldsymbol{1} - \boldsymbol{1}\|_\infty \le \epsilon$ guarantees that $\|G_{\min} - X_1^{(n)}\|_\infty \le \epsilon$.

The overall cost of Algorithm 8.5 is $O(m^3n_{\max} + m^2n_{\max}\log n_{\max})$ ops where $n_{\max} = 2^s$, and s is the maximum number of steps needed by the algorithm to arrive at completion. Therefore if the number s of doubling steps needed to arrive at a reliable approximation is large, then the complexity of the algorithm is not negligible. Decay properties of the coefficients of $I - z^{-1}A(z)$ and of its inverse might help in order to reduce the complexity of this computation even though this reduction is only minor with respect to the overall complexity. An interesting

Algorithm 8.4 Doubling method: single step

INPUT: An integer $n \geq 1$, the matrices A_i, $i = -1, \ldots, 2n$, the first and the last block columns $\boldsymbol{c}_n^{(1)}$, $\boldsymbol{c}_n^{(2)}$ and the first and the last block rows $(\boldsymbol{r}_n^{(1)})^{\mathrm{T}}$ and $(\boldsymbol{r}_n^{(2)})^{\mathrm{T}}$, respectively, of the matrix H_n^{-1}.

OUTPUT: The first and the last block columns $\boldsymbol{c}_{2n}^{(1)}$, $\boldsymbol{c}_{2n}^{(2)}$ and the first and the last block rows $(\boldsymbol{r}_{2n}^{(1)})^{\mathrm{T}}$ and $(\boldsymbol{r}_{2n}^{(2)})^{\mathrm{T}}$, respectively, of the matrix H_{2n}^{-1}.

COMPUTATION:

1. Compute the block vectors $\boldsymbol{u}^{(1)} = T_n \boldsymbol{c}_n^{(1)}$, $\boldsymbol{u}^{(2)} = T_n \boldsymbol{c}_n^{(2)}$.
2. Compute the $m \times m$ matrices $W_n = A_{-1}(I - (\boldsymbol{r}_n^{(2)})^{\mathrm{T}} \boldsymbol{u}^{(1)} A_{-1})^{-1}$, $Q_1 = W_n(\boldsymbol{l}_n^{\mathrm{T}} \boldsymbol{c}_n^{(1)})$, $Q_2 = W_n((\boldsymbol{r}_n^{(2)})^{\mathrm{T}} \boldsymbol{u}^{(2)})$, $Q_3 = (\boldsymbol{r}_n^{(1)})^{\mathrm{T}} \boldsymbol{u}^{(1)} W_n$, $Q_4 = (\boldsymbol{l}_n^{\mathrm{T}} \boldsymbol{c}_n^{(1)}) W_n$.
3. Compute the row vectors $(\boldsymbol{v}^{(1)})^{\mathrm{T}} = (\boldsymbol{r}_n^{(1)})^{\mathrm{T}} T_n$, $(\boldsymbol{v}^{(2)})^{\mathrm{T}} = (\boldsymbol{r}_n^{(2)})^{\mathrm{T}} T_n$.
4. Compute the block vectors $\boldsymbol{y}^{(1)} = H_n^{-1} \boldsymbol{u}^{(1)}$, $\boldsymbol{y}^{(2)} = H_n^{-1} \boldsymbol{u}^{(2)}$, $(\boldsymbol{t}^{(1)})^{\mathrm{T}} = (\boldsymbol{v}^{(1)})^{\mathrm{T}} H_n^{-1}$, $(\boldsymbol{t}^{(2)})^{\mathrm{T}} = (\boldsymbol{v}^{(2)})^{\mathrm{T}} H_n^{-1}$, by using the representation (8.39) and Algorithms 2.3 and 2.4 for computing the products of the block Toeplitz matrices and the block vectors involved in the formula.
5. Compute and output

$$\boldsymbol{c}_{2n}^{(1)} = \begin{bmatrix} \boldsymbol{c}_n^{(1)} \\ 0 \end{bmatrix} + \begin{bmatrix} -\boldsymbol{y}^{(1)} \\ \boldsymbol{c}_n^{(1)} \end{bmatrix} Q_1,$$

$$\boldsymbol{c}_{2n}^{(2)} = \begin{bmatrix} -\boldsymbol{y}^{(2)} \\ \boldsymbol{c}_n^{(2)} \end{bmatrix} + \begin{bmatrix} \boldsymbol{y}^{(1)} \\ -\boldsymbol{c}_n^{(1)} \end{bmatrix} Q_2,$$

$$(\boldsymbol{r}_{2n}^{(1)})^{\mathrm{T}} = [(\boldsymbol{r}_n^{(1)})^{\mathrm{T}}, -(\boldsymbol{t}^{(1)})^{\mathrm{T}}] - Q_3[(\boldsymbol{r}_n^{(2)})^{\mathrm{T}}, -(\boldsymbol{t}^{(2)})^{\mathrm{T}}],$$

$$(\boldsymbol{r}_{2n}^{(2)})^{\mathrm{T}} = [0, (\boldsymbol{r}_n^{(2)})^{\mathrm{T}}] + Q_4[(\boldsymbol{r}_n^{(2)})^{\mathrm{T}}, -(\boldsymbol{t}^{(2)})^{\mathrm{T}}].$$

feature of this algorithm is that we do not have to invert any $mn \times mn$ matrix for $n = 4, 8, \ldots$. The only inversions involved concerns H_2, which must be computed only at the initial recursive step, and the $m \times m$ matrix $I - (\boldsymbol{r}_n^{(2)})^{\mathrm{T}} \boldsymbol{u}^{(1)} A_{-1}$.

In the case of a QBD process where the matrix H_n is block tridiagonal, the equations (8.41) are much simplified. However, the algorithm still remains more expensive than logarithmic reduction or cyclic reduction.

8.5 Evaluation–interpolation techniques

Consider the general problem of approximating a finite number of block components of the vector $\boldsymbol{\pi}$ which solves the system

$$\boldsymbol{\pi}^{\mathrm{T}} P = \boldsymbol{\pi}^{\mathrm{T}}$$

where the matrix P is the infinite matrix (5.15) of Section 5.4 which we recall below

Algorithm 8.5 Doubling method: the case $\mu < 0$

INPUT: The $m \times m$ block elements A_i, $i = -1, 0, 1, \ldots$, defining the block Toeplitz block Hessenberg matrix in (7.47) with $\mu < 0$; an error bound $\epsilon > 0$.

OUTPUT: Approximation Y to the matrix $G_{\min}$ such that $\|G_{\min} - Y\|_\infty \le \epsilon$.

COMPUTATION:

1. Set $n = 2$, compute H_2^{-1} and set $\boldsymbol{c}_n^{(1)}$, $\boldsymbol{c}_n^{(2)}$, $(\boldsymbol{r}_n^{(1)})^{\mathrm{T}}$, $(\boldsymbol{r}_n^{(2)})^{\mathrm{T}}$ the first and the second block rows and block columns, respectively, of H_2^{-1}.
2. Compute $\boldsymbol{c}_{2n}^{(1)}$, $\boldsymbol{c}_{2n}^{(2)}$, $\boldsymbol{r}_{2n}^{(1)}$, $\boldsymbol{r}_{2n}^{(2)}$ by means of Algorithm 8.4.
3. Compute $Y = KA_{-1}$ where K is the first block component of $\boldsymbol{c}_{2n}^{(1)}$.
4. If $\|Y\boldsymbol{1} - \boldsymbol{1}\|_\infty > \epsilon$ then set $n = 2n$ and continue from stage 2. Otherwise output Y.

$$P = \begin{bmatrix} B_0 & B_1 & B_2 & B_3 & \ldots \\ B_{-1} & A_0 & A_1 & A_2 & \ldots \\ B_{-2} & A_{-1} & A_0 & A_1 & \ddots \\ B_{-3} & A_{-2} & A_{-1} & A_0 & \ddots \\ \vdots & \vdots & \ddots & \ddots & \ddots \end{bmatrix}. \tag{8.42}$$

This system, rewritten in the form $\boldsymbol{\pi}^{\mathrm{T}}(I - P) = 0$, can be solved by means of the weak canonical factorization of the function

$$I - S(z) = I - \sum_{i=-\infty}^{+\infty} z^i A_i$$

in light of the results of Section 3.5.

More precisely, assume that $S(z) = \sum_{i=-\infty}^{+\infty} z^i A_i$ is analytic for $r_1 < |z| < r_2$, where $r_1 < 1 < r_2$, and that there exists a weak canonical factorization

$$I - S(z) = U(z)L(z)$$

where $U(z) = \sum_{i=0}^{+\infty} z^i U_i$ is analytic for $|z| < 1$ and nonsingular for $|z| \le 1$ and $L(z) = \sum_{i=0}^{+\infty} z^{-i} L_{-i}$ is analytic for $|z| > 1$ and nonsingular for $|z| > 1$. We assume also that, if $L(z)$ is singular for some z of modulus 1, then $z = 1$ is the only zero of $\det L(z)$ and it is simple. Observe that if $\sum_{i=-\infty}^{+\infty} A_i$ is stochastic, one has $(I - S(1))\boldsymbol{1} = 0$, which implies $L(1)\boldsymbol{1} = 0$.

Since the matrix Laurent power series $S(z)$ is analytic, its coefficients A_i decay to zero so that they are negligible for $i < -N$ and $i > M$ for sufficiently large $N, M > 0$. Therefore for the sake of simplicity we assume that $S(z) = \sum_{i=-N}^{M} z^i A_i$. Here we consider the case where $L(1)$ is singular. If the latter condition is not satisfied, then the procedure is much simplified since we do not need to apply the shifting stage. We proceed according to the following steps:

1. Apply the shift technique of Section 3.6 and construct the matrix Laurent power series
$$I - \widetilde{S}(z) = (I - S(z))(I - z^{-1}Q)^{-1}$$
which admits a canonical factorization in light of the results of Section 3.6, where $Q = \mathbf{1}\boldsymbol{u}^{\mathrm{T}}$ and $\boldsymbol{u}$ is any vector such that $\boldsymbol{u}^{\mathrm{T}}\mathbf{1} = 1$.
2. Compute the canonical factorization $I - \widetilde{S}(z) = \widetilde{U}(z)\widetilde{L}(z)$ of the matrix Laurent power series $I - \widetilde{S}(z)$ by approximating the central K coefficients of the inverse of the matrix Laurent power series $I - \widetilde{S}(z)$ in light of Theorem 3.27, where $K > N$. Here, since the matrix Laurent power series $I - \widetilde{S}(z)$ is nonsingular for $|z| = 1$ we may apply Algorithm 3.2 based on the evaluation–interpolation technique at the Fourier points.
3. Compute the weak canonical factorization $I - S(z) = U(z)L(z)$ of $I - S(z)$ by means of the relations $U(z) = \widetilde{U}(z)$ and $L(z) = \widetilde{L}(z)(I - z^{-1}Q)^{-1}$, that is, compute $L_{-i} = \widetilde{L}_{-i} - \widetilde{L}_{-i+1}Q$, $i = 1, 2, \ldots, N$, $\widetilde{L}_0 = L_0$; set $U_i = \widetilde{U}_i$, $i = 0, 1, \ldots$ (compare with (3.51)).
4. Use the weak canonical factorization of $I - S(z)$ in order to solve the system (8.42) with the techniques of Section 3.5.

Now we describe in more detail the stages of the above computational scheme. The computation at stage 1 is simply performed in the following way.

1a. Choose any vector $\boldsymbol{u}$ such that $\boldsymbol{u}^{\mathrm{T}}\mathbf{1} = 1$;
1b. set $\widetilde{A}_i = A_i - (\sum_{j=-N}^{i} A_j)\mathbf{1}\boldsymbol{u}^{\mathrm{T}}$, for $i = -N, \ldots, -1$, and $\widetilde{A}_i = A_i + (I - \sum_{j=-N}^{i} A_j)\mathbf{1}\boldsymbol{u}^{\mathrm{T}}$, for $i = 0, 1, \ldots, M$;
1c. set $\widetilde{S}(z) = \sum_{i=-N}^{M} z^i \widetilde{A}_i$.

The computation at stage 2, i.e., the inversion by means of evaluation–interpolation, is slightly more involved and is described by Algorithm 8.6 which is a slight modification of Algorithm 3.2.

The evaluation part at stage 2 at each cycle of this algorithm may not be performed from scratch. In fact we may follow the strategy described in Remark 2.2 where it is shown how to update, with low cost, the discrete Fourier transform at $2h$ points of an assigned function, given the discrete Fourier transform at h points of the same function. Similarly we may do the same for the interpolation stage 4. We refer the reader to Remark 2.2 and to the entire Section 2.2 for more details in this regard and to related topics.

The cost of the algorithm amounts to $O(m^2 h \log h)$ ops for the evaluation and for the interpolation stages, and to $O(m^3 h)$ ops for the inversion of stage 3.

The computation of the canonical factorization of $I - \widetilde{S}(z)$ at stage 3 of the main scheme, once the central coefficients of $H(z) = (I - \widetilde{S}(z))^{-1}$ have been approximated, is performed in light of Theorem 3.27. More precisely, denoting by $H'(z)$ the approximation to $H(z)$ provided by Algorithm 8.6, and denoting by $[X_{-K+1}, X_{-K}, \ldots, X_0]$ the solution of the system

$$[X_{-K+1}, X_{-K}, \ldots, X_{-1}, X_0] T_K[H'(z)] = \boldsymbol{e}_K^{\mathrm{T}} \otimes I,$$

Algorithm 8.6 Matrix Laurent polynomial inversion

INPUT: Positive integers m, M, N, K, a positive real ϵ, and the $m \times m$ blocks $\widetilde{A}_{-M}, \ldots, \widetilde{A}_0, \ldots, \widetilde{A}_N$, of the matrix Laurent power series $\widetilde{S}(z) = \sum_{i=-N}^{M} z^i \widetilde{A}_i$.

OUTPUT: Approximations H'_i, $i = -K, \ldots, K$, of the central block components of $H(z) = (I - \widetilde{S}(z))^{-1} = \sum_{i=-\infty}^{+\infty} z^i H_i$.

COMPUTATION:

1. Let $h = 2^q$, $q = \max(\lceil \log_2(N + M) \rceil, \lceil \log_2 2K + 1 \rceil)$, $h_0 = h$.
2. Set $\omega_h = \cos(2\pi/h) + \mathrm{i} \sin(2\pi/h)$ and compute $\boldsymbol{v} = (V_i)_{i=1,h}$ such that $V_i = \widetilde{S}(\omega_h^{i-1})$, $i = 1, \ldots, h$.
3. Compute $\boldsymbol{w} = (W_i)$, where $W_i = (I - V_i)^{-1}$, $i = 1, \ldots, h$, which provide the values of $H(\omega_h^{i-1})$.
4. Compute $\boldsymbol{y} = (Y_i) = \mathrm{DFT}_h(\boldsymbol{w})$.
5. If $h = h_0$ then set $H_i^{(old)} = Y_{i+1}$, $i = 0, \ldots, K$, $H_i^{(old)} = Y_{h+i+1}$, for $i = -1, \ldots, -K$, $h = 2h$ and continue from stage 2.
6. Set $H'_i = Y_{i+1}$, $i = 0, \ldots, K$, $H'_i = Y_{h+i+1}$, for $i = -1, \ldots, -K$.
7. If $\max_{-K \le i \le K} \|H'_i - H_i^{(old)}\|_\infty \le \epsilon$ then output H'_i, $i = -K, \ldots, K$. Otherwise set $h = 2h$, $H_i^{(old)} = H'_i$, $i = -K, \ldots, K$ and continue from stage 2.

where $\boldsymbol{e}_K^{\mathrm{T}}$ is the last row of the $K \times K$ identity matrix, $T_K[H'(z)] = (H'_{j-i})_{i,j=1,K}$, and $H'(z) = \sum_{i=-K}^{K} z^i H'_i$, we have that the matrices $\widetilde{L}'_0 = I$, $\widetilde{U}'_0 = X_0$, $\widetilde{L}'_i = X_0^{-1} X_i$, $i = -1, \ldots, -N$, provide approximations to the blocks $\widetilde{U}_0$, $\widetilde{L}_i$, $i = 0, \ldots, -N$. This computation requires the solution of a $K \times K$ block Toeplitz system for which we may use any of the available direct or iterative algorithms (see Section 2.4). The choice of the algorithm depends on the size of the blocks and on the size of the block matrix. In certain cases the customary inversion algorithm based on Gaussian elimination is more convenient.

In order to compute the first K coefficients of $\widetilde{U}(z)$ we rewrite the equation $I - \widetilde{S}(z) = \widetilde{U}(z)\widetilde{L}(z)$ in matrix form as

$$[\widetilde{U}_0, \widetilde{U}_1, \ldots] \begin{bmatrix} I & & & & 0 \\ \widetilde{L}_{-1} & I & & & \\ \widetilde{L}_{-2} & \widetilde{L}_{-1} & I & & \\ \vdots & \ddots & \ddots & \ddots & \\ \widetilde{L}_{-N} & \ddots & \ddots & \ddots & \ddots \\ 0 & \ddots & \ddots & \ddots & \ddots \\ \vdots & \ddots & \ddots & \ddots & \ddots \end{bmatrix} = [H_0, H_1, H_2, \ldots]$$

and observe that for the decay properties of the coefficients of $\widetilde{U}(z)$, there is a positive integer q such that $\widetilde{U}_i$ is negligible for $i > q$. Therefore, assuming $U_i = 0$

for $i > q$ the above infinite system turns into a $q \times q$ (finite) block triangular block Toeplitz system which we may solve with the algorithms of Chapter 2 once we have replaced the matrices $\widetilde{L}_i$ with their computed approximations $\widetilde{L}'_i$, $i = 0, \ldots, -N$.

The cost of this latter computation amounts to $O(m^2 q \log q - m^3 q)$ operations.

Concerning the last stage of the main scheme, that is, computing the vector $\boldsymbol{\pi}$ once the weak canonical factorization of $I - S(z)$ has been computed, we describe the algorithm which relies on the results of Section 3.5. For the sake of simplicity we assume that $B_i = B_{-i} = 0$ for $i > d$. This assumption is no loss of generality since in practice the blocks B_i and B_{-i} have a decay property and are negligible for sufficiently large i.

Observe that in light of the results of Section 3.5, the system $\boldsymbol{\pi}^{\mathrm{T}}(I - P) = 0$ can be rewritten as

$$\boldsymbol{\pi}^{\mathrm{T}} \left[\begin{array}{c|ccc} I - B_0 & -B_1 & -B_2 & \ldots \\ \hline -B_{-1} & & & \\ -B_{-2} & & T_\infty[I - S(z)] & \\ \vdots & & & \end{array}\right] = 0$$

that is,

$$\begin{aligned} &\boldsymbol{\pi}_0^{\mathrm{T}} W = 0 \\ &[\boldsymbol{\pi}_1^{\mathrm{T}}, \boldsymbol{\pi}_2^{\mathrm{T}}, \ldots] T_\infty[I - S(z)] = \boldsymbol{\pi}_0^{\mathrm{T}} [B_1, B_2, \ldots] \\ &W = I - B_0 - [B_1, B_2, \ldots] T_\infty[I - S(z)]^{-1} \begin{bmatrix} B_{-1} \\ B_{-2} \\ \vdots \end{bmatrix}. \end{aligned}$$

In this way, the computation of the first K components $\boldsymbol{\pi}_i$, $i = 0, 1, \ldots, K-1$, of $\boldsymbol{\pi}$ is reduced to:

4a. solving the system

$$T_\infty[I - S(z)]\boldsymbol{y} = \begin{bmatrix} B_{-1} \\ B_{-2} \\ \vdots \end{bmatrix} \tag{8.43}$$

where $\boldsymbol{y} = (Y_i)_{i=1,2,\ldots}$;

4b. computing $W = I - B_0 - \sum_{i=1}^{d} B_i Y_i$;

4c. solving the homogeneous system $\boldsymbol{\pi}_0^{\mathrm{T}} W = 0$;

4d. solving the system

$$[\boldsymbol{\pi}_1^{\mathrm{T}}, \boldsymbol{\pi}_2^{\mathrm{T}}, \ldots] T_\infty[I - S(z)] = \boldsymbol{\pi}_0^{\mathrm{T}} [B_1, B_2, \ldots].$$

The most expensive parts in the above computation at stages 4a and 4d are much simplified in light of the weak canonical factorization

$$T_\infty[I - S(z)] = T_\infty[U(z)] T_\infty[L(z)].$$

In fact, the system (8.43), rewritten as

$$T_\infty[U(z)]T_\infty[L(z)]\boldsymbol{y} = \begin{bmatrix} B_{-1} \\ B_{-2} \\ \vdots \end{bmatrix}$$

can be solved in the following way

$$\begin{aligned} \boldsymbol{y} &= T_\infty[L(z)]^{-1}\boldsymbol{z} \\ \boldsymbol{z} &= T_\infty[U(z)]^{-1}\begin{bmatrix} B_{-1} \\ B_{-2} \\ \vdots \end{bmatrix}, \end{aligned} \tag{8.44}$$

and the system at stage 4.d, rewritten as

$$[\boldsymbol{\pi}_1^{\mathrm{T}}, \boldsymbol{\pi}_2^{\mathrm{T}}, \ldots]T_\infty[U(z)]T_\infty[L(z)] = \boldsymbol{\pi}_0^{\mathrm{T}}[B_1, B_2, \ldots],$$

can be solved by means of

$$[\boldsymbol{\pi}_1^{\mathrm{T}}, \boldsymbol{\pi}_2^{\mathrm{T}}, \ldots] = \boldsymbol{\pi}_0^{\mathrm{T}}[B_1, B_2, \ldots]T_\infty[L(z)]^{-1}T_\infty[U(z)]^{-1}. \tag{8.45}$$

More precisely, we perform the computation at stages 4a–4d according to the following steps:

4a'. compute the first d block elements $U_i^{(-1)}$, $i = 0, \ldots, d-1$, of the matrix power series $U(z)^{-1}$ by means of Algorithm 2.6, that is the block elements of the first block row of the matrix $T_d[U(z)]^{-1}$;

4b'. compute

$$\boldsymbol{z}_d = T_d[U(z)]^{-1}\begin{bmatrix} B_{-1} \\ B_{-2} \\ \vdots \\ B_{-d} \end{bmatrix};$$

by means of Algorithm 2.3;

4c'. compute $T_d[L(z)]^{-1}$ by means of Algorithm 2.5 and then the Toeplitz vector product $\boldsymbol{y}_d = T_d[L(z)]^{-1}\boldsymbol{z}_d$ which provides the first d components $Y_1, \ldots, Y_d$ of $\boldsymbol{y}$;

4d'. compute $W = I - B_0 - \sum_{i=1}^{d} B_i Y_i$.

Similarly, the solution of the system (8.45) can be computed in the following way:

4e'. compute $\boldsymbol{t}_d^{\mathrm{T}} = [B_1, \ldots, B_d]T_d[L(z)]^{-1}$;

4f'. compute $[\boldsymbol{\pi}_1^{\mathrm{T}}, \ldots, \boldsymbol{\pi}_K^{\mathrm{T}}] = \boldsymbol{\pi}_0^{\mathrm{T}}\boldsymbol{t}_d^{\mathrm{T}}T_K[U(z)]^{-1}$;

still relying on Algorithm 2.4 for the Toeplitz matrix vector product.

Observe that the components $\boldsymbol{\pi}_i$, $i = 1, \ldots, K$ are uniquely determined by $\boldsymbol{\pi}_0$ which is unique up to a scaling factor. In order to compute the vector $\boldsymbol{\pi} \geq 0$ such that $\boldsymbol{\pi}^{\mathrm{T}}\mathbf{1} = 1$ we have to choose a suitable scaling factor for $\boldsymbol{\pi}_0$. This can be done by relying on (8.45). Indeed, from (8.45) we deduce that $\sum_{i=1}^{+\infty} \boldsymbol{\pi}_i^{\mathrm{T}}\mathbf{1} = \boldsymbol{\pi}_0^{\mathrm{T}}\boldsymbol{t}_d^{\mathrm{T}}U(1)^{-1}\mathbf{1}$. Therefore $\boldsymbol{\pi}_0$ must be normalized so that $\boldsymbol{\pi}_0^{\mathrm{T}}(\mathbf{1}+\boldsymbol{t}_d^{\mathrm{T}}U(1)^{-1}\mathbf{1}) = 1$.

8.6 Invariant subspace method

The *invariant subspace* method consists in approximating the minimal nonnegative solution $G_{\min}$ of the matrix equation (8.1) by approximating the left invariant subspace of a suitable block companion matrix. In order to apply this approach, we need to assume that the matrix equation (8.1) is a polynomial matrix equation, i.e., $A_i = 0$ for $i > N$. Therefore the matrix power series $A(z)$ reduces to the matrix polynomial $A(z) = \sum_{i=-1}^{N-1} z^{i+1} A_i$.

Before describing the invariant subspace method, we recall the definition of matrix sign and the definition of left invariant subspace:

Definition 8.11 *Let M be an $m \times m$ real matrix, with no pure imaginary eigenvalues. Let $M = S(D+N)S^{-1}$ be the Jordan decomposition of M, where $D = \mathrm{Diag}(\lambda_1, \lambda_2, \ldots, \lambda_m)$ and N is nilpotent and commutes with D. Then the matrix sign of M is given by*

$$Z = \mathrm{Sign}(M) = S\ \mathrm{Diag}(\mathrm{Sign}(\lambda_1), \mathrm{Sign}(\lambda_2), \ldots, \mathrm{Sign}(\lambda_m))S^{-1},$$

where, for any complex number z with $\mathrm{real}(z) \neq 0$,

$$\mathrm{Sign}(z) = \begin{cases} 1 & \text{if } \mathrm{real}(z) > 0 \\ -1 & \text{if } \mathrm{real}(z) < 0. \end{cases}$$

Definition 8.12 *Let A be an $m \times m$ real matrix. Let $\mathcal{S}$ be a k-dimensional subspace of $\mathbb{R}^m$ such that $A\boldsymbol{x} \in \mathcal{S}$, for any $\boldsymbol{x} \in \mathcal{S}$. Let S be an $m \times k$ matrix, whose columns are a basis of $\mathcal{S}$. Denote with A_1 a $k \times k$ matrix such that $AS = SA_1$. Then the subspace $\mathcal{S}$ is called the (closed) left invariant subspace of A if the eigenvalues of A_1 are contained in the (closed) left half-plane of the complex plane, and there is no larger subspace for which this inclusion holds.*

The invariant subspace method consists in computing a left invariant subspace in the complex plane $\mathbb{C}$, i.e., it provides a splitting of the eigenvalues with respect to the imaginary axis. On the other hand, the eigenvalues of $G_{\min}$ are the roots of $zI - A(z)$ inside the closed unit disk, no matter what their position with respect to the imaginary axis. Therefore, in order to reduce the computation of $G_{\min}$ to the computation of a left invariant subspace, we have to apply the Cayley transform to the function $zI - A(z)$.

The Cayley transform $z(t)$ and its inverse $w(s)$ are defined by the functions

$$z(t) = (1+t)/(1-t), \quad w(s) = (s-1)/(s+1) \tag{8.46}$$

of a complex variable, where $t \neq 1$ and $s \neq -1$, respectively. The functions $z(t)$ and $w(s)$ respectively map the unit circle without the point 1 and without the point -1 into the imaginary axis, and they map the imaginary axis into the unit circle without the point -1 and the point 1, respectively. Moreover, $z(t)$ and $w(s)$ respectively map the open unit disk into $\mathbb{C}^+$ and into $\mathbb{C}^-$ and they map the complement of the closed unit disk into $\mathbb{C}^-$ and into $\mathbb{C}^+$, respectively.

By the properties of the Cayley transform the matrix polynomial $H(t) = \sum_{i=0}^{N} t^i H_i$, defined as

$$\begin{aligned} H(t) = \sum_{i=0}^{N} t^i H_i &= (1-t)^N (zI - A(z))|_{z=z(t)} \\ &= (1-t)^{N-1}(1+t)I - \sum_{i=0}^{N} (1-t)^{N-i}(1+t)^i A_{i-1}, \end{aligned}$$

is such that $\det H(t) = 0$ has exactly $m-1$ roots with imaginary part less than zero (which correspond to the roots of $zI - A(z)$ in the open unit disk) and a simple root at $t = 0$ (which corresponds to the root $z = 1$ of $zI - A(z)$). Moreover, $\det H_N \neq 0$, since $H_N = (-1)^N(-I - A(-1))$.

Since $G_{\min}$ is the unique solution of the matrix equation (8.1) with eigenvalues lying in the closed unit disk, by applying the inverse Cayley transform $w(s)$ to the matrix $G_{\min}$, we may easily verify that the matrix

$$\widehat{G} = (G_{\min} - I)(G_{\min} + I)^{-1}$$

is the unique solution of the matrix equation

$$X = H_0 + H_1 X + \cdots + H_N X^N$$

having eigenvalues with nonpositive real part.

Now, define the matrices $\widehat{H}_i = H_N^{-1} H_i$, for $i = 0, \ldots, N-1$, and consider the $N \times N$ block companion matrix

$$F = \begin{bmatrix} 0 & I & & 0 \\ \vdots & \ddots & \ddots & \\ 0 & \ldots & 0 & I \\ -\widehat{H}_0 & -\widehat{H}_1 & \ldots & -\widehat{H}_{N-1} \end{bmatrix}.$$

Let T be a $Nm \times m$ matrix whose columns are a basis of the closed left invariant subspace of the matrix F, and partition the matrix T as

$$T = \begin{bmatrix} T_1 \\ T_2 \\ \vdots \\ T_N \end{bmatrix},$$

where T_i for $i = 1, \ldots, N$, are $m \times m$ matrices. Then, in [1] it is shown that the matrix $G_{\min}$ is given by

$$G_{\min} = (T_1 + T_2)(T_1 - T_2)^{-1}. \tag{8.47}$$

Therefore, the computation of $G_{\min}$ is reduced to computing the closed left invariant subspace of F.

From the computational point of view, the computation of the closed invariant subspace of F is not trivial, since F has an eigenvalue equal to 0, and the sign of zero is not defined. This difficulty is overcome in [1] by observing that the closed left invariant subspace of F coincides with the left invariant subspace of the matrix

$$\widehat{F} = F - \frac{\boldsymbol{y}\boldsymbol{x}^{\mathrm{T}}}{\boldsymbol{x}^{\mathrm{T}}\boldsymbol{y}},$$

where

$$\boldsymbol{x}^{\mathrm{T}} = \begin{bmatrix} \boldsymbol{x}_0^{\mathrm{T}}\widehat{H}_1 & \boldsymbol{x}_0^{\mathrm{T}}\widehat{H}_2 & \cdots & \boldsymbol{x}_0^{\mathrm{T}}\widehat{H}_{N-1} & \boldsymbol{x}_0^{\mathrm{T}} \end{bmatrix}, \quad \boldsymbol{y} = \begin{bmatrix} \boldsymbol{y}_0 \\ 0 \\ \vdots \\ 0 \end{bmatrix},$$

and $\boldsymbol{x}_0$, $\boldsymbol{y}_0$ are two vectors such that $\boldsymbol{x}_0^{\mathrm{T}}\widehat{H}_0 = 0$, $\widehat{H}_0\boldsymbol{y}_0 = 0$. Now, a basis of the left invariant subspace of $\widehat{F}$, forming the columns of the matrix T, is given by m linearly independent columns of $I - S$, where S is the matrix sign of $\widehat{F}$.

In order to compute the matrix sign S we may use the Newton iteration (matrix sign function iteration)

$$\begin{cases} S_0 = \widehat{F} \\ S_{n+1} = \frac{1}{2}\left(S_n + S_n^{-1}\right), \quad n \geq 0, \end{cases} \tag{8.48}$$

with the stopping criterion

$$||S_{n+1} - S_n||_1 < \epsilon ||S_n||_1,$$

for a fixed small error bound ϵ. The above sequence quadratically converges to S. Once S is known, the matrix T is computed by means of the rank revealing QR decomposition of the matrix $I - S$. The first two block elements T_1 and T_2 of T allow us to recover $G_{\min}$ by means of (8.47).

The computational cost of this approach is dominated by the inversion of the $N \times N$ block matrix S_n at each step n of the matrix sign iteration. The matrix S_n belongs to the algebra generated by the block companion matrix F. This is a matrix with small displacement rank, and its inversion costs $O(m^3 N \log^2 N)$ ops. Finally we have to compute a QR factorization of an $N \times N$ block matrix; this computation requires $O(m^3 N^3)$ arithmetic operations, and this is the most expensive part of the overall computation for large N.

By denoting with ξ_i, $i = 1, \ldots, m-1$, the eigenvalues of $G_{\min}$ in the open unit disk, and by $\eta_i = w(\xi_i)$, $i = 1, \ldots, m-1$, the convergence speed of S_n to S depends on how fast the values $J^{(n)}(\eta_i)$ converge to ± 1, as n tends to infinity, where $J^{(n)}(t)$ denotes the composition n times with itself of Joukowski's function $J(t) = \frac{1}{2}(t + t^{-1})$. Defining $w_{n+1} = J(w_n)$, one has that $w_{n+1} \pm 1 = \frac{(w_n \pm 1)^2}{2w_n}$, so that one may expect slow convergence if $w_0 = w(\xi_i)$ is very close to zero or if w_0 is very large in modulus or if w_0 is very close to the imaginary axis. These three conditions correspond to eigenvalues of $G_{\min}$ being close to 1, to -1, or to the unit circle.

8.7 Bibliographic notes

The shifting technique at the basis of convergence acceleration was introduced in [59] for QBD processes. It has been generalized to M/G/1-type Markov chains in [15] and used to prove the quadratic convergence of logarithmic reduction for null recurrent QBDs in [57]. The results concerning the convergence of shifted cyclic reduction for M/G/1-type Markov chains and the applicability of shifted cyclic reduction for QBD processes were introduced in [26].

The combination of functional iterations and cyclic reduction has been proposed and analyzed in [10].

The divide-and-conquer method has been introduced by W. J. Stewart [106] in order to devise an efficient doubling method for solving general block Hessenberg systems. It has been applied by Latouche and Stewart [80] for computing $G_{\min}$, and improved by Bini and Meini in [20] by exploiting the Toeplitz structure of the block Hessenberg matrices. The convergence results presented in Section 8.4 are new. A survey on displacement structure, cyclic reduction and divide-and-conquer method can be found in [23].

The shifting technique has been combined with the evaluation–interpolation strategy in [24] where the application to Markov chains with limited displacement is analyzed in detail.

The invariant subspace method has been applied to the solution of M/G/1-type Markov chains by N. Akar and K. Sohraby in [1]. Comparisons among the invariant subspace method, logarithmic reduction, and cyclic reduction, in terms of accuracy and speed of convergence, are performed in [89] and [16].

Other methods for computing $G_{\min}$, which are not considered in this book, are the so-called "spectral methods", which consist of computing the eigenvalues (with their multiplicities) and the right (generalized) eigenvectors of $G_{\min}$ by approximating the zeros of $\det(zI-A(z))$ in the unit disk, and the corresponding null vectors (or Jordan chains), and then in recovering $G_{\min}$ by means of its spectral decomposition. A drawback of this approach is that, besides being no trivial to compute the zeros of $\det(zI - A(z))$ in the unit disk, the matrix of the Jordan chains of $G_{\min}$ must be computed together with its inverse. This task can be numerically ill-conditioned especially when multiple zeros corresponding to large Jordan blocks are encountered. We refer to [41; 44] and to the references cited therein for more details on these methods.

9

SPECIALIZED STRUCTURES

9.1 Introduction

In this chapter we specialize the numerical methods introduced in the previous chapters to Markov chains with specific structures.

In Section 9.2 we consider Markov chains with limited displacement (non-skip-free Markov chains, or simply NSF Markov chains), and we exploit the structure of the blocks defining the probability transition matrix, to reduce the computational cost of linearly convergent functional iterations and of the cyclic reduction method.

In Section 9.3 we consider the reduction of an M/G/1-type Markov chain to a QBD performed along the same lines as Section 5.7. We perform a detailed analysis of this particular structure and then apply the cyclic reduction method, combined with the concept of displacement rank, to solve the associated quadratic matrix equation.

Tree-like stochastic processes, introduced in Section 5.8 are the subject analyzed in the last section. For these special processes we reduce the computation of the steady state vector $\boldsymbol{\pi}$ to the solution of a nonlinear matrix equation which is expressed in rational form. For this equation we present and analyze different solution methods based on classical functional iterations, cyclic reduction and Newton's iteration.

9.2 Markov chains with limited displacement

For Markov chains with limited displacement where the transition matrix has the generalized Hessenberg structure (5.29), we are interested in the solution of the matrix equation

$$\mathcal{X} = \sum_{i=-1}^{+\infty} \mathcal{A}_i \mathcal{X}^{i+1} \tag{9.1}$$

where the coefficients $\mathcal{A}_i$, defined in (5.31), are obtained by the reblocking (5.30) of the transition matrix.

The peculiarity of the coefficients $\mathcal{A}_i$ allows one to design specific algorithms which have a computational cost per step lower than the cost of the general methods of Section 6.2. A first important feature concerns the minimal solution $\mathcal{G}_{\min}$ which has the form

$$\mathcal{G}_{\min} = C(\boldsymbol{h}^{\mathrm{T}})^N,$$

where $C(\boldsymbol{h}^{\mathrm{T}})$ is the block companion matrix associated with the block row vector $\boldsymbol{h}^{\mathrm{T}}$ defined on page 71 (see Theorem 5.15).

In fact, $\mathcal{G}_{\min}$ is uniquely defined in terms of its first block row which coincides with $\boldsymbol{h}^{\mathrm{T}}$ so that the computation of $\boldsymbol{h}^{\mathrm{T}}$ is sufficient to determine all its elements.

9.2.1 Functional iterations

Since $\mathcal{G}_{\min}$ is defined by its first block row $\boldsymbol{h}^{\mathrm{T}}$, we may modify the functional iterations of Section 6.2 by looking at the problem in terms of a fixed point of a vector function instead of a fixed point of a matrix function. This allows one to generate a sequence of vectors converging to $\boldsymbol{h}^{\mathrm{T}}$, this is computationally simpler than generating a matrix sequence.

Let us introduce the function

$$\delta(\boldsymbol{x}^{\mathrm{T}}) = (\boldsymbol{e}_1^{\mathrm{T}} \otimes I_m)\mathcal{A}(C(\boldsymbol{x}^{\mathrm{T}})^N),$$

for $\boldsymbol{x}^{\mathrm{T}} \in \mathbb{R}^{m \times mN}$, where $\mathcal{A}(\mathcal{X})$ is the matrix valued function

$$\mathcal{A}(\mathcal{X}) = \sum_{i=-1}^{+\infty} \mathcal{A}_i \mathcal{X}^{i+1}.$$

Observe that $\delta(\boldsymbol{x}^{\mathrm{T}})$ is the first block row of the matrix $\mathcal{A}(C(\boldsymbol{x}^{\mathrm{T}})^N)$. We may modify the natural iteration

$$\mathcal{X}_{n+1} = \mathcal{A}(\mathcal{X}_n), \quad n \geq 0, \tag{9.2}$$

turning it into a more computationally simple one. For this purpose, let $\mathcal{Y}_n$ be a matrix of the form $C(\boldsymbol{x}_n^{\mathrm{T}})^N$ for a suitable block vector $\boldsymbol{x}_n^{\mathrm{T}}$ and consider the matrix

$$W = \mathcal{A}(\mathcal{Y}_n).$$

We wish to approximate W with a matrix of the form $C(\boldsymbol{x}_{n+1}^{\mathrm{T}})^N$. For this purpose, we define

$$\boldsymbol{x}_{n+1}^{\mathrm{T}} = (\boldsymbol{e}_1^{\mathrm{T}} \otimes I_m)W = \delta(\boldsymbol{x}_n^{\mathrm{T}})$$

and we take $\mathcal{Y}_{n+1} = C(\boldsymbol{x}_{n+1}^{\mathrm{T}})^N$ as an approximation of W. In this way we generate the sequence of vectors

$$\boldsymbol{x}_{n+1}^{\mathrm{T}} = \delta(\boldsymbol{x}_n^{\mathrm{T}}), \quad n \geq 0, \tag{9.3}$$

which are a sort of "projection" of the sequence $\mathcal{X}_{n+1} = \mathcal{A}(\mathcal{X}_n)$ on the set of the N-th powers of all block companion matrices.

We have the following result which has been proved in [42]:

Theorem 9.1 *For $\boldsymbol{x}_0 = 0$, the sequence $\{\boldsymbol{x}_n^{\mathrm{T}}\}_{n\in\mathbb{N}}$ generated by (9.3) converges monotonically to the block vector $\boldsymbol{h}^{\mathrm{T}}$ such that $\mathcal{G}_{\min} = C(\boldsymbol{h}^{\mathrm{T}})^N$. Moreover, let $\{\mathcal{X}_n\}_{n\geq 0}$ be the sequence generated by the natural iteration (9.2) with $\mathcal{X}_0 = 0$. Then*

$$0 \leq \mathcal{X}_n \leq C(\boldsymbol{x}_n^{\mathrm{T}})^N \leq \mathcal{G}_{\min}, \quad n \geq 0.$$

According to the previous theorem the sequence $\{C(\boldsymbol{x}_n^{\mathrm{T}})^N\}_{n\geq 0}$ provides a better approximation to $\mathcal{G}_{\min}$ with respect to the sequence $\{\mathcal{X}_n\}_{n\geq 0}$.

In order to design an efficient algorithm to compute $\boldsymbol{x}_{n+1}^{\mathrm{T}}$ given $\boldsymbol{x}_n^{\mathrm{T}}$ we have to state some properties concerning block companion matrices. The following lemma can be verified by direct inspection.

Lemma 9.2 *Let $\boldsymbol{r}^{\mathrm{T}}$ an N-dimensional block row vector. One has*

$$C(\boldsymbol{r}^{\mathrm{T}})^j = \begin{bmatrix} \boldsymbol{e}_{j+1}^{\mathrm{T}} \otimes I_m \\ \vdots \\ \boldsymbol{e}_N^{\mathrm{T}} \otimes I_m \\ \boldsymbol{r}^{\mathrm{T}} \\ \vdots \\ \boldsymbol{r}^{\mathrm{T}} C(\boldsymbol{r}^{\mathrm{T}})^{j-1} \end{bmatrix}, \quad j = 1, \ldots, N-1$$

and

$$C(\boldsymbol{r}^{\mathrm{T}})^j = \begin{bmatrix} \boldsymbol{r}^{\mathrm{T}} C(\boldsymbol{r}^{\mathrm{T}})^{j-N} \\ \vdots \\ \boldsymbol{r}^{\mathrm{T}} C(\boldsymbol{r}^{\mathrm{T}})^{j-1} \end{bmatrix}, \quad j = N, N+1, \ldots$$

From Lemma 9.2 we directly obtain the following.

Theorem 9.3 *One has*

$$\delta(\boldsymbol{r}^{\mathrm{T}}) = \sum_{i=0}^{N-1} A_{i-N}(\boldsymbol{e}_{i+1}^{\mathrm{T}} \otimes I_m) + \sum_{i=N}^{+\infty} A_{i-N} \boldsymbol{r}^{\mathrm{T}} C(\boldsymbol{r}^{\mathrm{T}})^{i-N}.$$

Proof By definition $\delta(\boldsymbol{r}^{\mathrm{T}}) = (\boldsymbol{e}_1^{\mathrm{T}} \otimes I_m)\mathcal{A}(C(\boldsymbol{r}^{\mathrm{T}})^N)$. Replacing $C(\boldsymbol{r}^{\mathrm{T}})^N$ with the expression given in Lemma 9.2 yields

$$\delta(\boldsymbol{r}^{\mathrm{T}}) = [A_{-N}, A_{-N+1}, \ldots] \begin{bmatrix} \boldsymbol{e}_1^{\mathrm{T}} \otimes I_m \\ \vdots \\ \boldsymbol{e}_N^{\mathrm{T}} \otimes I_m \\ \boldsymbol{r}^{\mathrm{T}} \\ \boldsymbol{r}^{\mathrm{T}} C(\boldsymbol{r}^{\mathrm{T}}) \\ \boldsymbol{r}^{\mathrm{T}} C(\boldsymbol{r}^{\mathrm{T}})^2 \\ \vdots \end{bmatrix}$$

□

According to the above result, the computation of $\delta(\boldsymbol{r}^{\mathrm{T}})$ can be carried out once the vectors $\boldsymbol{r}_j^{\mathrm{T}} = \boldsymbol{r}^{\mathrm{T}} C(\boldsymbol{r}^{\mathrm{T}})^j$, $j = 0, 1, 2, \ldots$, are computed. The latter computation can be performed by means of the following relations

$$\boldsymbol{r}_{j+1}^{\mathrm{T}} = \boldsymbol{r}_j^{\mathrm{T}} C(\boldsymbol{r}^{\mathrm{T}}), \quad j = 0, 1, \ldots, \tag{9.4}$$

starting from $\boldsymbol{r}_0^{\mathrm{T}} = \boldsymbol{r}^{\mathrm{T}}$, which requires N multiplications and $N-1$ additions of $m \times m$ matrices.

Algorithm 9.1 Functional iteration for NSF problems

INPUT: Positive integers k, N and the $m \times m$ matrix coefficients A_i, $i = -N, \ldots, k$, of the matrix polynomial $A(z) = \sum_{i=-N}^{k} z^{i+N} A_i$, defining a non-skip-free M/G/1-type Markov chain; a positive ϵ.

OUTPUT: An approximation $\boldsymbol{y}^{\mathrm{T}}$ of the first block row of the minimal nonnegative solution $\mathcal{G}_{\min}$ of the equation $\mathcal{X} = \mathcal{A}(\mathcal{X})$ such that $\|\boldsymbol{y}^{\mathrm{T}} - \delta(\boldsymbol{y}^{\mathrm{T}})\|_\infty \leq \epsilon$.

COMPUTATION: Set $n = 0$, $\boldsymbol{x} = 0$.

1. Compute $\boldsymbol{x}^{\mathrm{T}} C(\boldsymbol{x}^{\mathrm{T}})^j$, for $j = 1, \ldots, k$, by means of (9.4).
2. Compute

$$\boldsymbol{y}^{\mathrm{T}} = \delta(\boldsymbol{x}^{\mathrm{T}}) = \sum_{i=0}^{N-1} A_{i-N}(\boldsymbol{e}_{i+1}^{\mathrm{T}} \otimes I_m) + \sum_{i=N}^{k+N} A_{i-N} \boldsymbol{x}^{\mathrm{T}} C(\boldsymbol{x}^{\mathrm{T}})^{i-N}.$$

3. If $\|\boldsymbol{y}^{\mathrm{T}} - \boldsymbol{x}^{\mathrm{T}}\|_\infty \leq \epsilon$ then output $\boldsymbol{y} = \boldsymbol{x}$. Otherwise set $\boldsymbol{x} = \boldsymbol{y}$ and repeat from stage 1.

The resulting method for approximating the first block row $\boldsymbol{h}^{\mathrm{T}}$ of $\mathcal{G}_{\min}$ is summarized in Algorithm 9.1.

The overall cost of computing $\boldsymbol{y}^{\mathrm{T}}$ given $\boldsymbol{x}^{\mathrm{T}}$ is about $4m^3Nk$ arithmetic operations while the cost of performing one step of the natural iteration (6.6) to the reblocked equation (9.1) is $2m^3N^2k$. This provides an acceleration by a factor of N in the computational cost per step.

Even for this algorithm, all the additions are performed with nonnegative numbers so that no cancellation is encountered in this computation. This ensures the numerical stability of the algorithm.

9.2.2 Cyclic reduction

In this section we analyze the behavior of cyclic reduction when applied to M/G/1-type Markov chains which are derived from a limited displacement process.

The main issues to be discussed here are the applicability and the convergence of cyclic reduction and the analysis of the structural properties of the matrix power series $\mathcal{A}^{(n)}(z) = \sum_{i=-1}^{+\infty} z^{i+1} \mathcal{A}_i^{(n)}$, and $\widehat{\mathcal{A}}^{(n)}(z) = \sum_{i=0}^{+\infty} z^i \widehat{\mathcal{A}}_i^{(n)}$, generated by applying (7.38) to the matrix power series $\mathcal{A}(z) = \sum_{i=-1}^{+\infty} z^{i+1} \mathcal{A}_i$ where $\mathcal{A}_i$, $i = -1, \ldots$, are $N \times N$ block matrices with $m \times m$ blocks defined in (5.31).

In fact, relying on these properties we will show that the cyclic reduction step can be performed with a computational cost and with a memory storage which are much less than the costs and the storage needed by cyclic reduction applied to an M/G/1-type Markov chain with $mN \times mN$ blocks.

In order to prove that cyclic reduction can be carried out, we may rely on Theorem 7.8. The only hypothesis that we have to assume is that Condition 4.20 is satisfied by the reblocked matrix (5.30). Under this assumption cyclic reduction applied to (5.30) can be carried out without breakdown.

In order to analyze the convergence properties of cyclic reduction applied to the matrix power series $\mathcal{A}(z)$, we rely on Theorems 5.17 and 7.13:

Theorem 9.4 *Assume that $A(z) = \sum_{i=-N}^{+\infty} z^{i+N} A_i$ is analytic for $|z| < r$, $r > 1$ and the bi-infinite Markov chain associated with the transition matrix $T_{\pm\infty}[\mathcal{A}(z)]$ satisfies Condition 4.20 and that the drift of $\mathcal{A}(z)$ is negative. If there exists a zero ζ of $\det(z^N I - A(z))$ of modulus greater than 1 then there exists a root ξ of $z^N I - A(z)$ of smallest modulus among the roots of modulus greater than 1, such that ξ is real and simple, $1 < \xi < r$. Moreover, for any matrix norm $\|\cdot\|$ and for any $\epsilon > 0$ such that $\eta^N + \epsilon < 1 < \xi^N - \epsilon$, where $\eta = \max\{|z| : \ |z| < 1, \det(z^N I - A(z)) = 0\}$, there exists a positive γ such that:*

1. *for any $n \geq 0$*

$$\|\widehat{\mathcal{A}}_i^{(n)}\| \leq \gamma \xi^{N2^n} (\xi^N - \epsilon)^{-(i+1)2^n}, \quad i \geq 1$$
$$\|\mathcal{A}_i^{(n)}\| \leq \gamma \xi^{N2^n} (\xi^N - \epsilon)^{-(i+1)2^n}, \quad i \geq 1;$$

2. *there exist $\mathcal{A}_0^{(\infty)} = \lim_n \mathcal{A}_0^{(n)}$, $\widehat{\mathcal{A}}_0^{(\infty)} = \lim_n \widehat{\mathcal{A}}_0^{(n)}$, $\mathcal{A}_{-1}^{(\infty)} = \lim_n \mathcal{A}_{-1}^{(n)}$, where $\mathcal{A}_{-1}^{(\infty)} = (I - \mathcal{A}_0^{(\infty)})\mathbf{1}\boldsymbol{g}^{\mathrm{T}}$, $\widehat{\mathcal{A}}_0^{(\infty)} = \sum_{i=0}^{+\infty} A_i G_{\min}^i$, and $\boldsymbol{g} \geq 0$ is such that $\boldsymbol{g}^{\mathrm{T}} \mathcal{G}_{\min} = \boldsymbol{g}^{\mathrm{T}}$, $\boldsymbol{g}^{\mathrm{T}}\mathbf{1} = 1$; moreover,*

$$\|\widehat{\mathcal{A}}_0^{(n)} - \widehat{\mathcal{A}}_0^{(\infty)}\| \leq \gamma \xi^{N2^n} (\xi^N - \epsilon)^{-2^{n+1}};$$

3. *the sequence $\{\varphi^{(n)}(z)\}_n$, where $\varphi^{(n)}(z) = zI - \mathcal{A}^{(n)}(z)$ for $n \geq 0$, uniformly converges to $\varphi^{(\infty)}(z) = -\mathcal{A}_{-1}^{(\infty)} - z(\mathcal{A}_0^{(\infty)} - I)$ over any compact subset of the disk $\{z \in \mathbb{C} : \ |z| < \xi^N\}$; moreover $\rho(\widehat{\mathcal{A}}_0^{(\infty)}) \leq \rho(\mathcal{A}_0^{(\infty)}) < 1$;*
4. *for any $n \geq 0$*

$$\|\mathcal{A}_0^{(n)} - \mathcal{A}_0^{(\infty)}\| \leq \gamma \xi^{N2^n} (\xi^N - \epsilon)^{-2^{n+1}},$$
$$\|\mathcal{A}_{-1}^{(n)} - \mathcal{A}_{-1}^{(\infty)}\| \leq \gamma \left(\xi^{N2^n} (\xi^N - \epsilon)^{-2^{n+1}} + (\eta^N + \epsilon)^{2^n} \right);$$

5. *for any $n \geq 0$*

$$\|\mathcal{G}_{\min} - G^{(n)}\| \leq \gamma \xi^{-N2^n} (\xi^N - \epsilon)^{-2^{n+1}}$$

where $G^{(n)} = (I - \widehat{\mathcal{A}}_0^{(n)})^{-1} \mathcal{A}_{-1}$.

Proof Since there exists a zero ζ of $\det(z^N I - A(z))$ of modulus greater than 1, then there exists a zero of $\det(zI - \mathcal{A}(z))$ of modulus greater than 1 by Theorem 5.17. From Theorem 4.11 there exists the root of $zI - \mathcal{A}(z)$ of minimum modulus σ among the roots of modulus greater than 1, and $\rho(\mathcal{A}(\sigma)) = \sigma$. By Theorem 5.17, $\sigma = \xi^N$ where ξ is the zero of minimum modulus of $\det(z^N I - A(z))$ among the zeros of modulus greater than 1. Therefore we may apply Theorem 7.13 to $\mathcal{A}(z)$ and the result follows. □

9.2.3 Displacement structure of cyclic reduction

Now we analyze the structure of the matrix power series $\mathcal{A}^{(n)}(z)$ relying on the structured matrix tools of Chapter 2.

Recall that $\mathcal{A}_i$, $i \geq -1$, are $N \times N$ block Toeplitz matrices with $m \times m$ blocks and that the matrix function $\mathcal{A}(z)$ is a block z-circulant matrix, as we already observed in (3.26) (for the concept of z-circulant matrices see also Definition 2.9). This is the key property which will enable us to reduce the amount of memory space and the computational cost of performing a single step of cyclic reduction.

In order to show this, let us consider cyclic reduction applied to the matrix power series $\mathcal{A}(z)$ expressed in functional form. We recall that for $\varphi^{(0)}(z) = \varphi(z) = zI - \mathcal{A}(z)$ one has

$$\varphi^{(n)}(z) = zI - \mathcal{A}^{(n)}(z), \quad \psi^{(n)}(z) = (z^{-1}\varphi^{(n)}(z))^{-1}, \quad n = 0, 1, \ldots,$$

where

$$\psi^{(n+1)}(z^2) = \frac{1}{2}(\psi^{(n)}(z) + \psi^{(n)}(-z)), \quad n = 0, 1, \ldots, \tag{9.5}$$

which is valid for any z for which $z^{-1}\varphi^{(n)}(z)$ is invertible (see Section 7.4).

We have the following result.

Theorem 9.5 *The matrix Laurent power series $\psi^{(n)}(z)$, $n \geq 0$, are block Toeplitz matrices for any z for which they are defined.*

Proof Recall that $\varphi^{(0)}(z)$ is a block z-circulant matrix therefore its inverse $\varphi^{(0)}(z)^{-1}$ is still block z-circulant since block z-circulant matrices are closed under inversion (see Section 2.3.2). Therefore also $\psi^{(0)}(z)$ is block z-circulant, and in particular is block Toeplitz since block z-circulant matrices are particular instances of block Toeplitz matrices.

From (9.5) we deduce that $\psi^{(1)}(z^2)$ is a sum of two block Toeplitz matrices and therefore is block Toeplitz for any z for which it is defined. This implies that $\psi^{(1)}(z)$ is Toeplitz as well. By means of an induction argument we deduce that $\psi^{(n)}(z)$ is block Toeplitz since it is the sum of two block Toeplitz matrices in light of (9.5). □

This fact implies that for any n and for any z for which it is defined, the matrix $\varphi^{(n)}(z) = z^{-1}\psi^{(n)}(z)^{-1}$, being the inverse of a block Toeplitz matrix, has block displacement rank at most 2 (see Section 2.4). This property enables us to represent each matrix $\varphi^{(n)}(z)$ in terms of a few block vectors and to implement the cyclic reduction algorithm in a fast way, despite the possibly large size of the blocks $\mathcal{A}_i$, by relying on the algorithms described in Chapter 2. For this purpose we have to analyze the displacement structure of the matrix power series $\varphi^{(n)}(z)$ for $n \geq 0$.

Now we give a detailed description of the displacement representation of the matrix power series $\varphi^{(n)}(z)$ and $\widehat{\varphi}^{(n)}(z)$ generated by cyclic reduction applied to a non-skip-free Markov chain. More specifically we show that also $\widehat{\varphi}^{(n)}(z)$ has a constant block displacement rank and we provide explicit formulae relating the displacement representation of $\varphi^{(n)}(z)$ and $\widehat{\varphi}^{(n)}(z)$ to the displacement representation of $\varphi^{(n+1)}(z)$ and $\widehat{\varphi}^{(n+1)}(z)$.

These properties allow us to devise an algorithm, based on FFT, that performs a single cyclic reduction step in $O(d\mu(N,m) - (m^3 + m^2\log(dN))dN)$ arithmetic operations, where d is the numerical degree of the matrix power series $\mathcal{A}(z)$, and $\mu(N,m)$ is the cost of inverting an $N \times N$ block matrix with $m \times m$ blocks having constant block displacement rank. This bound compares favourably with the estimate $O(N^3m^3d + (Nm)^2 d\log d)$ that we would obtain without exploiting the Toeplitz structure of the blocks $\mathcal{A}_i$, that is, by treating $\mathcal{A}_i$ as general $mN \times mN$ matrices.

Let Z be the lower shift matrix of (2.13), set $\mathcal{Z} = Z \otimes I_m$, and denote with $\Delta_1(A) = A\mathcal{Z} - \mathcal{Z}A$, $\Delta_2(A) = A\mathcal{Z}^{\mathrm{T}} - \mathcal{Z}^{\mathrm{T}}A$ the associated displacement operators.

For the sake of notational simplicity here and throughout the section we use $\boldsymbol{e}_1$ and $\boldsymbol{e}_N$ instead of $\boldsymbol{e}_1 \otimes I_m$ and $\boldsymbol{e}_N \otimes I_m$, respectively. We prove the following result.

Theorem 9.6 *For the matrix power series $\varphi^{(n)}(z) = zI - \mathcal{A}^{(n)}(z)$ generated after n steps of cyclic reduction one has*

$$\Delta_1(\varphi^{(n)}(z)) = -z^{-1}\varphi^{(n)}(z)\left(\boldsymbol{e}_1\boldsymbol{e}_1^{\mathrm{T}}\psi^{(n)}(z)\mathcal{Z} - \mathcal{Z}\psi^{(n)}(z)\boldsymbol{e}_N\boldsymbol{e}_N^{\mathrm{T}}\right)\varphi^{(n)}(z),$$
$$\Delta_2(\varphi^{(n)}(z)) = -z^{-1}\varphi^{(n)}(z)\left(\boldsymbol{e}_N\boldsymbol{e}_N^{\mathrm{T}}\psi^{(n)}(z)\mathcal{Z}^{\mathrm{T}} - \mathcal{Z}^{\mathrm{T}}\psi^{(n)}(z)\boldsymbol{e}_1\boldsymbol{e}_1^{\mathrm{T}}\right)\varphi^{(n)}(z).$$

where $\psi^{(n)}(z) = (z^{-1}\varphi^{(n)}(z))^{-1}$.

Proof By Theorem 9.5, $\psi^{(n)}(z) = z\varphi^{(n)}(z)^{-1}$ is a Toeplitz matrix for any value of z for which $z^{-1}\varphi^{(n)}(z)$ is invertible. Therefore, from (2.18) one has

$$\Delta_1(\psi^{(n)}(z)) = \boldsymbol{e}_1\boldsymbol{e}_1^{\mathrm{T}}\psi^{(n)}(z)\mathcal{Z} - \mathcal{Z}\psi^{(n)}(z)\boldsymbol{e}_N\boldsymbol{e}_N^{\mathrm{T}},$$

and similarly $\Delta_2(\psi^{(n)}(z)) = \boldsymbol{e}_N\boldsymbol{e}_N^{\mathrm{T}}\psi^{(n)}(z)\mathcal{Z}^{\mathrm{T}} - \mathcal{Z}^{\mathrm{T}}\psi^{(n)}(z)\boldsymbol{e}_1\boldsymbol{e}_1^{\mathrm{T}}$. Now, from (2.19) one has $\Delta(A^{-1}) = -A^{-1}\Delta(A)A^{-1}$, for $\Delta = \Delta_1, \Delta_2$. Since $\varphi^{(n)}(z) = z\psi^{(n)}(z)^{-1}$, one finds that

$$\Delta(\varphi^{(n)}(z)) = z\Delta(\psi^{(n)}(z)^{-1}) = -z\psi^{(n)}(z)^{-1}\Delta(\psi^{(n)}(z))\psi^{(n)}(z)^{-1},$$

and we may conclude that

$$\Delta_1(\varphi^{(n)}(z)) = -z^{-1}\varphi^{(n)}(z)\left(\boldsymbol{e}_1\boldsymbol{e}_1^{\mathrm{T}}\psi^{(n)}(z)\mathcal{Z} - \mathcal{Z}\psi^{(n)}(z)\boldsymbol{e}_N\boldsymbol{e}_N^{\mathrm{T}}\right)\varphi^{(n)}(z),$$
$$\Delta_2(\varphi^{(n)}(z)) = -z^{-1}\varphi^{(n)}(z)\left(\boldsymbol{e}_N\boldsymbol{e}_N^{\mathrm{T}}\psi^{(n)}(z)\mathcal{Z}^{\mathrm{T}} - \mathcal{Z}^{\mathrm{T}}\psi^{(n)}(z)\boldsymbol{e}_1\boldsymbol{e}_1^{\mathrm{T}}\right)\varphi^{(n)}(z).$$

□

The above result turns into simple expressions relating the displacement of $\varphi^{(n)}(z)$ with the displacement of $\varphi^{(n+1)}(z)$.

Corollary 9.7 *It holds*

$$\begin{aligned}\Delta_1(\varphi^{(n)}(z)) &= -z^{-1}(\varphi^{(n)}(z)\boldsymbol{e}_1\boldsymbol{u}^{(n)}(z)^{\mathrm{T}} - \boldsymbol{v}^{(n)}(z)\boldsymbol{e}_N^{\mathrm{T}}\varphi^{(n)}(z)),\\ \Delta_2(\varphi^{(n)}(z)) &= -z^{-1}(\varphi^{(n)}(z)\boldsymbol{e}_N\boldsymbol{u}^{(n)}(z)^{\mathrm{T}} - \boldsymbol{v}^{(n)}(z)\boldsymbol{e}_1^{\mathrm{T}}\varphi^{(n)}(z)),\end{aligned} \tag{9.6}$$

where $\boldsymbol{v}^{(n)}(z)$ is a block column vector power series, and $\boldsymbol{u}^{(n)}(z)^{\mathrm{T}}$ is a block row vector power series, such that

$$\begin{aligned}\boldsymbol{u}^{(n+1)}(z^2)^{\mathrm{T}} &= -\left[\boldsymbol{u}^{(n)}(z)^{\mathrm{T}}\varphi_{\mathrm{odd}}^{(n)}(z^2)^{-1}\varphi^{(n)}(-z)\right]_{\mathrm{even}}\\ \boldsymbol{v}^{(n+1)}(z^2) &= -\left[\varphi^{(n)}(-z)\varphi_{\mathrm{odd}}^{(n)}(z^2)^{-1}\boldsymbol{v}^{(n)}(z)\right]_{\mathrm{even}}\end{aligned} \tag{9.7}$$

for $n \geq 0$, where for Δ_1 the initial conditions are $\boldsymbol{u}^{(0)}(z)^{\mathrm{T}} = \boldsymbol{e}_N^{\mathrm{T}}$, $\boldsymbol{v}^{(0)}(z) = \boldsymbol{e}_1$, while for Δ_2, the initial conditions are $\boldsymbol{u}^{(0)}(z)^{\mathrm{T}} = z^2\boldsymbol{e}_1^{\mathrm{T}}$, $\boldsymbol{v}^{(0)}(z) = z^2\boldsymbol{e}_N$.

Proof Consider the case of the operator Δ_1. The first equation of (9.6) follows from Theorem 9.6 with $\boldsymbol{u}^{(n)}(z)^{\mathrm{T}} = \boldsymbol{e}_1^{\mathrm{T}}\psi^{(n)}(z)\mathcal{Z}\varphi^{(n)}(z)$ and with $\boldsymbol{v}^{(n)}(z) = \varphi^{(n)}(z)\mathcal{Z}\psi^{(n)}(z)\boldsymbol{e}_N$. Therefore, from (7.40) and (7.42)

$$\begin{aligned}\boldsymbol{u}^{(n+1)}(z^2)^{\mathrm{T}} =& \boldsymbol{e}_1^{\mathrm{T}}\psi^{(n+1)}(z^2)\mathcal{Z}\varphi^{(n+1)}(z^2)\\ =& -\frac{1}{2}\boldsymbol{e}_1^{\mathrm{T}}(\psi^{(n)}(z)+\psi^{(n)}(-z))\mathcal{Z}\varphi^{(n)}(z)\varphi_{\mathrm{odd}}^{(n)}(z^2)^{-1}\varphi^{(n)}(-z)\\ =& -\frac{1}{2}\boldsymbol{e}_1^{\mathrm{T}}\left(\psi^{(n)}(z)\mathcal{Z}\varphi^{(n)}(z)K(z)+\psi^{(n)}(-z)\mathcal{Z}\varphi^{(n)}(-z)K(-z)\right)\\ =& -\boldsymbol{e}_1^{\mathrm{T}}\left[\boldsymbol{u}^{(n)}(z)^{\mathrm{T}}K(z)\right]_{\mathrm{even}}\end{aligned}$$

with $K(z) = (\varphi_{\mathrm{odd}}^{(n)}(z^2))^{-1}\varphi^{(n)}(-z)$, where we used the fact that $\varphi(z)K(z) = \varphi(-z)K(-z)$. With a similar argument we may prove the equation relating $\boldsymbol{v}^{(n+1)}(z)$ and $\boldsymbol{v}^{(n)}(z)$. Concerning the initial conditions we observe that $\boldsymbol{u}^{(0)}(z)^{\mathrm{T}} = \boldsymbol{e}_N^{\mathrm{T}} - (\boldsymbol{e}_1^{\mathrm{T}}\varphi(z)^{-1}\boldsymbol{e}_1)\boldsymbol{e}_N^{\mathrm{T}}\varphi(z)$ and $\boldsymbol{v}^{(0)}(z) = \boldsymbol{e}_1 - \varphi(z)\boldsymbol{e}_1(\boldsymbol{e}_1^{\mathrm{T}}\varphi(z)^{-1}\boldsymbol{e}_1)$ and

$$\Delta_1(\varphi^{(0)}(z)) = \boldsymbol{e}_1\boldsymbol{e}_1^{\mathrm{T}}\varphi(z)\mathcal{Z} - \varphi(z)\mathcal{Z}\boldsymbol{e}_N\boldsymbol{e}_N^{\mathrm{T}} = -z^{-1}(\varphi(z)\boldsymbol{e}_1\boldsymbol{e}_N^{\mathrm{T}} - \boldsymbol{e}_1\boldsymbol{e}_N^{\mathrm{T}}\varphi(z))$$

where the latter equality holds since $\varphi(z)$ is a z-circulant matrix. Therefore, the vectors $(\boldsymbol{e}_1^{\mathrm{T}}\varphi(z)^{-1}\boldsymbol{e}_1)\boldsymbol{e}_N^{\mathrm{T}}\varphi(z)$ and $\varphi(z)\boldsymbol{e}_1(\boldsymbol{e}_1^{\mathrm{T}}\varphi(z)^{-1}\boldsymbol{e}_1)$ do not appear in the expression of Δ_1. By using an inductive argument it is possible to prove that

$$\begin{aligned}\boldsymbol{u}^{(n)}(z)^{\mathrm{T}} &= \boldsymbol{p}^{(n)}(z)^{\mathrm{T}} - W^{(n)}(z)\boldsymbol{e}_N^{\mathrm{T}}\varphi^{(n)}(z),\\ \boldsymbol{v}^{(n)}(z) &= \boldsymbol{q}^{(n)}(z) - \varphi^{(n)}(z)\boldsymbol{e}_1 W^{(n)}(z),\end{aligned}$$

where $\boldsymbol{p}^{(0)}(z) = \boldsymbol{e}_N$, $\boldsymbol{q}^{(0)}(z) = \boldsymbol{e}_1$. Moreover $\boldsymbol{p}^{(n)}(z)$ and $\boldsymbol{q}^{(n)}(z)$ are related to $p^{(n+1)}(z)$ and $\boldsymbol{q}^{(n+1)}(z)$, respectively, by (9.7) where $\boldsymbol{u}^{(n)}$, $\boldsymbol{u}^{(n+1)}$, $\boldsymbol{v}^{(n)}$, $\boldsymbol{v}^{(n+1)}$, are replaced by $\boldsymbol{p}^{(n)}$, $\boldsymbol{p}^{(n+1)}$, $\boldsymbol{q}^{(n)}$, $\boldsymbol{q}^{(n+1)}$, and $\Delta_1(\varphi^{(n)}(z))$ is given by (9.6) with the same replacement. Therefore we may ignore the contribution of $W^{(n)}(z)$ and choose $\boldsymbol{u}^{(0)}(z) = \boldsymbol{e}_N$ and $\boldsymbol{v}^{(0)}(z) = \boldsymbol{e}_1$ as initial vectors. Similarly we may proceed for Δ_2. □

The results expressed in the above corollary are fundamental to provide displacement representations of the matrix $\varphi^{(n)}(z)$ in light of Theorem 2.12, where, following the notation of Section 2.3.4, we denote with $L(\boldsymbol{b})$ the block lower triangular block Toeplitz matrix whose first block column is the block vector $\boldsymbol{b}$ and with $U(\boldsymbol{r}^{\mathrm{T}})$ the block upper triangular block Toeplitz matrix whose first block row is the block row vector $\boldsymbol{r}^{\mathrm{T}}$.

Theorem 9.8 *We have*

$$\begin{aligned}\varphi^{(n)}(z) =& L(\varphi^{(n)}(z)\boldsymbol{e}_1)U(\boldsymbol{e}_1^{\mathrm{T}} - z^{-1}\boldsymbol{u}^{(n)}(z)^{\mathrm{T}}\mathcal{Z}^{\mathrm{T}})\\ &+ z^{-1}L(\boldsymbol{v}^{(n)}(z))U(\boldsymbol{e}_N^{\mathrm{T}}\varphi^{(n)}(z)\mathcal{Z}^{\mathrm{T}}),\end{aligned}$$

where the matrix power series $\boldsymbol{u}^{(n)}(z)^{\mathrm{T}}$ and $\boldsymbol{v}^{(n)}(z)$ are defined in (9.7) with the initial conditions $\boldsymbol{u}^{(0)}(z)^{\mathrm{T}} = \boldsymbol{e}_N^{\mathrm{T}}$, $\boldsymbol{v}^{(0)}(z) = \boldsymbol{e}_1$. Moreover,

$$\begin{aligned}\varphi^{(n)}(z) =& z^{-1}L(\mathcal{Z}\varphi^{(n)}(z)\boldsymbol{e}_N)U(\boldsymbol{u}^{(n)}(z)^{\mathrm{T}})\\ &+ L(\boldsymbol{e}_1 - z^{-1}\mathcal{Z}\boldsymbol{v}^{(n)}(z))U(\boldsymbol{e}_1^{\mathrm{T}}\varphi^{(n)}(z)),\end{aligned} \tag{9.8}$$

where the matrix power series $\boldsymbol{u}^{(n)}(z)^{\mathrm{T}}$ and $\boldsymbol{v}^{(n)}(z)$ are defined in (9.7) with the initial conditions $\boldsymbol{u}^{(0)}(z)^{\mathrm{T}} = z^2\boldsymbol{e}_1^{\mathrm{T}}$, $\boldsymbol{v}^{(0)}(z) = z^2\boldsymbol{e}_N$.

Proof The first representation is obtained by applying Theorem 2.14 to the first equation of (9.6). The second representation is obtained by applying Theorem 2.15 to the second equation of (9.6). □

We may provide a similar result concerning the displacement of the function $\widehat{\mathcal{A}}^{(n)}(z)$ generated by cyclic reduction according to (7.38). In the following we describe this property by using the general notation Δ for the displacement operator and by specifying it as Δ_1 or Δ_2 according to the needs.

Theorem 9.9 *For the function $\widehat{\mathcal{A}}^{(n)}(z)$ generated by cyclic reduction according to (7.38) we have*

$$\begin{aligned}\Delta(\widehat{\mathcal{A}}^{(n)}(z)) = \Big(&(I - \widehat{\mathcal{A}}^{(n)}(z))\Delta(\psi^{(n)}(z)) - (\Delta(\mathcal{A}_{-1})T^{(n)}(z)\\ &- \mathcal{A}_{-1}\Delta(T^{(n)}(z)))\Big)\,\psi^{(n)}(z)^{-1}\end{aligned} \tag{9.9}$$

where we denote $T^{(n)}(z) = \sum_{i=-\infty}^{+\infty} z^i H_{i\cdot 2^n+1}$, $\psi(z) = \sum_{i=-\infty}^{+\infty} z^i H_i$, $\psi^{(n)}(z) = \sum_{i=-\infty}^{+\infty} z^i H_{i\cdot 2^n}$. Moreover, if $\Delta = \Delta_1$ then $\widehat{\mathcal{A}}^{(n)}(z)$ has displacement rank at most 4 and if $\Delta = \Delta_2$ then $\widehat{\mathcal{A}}^{(n)}(z)$ has displacement rank at most 3.

Proof Set $V^{(n)}(z) = \widehat{\mathcal{A}}^{(n)}(z)\psi^{(n)}(z)$ and apply the displacement operator Δ to both sides of the latter equation. From the property (2.21) valid for both $\Delta = \Delta_1$ and $\Delta = \Delta_2$ one obtains $\Delta(V^{(n)}) = \Delta(\widehat{\mathcal{A}}^{(n)}(z))\psi^{(n)}(z) + \widehat{\mathcal{A}}^{(n)}(z)\Delta(\psi^{(n)}(z))$ whence

$$\Delta(\widehat{\mathcal{A}}^{(n)}(z)) = \Big(\Delta(V^{(n)}(z)) - \widehat{\mathcal{A}}^{(n)}(z)\Delta(\psi^{(n)}(z))\Big)\,\psi^{(n)}(z)^{-1}. \tag{9.10}$$

Moreover, from (7.45) one has $V^{(n+1)}(z^2) = [V^{(n)}(z)]_{\text{even}}$ that is,

$$V^{(n)}(z) = \sum_{i=-\infty}^{+\infty} z^i V_{i\cdot 2^n} \tag{9.11}$$

where $V^{(0)}(z) = \sum_{i=-\infty}^{+\infty} z^i V_i$. Now, recall that $\widehat{\mathcal{A}}^{(0)}(z) = z^{-1}(\mathcal{A}(z) - \mathcal{A}_{-1})$ and $\psi(z) = z(zI - \mathcal{A}(z))^{-1}$, so that

$$\begin{aligned} V^{(0)}(z) =& (\mathcal{A}(z) - \mathcal{A}_{-1})(zI - \mathcal{A}(z))^{-1} \\ =& (\mathcal{A}(z) - zI + zI - \mathcal{A}_{-1})(zI - \mathcal{A}(z))^{-1} \\ =& -I + (I - z^{-1}\mathcal{A}_{-1})\psi(z). \end{aligned}$$

From (9.11) one has

$$\begin{aligned} V^{(n)}(z) &= -I + \psi^{(n)}(z) - \mathcal{A}_{-1}T^{(n)}(z), \\ T^{(n)}(z) &= \sum_{i=-\infty}^{+\infty} z^i H_{i\cdot 2^n+1}. \end{aligned}$$

Whence

$$\Delta(V^{(n)}(z)) = \left(\Delta(\psi^{(n)}(z)) - \Delta(\mathcal{A}_{-1})T^{(n)}(z) - \mathcal{A}_{-1}\Delta(T^{(n)}(z))\right). \tag{9.12}$$

Therefore, from (9.10) and (9.12) we immediately deduce (9.9). The properties of the displacement rank with $\Delta = \Delta_1$ and $\Delta = \Delta_2$ follow from (9.9) since $\psi^{(n)}(z)$, $T^{(n)}(z)$ and $\mathcal{A}_{-1}$ are Toeplitz and since $\Delta_2(\mathcal{A}_{-1}) = 0$. □

A more detailed representation of $\Delta(\widehat{\mathcal{A}}^{(n)}(z))$, which is particularly useful from the algorithmic point of view, is reported in the following theorem where, in particular, it is shown that the displacement rank of $\widehat{\mathcal{A}}^{(n)}(z)$ is at most 3 for both the operators Δ_1 and Δ_2.

Theorem 9.10 *For the matrix power series $\widehat{\mathcal{A}}^{(n)}(z)$ we have*

$$\begin{aligned} \Delta_1(\widehat{\mathcal{A}}^{(n)}(z)) =& \mathcal{Z}\widehat{\boldsymbol{v}}^{(n)}(z)\boldsymbol{e}_1^{\mathrm{T}}\varphi^{(n)}(z)\mathcal{Z} - \mathcal{Z}\widehat{\mathcal{A}}^{(n)}(z)\boldsymbol{e}_N(\boldsymbol{e}_N^{\mathrm{T}} - z^{-1}\boldsymbol{u}^{(n)}(z)^{\mathrm{T}}\mathcal{Z}) \\ &+ \boldsymbol{e}_1\boldsymbol{e}_1^{\mathrm{T}}\widehat{\mathcal{A}}^{(n)}(z)\mathcal{Z}, \end{aligned}$$

and

$$\Delta_2(\widehat{\mathcal{A}}^{(n)}(z)) = -\widehat{\boldsymbol{v}}^{(n)}(z)\boldsymbol{e}_1^{\mathrm{T}}\varphi^{(n)}(z) + z^{-1}(I - \widehat{\mathcal{A}}^{(n)}(z))\boldsymbol{e}_N\boldsymbol{u}^{(n)}(z)^{\mathrm{T}} - \mathcal{A}_{-1}\boldsymbol{e}_N\boldsymbol{e}_1^{\mathrm{T}} \tag{9.13}$$

where

$$\widehat{\boldsymbol{v}}^{(n+1)}(z) = \widehat{\boldsymbol{v}}_{\text{odd}}^{(n)}(z) + z^{-1}\widehat{\mathcal{A}}_{\text{odd}}^{(n)}(z)\varphi_{\text{odd}}^{(n)}(z)^{-1}\boldsymbol{v}_{\text{even}}^{(n)}(z), \tag{9.14}$$

with $\widehat{\boldsymbol{v}}^{(0)}(z) = \boldsymbol{e}_N$, and $\boldsymbol{u}^{(n)}(z)^{\mathrm{T}}$ and $\boldsymbol{v}^{(n)}(z)$ are defined in (9.7) with the initial conditions $\boldsymbol{u}^{(0)}(z)^{\mathrm{T}} = z^2\boldsymbol{e}_1^{\mathrm{T}}$ and $\boldsymbol{v}^{(0)}(z) = z^2\boldsymbol{e}_N$.

Proof The equation for Δ_2 is proved by induction in n. For $n = 0$ it can be proved by a direct inspection. For the inductive step, relying on the first equation of (7.44) and on Theorem 2.15, we obtain that

$$\begin{aligned}\Delta_2(\widehat{\mathcal{A}}^{(n+1)}(z^2)) =& \Delta_2(\widehat{\mathcal{A}}^{(n)}(z)) - \Delta_2(\widehat{\mathcal{A}}^{(n)}_{\text{odd}}(z^2))\varphi^{(n)}_{\text{odd}}(z^2)^{-1}\varphi^{(n)}(z)\\ &+ \widehat{\mathcal{A}}^{(n)}_{\text{odd}}(z^2)\varphi^{(n)}_{\text{odd}}(z^2)^{-1}\Delta_2(\varphi^{(n)}_{\text{odd}}(z^2))\varphi^{(n)}_{\text{odd}}(z^2)^{-1}\varphi^{(n)}(z)\\ &- \widehat{\mathcal{A}}^{(n)}_{\text{odd}}(z^2)\varphi^{(n)}_{\text{odd}}(z^2)^{-1}\Delta_2(\varphi^{(n)}(z)).\end{aligned}$$

The inductive step is proved by substituting

$$\begin{aligned}\Delta_2(\widehat{\mathcal{A}}^{(n)}_{\text{odd}}(z^2)) &= \frac{1}{2z}\left(\Delta_2(\widehat{\mathcal{A}}^{(n)}(z)) - \Delta_2(\widehat{\mathcal{A}}^{(n)}(-z))\right)\\ \Delta_2(\varphi^{(n)}_{\text{odd}}(z^2)) &= \frac{1}{2z}\left(\Delta_2(\varphi^{(n)}(z)) - \Delta_2(\varphi^{(n)}(-z))\right)\end{aligned}$$

in the above formula and by replacing $\Delta_2(\varphi^{(n)}(z))$ with the second equation in (9.6) and $\Delta_2(\widehat{\mathcal{A}}^{(n)}(z))$ with the equation (9.13). The equation for Δ_1 is obtained by formally substituting the equation for Δ_2 in the expression

$$\Delta_1(A) = -\mathcal{Z}\Delta_2(A)\mathcal{Z} - \mathcal{Z}A\boldsymbol{e}_N\boldsymbol{e}_N^{\mathrm{T}} + \boldsymbol{e}_1\boldsymbol{e}_1^{\mathrm{T}}A\mathcal{Z},$$

which holds by Theorem 2.16. □

From the above result we deduce the following displacement representations for the matrix power series $\widehat{\mathcal{A}}^{(n)}(z)$ in light of Theorems 2.14 and 2.15.

Theorem 9.11 *We have*

$$\begin{aligned}\widehat{\mathcal{A}}^{(n)}(z) =& L(\widehat{\mathcal{A}}^{(n)}(z)\boldsymbol{e}_1) + L(\mathcal{Z}\widehat{\boldsymbol{v}}^{(n)}(z))U(\boldsymbol{e}_1^{\mathrm{T}}\varphi^{(n)}(z)\mathcal{Z}\mathcal{Z}^{\mathrm{T}})\\ &+ z^{-1}L(\mathcal{Z}\widehat{\mathcal{A}}^{(n)}(z)\boldsymbol{e}_N)U(\boldsymbol{u}^{(n)}(z)^{\mathrm{T}}\mathcal{Z}\mathcal{Z}^{\mathrm{T}}) + U(\boldsymbol{e}_1^{\mathrm{T}}\widehat{\mathcal{A}}^{(n)}(z)\mathcal{Z}\mathcal{Z}^{\mathrm{T}}).\end{aligned}$$

Moreover,

$$\begin{aligned}\widehat{\mathcal{A}}^{(n)}(z) =& U(\boldsymbol{e}_1^{\mathrm{T}}\widehat{\mathcal{A}}^{(n)}(z)) + L(\mathcal{Z}\widehat{\boldsymbol{v}}^{(n)}(z))U(\boldsymbol{e}_1^{\mathrm{T}}\varphi^{(n)}(z))\\ &+ z^{-1}L(\mathcal{Z}\widehat{\mathcal{A}}^{(n)}(z)\boldsymbol{e}_N)U(\boldsymbol{u}^{(n)}(z)^{\mathrm{T}}) + L(\mathcal{Z}\mathcal{A}_{-1}\boldsymbol{e}_N)\end{aligned} \tag{9.15}$$

where $\widehat{\boldsymbol{v}}^{(n)}(z)$ *is defined in* (9.14) *with* $\widehat{\boldsymbol{v}}^{(0)}(z) = \boldsymbol{e}_N$, *and* $\boldsymbol{u}^{(n)}(z)^{\mathrm{T}}$ *is defined in* (9.7) *with* $\boldsymbol{u}^{(0)}(z)^{\mathrm{T}} = z^2\boldsymbol{e}_1^{\mathrm{T}}$, *and* $\mathcal{Z}\mathcal{Z}^{\mathrm{T}} = I - \boldsymbol{e}_1\boldsymbol{e}_1^{\mathrm{T}}$.

Proof The proof simply follows from Theorems 2.14 and 2.15 and from Theorem 9.10. □

From the above results it follows that, using the displacement representations (9.8), (9.15), given by the displacement operator Δ_2, at each step n of cyclic reduction only the seven block vector power series

$$\begin{aligned}&\boldsymbol{u}^{(n)}(z)^{\mathrm{T}},\quad \boldsymbol{v}^{(n)}(z),\quad \widehat{\boldsymbol{v}}^{(n)}(z),\\ &\boldsymbol{r}^{(n)}(z)^{\mathrm{T}} = \boldsymbol{e}_1^{\mathrm{T}}\varphi^{(n)}(z),\quad \boldsymbol{c}^{(n)}(z) = \varphi^{(n)}(z)\boldsymbol{e}_N,\\ &\widehat{\boldsymbol{r}}^{(n)}(z)^{\mathrm{T}} = \boldsymbol{e}_1^{\mathrm{T}}\widehat{\mathcal{A}}^{(n)}(z),\quad \widehat{\boldsymbol{c}}^{(n)}(z) = \widehat{\mathcal{A}}^{(n)}(z)\boldsymbol{e}_N,\end{aligned} \tag{9.16}$$

need to be computed in order to represent the matrix power series $\varphi^{(n)}(z)$ and $\widehat{\mathcal{A}}^{(n)}(z)$.

To sum up, the functions $\varphi^{(n)}(z)$ and $\widehat{\mathcal{A}}^{(n)}(z)$ are defined by the equations

$$\begin{aligned}\varphi^{(n)}(z) =& z^{-1}L(\mathcal{Z}\boldsymbol{c}^{(n)}(z))U(\boldsymbol{u}^{(n)}(z)^{\mathrm{T}}) + L(\boldsymbol{e}_1 - z^{-1}\mathcal{Z}\boldsymbol{v}^{(n)}(z))U(\boldsymbol{r}^{(n)}(z)^{\mathrm{T}}),\\ \widehat{\mathcal{A}}^{(n)}(z) =& U(\widehat{\boldsymbol{r}}^{(n)}(z)^{\mathrm{T}}) + L(\mathcal{Z}\widehat{\boldsymbol{v}}^{(n)}(z))U(\boldsymbol{r}^{(n)}(z)^{\mathrm{T}})\\ &+ z^{-1}L(\mathcal{Z}\widehat{\boldsymbol{c}}^{(n)}(z))U(\boldsymbol{u}^{(n)}(z)^{\mathrm{T}}) + L(\boldsymbol{w}),\\ \boldsymbol{w} =& \mathcal{Z}\mathcal{A}_{-1}\boldsymbol{e}_N.\end{aligned} \tag{9.17}$$

The vectors defining $\varphi^{(n)}(z)$ and $\widehat{\mathcal{A}}^{(n)}(z)$ are related by the following equations

$$\begin{aligned}\boldsymbol{u}^{(n+1)}(z^2)^{\mathrm{T}} &= -\left[\boldsymbol{u}^{(n)}(z)^{\mathrm{T}}\varphi^{(n)}_{\mathrm{odd}}(z^2)^{-1}\varphi^{(n)}(-z)\right]_{\mathrm{even}}\\ \boldsymbol{v}^{(n+1)}(z^2) &= -\left[\varphi^{(n)}(-z)\varphi^{(n)}_{\mathrm{odd}}(z^2)^{-1}\boldsymbol{v}^{(n)}(z)\right]_{\mathrm{even}}\\ \widehat{\boldsymbol{v}}^{(n+1)}(z) &= \widehat{\boldsymbol{v}}^{(n)}_{\mathrm{odd}}(z) + z^{-1}\widehat{\mathcal{A}}^{(n)}_{\mathrm{odd}}(z)\varphi^{(n)}_{\mathrm{odd}}(z)^{-1}\boldsymbol{v}^{(n)}_{\mathrm{even}}(z),\\ \boldsymbol{r}^{(n+1)}(z)^{\mathrm{T}} &= z\boldsymbol{r}^{(n)}_{\mathrm{odd}}(z)^{\mathrm{T}} - \boldsymbol{r}^{(n)}_{\mathrm{even}}(z)^{\mathrm{T}}\varphi^{(n)}_{\mathrm{odd}}(z)^{-1}\varphi^{(n)}_{\mathrm{even}}(z)\\ \boldsymbol{c}^{(n+1)}(z) &= z\boldsymbol{c}^{(n)}_{\mathrm{odd}}(z) - \varphi^{(n)}_{\mathrm{even}}(z)\varphi^{(n)}_{\mathrm{odd}}(z)^{-1}\boldsymbol{c}^{(n)}_{\mathrm{even}}(z)\\ \widehat{\boldsymbol{r}}^{(n+1)}(z)^{\mathrm{T}} &= \widehat{\boldsymbol{r}}^{(n)}_{\mathrm{even}}(z)^{\mathrm{T}} - \widehat{\boldsymbol{r}}^{(n)}_{\mathrm{odd}}(z)^{\mathrm{T}}\varphi^{(n)}_{\mathrm{odd}}(z)^{-1}\varphi^{(n)}_{\mathrm{even}}(z)\\ \widehat{\boldsymbol{c}}^{(n+1)}(z) &= \widehat{\boldsymbol{c}}^{(n)}_{\mathrm{even}}(z) - \widehat{\mathcal{A}}^{(n)}_{\mathrm{odd}}(z)\varphi^{(n)}_{\mathrm{odd}}(z)^{-1}\boldsymbol{c}^{(n)}_{\mathrm{even}}(z)\end{aligned} \tag{9.18}$$

with the initial conditions

$$\begin{aligned}&\boldsymbol{u}^{(0)}(z)^{\mathrm{T}} = z^2\boldsymbol{e}_1^{\mathrm{T}},\ \ \boldsymbol{v}^{(0)}(z) = z^2\boldsymbol{e}_N,\ \ \widehat{\boldsymbol{v}}^{(0)}(z) = \boldsymbol{e}_N,\\ &\boldsymbol{r}^{(0)}(z)^{\mathrm{T}} = \boldsymbol{e}_1^{\mathrm{T}}\varphi^{(0)}(z),\ \ \boldsymbol{c}^{(0)}(z) = \varphi^{(0)}(z)\boldsymbol{e}_N,\\ &\widehat{\boldsymbol{r}}^{(0)}(z)^{\mathrm{T}} = \boldsymbol{e}_1^{\mathrm{T}}\widehat{\mathcal{A}}^{(0)}(z),\ \ \widehat{\boldsymbol{c}}^{(0)}(z) = \widehat{\mathcal{A}}^{(0)}(z)\boldsymbol{e}_N.\end{aligned} \tag{9.19}$$

The above relations can be implemented in order to update the vectors defining the matrix power series $\varphi^{(n)}(z)$ and $\widehat{\mathcal{A}}^{(n)}(z)$. One difficulty that we may encounter in this computation is the fact that the coefficients of the vectors involved are matrix power series. We may overcome this difficulty if we know an upper bound d on the numerical degrees of the block vector power series at the step $n+1$, since in this case it is sufficient to apply the point-wise technique of evaluation interpolation at the set $\mathcal{F}_d = \{\omega_d^i,\quad i = 0,\ldots,d-1\}$, $\omega_d = \cos(2\pi/d)+\mathrm{i}\sin(2\pi/d)$, of the d roots of 1. That is, we evaluate all the vector power series at step n for $z \in \mathcal{F}_d$, we apply equations (9.18) for each value of $z \in \mathcal{F}_d$ and obtain the values of the new vectors at step $n+1$ at the same points $z \in \mathcal{F}_d$. Then it is sufficient to interpolate to these values in order to get an approximation of the coefficients of the vectors. We refer the reader to Section 3.1 for more details of this technique, in particular concerning the error introduced with this approximation.

In this way we turned the functional relations (9.18) into d independent matrix relations obtained by replacing z with the interpolation points ω_d^i, $i = 0, \dots, d-1$, for which it is possible to apply the Toeplitz matrix machinery described in Chapter 2.

The main operations involved in the matrix computation obtained for each value of $z \in \mathcal{F}_d$, are

(a1) Computing, for $z \in \mathcal{F}_d$, the displacement representation of the matrix $\varphi_{\mathrm{odd}}^{(n)}(z)^{-1}$ from the displacement representation of the matrix $\varphi_{\mathrm{odd}}^{(n)}(z)$ by means of the relation $\Delta(\varphi_{\mathrm{odd}}^{(n)}(z)^{-1}) = -\varphi_{\mathrm{odd}}^{(n)}(z)^{-1}\Delta(\varphi_{\mathrm{odd}}^{(n)}(z))\varphi_{\mathrm{odd}}^{(n)}(z)^{-1}$, where $\Delta(\varphi_{\mathrm{odd}}^{(n)}(z))$ is expressed as the sum of four terms of the kind $\boldsymbol{p}(z)\boldsymbol{q}(z)^{\mathrm{T}}$, for pairs of block column and block row vector power series $\boldsymbol{p}(z)$ and $\boldsymbol{q}(z)$. This computation requires solving eight block systems with the matrix $\varphi_{\mathrm{odd}}^{(n)}(z)$. The Toeplitz like systems can be solved either by means of general methods like pivoted Gaussian elimination or relying on fast techniques like the conjugate gradient method which are based on FFT (see Section 2.3.3).

(a2) Once the displacement representation of $\varphi_{\mathrm{odd}}^{(n)}(z)^{-1}$ has been computed, the implementation of (9.18) is reduced to computing a finite number of products of block triangular Toeplitz matrices and block vectors. For this computation we rely on Algorithms 2.3, 2.4.

We synthesize this computation in Algorithm 9.2 where, for simplicity, we assume that an upper bound d to the numerical degrees of the block vector power series at step $n+1$ is available in input.

Concerning the computational cost of this algorithm we observe that except for the computation of the displacement representation of the matrix $\varphi_{\mathrm{odd}}^{(n)}(z)$ at a single value $z \in \mathcal{F}_d$, whose cost is denoted with $\mu(N, m)$, the remaining part of the algorithm is reduced to computing a finite number of products of $N \times N$ block triangular block Toeplitz matrices and block vectors having $m \times m$ blocks. According to the estimates of Section 2.3.3, this cost is $O(m^3N + m^2N\log N)$. Therefore the overall cost of the algorithm has the order of

$$d\mu(N, m) + d(m^3N + m^2N\log N)$$

arithmetic operations.

From the convergence properties in the positive recurrent case, and from (9.17) we deduce that the vectors $\boldsymbol{r}^{(n)}(z)^{\mathrm{T}}$ and $\boldsymbol{c}^{(n)}(z)$ converge to matrix polynomials of degree at most one, while $\widehat{\boldsymbol{r}}^{(n)}(z)^{\mathrm{T}}$ and $\widehat{\boldsymbol{c}}^{(n)}(z)$ converge to constant matrices.

These convergence properties imply also that the numerical degree of the matrix vector power series encountered at each step becomes 1 after some steps of cyclic reduction. Therefore the number d of interpolation points needed in the evaluation–interpolation technique, is bounded from above by a constant independent of the number of steps.

Since $\lim_n(I - \widehat{\mathcal{A}}_0^{(n)})^{-1}\mathcal{A}_{-1} = \mathcal{G}_{\min}$, the first block row of $\mathcal{G}_{\min}$ is given by $\boldsymbol{h}^{\mathrm{T}} = \boldsymbol{e}_1^{\mathrm{T}}(I - \widehat{\mathcal{A}}_0^{(\infty)})^{-1}\mathcal{A}_{-1}$ where $\widehat{\mathcal{A}}_0^{(\infty)} = \lim_n \widehat{\mathcal{A}}_0^{(n)}$. Therefore, once the dis-

Algorithm 9.2 One step of cyclic reduction for NSF problems

INPUT: The block vector $\boldsymbol{w} = \mathcal{Z}\mathcal{A}_{-1}\boldsymbol{e}_N$, the matrix coefficients of the block vector power series $\boldsymbol{u}^{(n)}(z)^{\mathrm{T}}$, $\boldsymbol{v}^{(n)}(z)$, $\widehat{\boldsymbol{v}}^{(n)}(z)$, $\boldsymbol{r}^{(n)}(z)^{\mathrm{T}}$, $\boldsymbol{c}^{(n)}(z)$, $\widehat{\boldsymbol{r}}^{(n)}(z)^{\mathrm{T}}$, $\widehat{\boldsymbol{c}}^{(n)}(z)$, which define the matrix power series $\varphi^{(n)}(z)$ and $\widehat{\mathcal{A}}^{(n)}(z)$ according to (9.17) and (9.15), respectively at the step n; an upper bound d on the numerical degree of the block vector power series at the step $n+1$, where d is an integer power of 2.

OUTPUT: An approximation to the first d block coefficients of the block vector power series $\boldsymbol{u}^{(n+1)}(z)^{\mathrm{T}}$, $\boldsymbol{v}^{(n+1)}(z)$, $\widehat{\boldsymbol{v}}^{(n+1)}(z)$, $\boldsymbol{r}^{(n+1)}(z)^{\mathrm{T}}$, $\boldsymbol{c}^{(n+1)}(z)$, $\widehat{\boldsymbol{r}}^{(n+1)}(z)^{\mathrm{T}}$, $\widehat{\boldsymbol{c}}^{(n+1)}(z)$, which define the matrix power series $\varphi^{(n+1)}(z)$ and $\widehat{\mathcal{A}}^{(n+1)}(z)$.

COMPUTATION:

1. Evaluate the functions $\widehat{\boldsymbol{v}}^{(n)}_{\mathrm{odd}}(z)$ and $\widehat{\boldsymbol{c}}^{(n)}_{\mathrm{even}}(z)$ for $z \in \mathcal{F}_d$, evaluate the functions $\boldsymbol{u}^{(n)}(z)^{\mathrm{T}}$, $\boldsymbol{v}^{(n)}(z)$, $\boldsymbol{r}^{(n)}(z)^{\mathrm{T}}$, $\boldsymbol{c}^{(n)}(z)$ and $\widehat{\boldsymbol{r}}^{(n)}(z)^{\mathrm{T}}$ for $z \in \mathcal{F}_{2d}$, so that the values of the odd and of the even parts of such functions at $z \in \mathcal{F}_d$ are available.
2. For each value of $z = \omega_d^i$, $i = 0, \ldots, d-1$, compute a displacement representation of $\varphi^{(n)}_{\mathrm{odd}}(z)^{-1}$ by following the lines described in part (a1) above.
3. For each value of $z \in \mathcal{F}_d$, by using the displacement representation of $\varphi^{(n)}_{\mathrm{odd}}(z)^{-1}$, $\varphi^{(n)}(z)$, $\widehat{\mathcal{A}}^{(n)}_{\mathrm{odd}}(z)$, $\widehat{\mathcal{A}}^{(n)}_{\mathrm{even}}(z)$ compute the values at $z \in \mathcal{F}_d$, of the vectors at step $n+1$ by applying equations (9.18).
4. Interpolate to the values obtained in this way and recover an approximation of the first d block coefficients of the series $\boldsymbol{u}^{(n+1)}(z)^{\mathrm{T}}$, $\boldsymbol{v}^{(n+1)}(z)$, $\widehat{\boldsymbol{v}}^{(n+1)}(z)$, $\boldsymbol{r}^{(n+1)}(z)^{\mathrm{T}}$, $\boldsymbol{c}^{(n+1)}(z)$, $\widehat{\boldsymbol{r}}^{(n+1)}(z)^{\mathrm{T}}$, $\widehat{\boldsymbol{c}}^{(n+1)}(z)$.

placement representation of $\widehat{\mathcal{A}}_0^{(\infty)}$ has been approximated by means of cyclic reduction, it is sufficient to solve a Toeplitz like system in order to approximate the first block row of $\mathcal{G}_{\min}$.

In the case $m = 1$, where the blocks A_i reduce to scalar elements, the formulae displayed in Theorem 9.11 can be substantially simplified. In fact, by using induction on n and relying on (9.5) we may prove the following.

Theorem 9.12 *If $m = 1$, then the matrix polynomial $\mathcal{A}^{(n)}(z)$ satisfies the following relation*

$$\mathcal{A}^{(n)}(z)^{\mathrm{T}} = J\mathcal{A}^{(n)}(z)J,$$

where J denotes the $N \times N$ permutation (reversion) matrix having unit elements along the anti-diagonal.

Whence, the vectors $\boldsymbol{u}^{(n)}(z)$, $\boldsymbol{v}^{(n)}(z)$, $\boldsymbol{r}^{(n)}(z)$, $\boldsymbol{c}^{(n)}(z)$ are related, as stated by the following.

Theorem 9.13 *If $m = 1$, then the vectors $\boldsymbol{u}^{(n)}(z)$, $\boldsymbol{v}^{(n)}(z)$, $\boldsymbol{r}^{(n)}(z)$, $\boldsymbol{c}^{(n)}(z)$ are such that $\boldsymbol{v}^{(n)}(z) = J\boldsymbol{u}^{(n)}(z)$, $\boldsymbol{c}^{(n)}(z) = J\boldsymbol{r}^{(n)}(z)$, thus*

$$\Delta_2(\varphi^{(n)}(z)) = -z^{-1}(J\boldsymbol{r}^{(n)}(z)\boldsymbol{u}^{(n)}(z)^{\mathrm{T}} - J\boldsymbol{u}^{(n)}(z)\boldsymbol{r}^{(n)}(z)^{\mathrm{T}})$$

9.2.4 *The banded case*

In the case where $A_i = 0$ for $|i| > N$, the matrix (5.29) is block banded and the reblocked matrix (5.30) is block tridiagonal, therefore the power series $\varphi(z)$ turns into a quadratic polynomial $\varphi(z) = -\mathcal{A}_{-1} + z(I - \mathcal{A}_0) - z^2\mathcal{A}_1$. In this case the results concerning the structural properties of the blocks generated by cyclic reduction are further simplified, in fact, we can give explicit expressions of the blocks $\mathcal{A}_i^{(n)}$, $i = -1, 0, 1$, $\widehat{\mathcal{A}}_0^{(n)}$. Moreover, since $\mathcal{A}^{(n)}(z) = zI - \varphi^{(n)}(z)$ is a matrix polynomial of degree 2 having displacement rank at most 2, then we may prove that its coefficients have bounded displacement rank. We need the following.

Lemma 9.14 *The matrix power series $\boldsymbol{u}^{(n)}(z)$, $\boldsymbol{v}^{(n)}(z)$, and $\widehat{\boldsymbol{v}}^{(n)}(z)$, have the form*

$$\begin{aligned}
\boldsymbol{u}^{(n)}(z)^{\mathrm{T}} &= z(\boldsymbol{u}_0^{(n)})^{\mathrm{T}} + z^2(\boldsymbol{u}_1^{(n)})^{\mathrm{T}},\\
\boldsymbol{v}^{(n)}(z) &= z\boldsymbol{v}_0^{(n)} + z^2\boldsymbol{v}_1^{(n)},\\
\widehat{\boldsymbol{v}}^{(n)}(z) &= \boldsymbol{v}_1^{(n)},
\end{aligned}$$

where $\boldsymbol{u}_0^{(0)} = \boldsymbol{v}_0^{(0)} = 0$, $\boldsymbol{u}_1^{(0)} = \boldsymbol{e}_1$, $\boldsymbol{v}_1^{(0)} = \boldsymbol{e}_N$ and

$$\begin{aligned}
(\boldsymbol{u}_0^{(n+1)})^{\mathrm{T}} &= (\boldsymbol{u}_0^{(n)})^{\mathrm{T}} + (\boldsymbol{u}_1^{(n)})^{\mathrm{T}}(I - \mathcal{A}_0^{(n)})^{-1}\mathcal{A}_{-1}^{(n)}\\
(\boldsymbol{u}_1^{(n+1)})^{\mathrm{T}} &= (\boldsymbol{u}_1^{(n)})^{\mathrm{T}}(I - \mathcal{A}_0^{(n)})^{-1}\mathcal{A}_1^{(n)}\\
\boldsymbol{v}_0^{(n+1)} &= \boldsymbol{v}_0^{(n)} + \mathcal{A}_{-1}^{(n)}(I - \mathcal{A}_0^{(n)})^{-1}\boldsymbol{v}_1^{(n)}\\
\boldsymbol{v}_1^{(n+1)} &= \mathcal{A}_1^{(n)}(I - \mathcal{A}_0^{(n)})^{-1}\boldsymbol{v}_1^{(n)}, \quad n \ge 0.
\end{aligned}$$

Moreover, one has $c^{(n)}(z) = \boldsymbol{c}_0^{(n)} + z\boldsymbol{c}_1^{(n)} + z^2\boldsymbol{c}_2^{(n)}$ and $\boldsymbol{r}^{(n)}(z)^{\mathrm{T}} = (\boldsymbol{r}_0^{(n)})^{\mathrm{T}} + z(\boldsymbol{r}_1^{(n)})^{\mathrm{T}} + z^2(\boldsymbol{r}_2^{(n)})^{\mathrm{T}}$, where

$$\begin{aligned}
\boldsymbol{c}_0^{(n+1)} &= \mathcal{A}_{-1}^{(n)}(I - \mathcal{A}_0^{(n)})^{-1}\boldsymbol{c}_0^{(n)}\\
\boldsymbol{c}_1^{(n+1)} &= \boldsymbol{c}_1^{(n)} + \mathcal{A}_{-1}^{(n)}(I - \mathcal{A}_0^{(n)})^{-1}\boldsymbol{c}_2^{(n)} + \mathcal{A}_1^{(n)}(I - \mathcal{A}_0^{(n)})^{-1}\boldsymbol{c}_0^{(n)}\\
\boldsymbol{c}_2^{(n+1)} &= \mathcal{A}_1^{(n)}(I - \mathcal{A}_0^{(n)})^{-1}\boldsymbol{c}_2^{(n)}\\
(\boldsymbol{r}_0^{(n+1)})^{\mathrm{T}} &= (\boldsymbol{r}_0^{(n)})^{\mathrm{T}}(I - \mathcal{A}_0^{(n)})^{-1}\mathcal{A}_{-1}^{(n)}\\
(\boldsymbol{r}_1^{(n+1)})^{\mathrm{T}} &= (\boldsymbol{r}_1^{(n)})^{\mathrm{T}} + (\boldsymbol{r}_0^{(n)})^{\mathrm{T}}(I - \mathcal{A}_0^{(n)})^{-1}\mathcal{A}_1^{(n)} + (\boldsymbol{r}_2^{(n)})^{\mathrm{T}}(I - \mathcal{A}_0^{(n)})^{-1}\mathcal{A}_{-1}^{(n)}\\
(\boldsymbol{r}_2^{(n+1)})^{\mathrm{T}} &= (\boldsymbol{r}_2^{(n)})^{\mathrm{T}}(I - \mathcal{A}_0^{(n)})^{-1}\mathcal{A}_1^{(n)} \quad n \ge 0,
\end{aligned}$$

and $\boldsymbol{c}_0^{(0)} = -\mathcal{A}_{-1}\boldsymbol{e}_N$, $\boldsymbol{c}_1^{(0)} = (I - \mathcal{A}_0)\boldsymbol{e}_N$, $\boldsymbol{c}_2^{(0)} = -\boldsymbol{e}_N A_N$, $(\boldsymbol{r}_0^{(0)})^{\mathrm{T}} = -\boldsymbol{e}_1^{\mathrm{T}}\mathcal{A}_{-1}$, $(\boldsymbol{r}_1^{(0)})^{\mathrm{T}} = \boldsymbol{e}_1^{\mathrm{T}}(I - \mathcal{A}_0)$, $(\boldsymbol{r}_2^{(0)})^{\mathrm{T}} = -A_N\boldsymbol{e}_1^{\mathrm{T}}$.

Proof We proceed by induction. For $n = 0$ the properties are trivial. We prove the inductive step only for $\boldsymbol{u}^{(n)}(z)$, the other relations can be proved similarly. From the first equation of (9.18) we have

$$\begin{aligned}
\boldsymbol{u}^{(n+1)}(z^2)^{\mathrm{T}} = & - \left[(z(\boldsymbol{u}_0^{(n)})^{\mathrm{T}} + z^2(\boldsymbol{u}_1^{(n)})^{\mathrm{T}})(I - \mathcal{A}_0^{(n)})^{-1} \right. \\
& \left. (-\mathcal{A}_{-1}^{(n)} - z(I - \mathcal{A}_0^{(n)}) - z^2\mathcal{A}_1^{(n)})\right]_{\text{even}} \\
= & \; z^2((\boldsymbol{u}_0^{(n)})^{\mathrm{T}} + (\boldsymbol{u}_1^{(n)})^{\mathrm{T}}(I - \mathcal{A}_0^{(n)})^{-1}\mathcal{A}_{-1}^{(n)}) \\
& + z^4(\boldsymbol{u}_1^{(n)})^{\mathrm{T}}(I - \mathcal{A}_0^{(n)})^{-1}\mathcal{A}_1^{(n)}.
\end{aligned}$$

□

A nice consequence of the above lemma is that

$$\begin{aligned}
\boldsymbol{c}_2^{(n)} &= -\boldsymbol{v}_1^{(n)} A_N \\
(\boldsymbol{r}_2^{(n)})^{\mathrm{T}} &= -A_N(\boldsymbol{u}_1^{(n)})^{\mathrm{T}}
\end{aligned} \tag{9.20}$$

for any $n \geq 0$. Therefore, the vectors $\boldsymbol{c}_2^{(n)}$ and $\boldsymbol{r}_2^{(n)}$ can be expressed in terms of $\boldsymbol{v}_1^{(n)}$ and $\boldsymbol{u}_1^{(n)}$, respectively. By comparing the terms of the same degree in (9.6) and (9.13) and by using (9.20), we immediately obtain the following.

Theorem 9.15 *If $\mathcal{A}_i = 0$ for $i > 1$, then for the matrices $\mathcal{A}_i^{(n)}$, $i = -1, 0, 1$, and $\widehat{\mathcal{A}}_0^{(n)}$, $n \geq 0$, generated by cyclic reduction (7.17) we have, for $\Delta = \Delta_2$,*

$$\begin{aligned}
\Delta(\mathcal{A}_{-1}^{(n)}) &= \boldsymbol{c}_0^{(n)}(\boldsymbol{u}_0^{(n)})^{\mathrm{T}} - \boldsymbol{v}_0^{(n)}(\boldsymbol{r}_0^{(n)})^{\mathrm{T}} \\
\Delta(\mathcal{A}_0^{(n)}) &= \boldsymbol{c}_0^{(n)}(\boldsymbol{u}_1^{(n)})^{\mathrm{T}} + \boldsymbol{c}_1^{(n)}(\boldsymbol{u}_0^{(n)})^{\mathrm{T}} - \boldsymbol{v}_0^{(n)}(\boldsymbol{r}_1^{(n)})^{\mathrm{T}} - \boldsymbol{v}_1^{(n)}(\boldsymbol{r}_0^{(n)})^{\mathrm{T}} \\
\Delta(\mathcal{A}_1^{(n)}) &= (\boldsymbol{c}_1^{(n)} + \boldsymbol{v}_0^{(n)}A_N)(\boldsymbol{u}_1^{(n)})^{\mathrm{T}} - \boldsymbol{v}_1^{(n)}((\boldsymbol{r}_1^{(n)})^{\mathrm{T}} + A_N(\boldsymbol{u}_0^{(n)})^{\mathrm{T}}) \\
\Delta(\widehat{\mathcal{A}}_0^{(n)}) &= -\boldsymbol{v}_1^{(n)}(\boldsymbol{r}_0^{(n)})^{\mathrm{T}} + (\boldsymbol{e}_N - \widehat{\boldsymbol{c}}_0^{(n)})(\boldsymbol{u}_0^{(n)})^{\mathrm{T}} - \mathcal{A}_{-1}\boldsymbol{e}_N\boldsymbol{e}_1^{\mathrm{T}},
\end{aligned}$$

where $\widehat{\boldsymbol{c}}_0^{(n)} = \widehat{\mathcal{A}}_0^{(n)}\boldsymbol{e}_N$, $n \geq 0$.

From the above theorem and by the results of Section 2.4 we obtain the following representation of the matrices generated by cyclic reduction.

Theorem 9.16 *At each step n of cyclic reduction, the matrices $\mathcal{A}_{-1}^{(n)}$, $\mathcal{A}_0^{(n)}$, $\mathcal{A}_1^{(n)}$, $\mathcal{K}^{(n)} = (I - \mathcal{A}_0^{(n)})^{-1}$, $\widehat{\mathcal{A}}_0^{(n)}$ can be rewritten as*

$$\begin{aligned}
\mathcal{A}_{-1}^{(n)} =& (L(\mathcal{Z}\boldsymbol{v}_0^{(n)}) - I)U((\boldsymbol{r}_0^{(n)})^{\mathrm{T}}) - L(\mathcal{Z}\boldsymbol{c}_0^{(n)})U((\boldsymbol{u}_0^{(n)})^{\mathrm{T}}), \\
\mathcal{A}_0^{(n)} =& (L(\mathcal{Z}\boldsymbol{v}_0^{(n)}) - I)U((\boldsymbol{r}_1^{(n)})^{\mathrm{T}}) - L(\mathcal{Z}\boldsymbol{c}_1^{(n)})U((\boldsymbol{u}_0^{(n)})^{\mathrm{T}}) \\
& - L(\mathcal{Z}\boldsymbol{c}_0^{(n)})U((\boldsymbol{u}_1^{(n)})^{\mathrm{T}}) + L(\mathcal{Z}\boldsymbol{v}_1^{(n)})U((\boldsymbol{r}_0^{(n)})^{\mathrm{T}}), \\
\mathcal{A}_1^{(n)} =& - U((\boldsymbol{r}_1^{(n)})^{\mathrm{T}}) + L(\mathcal{Z}\boldsymbol{v}_1^{(n)})U((\boldsymbol{r}_1^{(n)})^{\mathrm{T}} + A_N(\boldsymbol{u}_0^{(n)})^{\mathrm{T}}), \\
\mathcal{K}^{(n)} =& U(\boldsymbol{e}_1^{\mathrm{T}}\mathcal{K}^{(n)}) - L(\mathcal{Z}\mathcal{K}^{(n)}\boldsymbol{v}_0^{(n)}))U((\boldsymbol{r}_1^{(n)})^{\mathrm{T}}\mathcal{K}^{(n)}) \\
& + L(\mathcal{Z}\mathcal{K}^{(n)}\boldsymbol{c}_1^{(n)})U((\boldsymbol{u}_0^{(n)})^{\mathrm{T}}\mathcal{K}^{(n)}) + L(\mathcal{Z}\mathcal{K}^{(n)}\boldsymbol{c}_0^{(n)})U((\boldsymbol{u}_1^{(n)})^{\mathrm{T}}\mathcal{K}^{(n)}) \\
& - L(\mathcal{Z}\mathcal{K}^{(n)}\boldsymbol{v}_1^{(n)})U((\boldsymbol{r}_0^{(n)})^{\mathrm{T}}\mathcal{K}^{(n)}), \\
\widehat{\mathcal{A}}_0^{(n)} =& U((\widehat{\boldsymbol{r}}_0^{(n)})^{\mathrm{T}}) + L(\mathcal{Z}\boldsymbol{v}_1^{(n)})U((\boldsymbol{r}_0^{(n)})^{\mathrm{T}}) + L(\mathcal{Z}\widehat{\boldsymbol{c}}_0^{(n)})U((\boldsymbol{u}_0^{(n)})^{\mathrm{T}}) + L(\boldsymbol{w})
\end{aligned}$$

where $\boldsymbol{w} = \mathcal{Z}\mathcal{A}_{-1}\boldsymbol{e}_N$ and $(\widehat{\boldsymbol{r}}_0^{(n)})^{\mathrm{T}} = \boldsymbol{e}_1^{\mathrm{T}}\widehat{\mathcal{A}}_0^{(n)}$.

Below we synthesize the relations between the vectors defining the matrices at two subsequent steps of cyclic reduction:

$$
\begin{aligned}
(\boldsymbol{u}_0^{(n+1)})^{\mathrm{T}} &= (\boldsymbol{u}_0^{(n)})^{\mathrm{T}} + (\boldsymbol{u}_1^{(n)})^{\mathrm{T}}(I - \mathcal{A}_0^{(n)})^{-1}\mathcal{A}_{-1}^{(n)} \\
(\boldsymbol{u}_1^{(n+1)})^{\mathrm{T}} &= (\boldsymbol{u}_1^{(n)})^{\mathrm{T}}(I - \mathcal{A}_0^{(n)})^{-1}\mathcal{A}_1^{(n)} \\
\boldsymbol{v}_0^{(n+1)} &= \boldsymbol{v}_0^{(n)} + \mathcal{A}_{-1}^{(n)}(I - \mathcal{A}_0^{(n)})^{-1}\boldsymbol{v}_1^{(n)} \\
\boldsymbol{v}_1^{(n+1)} &= \mathcal{A}_1^{(n)}(I - \mathcal{A}_0^{(n)})^{-1}\boldsymbol{v}_1^{(n)} \\
\boldsymbol{c}_0^{(n+1)} &= \mathcal{A}_{-1}^{(n)}(I - \mathcal{A}_0^{(n)})^{-1}\boldsymbol{c}_0^{(n)} \\
\boldsymbol{c}_1^{(n+1)} &= \boldsymbol{c}_1^{(n)} - \mathcal{A}_{-1}^{(n)}(I - \mathcal{A}_0^{(n)})^{-1}\boldsymbol{v}_1^{(n)}A_N + \mathcal{A}_1^{(n)}(I - \mathcal{A}_0^{(n)})^{-1}\boldsymbol{c}_0^{(n)} \\
(\boldsymbol{r}_0^{(n+1)})^{\mathrm{T}} &= (\boldsymbol{r}_0^{(n)})^{\mathrm{T}}(I - \mathcal{A}_0^{(n)})^{-1}\mathcal{A}_{-1}^{(n)} \\
(\boldsymbol{r}_1^{(n+1)})^{\mathrm{T}} &= (\boldsymbol{r}_1^{(n)})^{\mathrm{T}} + (\boldsymbol{r}_0^{(n)})^{\mathrm{T}}(I - \mathcal{A}_0^{(n)})^{-1}\mathcal{A}_1^{(n)} - A_N(\boldsymbol{u}_1^{(n)})^{\mathrm{T}}(I - \mathcal{A}_0^{(n)})^{-1}\mathcal{A}_{-1}^{(n)} \\
(\widehat{\boldsymbol{r}}_0^{(n+1)})^{\mathrm{T}} &= (\widehat{\boldsymbol{r}}_0^{(n)})^{\mathrm{T}} + A_N(\boldsymbol{u}_1^{(n)})^{\mathrm{T}}(I - \mathcal{A}_0^{(n)})^{-1}\mathcal{A}_{-1}^{(n)} \\
\widehat{\boldsymbol{c}}_0^{(n+1)} &= \widehat{\boldsymbol{c}}_0^{(n)} - \mathcal{A}_1^{(n)}(I - \mathcal{A}_0^{(n)})^{-1}\boldsymbol{c}_0^{(n)}
\end{aligned}
\tag{9.21}
$$

for $n \geq 0$, where $\boldsymbol{u}_0^{(0)} = \boldsymbol{v}_0^{(0)} = 0$, $\boldsymbol{u}_1^{(0)} = \boldsymbol{e}_N$, $\boldsymbol{v}_1^{(0)} = \boldsymbol{e}_1$, $\boldsymbol{c}_0^{(0)} = -\mathcal{A}_{-1}\boldsymbol{e}_N$, $\boldsymbol{c}_1^{(0)} = (I - \mathcal{A}_0)\boldsymbol{e}_N$, $(\boldsymbol{r}_0^{(0)})^{\mathrm{T}} = -\boldsymbol{e}_1^{\mathrm{T}}\mathcal{A}_{-1}$, $(\boldsymbol{r}_1^{(0)})^{\mathrm{T}} = \boldsymbol{e}_1^{\mathrm{T}}(I - \mathcal{A}_0)$, $(\widehat{\boldsymbol{r}}_0^{(0)})^{\mathrm{T}} = \boldsymbol{e}_1^{\mathrm{T}}\mathcal{A}_0$, $\widehat{\boldsymbol{c}}_0^{(0)} = \mathcal{A}_0\boldsymbol{e}_N$.

Algorithm 9.3 One step of cyclic reduction for NSF problems: banded case

INPUT: The block vectors $\widehat{\boldsymbol{c}}_0^{(n)}$, $\widehat{\boldsymbol{r}}_0^{(n)}$, $\boldsymbol{u}_i^{(n)}$, $\boldsymbol{v}_i^{(n)}$, $\boldsymbol{c}_i^{(n)}$, $\boldsymbol{r}_i^{(n)}$, for $i = 0, 1$, which define the matrices $\mathcal{A}_i^{(n)}$, $i = -1, 0, 1$ and $\widehat{\mathcal{A}}_0^{(n)}$ according to Theorem 9.16.

OUTPUT: The block vectors $\widehat{\boldsymbol{c}}_0^{(n+1)}$, $\widehat{\boldsymbol{r}}_0^{(n+1)}$, $\boldsymbol{u}_i^{(n+1)}$, $\boldsymbol{v}_i^{(n+1)}$, $\boldsymbol{c}_i^{(n+1)}$, $\boldsymbol{r}_i^{(n+1)}$, for $i = 0, 1$, which define the matrices $\mathcal{A}_i^{(n+1)}$, $i = -1, 0, 1$ and $\widehat{\mathcal{A}}_0^{(n+1)}$.

COMPUTATION:

1. Compute the diplacement representation of $(I - \mathcal{A}_0^{(n)})^{-1}$ given in Theorem 9.16. This computation requires solving six block systems with the matrix $I - \mathcal{A}_0^{(n)}$. These Toeplitz like systems can be solved either by means of general methods like pivoted Gaussian elimination or relying on fast techniques like the conjugate gradient method which are based on FFT (compare with Section 2.3.3).
2. By using the displacement representation of $\mathcal{A}_{-1}^{(n)}$, $(I - \mathcal{A}_0^{(n)})^{-1}$ and $\mathcal{A}_1^{(n)}$, compute the block vectors $(\boldsymbol{u}_i^{(n+1)})^{\mathrm{T}}$, $\boldsymbol{v}_i^{(n+1)}$, $\boldsymbol{c}_i^{(n+1)}$, $(\boldsymbol{r}_i^{(n+1)})^{\mathrm{T}}$, $i = 0, 1$, and $\widehat{\boldsymbol{c}}_0^{(n+1)}$, $(\widehat{\boldsymbol{r}}_0^{(n+1)})^{\mathrm{T}}$, by means of relations (9.21). This latter computation consists in computing a finite number of products of block triangular Toeplitz matrices and block vectors, which can be performed by using Algorithm 2.3.

The computational cost per step of Algorithm 9.3 is reduced to $O(\mu(N, m) + m^2 N \log N + m^3 N)$ operations.

According to the convergence theorems of Section 7.3, if the drift μ is nonzero, then $\mathcal{G}_{\min} = \lim_n (I - \widehat{\mathcal{A}}_0^{(n)})^{-1}\mathcal{A}_{-1}$, therefore the first block row of $\mathcal{G}_{\min}$ can be approximated by solving the Toeplitz like linear system $\boldsymbol{x}^{\mathrm{T}}(I - \widehat{\mathcal{A}}_0^{(n)}) = \boldsymbol{e}_1^{\mathrm{T}}$, for a sufficiently large n, and by computing the product $\boldsymbol{x}^{\mathrm{T}}\mathcal{A}_{-1}$.

The cyclic reduction algorithm particularly shows its effectiveness, with respect to the linearly convergent method of Section 9.2.1, when the block dimension N is large; in fact, the computational cost of one step of the latter method is $O(m^3N)$, and many iterations may need to be computed in order to reach a good approximation of $\mathcal{G}_{\min}$.

9.2.5 Acceleration techniques

The shift technique, introduced in Section 3.6 and exploited in Section 8.2 for M/G/1-type Markov chains, can be successfully applied to non-skip-free problems. Indeed, in this case we consider the modified matrix Laurent power series $I - \widetilde{S}(z) = (I - S(z))(I - z^{-1}Q)^{-1}$, where $Q = \boldsymbol{1}\boldsymbol{u}^{\mathrm{T}}$, $\boldsymbol{u}^{\mathrm{T}} \geq 0$, $\boldsymbol{u}^{\mathrm{T}}\boldsymbol{1} = 1$, and $\widetilde{S}(z) = \sum_{i=-N}^{+\infty} z^i \widetilde{A}_i$, $S(z) = \sum_{i=-N}^{+\infty} z^i A_i$.

Therefore, instead of applying the reblocking technique of Section 5.5 to the function $I - S(z)$ in order compute its weak canonical factorization, we may apply the same technique to the new function $I - \widetilde{S}(z)$ in order to compute a canonical factorization. Indeed, for the function $I - \widetilde{\mathcal{S}}(z)$ obtained by the reblocking of $I - \widetilde{S}(z)$ the convergence of iterative methods like cyclic reduction is faster than for the function $I - \mathcal{S}(z)$, obtained by reblocking $I - S(z)$ in light of Theorem 8.3.

Alternatively, we may apply either the functional iteration methods of Section 6.2 to the function $I - \widetilde{\mathcal{S}(z)}$ or the functional iteration of Section 9.2.1 to the function $I - \widetilde{S}(z)$. However, the convergence analysis of these methods applied to the shifted problem has not yet been performed.

9.3 Solving M/G/1-type Markov chains through a QBD

Consider the M/G/1-type Markov chain (4.3) and consider the QBD process obtained by embedding the M/G/1-type Markov chain, as in Section 5.7.

By Theorem 5.21 the computation of the minimal nonnegative solution $G_{\min}$ of $X = \sum_{i=-1}^{+\infty} A_i X^{i+1}$ can be reduced to the computation of the minimal nonnegative solution $\mathcal{G}_{\min}$ of the matrix equation

$$X = \mathcal{A}_{-1} + \mathcal{A}_0 X + \mathcal{A}_1 X^2, \tag{9.22}$$

where the matrices $\mathcal{A}_i$, $i = -1, 0, 1$, are defined in (5.49), (5.50) and (5.51), respectively. In fact, the matrix $\mathcal{G}_{\min}$ has the structure

$$\mathcal{G}_{\min} = \begin{bmatrix} G_{\min} & 0 & 0 & \dots \\ G_{\min}^2 & 0 & 0 & \dots \\ G_{\min}^3 & 0 & 0 & \dots \\ \vdots & \vdots & \vdots & \end{bmatrix}.$$

Therefore, algorithms designed for the QBD case can be applied in order to solve general M/G/1-type Markov chains. In particular we may consider quadratically convergent methods like the logarithmic reduction or the cyclic reduction algorithm. The main issue in the application of these methods is the exploitation of the structures of the blocks $\mathcal{A}_i$, $i=-1,0,1$, and of the solution $\mathcal{G}_{\min}$ that must be taken into account if we want to arrive at an efficient design of the solution algorithm.

Let us denote by $\mathcal{A}'_i$, $i=-1,0,1$, and $\mathcal{G}'_{\min}$ the block $N\times N$ matrices obtained by truncating $\mathcal{A}_i$, $i=-1,0,1$, and $\mathcal{G}_{\min}$ at the block size N, respectively. Then it is easy to verify that, if $A_i=0$, for $i>N-1$ then $\mathcal{G}'_{\min}=\mathcal{A}'_{-1}+\mathcal{A}'_0\mathcal{G}'_{\min}+\mathcal{A}'_1(\mathcal{G}'_{\min})^2$. Since $\sum_{i=-1}^{+\infty}A_i$ is stochastic, the matrices A_i are negligible for i large enough. In practice, if the elements of A_i are such that $\mathbf{1}-\sum_{i=-1}^{N-1}A_i\mathbf{1}<\epsilon\mathbf{1}$, where ϵ is the machine precision of the floating point arithmetic used in the computation, then we may assume that $A_i=0$ for $i>N-1$. In this way without loss of generality, for practical applications, we may assume that the blocks $\mathcal{A}_i$, $i=-1,0,1$, of (5.49), (5.50) and (5.51), have finite block size N. Here and hereafter we make this assumption.

Let $\mathcal{A}_i^{(n)}$, $i=-1,0,1$, $\widehat{\mathcal{A}}_0^{(n)}$, $n\ge 0$, be the blocks generated by cyclic reduction (7.17), with $\mathcal{A}_i^{(0)}=\mathcal{A}_i$, $i=-1,0,1$, $\widehat{\mathcal{A}}_0^{(0)}=\mathcal{A}_0$.

For notational simplicity we denote by $\boldsymbol{e}_1$ the block vector $\boldsymbol{e}_1\otimes I_m$, and recall that $\mathcal{Z}$ denotes the $N\times N$ block lower shift matrix $\mathcal{Z}=Z\otimes I_m$.

Concerning the structure properties of the matrices generated by cyclic reduction we have the following first result, which does not involve any diplacement operator.

Theorem 9.17 *The matrices $\mathcal{A}_i^{(n)}$, $i=-1,0,1$, and $\widehat{\mathcal{A}}_0^{(n)}$, satisfy the following relations:*

$$\begin{aligned}
&\widehat{\mathcal{A}}_0^{(n)}=\boldsymbol{e}_1(\boldsymbol{v}^{(0)})^{\mathrm{T}}+\mathcal{Z}\boldsymbol{u}^{(n)}\boldsymbol{e}_1^{\mathrm{T}},\\
&\mathcal{A}_0^{(n)}=\boldsymbol{e}_1(\boldsymbol{v}^{(n)})^{\mathrm{T}}+\mathcal{Z}\boldsymbol{u}^{(n)}\boldsymbol{e}_1^{\mathrm{T}},\\
&\mathcal{A}_1^{(n+1)}=\mathcal{A}_1^{(n)}(I+\widehat{\boldsymbol{u}}^{(n)}S^{(n)}(\boldsymbol{v}^{(n)})^{\mathrm{T}})\mathcal{A}_1^{(n)},\\
&\mathcal{A}_{-1}^{(n)}=\boldsymbol{e}_1A_{-1}^{(n)}\boldsymbol{e}_1^{\mathrm{T}},\\
&(I-\mathcal{A}_0^{(n)})^{-1}=I+\left[\boldsymbol{e}_1\middle|\mathcal{Z}\boldsymbol{u}^{(n)}\right]\begin{bmatrix}S^{(n)} & T^{(n)}\\ S^{(n)} & I+T^{(n)}\end{bmatrix}\begin{bmatrix}(\boldsymbol{v}^{(n)})^{\mathrm{T}}\\ \boldsymbol{e}_1^{\mathrm{T}}\end{bmatrix},\quad n\ge 0,
\end{aligned}\tag{9.23}$$

where $A_{-1}^{(n)}$ is an $m\times m$ matrix, $\boldsymbol{u}^{(n)}$, $\widehat{\boldsymbol{u}}^{(n)}$, $\boldsymbol{v}^{(n)}$ are block vectors such that $\boldsymbol{u}^{(n)}=\mathcal{Z}^{\mathrm{T}}\widehat{\boldsymbol{u}}^{(n)}$ and

$$\begin{aligned}
&\mathcal{A}_1^{(0)}=\mathcal{Z},\ A_{-1}^{(0)}=A_{-1},\ \widehat{\boldsymbol{u}}^{(0)}=\boldsymbol{e}_1,\ (\boldsymbol{v}^{(0)})^{\mathrm{T}}=[A_0,\ldots,A_{N-1}],\\
&A_{-1}^{(n+1)}=A_{-1}^{(n)}(S^{(n)}V_1^{(n)}+T^{(n)}+I)A_{-1}^{(n)},\\
&\widehat{\boldsymbol{u}}^{(n+1)}=\widehat{\boldsymbol{u}}^{(n)}+\mathcal{A}_1^{(n)}\widehat{\boldsymbol{u}}^{(n)}(S^{(n)}V_1^{(n)}+T^{(n)}+I)A_{-1}^{(n)},\\
&(\boldsymbol{v}^{(n+1)})^{\mathrm{T}}=(\boldsymbol{v}^{(n)})^{\mathrm{T}}+A_{-1}^{(n)}S^{(n)}(\boldsymbol{v}^{(n)})^{\mathrm{T}}\mathcal{A}_1^{(n)},\ n\ge 0,
\end{aligned}\tag{9.24}$$

with

$$S^{(n)}=(I-(\boldsymbol{v}^{(n)})^{\mathrm{T}}\widehat{\boldsymbol{u}}^{(n)})^{-1},\quad T^{(n)}=S^{(n)}(\boldsymbol{v}^{(n)})^{\mathrm{T}}\mathcal{Z}\boldsymbol{u}^{(n)},\tag{9.25}$$

and $V_1^{(n)}$ is the first block component of the block row vector $(\boldsymbol{v}^{(n)})^{\mathrm{T}}$.

Proof Proceed by induction on n. For $n = 0$ the theorem trivially holds. In the general case, by applying (7.17) we immediately obtain the representation of $\mathcal{A}_i^{(n)}$, $i = -1, 0, 1$. The representation of $(I - \mathcal{A}_0^{(n)})^{-1}$ follows from the Sherman–Morrison–Woodbury formula (formula (A.2) in the appendix) applied to $I - \mathcal{A}_0^{(n)}$. □

From the above theorem it follows that the blocks $\widehat{\mathcal{A}}_0^{(n)}$, $\mathcal{A}_i^{(n)}$, $i = -1, 0$ still retain the sparsity properties of $\widehat{\mathcal{A}}_0^{(0)}$, $\mathcal{A}_i^{(0)}$, $i = -1, 0$, respectively. More precisely, the matrices $\widehat{\mathcal{A}}_0^{(n)}$ and $\mathcal{A}_0^{(n)}$ have null block elements except for the blocks in the first block column and in the first block row; and the matrices $\mathcal{A}_{-1}^{(n)}$ have null block elements except for the one in position (1,1). At first glance, the matrices $\mathcal{A}_1^{(n)}$ have neither sparsity nor structural properties; indeed, it is easy to verify that if $n < \log_2 N$, then $\mathcal{A}_1^{(n)}$ has the following structure

$$\mathcal{A}_1^{(n)} = \begin{bmatrix} 0 & 0 & \dots & 0 \\ * & * & \dots & * \\ \vdots & \vdots & & \vdots \\ * & * & \dots & * \\ I & & & 0 \\ & \ddots & & \\ 0 & & I & \end{bmatrix}$$

where $*$ denotes a possibly nonzero block element and where the number of nonzero block rows above the identity matrices is $2^n - 1$; if $n \geq \log_2 N$, then $\mathcal{A}_1^{(n)}$ seems to be a general $N \times N$ block matrix without any structural properties. The lack of structure of $\mathcal{A}_1^{(n)}$ would lead to an inefficient implementation of the cyclic reduction algorithm; in fact each step of cyclic reduction would cost $O(N^3 m^3)$ ops.

Fortunately, it is possible to prove that the matrix $\mathcal{A}_1^{(n)}$ has a displacement structure that allows one to implement one step of cyclic reduction with $O(m^2 N \log N + m^3 N)$ ops. This property is not trivial and its proof relies on the tools introduced in Section 2.4.

Consider the matrix polynomial $\varphi_R(z) = -\mathcal{A}_1 + z(I - \mathcal{A}_0) - z^2 \mathcal{A}_{-1}$ obtained by reverting the order of the coefficients of $\varphi(z)$. Observe that cyclic reduction applied to $\varphi_R(z)$ generates the polynomials $\varphi_R^{(n)}(z)$ which are obtained by reverting the order of the coefficients of the polynomials $\varphi^{(n)}(z)$ generated by cyclic reduction applied to $\varphi^{(n)}(z)$. Observe that

$$\varphi_R(z) = \begin{bmatrix} zI - zA_0 - z^2 A_{-1} & -zA_1 & -zA_2 & \dots & -zA_{N-1} \\ -I & zI & & & 0 \\ & -I & zI & & \\ & & \ddots & \ddots & \\ 0 & & & -I & zI \end{bmatrix}. \tag{9.26}$$

The matrix functions $\varphi_R^{(n)}(z) = -\mathcal{A}_1^{(n)} + z(I - \mathcal{A}_0^{(n)}) - z^2\mathcal{A}_{-1}^{(n)}$ have displacement rank at most 3, as shown by the following.

Theorem 9.18 *For the operator $\Delta(A) = \mathcal{Z}A - A\mathcal{Z}$, the matrix polynomials $\varphi_R^{(n)}(z)$ generated by applying cyclic reduction to $\varphi_R^{(0)}(z) = \varphi_R(z)$ are such that*

$$\begin{aligned}\Delta(\varphi_R^{(n)}(z)) =&(\boldsymbol{s}_1^{(n)} + \boldsymbol{e}_N + z\boldsymbol{s}_2^{(n)})\boldsymbol{e}_1^{\mathrm{T}}\varphi_R^{(n)}(z) + (\boldsymbol{s}_1^{(n)} + z\boldsymbol{s}_2^{(n)})\boldsymbol{\gamma}^{\mathrm{T}}\varphi_R^{(n)}(z)\\ &+ \varphi_R^{(n)}(z)\boldsymbol{e}_1((\boldsymbol{r}_1^{(n)})^{\mathrm{T}} + z(\boldsymbol{r}_2^{(n)})^{\mathrm{T}})\end{aligned} \tag{9.27}$$

where

$$\begin{aligned}\boldsymbol{\gamma} &= [0, A_1, A_2, \ldots, A_{N-1}],\\ \boldsymbol{s}_2^{(n)} &= \boldsymbol{e}_1 W^{(n)},\\ (\boldsymbol{r}_2^{(n)})^{\mathrm{T}} &= H^{(n)}\boldsymbol{e}_1^{\mathrm{T}}, \quad n \geq 0,\end{aligned}$$

and the $m \times m$ matrices $W^{(n)}$, $H^{(n)}$, and the block vectors $\boldsymbol{s}_2^{(n)}$, $(\boldsymbol{r}_2^{(n)})^{\mathrm{T}}$ are recursively defined by

$$\begin{aligned}\boldsymbol{s}_1^{(n+1)} &= \boldsymbol{s}_1^{(n)} + \mathcal{A}_1^{(n)}(I - \mathcal{A}_0^{(n)})^{-1}\boldsymbol{e}_1 W^{(n)},\\ (\boldsymbol{r}_1^{(n+1)})^{\mathrm{T}} &= (\boldsymbol{r}_1^{(n)})^{\mathrm{T}} + H^{(n)}\boldsymbol{e}_1^{\mathrm{T}}(I - \mathcal{A}_0^{(n)})^{-1}\mathcal{A}_1^{(n)},\\ W^{(n+1)} &= \mathcal{A}_{-1}^{(n)}\boldsymbol{e}_1^{\mathrm{T}}(I - \mathcal{A}_0^{(n)})^{-1}\boldsymbol{e}_1 W^{(n)},\\ H^{(n+1)} &= H^{(n)} + \boldsymbol{e}_1^{\mathrm{T}}(I - \mathcal{A}_0^{(n)})^{-1}\boldsymbol{e}_1\mathcal{A}_{-1}^{(n)}, \quad n \geq 0,\end{aligned} \tag{9.28}$$

with the initial conditions

$$\boldsymbol{s}_1^{(0)} = 0, \quad \boldsymbol{r}_1^{(0)} = 0, \quad W^{(0)} = -I, \quad H^{(0)} = I.$$

Proof We prove (9.27) by induction on n. Let $n = 0$, and observe that, from (9.26), one has

$$[A_1, \ldots, A_{N-1}, 0] = z\boldsymbol{\gamma}^{\mathrm{T}} - \boldsymbol{\gamma}^{\mathrm{T}}\varphi_R^{(0)}(z),$$

which can be rewritten as

$$[A_1, \ldots, A_{N-1}, 0] = -\boldsymbol{e}_1^{\mathrm{T}}\varphi_R^{(0)}(z) + \boldsymbol{e}_1(zI - zA_0 - z^2A_{-1})\boldsymbol{e}_1^{\mathrm{T}} - \boldsymbol{\gamma}^{\mathrm{T}}\varphi_R^{(0)}(z). \tag{9.29}$$

By applying the displacement operator Δ to $\varphi_R^{(0)}(z)$ we find that

$$\Delta(\varphi_R^{(0)}(z)) = \boldsymbol{e}_N\boldsymbol{e}_1^{\mathrm{T}}\varphi_R^{(0)}(z) - z\boldsymbol{e}_N\boldsymbol{e}_1^{\mathrm{T}} + z\boldsymbol{e}_1[A_1, \ldots, A_{N-1}, 0].$$

By replacing in the above equation the expression of the block vector (9.29), we find that

$$\Delta(\varphi_R^{(0)}(z)) = (\boldsymbol{e}_N - z\boldsymbol{e}_1)\boldsymbol{e}_1^{\mathrm{T}}\varphi_R^{(0)}(z) - z\boldsymbol{e}_1\boldsymbol{\gamma}^{\mathrm{T}}\varphi_R^{(0)}(0) + z\varphi_R^{(0)}(z)\boldsymbol{e}_1\boldsymbol{e}_1^{\mathrm{T}}.$$

Therefore (9.27) holds for $n = 0$, with $\boldsymbol{s}_1^{(0)} = 0$, $\boldsymbol{r}_1^{(0)} = 0$, $W^{(0)} = -I$, $H^{(0)} = I$.

Now assume that (9.27) holds for an $n \geq 0$. Let us denote $B(z) = \varphi_R^{(n)}(z)^{-1} - \varphi_R^{(n)}(-z)^{-1}$ and observe that

$$\begin{aligned} B(z)^{-1} &= -(2z)^{-1}\varphi_R^{(n)}(z)\left(\frac{\varphi_R^{(n)}(z) - \varphi_R^{(n)}(-z)}{2z}\right)^{-1}\varphi_R^{(n)}(-z) \\ &= -(2z)^{-1}\varphi_R^{(n)}(z)(I - \mathcal{A}_0^{(n)})^{-1}\varphi_R^{(n)}(-z), \end{aligned} \tag{9.30}$$

and since $B(z) = -B(-z)$,

$$B(z)^{-1} = -(2z)^{-1}\varphi_R^{(n)}(-z)(I - \mathcal{A}_0^{(n)})^{-1}\varphi_R^{(n)}(z). \tag{9.31}$$

Since, $\varphi_R^{(n+1)}(z^2) = 2zB(z)^{-1}$ (see (7.40)) and $\Delta(A^{-1}) = -A^{-1}\Delta(A)A^{-1}$, $\Delta(A+B) = \Delta(A) + \Delta(B)$ (see Section 2.4), we have:

$$\begin{aligned} \Delta(\varphi_R^{(n+1)}(z^2)) =& 2z\Delta(B(z)^{-1}) \\ =& -2zB(z)^{-1}\Delta(B(z))B(z)^{-1} \\ =& -2zB(z)^{-1}(-\varphi_R^{(n)}(z)^{-1}\Delta(\varphi_R^{(n)}(z))\varphi_R^{(n)}(z)^{-1} \\ &+ \varphi_R^{(n)}(-z)^{-1}\Delta(\varphi_R^{(n)}(-z))\varphi_R^{(n)}(-z)^{-1})B(z)^{-1}; \end{aligned}$$

by substituting relation (9.27) in the above equation we obtain

$$\begin{aligned} z^{-1}\Delta(\varphi_R^{(n+1)}(z^2)) =& 2B(z)^{-1}\varphi_R^{(n)}(z)^{-1}(\boldsymbol{s}_1^{(n)} + \boldsymbol{e}_N + z\boldsymbol{s}_2^{(n)})\boldsymbol{e}_1^{\mathrm{T}}B(z)^{-1} \\ &- 2B(z)^{-1}\varphi_R^{(n)}(-z)^{-1}(\boldsymbol{s}_1^{(n)} + \boldsymbol{e}_N - z\boldsymbol{s}_2^{(n)})\boldsymbol{e}_1^{\mathrm{T}}B(z)^{-1} \\ &+ 2B(z)^{-1}\varphi_R^{(n)}(z)^{-1}(\boldsymbol{s}_1^{(n)} + z\boldsymbol{s}_2^{(n)})\boldsymbol{\gamma}^{\mathrm{T}}B(z)^{-1} \\ &- 2B(z)^{-1}\varphi_R^{(n)}(-z)^{-1}(\boldsymbol{s}_1^{(n)} - z\boldsymbol{s}_2^{(n)})\boldsymbol{\gamma}^{\mathrm{T}}B(z)^{-1} \\ &+ 2B(z)^{-1}\boldsymbol{e}_1(\boldsymbol{r}_1^{(n)} + z\boldsymbol{r}_2^{(n)})^{\mathrm{T}}\varphi_R^{(n)}(z)^{-1}B(z)^{-1} \\ &- 2B(z)^{-1}\boldsymbol{e}_1(\boldsymbol{r}_1^{(n)} - z\boldsymbol{r}_2^{(n)})^{\mathrm{T}}\varphi_R^{(n)}(-z)^{-1}B(z)^{-1}. \end{aligned}$$

Since

$$\begin{aligned} 2z\boldsymbol{e}_1^{\mathrm{T}}B(z)^{-1} &= \boldsymbol{e}_1^{\mathrm{T}}\varphi_R^{(n+1)}(z^2), \\ 2z\boldsymbol{\gamma}^{\mathrm{T}}B(z)^{-1} &= \boldsymbol{\gamma}^{\mathrm{T}}\varphi_R^{(n+1)}(z^2), \\ 2zB(z)^{-1}\boldsymbol{e}_1 &= \varphi_R^{(n+1)}(z^2)\boldsymbol{e}_1, \end{aligned}$$

we obtain

$$
\begin{aligned}
\Delta(\varphi_R^{(n+1)}(z^2)) =& B(z)^{-1}\Big[(\varphi_R^{(n)}(z)^{-1} - \varphi_R^{(n)}(-z)^{-1})(\boldsymbol{s}_1^{(n)} + \boldsymbol{e}_N) \\
&+ z(\varphi_R^{(n)}(z)^{-1} + \varphi_R^{(n)}(-z)^{-1})\boldsymbol{s}_2^{(n)}\Big]\,\boldsymbol{e}_1^{\mathrm{T}}\varphi_R^{(n-1)}(z^2) \\
&+ B(z)^{-1}\Big[(\varphi_R^{(n)}(z)^{-1} - \varphi_R^{(n)}(-z)^{-1})\boldsymbol{s}_1^{(n)} \\
&+ z(\varphi_R^{(n)}(z)^{-1} + \varphi_R^{(n)}(-z)^{-1})\boldsymbol{s}_2^{(n)}\Big]\,\boldsymbol{\gamma}^{\mathrm{T}}\varphi_R^{(n+1)}(z^2) \\
&+ \varphi_R^{(n+1)}(z^2)\boldsymbol{e}_1\Big[(\boldsymbol{r}_1^{(n)})^{\mathrm{T}}(\varphi_R^{(n)}(z)^{-1} - \varphi_R^{(n)}(-z)^{-1}) \\
&+ z(\boldsymbol{r}_2^{(n)})^{\mathrm{T}}(\varphi_R^{(n)}(z)^{-1} + \varphi_R^{(n)}(-z)^{-1})\Big]\,B(z)^{-1}.
\end{aligned}
\tag{9.32}
$$

By using the definition of $B(z)$, from the above equation we have

$$
B(z)^{-1}(\varphi_R^{(n)}(z)^{-1} - \varphi_R^{(n)}(-z)^{-1}) = I,
$$

and by (9.30), (9.31)

$$
\begin{aligned}
B(z)^{-1}(\varphi_R^{(n)}(z)^{-1} + \varphi_R^{(n)}(-z)^{-1}) &= -(\varphi_R^{(n)}(z) + \varphi_R^{(n)}(-z))(I - \mathcal{A}_0^{(n)})^{-1}/(2z) \\
&= (\mathcal{A}_1^{(n)} + z\mathcal{A}_{-1}^{(n)})(I - \mathcal{A}_0^{(n)})^{-1}/z.
\end{aligned}
$$

Therefore (9.32) can be rewritten as

$$
\begin{aligned}
\Delta(\varphi_R^{(n+1)}(z)) =& (\boldsymbol{s}_1^{(n+1)} + \boldsymbol{e}_N + z\boldsymbol{s}_2^{(n+1)})\boldsymbol{e}_1^{\mathrm{T}}\varphi_R^{(n+1)}(z) \\
&+ (\boldsymbol{s}_1^{(n+1)} + z\boldsymbol{s}_2^{(n+1)})\boldsymbol{\gamma}^{\mathrm{T}}\varphi_R^{(n+1)}(z) \\
&+ \varphi_R^{(n+1)}(z)\boldsymbol{e}_1((\boldsymbol{r}_1^{(n+1)})^{\mathrm{T}} + z(\boldsymbol{r}_2^{(n+1)})^{\mathrm{T}})
\end{aligned}
$$

where

$$
\begin{aligned}
\boldsymbol{s}_1^{(n+1)} &= \boldsymbol{s}_1^{(n)} + \mathcal{A}_1^{(n)}(I - \mathcal{A}_0^{(n)})^{-1}\boldsymbol{s}_2^{(n)}, \\
\boldsymbol{s}_2^{(n+1)} &= \mathcal{A}_{-1}^{(n)}(I - \mathcal{A}_0^{(n)})^{-1}\boldsymbol{s}_2^{(n)}, \\
(\boldsymbol{r}_1^{(n+1)})^{\mathrm{T}} &= (\boldsymbol{r}_1^{(n)})^{\mathrm{T}} + (\boldsymbol{r}_2^{(n)})^{\mathrm{T}}(I - \mathcal{A}_0^{(n)})^{-1}\mathcal{A}_1^{(n)}, \\
(\boldsymbol{r}_2^{(n+1)})^{\mathrm{T}} &= (\boldsymbol{r}_2^{(n)})^{\mathrm{T}}(I - \mathcal{A}_0^{(n)})^{-1}\mathcal{A}_{-1}^{(n)}.
\end{aligned}
$$

By using the above relations and the equation $\mathcal{A}_{-1}^{(n)} = \boldsymbol{e}_1 A_{-1}^{(n)}\boldsymbol{e}_1^{\mathrm{T}}$, we may easily show by induction on n that

$$
\boldsymbol{s}_2^{(n)} = \boldsymbol{e}_1 W^{(n)}, \quad (\boldsymbol{r}_2^{(n)})^{\mathrm{T}} = H^{(n)}\boldsymbol{e}_1^{\mathrm{T}},
$$

where the matrices $W^{(n)}$, $H^{(n)}$, $n \geq 0$, are recursively defined as in (9.28). □

From the theorem above the displacement representation of $\mathcal{A}_1^{(n)}$ immediately follows:

Corollary 9.19 *For the blocks $\mathcal{A}_1^{(n)}$, $n \geq 0$, one has*

$$\Delta(\mathcal{A}_1^{(n)}) = \boldsymbol{s}_1^{(n)}\boldsymbol{\gamma}^{\mathrm{T}}\mathcal{A}_1^{(n)} + \mathcal{A}_1^{(n)}\boldsymbol{e}_1(\boldsymbol{r}_1^{(n)})^{\mathrm{T}}$$

where the vectors $\boldsymbol{s}_1^{(n)}$, $(\boldsymbol{r}_1^{(n)})^{\mathrm{T}}$, and $\boldsymbol{\gamma}^{\mathrm{T}}$ are defined in Theorem 9.18.

Proof By comparing the block coefficients of zero degree in (9.27), we obtain

$$\Delta(\mathcal{A}_1^{(n)}) = (\boldsymbol{s}_1^{(n)} + \boldsymbol{e}_N)\boldsymbol{e}_1^{\mathrm{T}}\mathcal{A}_1^{(n)} + \boldsymbol{s}_1^{(n)}\boldsymbol{\gamma}^{\mathrm{T}}\mathcal{A}_1^{(n)} + \mathcal{A}_1^{(n)}\boldsymbol{e}_1(\boldsymbol{r}_1^{(n)})^{\mathrm{T}}. \tag{9.33}$$

Since $\boldsymbol{e}_1^{\mathrm{T}}\mathcal{A}_1^{(0)} = 0$, we may easily verify by induction on n that $\boldsymbol{e}_1^{\mathrm{T}}\mathcal{A}_1^{(n)} = 0$ for any $n \geq 0$, therefore the first term in the right-hand side of (9.33) is the null matrix. □

As a consequence of the above result we find a displacement representation of the matrix $\mathcal{A}_1^{(n)}$

$$\mathcal{A}_1^{(n)} = L(\boldsymbol{c}^{(n)})(I + U((\boldsymbol{r}_1^{(n)})^{\mathrm{T}}\mathcal{Z}^{\mathrm{T}})) + L(\boldsymbol{s}_1^{(n)})U((\boldsymbol{w}^{(n)})^{\mathrm{T}}\mathcal{Z}^{\mathrm{T}}), \tag{9.34}$$

where $\boldsymbol{c}^{(n)} = \mathcal{A}_1^{(n)}\boldsymbol{e}_1$ and $(\boldsymbol{w}^{(n)})^{\mathrm{T}} = \boldsymbol{\gamma}^{\mathrm{T}}\mathcal{A}_1^{(n)}$.

Concerning the convergence properties of cyclic reduction, we observe that the roots of $\varphi(z)$ coincide with the roots of $zI - \sum_{i=-1}^{N-1} z^{i+1}A_i$. Therefore, if the original Markov chain is positive recurrent or transient, then also the associated QBD obtained by means of the embedding technique is positive recurrent or transient. Assume that the bi-infinite Markov chain associated with $T_{\pm\infty}[\mathcal{A}(z)]$ satisfies Condition 4.20. Therefore the assumptions of Theorems 7.8 and 7.13 are satisfied so that cyclic reduction can be applied to the QBD problem and convergence is guaranteed.

Now we are ready to describe the main algorithm for performing one step of cyclic reduction. This is summarized in Algorithm 9.4.

The computational cost of stage 1 is $O(m^3N)$ ops; at stage 2 we have to perform products between matrices with block displacement rank at most 2, and block vectors that can be computed in $O(m^3N + m^2N\log N)$ ops; similar computations are involved in stage 4 with the same cost; the cost of stage 3 is $O(m^3)$ ops. Thus, the overall computational cost of one step of cyclic reduction is $O(m^3N + m^2N\log N)$.

9.4 Tree-like stochastic processes

In this section we design algorithms based on functional iterations for solving matrix equations of the kind

$$X + \sum_{i=1}^{d} A_iX^{-1}D_i = C, \tag{9.35}$$

encountered in the analysis of tree-like stochastic processes (see Section 5.8). We recall that, according to the assumptions of Section 5.8, the $m \times m$ matrices C,

---

Algorithm 9.4 Cyclic reduction step for the embedded problem

INPUT: The block vectors $\widehat{\boldsymbol{u}}^{(n)}$, $\boldsymbol{v}^{(n)}$, $\boldsymbol{s}_1^{(n)}$, $\boldsymbol{r}_1^{(n)}$, $\boldsymbol{v}^{(n)}$, $(\boldsymbol{w}^{(n)})^{\mathrm{T}} = \boldsymbol{\gamma}^{\mathrm{T}}\mathcal{A}_1^{(n)}$, $\boldsymbol{c}^{(n)} = \mathcal{A}_1^{(n)}\boldsymbol{e}_1$; the matrices $A_{-1}^{(n)}$, $W^{(n)}$ and $H^{(n)}$;

OUTPUT: The block vectors $\widehat{\boldsymbol{u}}^{(n+1)}$, $\boldsymbol{v}^{(n-1)}$, $\boldsymbol{s}_1^{(n+1)}$, $\boldsymbol{r}_1^{(n+1)}$, $\boldsymbol{v}^{(n+1)}$, $(\boldsymbol{w}^{(n+1)})^{\mathrm{T}} = \boldsymbol{\gamma}^{\mathrm{T}}\mathcal{A}_1^{(n+1)}$, $\boldsymbol{c}^{(n+1)} = \mathcal{A}_1^{(n+1)}\boldsymbol{e}_1$; the matrices $A_{-1}^{(n+1)}$, $W^{(n+1)}$ and $H^{(n+1)}$;

COMPUTATION:

1. Compute $S = (I - (\boldsymbol{v}^{(n)})^{\mathrm{T}}\widehat{\boldsymbol{u}}^{(n)})^{-1}$, $T = S(\boldsymbol{v}^{(n)})^{\mathrm{T}}\mathcal{Z}\mathcal{Z}^{\mathrm{T}}\widehat{\boldsymbol{u}}^{(n)}$.
2. Compute

$$\begin{aligned}
A_{-1}^{(n+1)} &= A_{-1}^{(n)}(SV_1 - T + I)A_{-1}^{(n)}\\
\widehat{\boldsymbol{u}}^{(n+1)} &= \widehat{\boldsymbol{u}}^{(n)} + \mathcal{A}_1^{(n)}\widehat{\boldsymbol{u}}^{(n)}(SV_1 + T + I)A_{-1}^{(n)},\\
(\boldsymbol{v}^{(n+1)})^{\mathrm{T}} &= (\boldsymbol{v}^{(n)})^{\mathrm{T}} + A_{-1}^{(n)}S(\boldsymbol{v}^{(n)})^{\mathrm{T}}\mathcal{A}_1^{(n)},
\end{aligned}$$

   where V_1 is the first block element of $(\boldsymbol{v}^{(n)})^{\mathrm{T}}$ and where the products $\mathcal{A}_1^{(n)}\widehat{\boldsymbol{u}}^{(n)}$ and $(\boldsymbol{v}^{(n)})^{\mathrm{T}}\mathcal{A}_1^{(n)}$ are computed by means of the displacement representation (9.34) of $\mathcal{A}_1^{(n)}$.
3. Compute

$$\begin{aligned}
W^{(n+1)} &= A_{-1}^{(n)}\boldsymbol{e}_1^{\mathrm{T}}(I - \mathcal{A}_0^{(n)})^{-1}\boldsymbol{e}_1 W^{(n)},\\
H^{(n+1)} &= H^{(n)} + \boldsymbol{e}_1^{\mathrm{T}}(I - \mathcal{A}_0^{(n)})^{-1}\boldsymbol{e}_1 A_{-1}^{(n)},
\end{aligned}$$

   where the matrix vector products involving the matrix $(I - \mathcal{A}_0^{(n)})^{-1}$ are computed by using the representation in (9.23).
4. Compute

$$\begin{aligned}
\boldsymbol{s}_1^{(n+1)} &= \boldsymbol{s}_1^{(n)} + \mathcal{A}_1^{(n)}(I - \mathcal{A}_0^{(n)})^{-1}\boldsymbol{e}_1 W^{(n)},\\
(\boldsymbol{r}_1^{(n+1)})^{\mathrm{T}} &= (\boldsymbol{r}_1^{(n)})^{\mathrm{T}} + H^{(n)}\boldsymbol{e}_1^{\mathrm{T}}(I - \mathcal{A}_0^{(n)})^{-1}\mathcal{A}_1^{(n)},\\
\boldsymbol{c}^{(n+1)} &= \mathcal{A}_1^{(n)}(I - \mathcal{A}_0^{(n)})^{-1}\boldsymbol{c}^{(n)},\\
(\boldsymbol{w}^{(n+1)})^{\mathrm{T}} &= (\boldsymbol{w}^{(n)})^{\mathrm{T}}(I - \mathcal{A}_0^{(n)})^{-1}\mathcal{A}_1^{(n)},
\end{aligned}$$

   where the matrix vector products involving the matrix $(I - \mathcal{A}_0^{(n)})^{-1}$ are computed by using the representation in (9.23), the matrix vector products involving the matrix $\mathcal{A}_1^{(n)}$ are computed by using the displacement representation (9.34).

A_i, and D_i are such that $C = I - B$, $B \geq 0$, $A_i, D_i \geq 0$, for $i = 1, \ldots, d$, B is substochastic, and the matrices $I + C + D_i + A_1 + \cdots + A_d$, $i = 1, \ldots, d$, are stochastic.

We define the matrices $G_i = (-S)^{-1}D_i$ for $i = 1, \ldots, d$ which have the following interpretation: $(G_i)_{k,k'}$ is the probability that, starting at time 0 from the state $(i;k)$ in $\mathcal{N}_i$, the tree-like process eventually moves to the root node and (k') is the first state visited there. If the process is positive recurrent, then G_i is stochastic for all i.

With this definition, we may rewrite (9.35) as a system of coupled equations:

$$S = C + \sum_{i=1}^{d} A_i G_i, \tag{9.36}$$

$$G_i = (-S)^{-1} D_i, \qquad \text{for } 1 \le i \le d. \tag{9.37}$$

9.4.1 Linearly convergent iterations

From (9.36) and (9.37) we may obtain S by fixed point iterations, as stated in the following theorem (see Latouche and Ramaswami [79, Section 14.3] and Yeung and Alfa [121, Section 8] for the proof).

Theorem 9.20 *The sequences $\{S_n\}_{n\ge0}$ and $\{G_{i,n}\}_{n\ge0}$, $i = 1, \ldots, d$, defined by*

$$S_n = C + \sum_{i=1}^{d} A_i G_{i,n}, \tag{9.38}$$

$$G_{i,n+1} = (-S_n)^{-1} D_i, \qquad \textit{for } 1 \le i \le d,\ n \ge 0, \tag{9.39}$$

with $G_{1,0} = \ldots = G_{d,0} = 0$, monotonically converge to S and G_i for $i = 1, \ldots, d$, respectively.

It is useful to note that $G_{i,n}$ is the matrix of first passage probabilities from $\mathcal{N}_i$ to $\mathcal{N}_\emptyset$ in a truncated process where transitions are not allowed beyond the level $\mathcal{S}_n$. Furthermore, it is shown in [121] that S is minimal in the following sense: $S = T - I$ where T is a substochastic matrix satisfyng the minimality property:

Lemma 9.21 *The matrix $T = S + I$ is the minimal nonnegative solution of the equation*

$$X = B + \sum_{i=1}^{d} A_i (I - X)^{-1} D_i.$$

Armed with these, we may define another converging sequence.

Theorem 9.22 *The sequences $\{\bar{S}_n\}_{n\ge0}$ and $\{\bar{G}_{i,n}\}_{n\ge0}$, $i = 1, \ldots, d$, defined by (9.38), (9.39) with $\bar{G}_{1,0} = \ldots = \bar{G}_{d,0} = I$ converge to S and G_i for $i = 1, \ldots, d$, respectively. Moreover, $\bar{G}_{i,n}$ is stochastic for all i and n.*

Proof The proof proceeds along the same lines as the proof of [79, Theorem 8.3.1] and we briefly outline it here.

First, we consider a tree-like process on the finite set of levels $\mathcal{N}_\emptyset \cup \mathcal{S}_1 \cup \cdots \cup \mathcal{S}_n$. We keep the same transition probabilities as in the original process except at the last level $\mathcal{S}_n$: if $\mathcal{N}_J$ is a leaf, that is, if J is of length $|J| = n$, it is not possible to move from $(J;k)$ to any of the states $(J+i;k')$, with $1 \le i \le d$; instead, the transition probability to (J,k') is now equal to $(B + A_1 + \cdots + A_d)_{k,k'}$.

By repeating verbatim the argument on [79, Page 180] we show that the new tree-like process is irreducible, for every $n \ge 1$. Moreover, one readily verifies that $\bar{G}_{i,n}$ is the matrix of first passage probabilities from $\mathcal{N}_i$ to $\mathcal{N}_\emptyset$ in this new process, from which we conclude that $\bar{S}_n$ is nonsingular and that $\bar{G}_{i,n}$ is a stochastic matrix for all n and i.

Since the set of stochastic matrices of order m is compact, the sequence $\{(\bar{G}_{1,n}, \ldots, \bar{G}_{d,n}) : n \ge 0\}$ has at least one accumulation point. We denote by $(G_1^*, \ldots, G_d^*)$ one such point and we choose a subset of indices $\{n_1, n_2, \ldots\}$ such that $\lim_{t\to\infty} \bar{G}_{i,n_t} = G_i^*$. By Theorem 9.20 we have that $\lim_{t\to\infty} G_{i,n_t} = G_i$. Note that both G_i^* and G_i are stochastic matrices.

Furthermore, $\bar{G}_{i,0} \ge G_{i,0}$ and we easily show by induction that $\bar{G}_{i,n} \ge G_{i,n}$ for all n, so that $G_i^* \ge G_i$. Since $G_i^*\mathbf{1} = G_i\mathbf{1} = \mathbf{1}$, necessarily $G_i^* = G_i$ and the sequence $\{(\bar{G}_{1,n}, \ldots, \bar{G}_{d,n}) : n \ge 0\}$ has only one accumulation point, which proves the theorem. □

Both the sequences $\{S_n\}_n$ and $\{\bar{S}_n\}_n$ are linearly convergent, and each step of the fixed point iteration requires one $m \times m$ matrix inversion and $2d$ matrix products. The sequence $\{\bar{S}_n\}_n$ converges faster than $\{S_n\}_n$ and we estimate the asymptotic rate of convergence in Theorem 9.24 below. Before doing so, however, we need to recall the following notation introduced in (2.1) of Section 2.1: $\mathrm{vec}(A)$ is the mn-dimensional vector obtained by arranging column-wise the elements of the $m \times n$ matrix A; $C = A \otimes B$ is the matrix having block elements $C_{i,j} = a_{i,j}B$, where $A = (a_{i,j})_{i,j}$. We make use of the fact that $Y = AXB$ if and only if $\mathrm{vec}(Y) = (B^{\mathrm{T}} \otimes A)\,\mathrm{vec}(X)$, where A, B, X, Y are matrices of compatible size (see (2.2)).

The following result relates the error at two subsequent steps:

Theorem 9.23 *Let $\{S_n\}_{n\ge 0}$ be the sequence defined by (9.38)–(9.39), with arbitrary initial approximations $G_{i,0}$, $i = 1, \ldots, d$, and define $\boldsymbol{\epsilon}_n = \mathrm{vec}(E_n)$, where $E_n = S - S_n$. The following relation holds:*

$$\boldsymbol{\epsilon}_{n+1} = \left(\sum_{i=1}^{d} ((-S_n)^{-1} D_i)^{\mathrm{T}} \otimes R_i \right) \boldsymbol{\epsilon}_n, \quad n \ge 0. \tag{9.40}$$

Proof By subtracting (9.38) from (9.36) and by observing that

$$G_i - G_{i,n+1} = S^{-1} E_n S_n^{-1} D_i, \quad i = 1, \ldots, d, \quad n \ge 0,$$

we obtain that

$$E_{n+1} = \sum_{i=1}^{d} R_i E_n (-S_n)^{-1} D_i, \quad n \ge 0,$$

from which (9.40) immediately follows. □

In the following, $\lambda(A)$ denotes the set of the eigenvalues of A. From Theorem 9.23 we derive the following estimate of the mean asymptotic rate of convergence.

Theorem 9.24 *The eigenvalues of the matrix* $R = \sum_{i=1}^{d} R_i$ *are such that*

$$\lambda(R) \subset \lambda\left(\sum_{i=1}^{d} G_i^{\mathrm{T}} \otimes R_i\right).$$

Moreover, if R *has a positive left eigenvector, corresponding to its spectral radius, then*

$$\rho(R) = \rho\left(\sum_{i=1}^{d} G_i^{\mathrm{T}} \otimes R_i\right)$$

and the sequences $\{S_n\}_{n\geq 0}$, $\{\bar{S}_n\}_{n\geq 0}$ *defined in* (9.38)–(9.39), *obtained with* $G_{1,0} = \ldots = G_{d,0} = 0$ *and* $\bar{G}_{1,0} = \ldots = \bar{G}_{d,0} = I$, *respectively, are such that*

$$\lim_{n\to\infty} \|\boldsymbol{\epsilon}_n\|^{1/n} = \rho(R), \tag{9.41}$$

$$\lim_{n\to\infty} \|\bar{\boldsymbol{e}}_n\|^{1/n} \leq \max\left\{|\sigma| : \sigma \in \lambda\left(\sum_{i=1}^{d} G_i^{\mathrm{T}} \otimes R_i\right) \setminus \lambda(R)\right\} \tag{9.42}$$

for any vector norm $||\cdot||$, *where* $\boldsymbol{\epsilon}_n = \mathrm{vec}(S - S_n)$, $\bar{\boldsymbol{e}}_n = \mathrm{vec}(S - \bar{S}_n)$.

Proof We proceed in a manner similar to Section 6.2 and define the orthogonal $m^2 \times m^2$ matrix $\Pi = [\Pi_1|\Pi_2]$, with $\Pi_1 = (1/\sqrt{m})\mathbf{1} \otimes I \in \mathbb{R}^{m^2\times m}$, and $\Pi_2 \in \mathbb{R}^{m^2\times(m^2-m)}$. Since $\mathbf{1}^{\mathrm{T}} G_i^{\mathrm{T}} = \mathbf{1}^{\mathrm{T}}$, we have that

$$\Pi_1^{\mathrm{T}} \left(\sum_{i=1}^{d} G_i^{\mathrm{T}} \otimes R_i\right) \Pi_2 = 0$$

and

$$\Pi_1^{\mathrm{T}} \left(\sum_{i=1}^{d} G_i^{\mathrm{T}} \otimes R_i\right) \Pi_1 = R,$$

that is,

$$\Pi^{\mathrm{T}} \left(\sum_{i=1}^{d} G_i^{\mathrm{T}} \otimes R_i\right) \Pi = \begin{bmatrix} R & 0 \\ H & K \end{bmatrix}, \tag{9.43}$$

where $H = \Pi_2^{\mathrm{T}} \left(\sum_{i=1}^{d} G_i^{\mathrm{T}} \otimes R_i\right) \Pi_1$ and $K = \Pi_2^{\mathrm{T}} \left(\sum_{i=1}^{d} G_i^{\mathrm{T}} \otimes R_i\right) \Pi_2$. Thus, it is clear that $\lambda(R) \subset \lambda\left(\sum_{i=1}^{d} G_i^{\mathrm{T}} \otimes R_i\right)$.

Let $\boldsymbol{v}^{\mathrm{T}} > 0$ be a left eigenvector of R corresponding to $\rho(R)$. Since G_i, $i = 1, \ldots, d$, are stochastic, it follows that

$$(\mathbf{1}^{\mathrm{T}} \otimes \boldsymbol{v}^{\mathrm{T}}) \left(\sum_{i=1}^{d} G_i^{\mathrm{T}} \otimes R_i \right) = \rho(R)(\mathbf{1}^{\mathrm{T}} \otimes \boldsymbol{v}^{\mathrm{T}}), \tag{9.44}$$

that is, $\mathbf{1}^{\mathrm{T}} \otimes \boldsymbol{v}^{\mathrm{T}}$ is a left eigenvector corresponding to the eigenvalue $\rho(R)$. Since $\sum_{i=1}^{d} G_i^{\mathrm{T}} \otimes R_i$ is a nonnegative matrix and $\mathbf{1}^{\mathrm{T}} \otimes \boldsymbol{v}^{\mathrm{T}}$ is a positive vector, it follows that $\rho\left(\sum_{i=1}^{d} G_i^{\mathrm{T}} \otimes R_i\right) = \rho(R)$ (see [116]).

The relation (9.41) is proved as follows. Since $\boldsymbol{v}$ has positive components, we may define the vector norm $||\cdot||$ such that $||\boldsymbol{x}|| = (\mathbf{1}^{\mathrm{T}} \otimes \boldsymbol{v}^{\mathrm{T}})|\boldsymbol{x}|$, where $\boldsymbol{x} \in \mathbb{R}^{m^2}$ and $|\boldsymbol{x}| = (|x_i|)_i$. Since the vectors $\boldsymbol{\epsilon}_n$ are nonnegative and since the sequences $\{(-S_n)^{-1}D_i\}_{n\geq 0}$ converge monotonically to G_i for $i = 1, \ldots, d$, we obtain from (9.40) that

$$\begin{aligned} \|\boldsymbol{\epsilon}_n\| &= (\mathbf{1}^{\mathrm{T}} \otimes \boldsymbol{v}^{\mathrm{T}})\boldsymbol{\epsilon}_n \leq (\mathbf{1}^{\mathrm{T}} \otimes \boldsymbol{v}^{\mathrm{T}}) \left(\sum_{i=1}^{d} G_i^{\mathrm{T}} \otimes R_i \right) \boldsymbol{\epsilon}_{n-1} \\ &= \rho(R)(\mathbf{1}^{\mathrm{T}} \otimes \boldsymbol{v}^{\mathrm{T}})\boldsymbol{\epsilon}_{n-1} = \rho(R)||\boldsymbol{\epsilon}_{n-1}||. \end{aligned}$$

Hence $\lim_{n\to\infty} \|\boldsymbol{\epsilon}_n\|^{1/n} \leq \rho(R)$. Let us now show the opposite inequality. From (9.40) one has

$$\begin{aligned} \|\boldsymbol{\epsilon}_n\| =&(\mathbf{1}^{\mathrm{T}} \otimes \boldsymbol{v}^{\mathrm{T}})\boldsymbol{\epsilon}_n \\ =&(\mathbf{1}^{\mathrm{T}} \otimes \boldsymbol{v}^{\mathrm{T}}) \left(\left(\sum_{i=1}^{d} G_i^{\mathrm{T}} \otimes R_i \right) - \left(\sum_{i=1}^{d} (G_i - (-S_{n-1})^{-1}D_i)^{\mathrm{T}} \otimes R_i \right) \right) \boldsymbol{\epsilon}_{n-1} \\ =&\rho(R)||\boldsymbol{\epsilon}_{n-1}|| - (\mathbf{1}^{\mathrm{T}} \otimes \boldsymbol{v}^{\mathrm{T}}) \left(\sum_{i=1}^{d} (G_i - (-S_{n-1})^{-1}D_i)^{\mathrm{T}} \otimes R_i \right) \boldsymbol{\epsilon}_{n-1}. \end{aligned}$$

Since the sequences $\{(-S_n)^{-1}D_i\}_{n\geq 0}$ converge monotonically to G_i for $i = 1, \ldots, d$, for any $\epsilon > 0$ there exists an integer n_0 such that

$$G_i - (-S_n)^{-1}D_i \leq \frac{\epsilon}{m}(\mathbf{1}\ \mathbf{1}^{\mathrm{T}})$$

for any $i = 1, \ldots, d$ and for any $n \geq n_0$. Thus, we obtain that for any $n \geq n_0$

$$\begin{aligned} \|\boldsymbol{\epsilon}_n\| \geq&\rho(R)||\boldsymbol{\epsilon}_{n-1}|| - (\epsilon/m)(\mathbf{1}^{\mathrm{T}} \otimes \boldsymbol{v}^{\mathrm{T}}) \left(\sum_{i=1}^{d} (\mathbf{1}\ \mathbf{1}^{\mathrm{T}}) \otimes R_i \right) \boldsymbol{\epsilon}_{n-1} \\ =&(1-\epsilon)\rho(R)||\boldsymbol{\epsilon}_{n-1}||, \end{aligned}$$

so that $\lim_{n\to\infty} \|\boldsymbol{\epsilon}_n\|^{1/n} \geq \rho(R)(1-\epsilon)$. Since ϵ is arbitrary, we deduce that $\lim_{n\to\infty} \|\boldsymbol{\epsilon}_n\|^{1/n} = \rho(R)$, and this equality holds for any vector norm for the equivalence of the norms (see Theorem A.1 in the Appendix).

We finally prove (9.42). Since the matrices $\bar{G}_{i,n}$ are stochastic for any i and n, it can be easily shown by induction on n that $\Pi_1^{\mathrm{T}}\boldsymbol{\epsilon}_n = \mathbf{0}$ for all n, and thus $\Pi^{\mathrm{T}}\boldsymbol{\epsilon}_n = \left[\frac{\mathbf{0}}{\boldsymbol{h}_n}\right]$, where $\boldsymbol{h}_n = \Pi_2^{\mathrm{T}}\boldsymbol{\epsilon}_n$. Since

$$\begin{aligned}\Pi^{\mathrm{T}}\boldsymbol{\epsilon}_n &= \Pi^{\mathrm{T}}\left(\sum_{i=1}^{d}((-S_{n-1})^{-1}D_i)^{\mathrm{T}} \otimes R_i\right)\Pi(\Pi^{\mathrm{T}}\boldsymbol{\epsilon}_{n-1})\\ &= \begin{bmatrix} R & 0 \\ H_{n-1} & K_{n-1}\end{bmatrix}(\Pi^{\mathrm{T}}\boldsymbol{\epsilon}_{n-1}),\end{aligned}$$

where

$$H_n = \Pi_2^{\mathrm{T}}\left(\sum_{i=1}^{d}((-S_n)^{-1}D_i)^{\mathrm{T}} \otimes R_i\right)\Pi_1$$

and

$$K_n = \Pi_2^{\mathrm{T}}\left(\sum_{i=1}^{d}((-S_n)^{-1}D_i)^{\mathrm{T}} \otimes R_i\right)\Pi_2,$$

one has $\boldsymbol{h}_n = K_{n-1}\boldsymbol{h}_{n-1}$, from which we deduce (9.42) by applying the same techniques as in [88]. □

From the above theorem it follows that the convergence of the sequence obtained with $G_{i,0} = I$ is generally faster than the one obtained with $G_{i,0} = 0$, when the Perron–Frobenius eigenvector $\boldsymbol{v}^{\mathrm{T}}$ of R is strictly positive.

Remark 9.25 The property that the Perron–Frobenius eigenvector $\boldsymbol{v}^{\mathrm{T}}$ of R is strictly positive is related to irreducibility characteristics of the stochastic process under study and does not seem to be a very restrictive assumption. For instance, for $d = 1$, if $A_1 + B + D_1$ is irreducible then $\boldsymbol{v}^{\mathrm{T}} > 0$ (Neuts [91, Lemma 1.3.2]). For a general tree-like process, such simple sufficient conditions are not as readily available.

We report in Algorithm 9.5 the method for solving tree-like stochastic processes based on functional iterations.

9.4.2 Newton's iterations

Another tool for solving tree-like stochastic processes is applying Newton's iteration. Define the matrices

$$G = \begin{bmatrix} G_1 \\ \vdots \\ G_d \end{bmatrix}, \quad D = \begin{bmatrix} D_1 \\ \vdots \\ D_d \end{bmatrix}, \quad A = \begin{bmatrix} A_1 & \cdots & A_d \end{bmatrix},$$

and the four matrix operators

$$\begin{aligned}
\mathcal{I} &: \mathbb{R}^{md\times m} \to \mathbb{R}^{md\times m} : \mathcal{I}X = X,\\
\mathcal{M} &: \mathbb{R}^{md\times m} \to \mathbb{R}^{m\times m} : \mathcal{M}X = C + AX,\\
\mathcal{L} &: \mathbb{R}^{m\times m} \to \mathbb{R}^{md\times m} : \mathcal{L}V = [I_d \otimes (-V)^{-1}]D,\\
\mathcal{F} &: \mathbb{R}^{md\times m} \to \mathbb{R}^{md\times m} : \mathcal{F}X = (\mathcal{I} - \mathcal{L}\mathcal{M})X.
\end{aligned}$$

Algorithm 9.5 Functional iteration for tree-like processes

INPUT: The $m \times m$ matrices C, A_i, and D_i are such that $C = I - B$, $B \geq 0$, $A_i, D_i \geq 0$, for $i = 1, \ldots, d$, B is substochastic, and the matrices $I + C + D_i + A_1 + \cdots + A_d$, $i = 1, \ldots, d$, are stochastic. An error bound ϵ.

OUTPUT: An approximation X to the minimal solution S of the matrix equation (9.35).

COMPUTATION:

1. Set $Y = C + \sum_{i=1}^{d} A_i$.
2. Compute $W_i = Y^{-1} D_i$, $i = 1, \ldots, d$; and set

$$X = C - \sum_{i=1}^{d} A_i W_i.$$

3. If $\|X - Y\|_\infty \leq \epsilon$, then output X; otherwise set $Y = X$ and continue from step 2.

The first is the identity operator. The operator $\mathcal{M}$ is defined for matrices X in $< 0, G >$, that is, for matrices such that $0 \leq X \leq G$, and the operator $\mathcal{L}$ is defined for matrices V in $< C, S >$.

It is not difficult to verify that if $C \leq V \leq S$, then $0 \leq -C^{-1} \leq -V^{-1} \leq -S^{-1}$, so that $\mathcal{L}V$ is in $< 0, G >$; similarly, it is clear that if X is in $< 0, G >$, then $\mathcal{M}X$ is in $< C, S >$. We eventually conclude that G is the minimal nonnegative solution of $\mathcal{F}G = 0$.

Newton's method yields the sequence of matrices

$$\widehat{G}_{n+1} = \widehat{G}_n - \mathcal{F}'(\widehat{G}_n)^{-1} \mathcal{F}\widehat{G}_n, \tag{9.45}$$

with $\widehat{G}_0 = 0$. As we show below, the chain rule applies, so that

$$\mathcal{F}'(X) = \mathcal{I} - \mathcal{L}'(\mathcal{M}X)\mathcal{M}'(X)$$

with

$$\begin{aligned} \mathcal{M}'(X)H &= AH \\ \mathcal{L}'(V)K &= \left[I \otimes V^{-1} K V^{-1}\right] D. \end{aligned}$$

To see that the chain rule applies, one needs to repeat nearly verbatim the argument of Section 6.3. We only repeat the salient steps here.

Lemma 9.26 *The operators $\mathcal{M}$ and $\mathcal{M}'$ are Lipschitz-continuous on $< 0, G >$ and $\mathcal{M}'(X)$ is the Fréchet derivative of $\mathcal{M}$. The norm of $\mathcal{M}'$ is uniformly bounded on $< 0, G >$.*

Proof The proof directly follows that of Lemma 6.24; it is much simpler here because $[\mathcal{M}'(X) - \mathcal{M}'(Y)]H \equiv 0$. □

Lemma 9.27 *The operator $\mathcal{L}'$ is Lipschitz-continuous and uniformly bounded in $< C, S >$.*

Proof We use the infinity norm for matrices.

If V and W are in $< C, S >$ and if $\|K\| = 1$, then

$$\begin{aligned}
&\|[\mathcal{L}'(V) - \mathcal{L}'(W)]K\| \\
&= \|[I \otimes V^{-1}KV^{-1} - I \otimes W^{-1}KW^{-1}]D\| \\
&\leq \|D\| \cdot \|V^{-1}KV^{-1} - W^{-1}KW^{-1}\| \\
&\leq \|D\|(\|(V^{-1} - W^{-1})KV^{-1}\| + \|W^{-1}K(V^{-1} - W^{-1})\|) \\
&\leq \|D\| \cdot \|W^{-1}\| \cdot \|V^{-1}\| \cdot \|V - W\|(\|V^{-1}\| + \|W^{-1}\|) \\
&\leq 2\|D\| \cdot \|S^{-1}\|^3\|V - W\|
\end{aligned}$$

since $0 \leq -V^{-1}, -W^{-1} \leq -S^{-1}$. The remainder of the proof is identical to that of Lemma 6.25. □

Since both $\mathcal{M}'$ and $\mathcal{L}'$ are Fréchet derivatives, the chain rule applies, $\mathcal{F}'(X) = \mathcal{I} - \mathcal{L}'(\mathcal{M}X)\mathcal{M}'(X)$ is the Fréchet-derivative of $\mathcal{F}$ at X by [95, Proposition 3.1.7] and Newton's sequence (9.45) may be written as

$$\widehat{G}_{n+1} = \widehat{G}_n - Z_n \tag{9.46}$$

with

$$Z_n = \left[\mathcal{I} - \mathcal{L}'(\mathcal{M}\widehat{G}_n)\mathcal{M}'(\widehat{G}_n)\right]^{-1} (\mathcal{I} - \mathcal{L}\mathcal{M})\,\widehat{G}_n.$$

If we define

$$\widehat{S}_n = \mathcal{M}\widehat{G}_n = C + A\widehat{G}_n,$$

we may write that Z_n is the solution of

$$\left[\mathcal{I} - \mathcal{L}'(\widehat{S}_n)\mathcal{M}'(\widehat{G}_n)\right] Z_n = \widehat{G}_n - \mathcal{L}\mathcal{M}\widehat{G}_n$$

and this may be written after a few algebraic manipulations as

$$Z_n - \left[I \otimes \widehat{S}_n^{-1}AZ_n\widehat{S}_n^{-1}\right] D = \widehat{G}_n + \left[I \otimes \widehat{S}_n^{-1}\right] D. \tag{9.47}$$

If we premultiply this equation by A and if we define

$$Y_n = AZ_n, \tag{9.48}$$

we find that Y_n is a solution of the equation

$$Y_n + \sum_{i=1}^{d} A_i\widehat{S}_n^{-1}Y_n(-\widehat{S}_n)^{-1}D_i = L_n, \tag{9.49}$$

where

$$L_n = \widehat{S}_n - C - \sum_{i=1}^{d} A_i \widehat{S}_n^{-1} D_i.$$

Now, with $AZ_n = A\widehat{G}_n - A\widehat{G}_{n+1} = \widehat{S}_n - \widehat{S}_{n+1}$, we have

$$\widehat{S}_{n+1} = \widehat{S}_n - Y_n.$$

Finally, we have by (9.46, 9.47, 9.48) that

$$\widehat{G}_{n+1} = -\left[I \otimes \widehat{S}_n^{-1} Y_n \widehat{S}_n^{-1}\right] D - \left[I \otimes \widehat{S}_n^{-1}\right] D,$$

so that

$$\widehat{G}_{i,n+1} = \widehat{S}_n^{-1}(Y_n(-\widehat{S}_n)^{-1} D_i - D_i).$$

In summary, Newton's method generates $d+1$ sequences $\{\widehat{S}_n\}_{n\geq 0}$, $\{\widehat{G}_{i,n}\}_{n\geq 0}$, $i = 1, \ldots, d$, recursively defined by

$$\begin{aligned}
&\widehat{S}_{n+1} = \widehat{S}_n - Y_n,\\
&\widehat{G}_{i,n+1} = \widehat{S}_n^{-1}(Y_n(-\widehat{S}_n)^{-1} D_i - D_i), \ i = 1, \ldots, d, \ n \geq 0,\\
&\widehat{S}_0 = C,
\end{aligned}$$

where Y_n solves (9.49). The above sequences quadratically converge to S and G_i, $i = 1, \ldots, d$, respectively.

The applicability of this method relies on the efficient solution of the linear matrix equation (9.49). Indeed, the computation of L_n and of the matrix coefficients in (9.49) requires $3d$ matrix products, while the computation of $\widehat{S}_{n+1}$ and $\widehat{G}_{i,n+1}$, $1 \leq i \leq d$, requires $2d$ more matrix products.

Equation (9.49) constitutes a linear system of m^2 equations in m^2 unknowns. The customary solution algorithms like Gaussian elimination would cost $O(m^6)$ ops. There are efficient direct methods in the literature for solving matrix equations of the form $H_1 Y K_1 + H_2 Y K_2 = L$, which are known as Sylvester equations; these methods consist in performing a Hessenberg–Schur orthogonal transformation of the matrix coefficients H_1, H_2, K_1, K_2, and in solving a quasi-triangular linear system for the overall cost $O(m^3)$ (see [45]). Unfortunately, the more general case $\sum_{i=1}^{r} H_i Y K_i = L$, where $r > 2$, cannot be solved with these methods.

To solve (9.49), we may apply two kinds of fixed point iterations. The first fixed point iteration consists in generating the sequence $\{Y_{n,h}\}_{h\geq 0}$ defined by

$$Y_{n,h+1} = L_n + \sum_{i=1}^{d} (A_i \widehat{S}_n^{-1}) Y_{n,h} (\widehat{S}_n^{-1} D_i), \quad h \geq 0, \tag{9.50}$$

with $Y_{n,0} = L_n$. The second one generates the sequence $\{Y_{n,h}\}_{h\geq 0}$ defined by

$$Y_{n,h+1} - (A_r \widehat{S}_n^{-1}) Y_{n,h+1} (\widehat{S}_n^{-1} D_r) = L_n + \sum_{\substack{i=1\\ i\neq r}}^{d} (A_i \widehat{S}_n^{-1}) Y_{n,h} (\widehat{S}_n^{-1} D_i), \quad h \geq 0, \tag{9.51}$$

with $Y_{n,0} = L_n$, and where r is such that $1 \leq r \leq d$.

Concerning the choice of r we observe that (9.49) can be viewed as a linear system with matrix $I-B_n$, $B_n = \sum_{i=1}^{d} \hat{S}_n^{-1} D_i \otimes A_i S_n^{-1}$. If $B_n \geq 0$ and $\rho(B_n) < 1$ then $I - B_n$ is a nonsingular M-matrix, therefore, for the properties of regular splittings (see Theorem A.15) the more convenient choice for r is the one for which $||A_r \widehat{S}_n^{-1}|| \cdot ||\widehat{S}^{-1} D_r||$ is maximum for a given norm $|| \cdot ||$.

From numerical experiments it seems that the functional iteration (9.51) is generally faster than (9.50), but each step requires the solution of a Sylvester matrix equation. In order to solve it, we can apply the algorithm of [45], where the Hessenberg–Schur orthogonal transformation of the matrix coefficients is done once and for all. In principle, the convergence of the sequences (9.50) and (9.51) is not guaranteed; in fact, examples can be constructed where (9.51) diverges.

We summarize the method based on Newton's iteration as Algorithm 9.6.

Algorithm 9.6 Newton's iteration for tree-like processes

INPUT: The $m \times m$ matrices C, A_i, and D_i are such that $C = I - B$, $B \geq 0$, $A_i, D_i \geq 0$, for $i = 1, \ldots, d$, B is substochastic, and the matrices $I + C + D_i + A_1 + \cdots + A_d$, $i = 1, \ldots, d$, are stochastic. An error bound ϵ.

OUTPUT: An approximation X to the minimal solution S of the matrix equation (9.35).

COMPUTATION:

1. Set $X = C$.
2. Compute $W_i = X^{-1} D_i$, $i = 1, \ldots, d$.
3. Compute $L = X - C + \sum_{i=1}^{d} A_i W_i$.
4. Compute Y solving the equation
$$Y + \sum_{i=1}^{d} A_i X^{-1} Y W_i = L$$
and set $X = X + Y$.
5. If $||Y||_\infty \leq \epsilon$, then output X; otherwise continue from step 2.

9.4.3 *Cyclic reduction*

Cyclic reduction can be combined with functional iterations for solving the equation (9.35). By writing (9.37) as $D_i + SG_i = 0$ and replacing S by the right-hand side of (9.36), we obtain the system

$$D_i + (C + \sum_{\substack{j=1 \\ j \neq i}}^{d} A_j G_j) G_i + A_i G_i^2 = 0, \qquad (9.52)$$

for $i = 1, \ldots, d$. If we define F_i as

$$F_i = C + \sum_{\substack{j=1 \\ j \neq i}}^{d} A_j G_j, \tag{9.53}$$

then (9.52) becomes

$$D_i + F_i G_i + A_i G_i^2 = 0, \quad 1 \leq i \leq d. \tag{9.54}$$

For each i, we recognize this as the equation which defines the matrix of first passage probabilities to lower levels in a QBD process characterized by the transition matrices A_i, F_i and D_i. Those quadratic equations may be solved by the cyclic reduction method of Section 7.3 or by means of logarithmic reduction of Section 7.2 which converge quadratically. Thus, we may determine the matrices G_i, $1 \leq i \leq d$, and the matrix S by the following iterative procedure: we define sequences of matrices $\{G_{i,n}\}_{n \geq 0}$, for $i = 1, \ldots, d$, such that, for $n \geq 1$ and for each i, $G_{i,n}$ is the minimal nonnegative solution of the quadratic matrix equation

$$D_i + F_{i,n} G_{i,n} + A_i G_{i,n}^2 = 0, \tag{9.55}$$

where

$$F_{i,n} = C + \sum_{j=1}^{i-1} A_j G_{j,n} + \sum_{j=i+1}^{d} A_j G_{j,n-1}. \tag{9.56}$$

Theorem 9.28 *The sequences $\{G'_{i,n}\}_{n \geq 0}$ obtained from (9.55), (9.56), starting with $G'_{1,0} = G'_{2,0} = \cdots = G'_{d,0} = 0$, monotonically converge to G_i, for $1 \leq i \leq d$. The sequences $\{\tilde{G}_{i,n}\}_{n \geq 0}$ obtained from (9.55), (9.56), starting with $\tilde{G}_{1,0} = \tilde{G}_{2,0} = \cdots = \tilde{G}_{d,0} = I$, converge to G_i, for $1 \leq i \leq d$. Moreover, $\tilde{G}_{i,n}$ is stochastic for all i and n.*

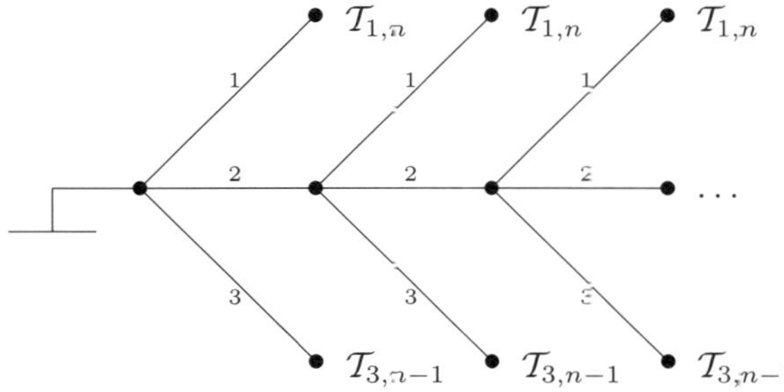

Fig. 9.1 A pictorial representation of the tree $\mathcal{T}_{2,n}$ for $d = 3$.

Proof We define a sequence of trees $\{\mathcal{T}_{i,n} : n \geq 0, 1 \leq i \leq d\}$ as follows (see Figure 9.1 for a graphical description). The trees $\mathcal{T}_{i,0}$ comprise a single node.

For $n \geq 1$, the root of $\mathcal{T}_{i,n}$ has d children; each child with index $j \leq i-1$ is the root of a tree which is isomorphic to $\mathcal{T}_{j,n}$; each child with index $j \geq i+1$ is the root of a tree which is isomorphic to $\mathcal{T}_{j,n-1}$; the child with index i is the root of a tree isomorphic to $\mathcal{T}_{i,n}$ itself. The transition probabilities are the same as in the original tree-like process, except at the leaf nodes where two cases are considered.

In the first case, for every node $\mathcal{N}_J$ which has no child, the transition probability from (J,k) to (J,k') is $B_{kk'}$. One proves by induction that for all $n \geq 1$ and all i, $G'_{i,n}$ is the matrix of first passage probabilities from $\mathcal{N}_i$ to $\mathcal{N}_\emptyset$ in $\mathcal{T}_{i,n}$. The detailed proof is similar to that of Theorem 9.20.

In the second case, for every node $\mathcal{N}_J$ which has no child, the transition probability from (J,k) to (J,k') is $(B+A_1+\cdots+A_d)_{kk'}$. One proves by induction that $\tilde{G}_{i,n}$ is the matrix of first passage probabilities from $\mathcal{N}_i$ to $\mathcal{N}_\emptyset$ in $\mathcal{T}_{i,n}$. Then we repeat the argument of Theorem 9.22. □

Let us now analyze the convergence speed of the sequences $\{\tilde{G}_{i,n}\}_n$ and $\{G'_{i,n}\}_n$. Define the matrices $E_{i,n} = G_i - G_{i,n}$, $i = 1,\ldots,d$, representing the errors at step n, and the corresponding vectors $\epsilon_{i,n} = \mathrm{vec}(E_{i,n})$. The errors at two subsequent steps are related as stated in the following theorem.

Theorem 9.29 *Let $\{G_{i,n}\}_n$, $i = 1,\ldots,d$, be the sequences generated by means of (9.55),(9.56), with arbitrary initial approximations $G_{i,0}$, $i = 1,\ldots,d$. Then, at each step n we have that*

$$\epsilon_{i,n} - \sum_{j=1}^{i} \left(G_{i,n}^{\mathrm{T}} \otimes (-S)^{-1}A_j\right)\epsilon_{j,n} = \sum_{j=i+1}^{d} \left(G_{i,n}^{\mathrm{T}} \otimes (-S)^{-1}A_j\right)\epsilon_{j,n-1}, \quad (9.57)$$

for $i = 1,\ldots,d$.

Proof From (9.54, 9.55) we obtain that

$$F_i G_i - F_{i,n}G_{i,n} + A_i(G_i^2 - G_{i,n}^2) = 0$$

for all i and n. By replacing in the latter equation F_i and $F_{i,n}$, respectively, by the expressions from (9.53) and (9.56), and $G_i^2 - G_{i,n}^2$ by $G_iE_{i,n} + E_{i,n}G_{i,n}$, we obtain that

$$\begin{aligned} CE_{i,n} &+ \sum_{j=1}^{i-1} A_j(G_jG_i - G_{j,n}G_{i,n}) + \sum_{j=i+1}^{d} A_j(G_jG_i - G_{j,n-1}G_{i,n}) + \\ &+ A_iG_iE_{i,n} + A_iE_{i,n}G_{i,n} = 0, \end{aligned}$$

from which, using

$$G_jG_i - G_{j,n}G_{i,n} = G_jE_{i,n} + E_{j,n}G_{i,n}$$

and

$$G_jG_i - G_{j,n-1}G_{i,n} = G_jE_{i,n} + E_{j,n-1}G_{i,n},$$

we arrive at

$$\left(C + \sum_{j=1}^{d} A_j G_j\right) E_{i,n} + \sum_{j=1}^{i} A_j E_{j,n} G_{i,n} + \sum_{j=i+1}^{d} A_j E_{j,n-1} G_{i,n} = 0.$$

We obtain (9.57) after multiplying this equation on the left by S^{-1}. □

The equation (9.57) may also be written as

$$\boldsymbol{f}_n = (I - H_n)^{-1} K_n \boldsymbol{f}_{n-1} \tag{9.58}$$

where $\boldsymbol{f}_n$ is the dm^2-dimensional vector made up by the vectors $\boldsymbol{\epsilon}_{i,n}$ ($i = 1, \ldots, d$), K_n is the $d \times d$ block lower triangular matrix with block elements

$$(K_n)_{i,j} = \begin{cases} G_{i,n}^{\mathrm{T}} \otimes \tilde{R}_j & i = 1, \ldots, d, \quad j = 1, \ldots, i \\ 0 & \text{otherwise,} \end{cases} \tag{9.59}$$

K_n is the $d \times d$ block strictly upper triangular matrix with block elements

$$(K_n)_{i,j} = \begin{cases} G_{i,n}^{\mathrm{T}} \otimes \tilde{R}_j & i = 1, \ldots, d, \quad j = i+1, \ldots, d \\ 0 & \text{otherwise,} \end{cases} \tag{9.60}$$

and $\tilde{R}_i = (-S)^{-1} A_i = S^{-1} R_i S$, $1 \leq i \leq d$.

Since $H_n + K_n = D_{1,n}(\mathbf{1}\mathbf{1}^{\mathrm{T}} \otimes I) D_{2,n}$ where

$$\begin{aligned} D_{1,n} &= \mathrm{Diag}\left(G_{1,n}^{\mathrm{T}} \otimes I, G_{2,n}^{\mathrm{T}} \otimes I, \ldots, G_{d,n}^{\mathrm{T}} \otimes I\right), \\ D_{2,n} &= \mathrm{Diag}\left(I \otimes \tilde{R}_1, I \otimes \tilde{R}_2, \ldots, I \otimes \tilde{R}_d\right), \end{aligned}$$

we conclude that $H_n + K_n$ has $(d-1)m^2$ eigenvalues equal to 0, and the remaining m^2 eigenvalues are the eigenvalues of $\sum_{i=1}^{d} G_{i,n}^{\mathrm{T}} \otimes \tilde{R}_i$. Since $R_i = S\tilde{R}_i S^{-1}$, these are the eigenvalues of $\sum_{i=1}^{d} G_{i,n}^{\mathrm{T}} \otimes R_i$.

Let H and K, respectively, be the $d \times d$ block lower and strictly upper triangular matrices with block elements

$$(H)_{i,j} = \begin{cases} G_i^{\mathrm{T}} \otimes \tilde{R}_j & i = 1, \ldots, d, \quad j = 1, \ldots, i \\ 0 & \text{otherwise,} \end{cases} \tag{9.61}$$

$$(K)_{i,j} = \begin{cases} G_i^{\mathrm{T}} \otimes \tilde{R}_j & i = 1, \ldots, d, \quad j = i+1, \ldots, d \\ 0 & \text{otherwise.} \end{cases} \tag{9.62}$$

The matrices H and K are the limits of $\{H_n\}_n$ and $\{K_n\}_n$, respectively, when $G_{i,0}$, $i = 1, \ldots, d$, are the null matrices, or the identity matrices. As for $H_n + K_n$, $H + K$ has $(d-1)m^2$ eigenvalues equal to 0, and the remaining m^2 eigenvalues are the eigenvalues of $\sum_{i=1}^{d} G_i^{\mathrm{T}} \otimes R_i$. By Theorem 9.24, we have $\lambda(R) \subset \lambda(H + K)$ and, if the Perron–Frobenius eigenvector of R is strictly positive, then $\rho(H+K) = \rho(R) < 1$. Since $H + K$ is nonnegative, $M = I - H - K$ is a nonsingular M-matrix and $N = (I - H)^{-1} K$ is the iteration matrix obtained by means of a regular splitting applied to M. From the properties of regular splittings (see Theorem A.15) it follows that $\rho(N) \leq \rho(H+K)$, and thus $\rho(N) \leq \rho(R)$. This inequality allows us to prove that the iteration defined by (9.55), (9.56) with $G'_{i,0} = 0$ is generally faster than the iteration (9.38), (9.39) starting with $G_{i,0} = 0$, $i = 1, \ldots, d$. This result is reported in the next theorem which shows also that the sequences $\{\tilde{G}_{i,n}\}_{n\geq 0}$, $i = 1, \ldots, d$, converge faster than the sequences $\{G'_{i,n}\}_{n\geq 0}$, $i = 1, \ldots, d$.

Theorem 9.30 *One has*

$$\lim_{n\to\infty} ||\boldsymbol{f}'_n||^{1/n} = \rho((I-H)^{-1}K)$$

for any vector norm $\|\cdot\|$, *and*

$$\lambda\left((I-\tilde{H})^{-1}\tilde{K}\right) \subset \lambda\left((I-H)^{-1}K\right),$$

where $\boldsymbol{f}'_n$ *is the* dm^2*-dimensional vector made up by the vectors* $\boldsymbol{\epsilon}'_{i,n} = \mathrm{vec}(G_i - G'_{i,n})$, $i = 1, \ldots, d$,

$$\tilde{H} = \begin{bmatrix} \tilde{R}_1 & & & 0 \\ \tilde{R}_1 & \tilde{R}_2 & & \\ \vdots & \vdots & \ddots & \\ \tilde{R}_1 & \tilde{R}_2 & \ldots & \tilde{R}_d \end{bmatrix} \quad \text{and} \quad \tilde{K} = \begin{bmatrix} 0 & \tilde{R}_2 & \tilde{R}_3 & \ldots & \tilde{R}_d \\ & 0 & \tilde{R}_3 & \ldots & \tilde{R}_d \\ & & \ddots & \ddots & \vdots \\ & & & 0 & \tilde{R}_d \\ 0 & & & & 0 \end{bmatrix}.$$

Moreover, if the Perron–Frobenius eigenvector of $(I-\tilde{H})^{-1}\tilde{K}$ *is strictly positive, then one has that*

$$\rho\left((I-\tilde{H})^{-1}\tilde{K}\right) = \rho\left((I-H)^{-1}K\right)$$

and

$$\lim_{n\to\infty} \|\widetilde{\boldsymbol{f}}_n\|^{1/n} \leq \max\left\{|\sigma| : \sigma \in \lambda\left((I-H)^{-1}K\right) \setminus \lambda\left((I-\tilde{H})^{-1}\tilde{K}\right)\right\}$$

for any vector norm $\|\cdot\|$, *where* $\widetilde{\boldsymbol{f}}_n$ *is the* dm^2*-dimensional vector made up by the vectors* $\widetilde{\boldsymbol{\epsilon}}_{i,n} = \mathrm{vec}(G_i - \tilde{G}_{i,n})$, $i = 1, \ldots, d$.

Proof Concerning the convergence of $\{\boldsymbol{f}'_n\}_n$, from the monotonicity of the sequences $\{G'_{i,n}\}_n$, from (9.59), (9.60) and (9.58) we deduce that

$$\boldsymbol{f}'_n \leq (I-H)^{-1}K\boldsymbol{f}'_{n-1} \leq ((I-H)^{-1}K)^n \boldsymbol{f}'_0$$

whence $\lim_{n\to\infty} ||\boldsymbol{f}'_n||^{1/n} \leq \rho((I-H)^{-1}K)$. For the opposite inequality observe that

$$\boldsymbol{f}'_n = (I-H)^{-1}K\boldsymbol{f}'_{n-1} - ((I-H)^{-1}K - (I-H_n)^{-1}K_n)\boldsymbol{f}'_{n-1}. \tag{9.63}$$

Let Q be any nonnegative matrix such that $(I-H)^{-1}K - \epsilon Q \geq 0$ for any $\epsilon > 0$ in a suitable neighborhood $\mathcal{U}$ of 0. Then for any positive $\epsilon \in \mathcal{U}$ there exists n_0

such that for any $n \geq n_0$ we have $(I-H)^{-1}K - (I-H_n)^{-1}K_n \leq \epsilon Q$. From the latter inequality and from (9.63) we deduce that $\boldsymbol{f}'_n \geq ((I-H)^{-1}K - \epsilon Q)\boldsymbol{f}'_{n-1} \geq ((I-H)^{-1}K-\epsilon Q)^n \boldsymbol{f}'_0$, hence $\lim_{n\to\infty} \|\boldsymbol{f}'_n\|^{1/n} \geq \rho((I-H)^{-1}K - \epsilon Q)$. For the arbitrariness of ϵ we obtain $\lim_{n\to\infty} \|\boldsymbol{f}'_n\|^{1/n} = \rho((I-H)^{-1}K)$. Concerning the remaining part of the theorem, let us define the $(dm^2)\times(dm^2)$ orthogonal matrix $\tilde{\Pi} = [\tilde{\Pi}_1|\tilde{\Pi}_2]$, where $\tilde{\Pi}_1 = I_d \otimes \Pi_1$, $\tilde{\Pi}_2 = I_d \otimes \Pi_2$, I_d is the $d \times d$ identity matrix, and $\Pi = [\Pi_1|\Pi_2]$ is the $(m^2)\times(m^2)$ orthogonal matrix defined in the proof of Theorem 9.24. Since $\tilde{G}_i$ is stochastic for any i, from (9.61), (9.62) it follows that $\tilde{\Pi}_1^{\mathrm{T}} H = \tilde{H}\tilde{\Pi}_1^{\mathrm{T}}$, and that $\tilde{\Pi}_1^{\mathrm{T}} K = \tilde{K}\tilde{\Pi}_1^{\mathrm{T}}$. Thus, we have that

$$\tilde{\Pi}^{\mathrm{T}}(I-H)^{-1}K\tilde{\Pi} = \begin{bmatrix} (I-\tilde{H})^{-1}\tilde{K} & 0 \\ T_1 & T_2 \end{bmatrix},$$

where $T_1 = \tilde{\Pi}_2^{\mathrm{T}}(I-H)^{-1}K\tilde{\Pi}_1$ and $T_2 = \tilde{\Pi}_2^{\mathrm{T}}(I-H)^{-1}K\tilde{\Pi}_2$, and we conclude that $\lambda\left((I-\tilde{H})^{-1}\tilde{K}\right) \subset \lambda\left((I-H)^{-1}K\right)$. If $\boldsymbol{u}$ is a positive Perron–Frobenius right eigenvector of $(I-\tilde{H})^{-1}\tilde{K}$, that is, $(I-\tilde{H})^{-1}\tilde{K}\boldsymbol{u} = \rho\boldsymbol{u}$ then $(I-H)^{-1}K\boldsymbol{v} = \rho\boldsymbol{v}$, where $\boldsymbol{v} = \tilde{\Pi}_1\boldsymbol{u}$. Since $\boldsymbol{v}$ is positive and $(I-H)^{-1}K$ is nonnegative, ρ is the spectral radius of $(I-H)^{-1}K$ [116]. Moreover, since $\tilde{G}_{i,n}$ is stochastic for any i and n, we have that $\tilde{\Pi}_1^{\mathrm{T}}\widetilde{\boldsymbol{f}}_n = \boldsymbol{0}$ for any n. Thus, as in the proof of Theorem 9.24, $\lim_{n\to\infty} \|\widetilde{\boldsymbol{f}}_n\|^{1/n} \leq \rho(T_2)$. □

Unfortunately, the result above does not allow us to conclude that, when one starts with $G_{i,0} = I$, $i = 1, \ldots, d$, the iteration defined by (9.55), (9.56) is generally faster than the iteration (9.38), (9.39). Nevertheless, we conjecture that the procedure described in this section does require fewer iterations than the ones in Section 9.4.1. Our argument is illustrated in Figure 9.2 where we show how the tree fills up in the case where $d = 2$ during the first two iterations of (9.55), (9.56). We represent at the top the first eight levels of a binary tree; underneath are the same levels for the trees $\mathcal{T}_{1,1}$, $\mathcal{T}_{2,1}$, $\mathcal{T}_{1,2}$ and $\mathcal{T}_{2,2}$. By contrast, only the first 2 levels are filled after the first two iterations of (9.38), (9.39).

Concerning the computational cost, the most expensive part at each iteration is the solution of the d quadratic equations (9.55); the computation of the coefficients $F_{i,n}$, $1 \leq i \leq d$, in (9.55) requires only d matrix products. If the quadratic matrix equations are solved by means of the cyclic reduction algorithm, which is quadratically convergent, the cost is one matrix inversion and six matrix products per step.

9.4.4 Numerical behavior

An experimental analysis of the fixed point iteration (FPI), the algorithm based on the reduction to quadratic equations (QE), solved by means of cyclic reduction, and Newton's method (NM), can be easily performed. We report here the results of some experiments carried out in [17]. Concerning NM, we have solved (9.49) by applying the fixed point iteration (9.51), with $r = 1$.

We consider a system similar to the M/M/1 queue in a random environment (see [79, Example 9.2.2]). The service rate is constant and the arrival rate depends

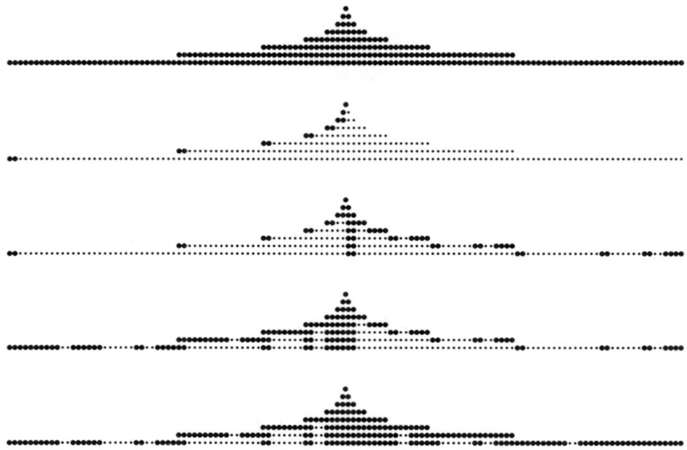

Fig. 9.2 Pictorial representation of the filling up of the nodes of the tree-like process, when $d = 2$. The first eight levels are depicted, for the whole tree and for the trees $\mathcal{T}_{1,1}$, $\mathcal{T}_{2,1}$, $\mathcal{T}_{1,2}$ and $\mathcal{T}_{2,2}$.

on the state of a Markovian environmental process. Each arrival has a label which indicates which direction is taken in the tree.

The parameters are as follows: the number of children is $d = 2$, the size of the blocks is m, $D_1 = \alpha I$, $D_2 = D_1$, $A_1 = \mathrm{Diag}(a, b, \ldots, b)$, $A_2 = bI$, $C = T - D_1 - A_1 - A_2$, $a = \rho f m$, $b = \rho(1 - f)m/(2m - 1)$, and $T = (t_{i,j})_{i,j=1,m}$, $t_{i,i} = -1$, for $i = 1, \ldots, m$, $t_{i,i+1} = 1$, for $i = 1, \ldots, m - 1$, $t_{m,1} = 1$, $t_{i,j} = 0$ elsewhere.

This system is stable if ρ (the stationary arrival rate) is less than α (the service rate). The parameter f, which ranges from 0 to 1, measures the fraction of arrivals which occur in phase 1; when f is small, the system behaves nearly like an ordinary M/M/1 queue, when f is large, the arrivals are very bursty: in phase 1 the system is strongly driven to higher levels in the direction of the children labeled "1", in all the other phases the system is strongly driven toward the root of the tree. The burstiness is higher for large values of m.

In Figure 9.3, we represent the CPU time as a function of the parameter f, needed to run the programs on a Pentium III at 500 MHz. It is interesting to observe the opposite monotonicity of FPI, with respect to QE and NM (which have the same behavior). The algorithm FPI is convenient when the number of iterations is not too large; otherwise QE is more convenient. QE and NM have the same behavior in terms of monotonicity.

We report in Table 9.1 the number of iterations and the residual error, for different values of the parameter f. It is interesting to observe the different behavior of the number of iterations for the three methods. The number of iterations of FPI grows with f. The number of iterations of QE decreases, as f

Algorithm 9.7 Cyclic reduction for tree-like processes

INPUT: The $m \times m$ matrices C, A_i, and D_i are such that $C = I - B$, $B \geq 0$, $A_i, D_i \geq 0$, for $i = 1, \ldots, d$, B is substochastic, and the matrices $I + C + D_i + A_1 + \cdots + A_d$, $i = 1, \ldots, d$, are stochastic. An error bound ϵ.

OUTPUT: Approximations W_i, $i = 1, \ldots, d$, to the minimal nonnegative solutions G_i, $i = 1, \ldots, d$, of the matrix equations (9.54).

COMPUTATION:

1. Set $W_i = I$, for $i = 1, \ldots, d$.
2. For $i = 1, \ldots, d$:
   (a) compute $F = C + \sum_{j=1}^{i-1} A_j W_j' + \sum_{j=i+1}^{d} A_j W_j$;
   (b) compute by means of Algorithm 7.1 or Algorithm 7.3 the minimal nonnegative solution W_i' of the matrix equation $X = -F^{-1}D_i - F^{-1}A_i X^2$.
3. If $\max_i \|W_i - W_i'\|_\infty \leq \epsilon$, then output $W_i = W_i'$, $i = 1, \ldots, d$; otherwise continue from step 2.

grows, and the number of inner iterations needed to solve the quadratic matrix equations is almost constant. The number of iterations of NM is almost constant, as f grows, while the number of inner iterations of (9.51) decreases.

9.5 Bibliographic notes

Markov chains with limited displacement have been analyzed by Gail, Hantler and Taylor in [42], where the classic functional iterations have been adapted to this specific case and where error bounds to the solution of the truncated matrix equation are proved. The analysis of cyclic reduction for NSF Markov chains is performed in [21] and in [22] in terms of solving a banded Toeplitz system, where the displacement rank structure is pointed out and exploited for designing fast algorithms.

The reduction of M/G/1-type Markov chains to infinite QBDs has been discovered by Ramaswami in [100]; the algorithmic analysis of cyclic reduction applied to the infinite QBD is performed in [25]

Functional iterations for tree-like stochastic processes are introduced and analyzed by Latouche and Ramaswami [79, Section 14.3] and Yeung and Alfa [121, Section 8]. The analysis of convergence of functional iterations is performed in [17]. Newton's iteration and the combination of functional iterations with cyclic reduction are introduced and analyzed in [17].

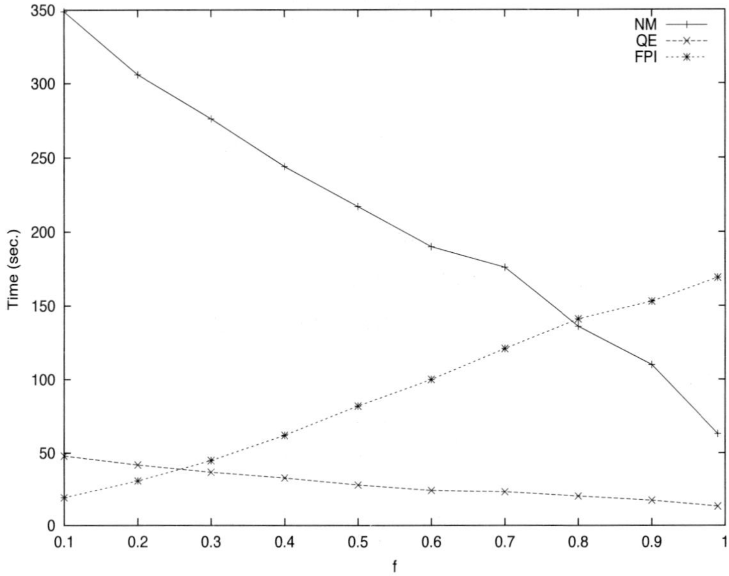

Fig. 9.3 $\alpha = 2$, $\rho = 1.8$, $m = 100$, CPU time

	FPI		QE		NM	
f	Steps	Residual	Steps	Residual	Steps	Residual
0.1	269	7.8e-15	25 (10)	3.6e-15	10 (101)	9.3e-16
0.2	445	8.0e-15	21 (11)	2.8e-15	10 (82)	5.4e-16
0.3	643	9.5e-15	17 (12)	2.6e-15	10 (69)	4.2e-16
0.4	873	9.1e-15	15 (12)	2.2e-15	10 (60)	3.7e-16
0.5	1133	9.7e-15	13 (13)	3.4e-15	10 (53)	5.8e-16
0.6	1379	9.0e-15	11 (13)	3.5e-15	10 (46)	6.7e-16
0.7	1664	9.7e-15	10 (14)	2.3e-15	10 (38)	1.0e-15
0.8	1927	1.0e-14	9 (14)	2.5e-15	10 (27)	8.9e-16
0.9	2153	1.0e-14	8 (15)	1.8e-15	10 (21)	8.9e-16
0.99	2411	9.5e-15	6 (15)	1.5e-15	11 (9)	1.0e-15

Table 9.1 $\alpha = 2$, $\rho = 1.8$, $m = 100$

APPENDIX

An application $\|\cdot\| : \mathbb{C}^n \to \mathbb{R}$ is called a *norm* or a *vector norm* if for any $\boldsymbol{x} \in \mathbb{C}^n$

$$
\begin{aligned}
&\|\boldsymbol{x}\| \geq 0 \\
&\|\boldsymbol{x}\| = 0 \text{ if and only if } \boldsymbol{x} = 0 \\
&\|\alpha \boldsymbol{x}\| = |\alpha| \|\boldsymbol{x}\|, \quad \text{for } \alpha \in \mathbb{C} \\
&\|\boldsymbol{x} + \boldsymbol{y}\| \leq \|\boldsymbol{x}\| + \|\boldsymbol{y}\|, \quad \text{for any } \boldsymbol{y} \in \mathbb{C}^n.
\end{aligned}
$$

Useful norms are:

$$
\begin{aligned}
\|\boldsymbol{x}\|_1 &= \sum_i |x_i| \\
\|\boldsymbol{x}\|_2 &= \sqrt{\sum_i |x_i|^2} \\
\|\boldsymbol{x}\|_\infty &= \max_i |x_i|
\end{aligned}
$$

They are called the 1-norm, the 2-norm (or the Euclidean norm) and the infinity-norm, respectively.

Similarly we may define a norm in the set of $n \times n$ complex matrices. Given a vector norm $\|\cdot\|$, the following application defines a norm in the set of $n \times n$ matrices, called the *induced matrix norm* or *operator norm* and is denoted with the same symbol $\|A\|$:

$$\|A\| = \max_{\|\boldsymbol{x}\|=1} \|A\boldsymbol{x}\|.$$

The operator norm induced by the 1-norm, the 2-norm and the infinity norm are given by

$$
\begin{aligned}
\|A\|_1 &= \max_j \sum_i |a_{i,j}|, \\
\|A\|_2 &= \sqrt{\rho(\bar{A}^{\mathrm{T}} A)}, \\
\|A\|_\infty &= \max_i \sum_j |a_{i,j}|,
\end{aligned}
$$

where $\rho(A)$ denotes the *spectral radius* of A, that is the maximum modulus of the eigenvalues of A and $\bar{A}$ is the conjugate of A.

The following property of norms is fundamental.

Theorem A.1. (Equivalence of norms) *For any pair of norms $\|\cdot\|'$, $\|\cdot\|''$ in $\mathbb{C}^n$ there exist two positive constants α, β such that for any $\boldsymbol{x} \in \mathbb{C}^n$*

$$\alpha\|\boldsymbol{x}\|' \leq \|\boldsymbol{x}\|'' \leq \beta\|\boldsymbol{x}\|'.$$

A similar result holds for matrix norms.

The following theorem relates operator norms and the spectral radius of a matrix.

Theorem A.2 *Let A be an $m \times m$ matrix. Then for any operator norm $\|\cdot\|$ one has*

$$\rho(A) \leq \|A\|.$$

Therefore, for any $\epsilon > 0$ there exists an operator norm $\|\cdot\|$ such that

$$\|A\| \leq \rho(A) + \epsilon.$$

If any eigenvalue of modulus $\rho(A)$ belongs to a Jordan block of dimension 1, then there exists an operator norm $\|\cdot\|$ such that $\|A\| = \rho(A)$. In particular, this holds if the eigenvalues of modulus $\rho(A)$ are simple.

The following result relates spectral radius with any matrix norm.

Theorem A.3 *Let A be an $m \times m$ matrix. Then for any matrix norm $\|\cdot\|$ and for any $q \in \mathbb{Z}$ one has*

$$\rho(A) = \lim_n \|A^{n+q}\|^{1/n}.$$

Corollary A.4 *Let A be an $m \times m$ matrix, then for any matrix norm $\|\cdot\|$ and for any $\epsilon > 0$ there exists $n_0 > 0$ such that*

$$\|A^n\| \leq (\rho(A) + \epsilon)^n, \quad \forall n \geq n_0.$$

Moreover there exists $\gamma > 0$ such that

$$\|A^n\| \leq \gamma(\rho(A) + \epsilon)^n \quad \text{for any } n \geq 0.$$

Proof From Theorem A.3 recall that $\lim_n \|A^n\|^{1/n} = \rho(A)$. Therefore, from the definition of limit, one has that for any $\epsilon > 0$ there exists n_0 such that

$$\|A^n\|^{1/n} \leq \rho(A) + \epsilon, \quad \forall n \geq n_0.$$

Whence $\|A^n\| \leq (\rho(A) + \epsilon)^n$, for any $n \geq n_0$. □

Theorem A.5 *Let $\|\cdot\|$ be any operator norm such that $\|A\| < 1$ where A is an $m \times m$ matrix. Then $I - A$ is nonsingular and*

$$\|(I - A)\|^{-1} \leq \frac{1}{1 - \|A\|}.$$

Definition A.6 *Consider the following partitioning of the $n \times n$ matrix A:*

$$A = \begin{bmatrix} A_{1,1} & A_{1,2} \\ A_{2,1} & A_{2,2} \end{bmatrix} \tag{A.1}$$

where $A_{1,1}$ is $m \times m$ and $A_{2,2}$ is $(n-m) \times (n-m)$. If $A_{1,1}$ is nonsingular then the matrix $S = A_{2,2} - A_{2,1}A_{1,1}^{-1}A_{1,2}$ is called the Schur complement of $A_{1,1}$ in A. If $A_{2,2}$ is nonsingular then the matrix $S' = A_{1,1} - A_{1,2}A_{2,2}^{-1}A_{2,1}$ is the Schur complement of $A_{2,2}$ in A.

Schur complements are related to block LU factorization:

Theorem A.7 *If $A_{1,1}$ is nonsingular then*

$$A = \begin{bmatrix} I & 0 \\ A_{2,1}A_{1,1}^{-1} & I \end{bmatrix} \begin{bmatrix} A_{1,1} & A_{1,2} \\ 0 & S \end{bmatrix}$$

Similarly, if $A_{2,2}$ is nonsingular then

$$A = \begin{bmatrix} S' & A_{1,2} \\ 0 & A_{2,2} \end{bmatrix} \begin{bmatrix} I & 0 \\ A_{2,2}^{-1}A_{2,1} & I \end{bmatrix}.$$

Moreover, if A and $A_{1,1}$ are nonsingular, then also S is nonsingular and the lower rightmost $(n-m) \times (n-m)$ submatrix of A^{-1} coincides with S^{-1}. Similarly, if A and $A_{2,2}$ are nonsingular, then also S' is nonsingular and the leading principal $m \times m$ submatrix of A^{-1} coincides with S'^{-1}.

The irreducibility of a matrix A can be defined in terms of the directed graph $G[A]$ associated with A, where an oriented arc connects the node i with the node j if $a_{i,j} \neq 0$.

Definition A.8 *A matrix A is irreducible if its directed graph is strongly connected, that is, for any pair i, j there exists a sequence of oriented arcs connecting the node i with the node j. A is said to be reducible if it is not irreducible.*

Theorem A.9 *An $m \times m$ matrix A is reducible if and only if there exists a permutation matrix Π such that $\Pi^{\mathrm{T}} A \Pi$ has the form*

$$\Pi^{\mathrm{T}} A \Pi = \begin{bmatrix} A_{1,1} & 0 \\ A_{2,1} & A_{2,2} \end{bmatrix}$$

where $A_{1,1}$ and $A_{2,2}$ are square blocks.

Theorem A.10. (Cayley–Hamilton theorem) *Let $p(\lambda) = \det(A - \lambda I) = \sum_{i=0}^{n} \lambda^i a_i$ the characteristic polynomial of the $n \times n$ matrix A. Then*

$$p(A) = \sum_{i=0}^{n} a_i A^i = 0.$$

Moreover, if A is nonsingular one has

$$A^{-1} = -a_0^{-1} \sum_{i=0}^{n-1} a_{i+1} A^i.$$

Theorem A.11 *Let A_i, $i = 0, \ldots, n-1$, be $m \times m$ matrices and consider the block companion matrix*

$$C(\boldsymbol{a}^{\mathrm{T}}) = \begin{bmatrix} 0 & I & 0 \ldots & & 0 \\ 0 & 0 & I & & \vdots \\ \vdots & & \ddots & \ddots & 0 \\ 0 & \ddots & \ddots & 0 & I \\ -A_0 & -A_1 & \ldots & \ldots & -A_{n-1} \end{bmatrix},$$

where $\boldsymbol{a}^{\mathrm{T}} = [-A_0, -A_1, \ldots, -A_{n-1}]$. Then the eigenvalues of $C(\boldsymbol{a}^{\mathrm{T}})$ are the solutions of the equation $\det(\sum_{i=0}^{n-1} \lambda^i A_i + \lambda^n I) = 0$.

M-matrices play an important role in Markov chains and in many other applications.

Definition A.12 *Let $B \geq 0$ be an $n \times n$ matrix. The matrix $A = \alpha I - B$ is an M-matrix, if $\rho(B) \leq \alpha$.*

M-matrices and nonnegative matrices are strictly related.

Theorem A.13 *The inverse of a nonsingular M-matrix is nonnegative.*

Regular splittings are at the basis of many iterative methods:

Definition A.14 *Let A be an M-matrix. The splitting $A = M - N$ of A is called a regular splitting if the matrices M and N are such that $N \geq 0$ and M is an M-matrix.*

Theorem A.15 *If $A = M_1 - N_1$ and $A = M_2 - N_2$ are two regular splittings of the M-matrix matrix A and if $N_1 \geq N_2$ and $\det M_1 \neq 0$, $\det M_2 \neq 0$ then $\rho(M_1^{-1}N_1) \geq \rho(M_2^{-1}N_2)$. If A is singular then $\rho(M_1^{-1}N_1) = \rho(M_2^{-1}N_2) = 1$.*

Lemma A.16 *Let $\{a_n\}_n$ be a real sequence such that $\lim_n a_n = a^*$. If $|a_{n+1} - a_n| \leq \gamma\sigma^{2^n}$ for some $\gamma > 0$ and $0 < \sigma < 1$, then there exists θ such that $|a_n - a^*| \leq \theta\sigma^{2^n}$.*

Proof Let $e_n = a_n - a^*$. Then we have $e_n = e_{n+1} + a_{n+1} - a_n$. Therefore, $|e_n| \leq |e_{n+1}| + \gamma\sigma^{2^n}$. Whence we deduce that

$$|e_n| \leq \gamma\sigma^{2^n} + \gamma\sigma^{2^{n+1}} + \gamma\sigma^{2^{n+2}} + \cdots \leq \gamma\sigma^{2^n}\frac{1}{1-\sigma}$$

The proof is completed by setting $\theta = \gamma/(1-\sigma)$. □

Inverses of matrices which differ by a low rank correction are closely related. Let $A = B + UV^{\mathrm{T}}$ where A and B are $n \times n$ matrices, U and V are $n \times k$ matrices with $k < n$. Then, if A and B are nonsingular then also the $k \times k$ matrix $I + V^{\mathrm{T}}B^{-1}U$ is nonsingular and the inverses of A and B are related by the Sherman–Morrison–Woodbury formula

$$A^{-1} = B^{-1} - B^{-1}U(I + V^{\mathrm{T}}B^{-1}U)^{-1}V^{\mathrm{T}}B^{-1}. \qquad \text{(A.2)}$$

Let A, B be $m \times n$ matrices and C, D be $n \times p$ matrices. The real part and the imaginary part of the product of the complex matrices $A + \mathrm{i}B$ and $C + \mathrm{i}D$ can be computed with three real matrix multiplications and five real matrix additions by means of the following formula

$$\begin{aligned} &(A + \mathrm{i}B)(C + \mathrm{i}D) = P_1 + P_2 - P_3 + \mathrm{i}(P_2 + P_3) \\ &P_1 = (A + B)(C - D), \quad P_2 = AD, \quad P_3 = BC. \end{aligned} \qquad \text{(A.3)}$$

Theorem A.17. (Monotone convergence theorem) *Let X be a measure space, and let $0 \le f_1 \le f_2 \le \cdots$ be a monotone increasing sequence of nonnegative measurable functions. Let f be the function defined almost everywhere by $f(x) = \lim_{n\to\infty} f_n(x)$. Then f is measurable, and $\lim\limits_{n\to\infty} \int_X f_n = \int_X f$.*

Concerning functions of a complex variable we have the following definitions and properties.

Definition A.18 *A function $f(z)$ defined in an open subset Ω of the complex plane $\mathbb{C}$ is analytic in Ω if for any $z \in \Omega$ there exists the first derivative $f'(z)$ in z. A function $f(z)$ defined on a nonempty open connected subset R of the complex plane is meromophic if at every point of R it is either analytic or has an isolated singularity that is at most a pole. A function $f(z)$ is entire if it is analytic at all the finite points of the complex plane.*

A function analytic in Ω has a power series representation for any $z_0 \in \Omega$ of the kind $f(z) = \sum_{i=0}^{+\infty} (z - z_0)^i a_i$ which is convergent for any $z \in \Omega$.

The following result is proved in [63] (see Theorem 4.10d).

Theorem A.19 *Let $\{f_n(z)\}_n$ be a sequence of analytic functions in a nonempty open connected set R in the complex plane. Assume that the sequence $\{f_n(z)\}$ converges uniformly to $f(z)$ on every compact subset of R. Let Γ be any Jordan curve in R whose interior belongs to R and let $f(z) \neq 0$ for $z \in \Gamma$. Then for all sufficiently large n, the functions $f_n(z)$ have the same number of zeros in the interior of Γ as $f(z)$.*

NOTATION

$\mathbb{R}$, $\mathbb{C}$: real and complex fields
$\mathbb{N}$, $\mathbb{Z}$: set of natural and integer numbers
$\mathbb{D} = \{z \in \mathbb{C} : \quad |z| < 1\}$: open unit disk
$\mathbb{T} = \{z \in \mathbb{C} : \quad |z| = 1\}$: unit circle
$\mathbb{A} = \{z \in \mathbb{C} : \quad r_1 < |z| < r_2\}$: open annulus
$\mathtt{i}$: imaginary unit
$T_N[A(z)]$: finite block Toeplitz matrix associated with the function $A(z)$
$T_\infty[A(z)]$: infinite block Toeplitz matrix associated with the function $A(z)$
$T_{\pm\infty}[A(z)]$: bi-infinite block Toeplitz matrix associated with the function $A(z)$
$\omega_n = \cos\frac{2\pi}{n} + \mathtt{i}\sin\frac{2\pi}{n}$: primitive n-th root of 1
Ω_n: Fourier matrix
DFT_n: discrete Fourier transform of order n
IDFT_n: inverse discrete Fourier transform of order n
$\mathbf{1}$: vector of elements equal to 1
$\boldsymbol{e}_i$: vector with unit ith component and with zero components elsewhere
$\|\boldsymbol{u}\|$: vector norm
$\|\boldsymbol{u}\|_\infty$: infinity norm
$\|\boldsymbol{u}\|_1$: 1-norm
$\|\boldsymbol{u}\|_2$: 2-norm (Euclidean norm)
$\doteq$: equality up to higher-order terms
$\dot{\leq}$: inequality up to higher-order terms
$\lceil x \rceil$: minimum integer greater than or equal to x
$\lfloor x \rfloor$: maximum integer less than or equal to x
$\ell^2(\mathbb{N})$, $\ell^2(\mathbb{Z})$: spaces of sequences with finite sum of squares of the moduli
$\ell^1(\mathbb{N})$, $\ell^1(\mathbb{Z})$: spaces of sequences with finite sum of moduli
$\ell^\infty(\mathbb{N})$, $\ell^\infty(\mathbb{Z})$: spaces of sequences with bounded values
$\mathcal{W}$: Wiener algebra
$\mathcal{W}_+$: Wiener algebra of matrix power series in z
$\mathcal{W}_-$: Wiener algebra of matrix power series in z^{-1}
$\|A(z)\|_*$: the infinity norm of $\sum_{i=-\infty}^{+\infty}|A_i|$, for $A(z) \in \mathcal{W}$
$f_{\mathrm{odd}}(z^2)$: the odd part of the function $f(z)$, i.e., $(f(z) - f(-z))/(2z)$
$f_{\mathrm{even}}(z^2)$: the even part of the function $f(z)$, i.e., $(f(z) + f(-z))/2$
$\otimes$: Kronecker product
$*$: Hadamard (component-wise) product
I_m: $m \times m$ identity matrix
I: identity matrix
Z: lower shift matrix
$\mathcal{Z}$: the matrix $Z \otimes I$

Δ: displacement operator
$\boldsymbol{a} = \mathrm{vec}(A)$: vector formed by column-wise arranging the matrix A
$A = \mathrm{vec}_m^{-1}(\boldsymbol{a})$: the $m \times n$ matrix such that $\boldsymbol{a} = \mathrm{vec}(A)$
P_A: submatrix of the $m \times m$ matrix P with indices in $A \subset \{1, \ldots, m\}$
$P_{A,B}$: submatrix of P with row indices in A and column indices in B
$\rho(A)$: spectral radius of A
$\lambda(A)$: set of eigenvalues of A
E: the matrix with all elements equal to 1
$\mathrm{E}[X]$: the expected value of the random variable X
$\mathrm{E}[X|A]$: conditional expectation
$\mathrm{P}[X = j]$: probability that the random variable X takes the value j
$\mathrm{P}[X = j|Y = i]$: conditional probability
$\mathbb{I}\{\}$: indicator function
$\boldsymbol{\pi}$: invariant probability vector
μ: drift of an M/G/1-type Markov chain

LIST OF ALGORITHMS

BIBLIOGRAPHY

[1] N. Akar and K. Sohraby. An invariant subspace approach in $M/G/1$ and $G/M/1$ type Markov chains. *Comm. Statist. Stochastic Models*, 13(3):381–416, 1997.

[2] P. Amodio and F. Mazzia. Backward error analysis of cyclic reduction for the solution of tridiagonal systems. *Math. Comp.*, 62(206):601–617, 1994.

[3] G. Anastasi, L. Lenzini, and B. Meini. Performance evaluation of a worst case model of the MetaRing MAC protocol with global fairness. *Perf. Eval.*, 29:127–151, 1997.

[4] S. Asmussen. *Applied probability and queues.* John Wiley, New York, 1987.

[5] S. Asmussen and V. Ramaswami. Probabilistic interpretations of some duality results for the matrix paradigms in queueing theory. *Comm. Statist. Stochastic Models*, 6:715–733, 1990.

[6] Z.-Z. Bai. A class of iteration methods based on the Moser formula for nonlinear equations in Markov chains. *Linear Algebra Appl.*, 266:219–241, 1997.

[7] S. Barnett. *Polynomials and linear control systems*, volume 77 of Monographs and Textbooks in Pure and Applied Mathematics. Marcel Dekker Inc., New York, 1983.

[8] P. Benner, V. Mehrmann, V. Sima, S. Van Huffel, and A. Varga. SLICOT–A Subroutine Library in Systems and Control Theory. Technical Report 97-3, NICONET, June 1997.

[9] A. Berman and R. J. Plemmons. *Nonnegative matrices in the mathematical sciences*, volume 9 of Classics in Applied Mathematics. Society for Industrial and Applied Mathematics (SIAM), Philadelphia, PA, 1994. Revised reprint of the 1979 original.

[10] D. Bini and L. Gemignani. Solving quadratic matrix equations and factoring polynomials: new fixed point iterations based on Schur complements of Toeplitz matrices. *Numerical Linear Algebra with Applications*, 2004, to appear.

[11] D. Bini and B. Meini. On the solution of a nonlinear matrix equation arising in queueing problems. *SIAM J. Matrix Anal. Appl.*, 17(4):906–926, 1996.

[12] D. Bini and V. Y. Pan. *Polynomial and matrix computations.* Volume 1. Progress in Theoretical Computer Science. Birkhäuser, Boston, MA, 1994.

[13] D. A. Bini, S. Chakravarthy, and B. Meini. A new algorithm for the design of finite capacity service units. In B. Plateau, W. J. Stewart, and M. Silva,

editors, *Numerical Solution of Markov Chains, 3rd International Workshop, Saragoza (Spain), Sept. 1999*, pages 247–260. Prensas Unversitarias de Zaragoza, 1999.

[14] D. A. Bini, L. Gemignani, and B. Meini. Computations with infinite Toeplitz matrices and polynomials. *Linear Algebra Appl.*, 343/344:21–61, 2002. Special issue on structured and infinite systems of linear equations.

[15] D. A. Bini, L. Gemignani, and B. Meini. Solving certain matrix equations by means of Toeplitz computations: algorithms and applications. In *Fast algorithms for structured matrices: theory and applications*, volume 323 of *Contemp. Math.*, pages 151–167. Amer. Math. Soc., Providence, RI, 2003.

[16] D. A. Bini, G. Latouche, and B. Meini. Solving matrix polynomial equations arising in queueing problems. *Linear Algebra Appl.*, 340:225–244, 2002.

[17] D. A. Bini, G. Latouche, and B. Meini. Solving nonlinear matrix equations arising in tree-like stochastic processes. *Linear Algebra Appl.*, 366:39–64, 2003.

[18] D. A. Bini and B. Meini. On cyclic reduction applied to a class of Toeplitz-like matrices arising in queueing problems. In W. J. Stewart, editor, *Computations with Markov Chains*, pages 21–38. Kluwer Academic Publishers, Boston, 1995.

[19] D. A. Bini and B. Meini. Improved cyclic reduction for solving queueing problems. *Numer. Algorithms*, 15(1):57–74, 1997.

[20] D. A. Bini and B. Meini. Inverting block Toeplitz matrices in block Hessenberg form by means of displacement operators: application to queueing problems. *Linear Algebra Appl.*, 272:1–16, 1998.

[21] D. A. Bini and B. Meini. Using displacement structure for solving non-skip-free M/G/1 type Markov chains. In A. Alfa and S. Chakravarthy, editors, *Advances in Matrix Analytic Methods for Stochastic Models—Proceedings of the 2nd international conference on matrix analytic methods*, pages 17–37. Notable Publications Inc, New Jersey, 1998.

[22] D. A. Bini and B. Meini. Effective methods for solving banded Toeplitz systems. *SIAM J. Matrix Anal. Appl.*, 20(3):700–719 (electronic), 1999.

[23] D. A. Bini and B. Meini. Fast algorithms with applications to Markov chains and queueing models. In *Fast reliable algorithms for matrices with structure*, pages 211–243. SIAM, Philadelphia, PA, 1999.

[24] D. A. Bini and B. Meini. Non-skip-free M/G/1-type Markov chains and Laurent matrix power series. *Linear Algebra Appl.*, 386:187–206, 2004. Special Issue on the Conference on the Numerical Solution of Markov Chains 2003, Urbana-Champaign, IL, 3–5 September 2003.

[25] D. A. Bini, B. Meini, and V. Ramaswami. Analyzing M/G/1 paradigms through QBDs: the role of the block structure in computing the matrix G. In G. Latouche and P. Taylor, editors, *Advances in Algorithmic Methods for Stochastic Models*, pages 73–86. Notable Publications, New Jersey, 2000. Proceedings of the Third Conference on Matrix Analytic Methods.

[26] D. A. Bini, B. Meini, and I. M. Spitkowsky. Shift techniques and canonical factorizations in the solution of M/G/1-type Markov chains. Technical report, Dipartimento di Matematica, Università di Pisa, 2004.

[27] A. Böttcher and B. Silbermann. *Introduction to large truncated Toeplitz matrices.* Universitext. Springer-Verlag, New York, 1999.

[28] B. L. Buzbee, G. H. Golub, and C. W. Nielson. On direct methods for solving Poisson's equations. *SIAM J. Numer. Anal.*, 7:627–656, 1970.

[29] S. L. Campbell and C. D. Meyer. *Generalized inverses of linear transformations.* Dover Publications, New York, 1991 Republication.

[30] E. Çinlar. *Introduction to stochastic processes.* Prentice-Hall, Englewood Cliffs, NJ, 1975.

[31] R. H. Chan and M. K. Ng. Conjugate gradient methods for Toeplitz systems. *SIAM Rev.*, 38(3):427–482, 1996.

[32] R. H. Chan and M. K. Ng. Iterative methods for linear systems with matrix structure. In *Fast reliable algorithms for matrices with structure*, pages 117–152. SIAM, Philadelphia, PA, 1999.

[33] K. L. Chung. *Markov chains with stationary transition probabilities.* Springer-Verlag, Berlin, second edition, 1967.

[34] R. E. Cline, R. J. Plemmons, and G. Worm. Generalized inverses of certain Toeplitz matrices. *Linear Algebra and Appl.*, 8:25–33, 1974.

[35] P. J. Davis. *Circulant matrices.* John Wiley & Sons, New York, Chichester, and Brisbane, 1979. A Wiley-Interscience Publication, Pure and Applied Mathematics.

[36] D. F. Elliott and K. R. Rao. *Fast transforms. Algorithms, analyses, applications..* Academic Press Inc, New York, 1982.

[37] P. Favati and B. Meini. Relaxed functional iteration techniques for the numerical solution of $M/G/1$ type Markov chains. *BIT*, 38(3):510–526, 1998.

[38] P. Favati and B. Meini. On functional iteration methods for solving nonlinear matrix equations arising in queueing problems. *IMA J. Numer. Anal.*, 19(1):39–49, 1999.

[39] M. Frigo and S. G. Johnson. FFTW: An adaptive software architecture for the FFT. In *Proc. 1998 IEEE Intl. Conf. Acoustics Speech and Signal Processing*, volume 3, pages 1381–1384. IEEE, 1998.

[40] H. R. Gail, S. L. Hantler, and B. A. Taylor. Solutions of the basic matrix equation for $M/G/1$ and $G/M/1$ type Markov chains. *Comm. Statist. Stochastic Models*, 10(1):1–43, 1994.

[41] H. R. Gail, S. L. Hantler, and B. A. Taylor. Spectral analysis of $M/G/1$ and $G/M/1$ type Markov chains. *Adv. in Appl. Probab.*, 28(1):114–165, 1996.

[42] H. R. Gail, S. L. Hantler, and B. A. Taylor. Non-skip-free $M/G/1$ and $G/M/1$ type Markov chains. *Adv. in Appl. Probab.*, 29(3):733–758, 1997.

[43] H. R. Gail, S. L. Hantler, and B. A. Taylor. Matrix-geometric invariant measures for $G/M/1$ type Markov chains. *Comm. Statist. Stochastic Models*, 14(3):537–569, 1998.

[44] H. R. Gail, S. L. Hantler, and B. A. Taylor. Use of characteristics roots for solving infinite state Markov chains. In W. K. Grassmann, editor, *Computational probability*, pages 205–255. Kluwer Academic Publishers, 2000.

[45] J. D. Gardiner, A. J. Laub, J. J. Amato, and C. B. Moler. Solution of the Sylvester matrix equation $AXB^{\mathrm{T}} + CXD^{\mathrm{T}} = E$. *ACM Trans. Math. Software*, 18(2):223–231, 1992.

[46] I. Gohberg, M. A. Kaashoek, and I. M. Spitkovsky. An overview of matrix factorization theory and operator applications. In *Factorization and integrable systems (Faro, 2000)*, volume 141 of *Oper. Theory Adv. Appl.*, pages 1–102. Birkhäuser, Basel, 2003.

[47] I. Gohberg, P. Lancaster, and L. Rodman. *Matrix polynomials.* Academic Press Inc, New York, 1982. Computer Science and Applied Mathematics.

[48] I. C. Gohberg and I. A. Fel′dman. *Convolution equations and projection methods for their solution.* American Mathematical Society, Providence, R.I., 1974. Translated from the Russian by F. M. Goldware, Translations of Mathematical Monographs, Vol. 41.

[49] I. C. Gohberg and M. G. Kreĭn. Systems of integral equations on a half line with kernels depending on the difference of arguments. *Amer. Math. Soc. Transl. (2)*, 14:217–287, 1960.

[50] G. H. Golub and C. F. Van Loan. *Matrix computations.* Johns Hopkins Studies in the Mathematical Sciences. Johns Hopkins University Press, Baltimore, MD, third edition, 1996.

[51] W. K. Grassmann and D. P. Heyman. Equilibrium distribution of block-structured Markov chains with repeating rows. *J. Appl. Probab.*, 27:557–576, 1990.

[52] W. K. Grassmann and D. P. Heyman. Computation of steady-state probabilities for infinite Markov chains with repeating rows. *ORSA J. Comput.*, 5:292–303, 1993.

[53] W. K. Grassmann, M. I. Taksar, and D. P. Heyman. Regenerative analysis and steady state distributions for Markov chains. *Oper. Res.*, 33(5):1107–1116, 1985.

[54] C. M. Grinstead and J. L. Snell. *Introduction to Probability.* AMS, Second Revised Edition, 1997.

[55] C.-H. Guo. On the numerical solution of a nonlinear matrix equation in Markov chains. *Linear Algebra Appl.*, 288(1-3):175–186, 1999.

[56] C.-H. Guo. Convergence analysis of the Latouche–Ramaswami algorithm for null recurrent quasi-birth-death processes. *SIAM J. Matrix Anal. Appl.*, 23(3):744–760 (electronic), 2001/2.

[57] C.-H. Guo. Comments on a shifted cyclic reduction algorithm for quasi-birth-death problems. *SIAM J. Matrix Anal. Appl.*, 24(4):1161–1166 (electronic), 2003.

[58] B. Haverkort. *Performance of computer communication systems: a model-based approach.* John Wiley, New York, 1998.

[59] C. He, B. Meini, and N. H. Rhee. A shifted cyclic reduction algorithm for quasi-birth-death problems. *SIAM J. Matrix Anal. Appl.*, 23(3):673–691 (electronic), 2001/2.

[60] Q.-M. He and A. S. Alfa. The $MMAP[K]/PH[K]/1$ queues with a last-come-first-served preemptive service discipline. *Queueing Systems Theory Appl.*, 29(2-4):269–291, 1998.

[61] Q.-M. He and A. S. Alfa. The discrete time MMAP[K]/PH[K]/1/LCFS-GPR queue and its variants. In G. Latouche and P. Taylor, editors, *Advances in algorithmic methods for stochastic models*, pages 167–190. Notable Publications, New Jersey, 2000. Proceedings of the Third Conference on Matrix Analytic Methods.

[62] G. Heinig and K. Rost. *Algebraic methods for Toeplitz-like matrices and operators*, volume 13 of *Operator theory: advances and applications.* Birkhäuser Verlag, Basel, 1984.

[63] P. Henrici. *Applied and computational complex analysis,* volume 1. Wiley Classics Library. John Wiley & Sons Inc., New York, 1988. Power series—integration—conformal mapping—location of zeros, Reprint of the 1974 original, A Wiley-Interscience Publication.

[64] N. J. Higham and H.-M. Kim. Numerical analysis of a quadratic matrix equation. *IMA J. Numer. Anal.*, 20(4):499–519, 2000.

[65] N. J. Higham and H.-M. Kim. Solving a quadratic matrix equation by Newton's method with exact line searches. *SIAM J. Matrix Anal. Appl.*, 23(2):303–316 (electronic), 2001.

[66] X.-Q. Jin. *Developments and applications of block Toeplitz iterative solvers,* volume 2 of *Combinatorics and Computer Science.* Kluwer Academic Publishers, Dordrecht, 2002.

[67] T. Kailath, S. Y. Kung, and M. Morf. Displacement ranks of a matrix. *Bull. Amer. Math. Soc. (N.S.)*, 1(5):769–773, 1979.

[68] T. Kailath, S. Y. Kung, and M. Morf. Displacement ranks of matrices and linear equations. *J. Math. Anal. Appl.*, 68(2):395–407, 1979.

[69] T. Kailath and A. H. Sayed. Displacement structure: theory and applications. *SIAM Rev.*, 37(3):297–386, 1995.

[70] T. Kailath and A. H. Sayed, editors. *Fast reliable algorithms for matrices with structure.* Society for Industrial and Applied Mathematics (SIAM), Philadelphia, PA, 1999.

[71] G. J. Kemeny and J. L. Snell. *Finite Markov chains.* Van Nostrand, Princeton, NJ, 1960.

[72] G. J. Kemeny, J. L. Snell, and A. W. Knapp. *Denumerable Markov chains.* Springer-Verlag, New York, second edition, 1976.

[73] M. G. Kreĭn. Integral equations on the half-line with a kernel depending on the difference of the arguments. *Uspehi Mat. Nauk*, 13(5):3–120, 1958.

[74] P. Lancaster and M. Tismenetsky. *The theory of matrices.* Computer Science and Applied Mathematics. Academic Press Inc., Orlando, FL, second edition, 1985.

[75] G. Latouche. Algorithms for infinite Markov chains with repeating columns. In *Linear algebra, Markov chains, and queueing models (Minneapolis, MN, 1992)*, volume 48 of *IMA Vol. Math. Appl.*, pages 231–265. Springer, New York, 1993.

[76] G. Latouche. Algorithms for evaluating the matrix G in Markov chains of $PH/G/1$ type. *Cahiers Centre Études Rech. Opér.*, 36:251–258, 1994. Hommage à Simone Huyberechts.

[77] G. Latouche. Newton's iteration for non-linear equations in Markov chains. *IMA J. Numer. Anal.*, 14(4):583–598, 1994.

[78] G. Latouche and V. Ramaswami. A logarithmic reduction algorithm for quasi-birth-death processes. *J. Appl. Probab.*, 30(3):650–674, 1993.

[79] G. Latouche and V. Ramaswami. *Introduction to matrix analytic methods in stochastic modeling.* ASA-SIAM Series on Statistics and Applied Probability. Society for Industrial and Applied Mathematics (SIAM), Philadelphia, PA, 1999.

[80] G. Latouche and G. Stewart. Numerical methods for M/G/1 type queues. In W. J. Stewart, editor, *Computations with Markov chains*, pages 571–581. Kluwer Academic Publishers, 1995.

[81] G. Latouche and P. Taylor. Truncation and augmentation of level-independent QBD processes. *Stochastic Process. Appl.*, 99(1):53–80, 2002.

[82] G. Latouche and P. G. Taylor. Level-phase independence for GI/M/1-type Markov chains. *J. Appl. Probab.*, 37:984–998, 2000.

[83] G. Latouche and P. G. Taylor. Drift conditions for matrix-analytic models. *Mathematics of Operations Research*, to appear.

[84] S.-Q. Li and H.-D. Sheng. Generalized folding-algorithm for sojourn time analysis of finite QBD processes and its queueing applications. *Comm. Statist. Stochastic Models*, 12(3):507–522, 1996.

[85] G. S. Litvinchuk and I. M. Spitkovsky. *Factorization of measurable matrix functions*, volume 25 of *Operator theory: advances and applications.* Birkhäuser Verlag, Basel, 1987. Translated from the Russian by Bernd Luderer, with a foreword by Bernd Silbermann.

[86] G. Louchard and G. Latouche. Geometric bounds on iterative approximations for nearly completely decomposable Markov chains. *J. Appl. Probab.*, 27:521–529, 1990.

[87] B. Meini. An improved FFT-based version of Ramaswami's formula. *Stochastic Models*, 13(2):223–238, 1997.

[88] B. Meini. New convergence results on functional iteration techniques for the numerical solution of $M/G/1$ type Markov chains *Numer. Math.*, 78(1):39–58, 1997.

[89] B. Meini. Solving QBD problems: the cyclic reduction algorithm versus the invariant subspace method. *Adv. Perf. Anal.*, 1:215–225, 1998.

[90] M. F. Neuts. Moment formulas for the Markov renewal branching process. *Adv. Appl. Prob.*, 8:690–711, 1976.

[91] M. F. Neuts. *Matrix-geometric solutions in stochastic models. An algorithmic approach*, volume 2 of Johns Hopkins Series in the Mathematical Sciences. Johns Hopkins University Press, Baltimore, MD, 1981.

[92] M. F. Neuts. *Structured stochastic matrices of $M/G/1$ type and their applications*, volume 5 of Probability: Pure and Applied. Marcel Dekker Inc., New York, 1989.

[93] J. R. Norris. *Markov chains.* Cambridge Series on Statistical and Probabilistic Mathematics. Cambridge University Press, Cambridge, 3rd edition, 1999.

[94] C. A. O'Cinneide. Relative-error bounds for the LU decomposition via the GTH algorithm. *Numer. Math.*, 73:507–519, 1996.

[95] J. M. Ortega and W. C. Rheinboldt. *Iterative solution of nonlinear equations in several variables*, volume 30 of Classics in Applied Mathematics. Society for Industrial and Applied Mathematics (SIAM), Philadelphia, PA, 2000. Reprint of the 1970 original.

[96] G. Rama Murthy, M. Kim, and E. J. Coyle. Equilibrium analysis of skip-free Markov chains: Nonlinear matrix equations. *Comm. Statist. Stochastic Models*, 7:547–571, 1991.

[97] V. Ramaswami. Nonlinear matrix equations in applied probability—solution techniques and open problems. *SIAM Rev.*, 30(2):256–263, 1988.

[98] V. Ramaswami. A stable recursion for the steady state vector in Markov chains of $M/G/1$ type. *Comm. Statist. Stochastic Models*, 4(1):183–188, 1988.

[99] V. Ramaswami. A duality theorem for the matrix paradigms in queueing theory. *Comm. Statist. Stochastic Models*, 6:151–161, 1990.

[100] V. Ramaswami. The generality of quasi birth-and-death processes. In A. S. Alfa and S. R. Chakravarthy, editors, *Advances in matrix analytic methods for stochastic models*, pages 93–113. Notable Publications, NJ, 1998.

[101] V. Ramaswami and P. G. Taylor. An operator-analytic approach to product-form networks. *Comm. Statist. Stochastic Models*, 12:121–142, 1996.

[102] S. I. Resnick. *Adventures in Stochastic Processes.* Birkhäuser, Cambridge, MA, 1992.

[103] Y. Saad. *Iterative methods for sparse linear systems.* Society for Industrial and Applied Mathematics, Philadelphia, PA, second edition, 2003.

[104] H. Schellhaas. On Ramaswami's algorithm for the computation of the steady state vector in Markov chains of M/G/1-type. *Comm. Statist. Stochastic Models*, 6:541–550, 1990.

[105] E. Seneta. *Non-negative matrices and Markov chains.* Springer-Verlag, New York, second edition, 1981.

[106] G. W. Stewart. On the solution of block Hessenberg systems. *Numer. Linear Algebra Appl.*, 2(3):287–296, 1995.

[107] W. J. Stewart. *Introduction to the numerical solution of Markov chains.* Princeton University Press, Princeton, NJ, 1994.

[108] D. Stirzaker. *Probability and random variables. A beginner's guide.* Cambridge University Press, 1999.

[109] P. N. Swarztrauber. Vectorizing the FFTs. In *Parallel computations*, volume 1 of *Comput. Tech.*, pages 51–83. Academic Press, Orlando, FL, 1982.

[110] T. Takine, B. Sengupta, and R. W. Yeung. A generalization of the matrix $M/G/1$ paradigm for Markov chains with a tree structure. *Comm. Statist. Stochastic Models*, 11(3):411–421, 1995.

[111] O. Toeplitz. Zur Theorie der quadratischen und bilinearen Formen von unendlichvielen Veränderlichen. *Math. Ann.*, 70:351–376, 1911.

[112] R. L. Tweedie. Operator-geometric stationary distributions for Markov chains, with applications to queueing models. *Adv. in Appl. Probab.*, 14:368–391, 1982.

[113] B. Van Houdt and C. Blondia. Stability and performance of stack algorithms for random access communication modeled as a tree structured QBD Markov chain. *Stoch. Models*, 17(3):247–270, 2001.

[114] B. Van Houdt and C. Blondia. Tree structured QBD Markov chains and tree-like QBD processes. *Stoch. Models*, 19(4):467–482, 2003.

[115] C. Van Loan. *Computational frameworks for the fast Fourier transform*, volume 10 of Frontiers in Applied Mathematics. Society for Industrial and Applied Mathematics (SIAM), Philadelphia, PA, 1992.

[116] R. S. Varga. *Matrix iterative analysis*, volume 27 of Springer Series in Computational Mathematics. Springer-Verlag, Berlin, expanded edition, 2000.

[117] P. Yalamov and V. Pavlov. Stability of the block cyclic reduction. *Linear Algebra Appl.*, 249:341–358, 1996.

[118] P. Y. Yalamov. On the stability of the cyclic reduction without back substitution for tridiagonal systems. *BIT*, 35(3):428–447, 1995.

[119] J. Ye and S.-Q. Li. Analysis of multi-media traffic queues with finite buffer and overload control—Part I: algorithm. In *Proc. IEEE Infocom 91, Bal Harbour*, pages 1464–1474. 1991.

[120] Q. Ye. On Latouche–Ramaswami's logarithmic reduction algorithm for quasi-birth-and-death processes. *Stoch. Models*, 18(3):449–467, 2002.

[121] R. W. Yeung and A. S. Alfa. The quasi-birth-death type Markov chain with a tree structure. *Comm. Statist. Stochastic Models*, 15(4):639–659,

1999.

[122] R. W. Yeung and B. Sengupta. Matrix product-form solutions for Markov chains with a tree structure. *Adv. in Appl. Probab.*, 26(4):965–987, 1994.

INDEX